VOLUME THREE HUNDRED AND FORTY

INTERNATIONAL REVIEW OF CELL AND MOLECULAR BIOLOGY

Mitochondria and Longevity

INTERNATIONAL REVIEW OF CELL AND MOLECULAR BIOLOGY

VOLUME THREE HUNDRED AND FORTY

INTERNATIONAL REVIEW OF CELL AND MOLECULAR BIOLOGY

Mitochondria and Longevity

Edited by

CARLOS LÓPEZ-OTÍN
Department of Biochemistry and Molecular Biology
Universidad de Oviedo, Asturias, Spain

LORENZO GALLUZZI
Weill Cornell Medical College, New York, NY, United States

ELSEVIER

ACADEMIC PRESS
An imprint of Elsevier

Academic Press is an imprint of Elsevier
125 London Wall, London EC2Y 5AS, United Kingdom
525 B Street, Suite 1650, San Diego, CA 92101, United States
50 Hampshire Street, 5th Floor, Cambridge, MA 02139, United States
The Boulevard, Langford Lane, Kidlington, Oxford OX5 1GB, United Kingdom

First edition 2018

Notices

Knowledge and best practice in this field are constantly changing. As new research and experience broaden our understanding, changes in research methods, professional practices, or medical treatment may become necessary.

Practitioners and researchers must always rely on their own experience and knowledge in evaluating and using any information, methods, compounds, or experiments described herein. In using such information or methods they should be mindful of their own safety and the safety of others, including parties for whom they have a professional responsibility.

To the fullest extent of the law, neither the Publisher nor the authors, contributors, or editors, assume any liability for any injury and/or damage to persons or property as a matter of products liability, negligence or otherwise, or from any use or operation of any methods, products, instructions, or ideas contained in the material herein.

ISBN: 978-0-12-815736-7
ISSN: 1937-6448

Publisher: Zoe Kruze
Acquisition Editor: Ashlie M. Jackman
Editorial Project Manager: Teresa Pons-Ferrer
Production Project Manager: James Selvam
Cover Designer: Victoria Pearson

Typeset by TNQ Technologies

CONTENTS

CONTRIBUTORS

Giacomo Baccolo
SYSBIO Centre for Systems Biology, Milano, Italy; Dipartimento di Biotecnologie e Bioscienze, Università di Milano-Bicocca, Milano, Italy

Clea Bárcena
Departamento de Bioquímica y Biología Molecular, Facultad de Medicina, Instituto Universitario de Oncología (IUOPA), Universidad de Oviedo, 33006-Oviedo, Spain

Massimo Bonora
Departments of Cell Biology and Gottesman Institute for Stem Cell & Regenerative Medicine Research, Albert Einstein College of Medicine, Bronx, NY, United States

Sabine Borchard
Institute of Molecular Toxicology and Pharmacology, Helmholtz Center Munich, German Research Center for Environmental Health, Neuherberg, Germany

Carles Cantó
Nestle Institute of Health Sciences NIHS, Lausanne, Switzerland; School of Life Sciences, Ecole Polytechnique Fédérale de Lausanne (EPFL), Lausanne, Switzerland

Coppola Damiano Pellegrino
Dipartimento di Biotecnologie e Bioscienze, Università di Milano-Bicocca, Milano, Italy

Agnieszka Dobrzyn
The Francis Crick Institute, London, United Kingdom

Pawel Dobrzyn
Department of Biochemistry, Nencki Institute of Experimental Biology, Polish Academy of Sciences, Warsaw, Poland

Jerzy Duszynski
Department of Biochemistry, Nencki Institute of Experimental Biology, Polish Academy of Sciences, Warsaw, Poland

Carlotta Giorgi
Department of Morphology Surgery and Experimental Medicine, Section of Pathology Oncology and Experimental Biology, Interdisciplinary Center for the Study of Inflammation (ICSI), Laboratory for Technologies of Advanced Therapies (LTTA), University of Ferrara, Ferrara, Italy

Paulina Jędrak
Department of Molecular Biology, University of Gdańsk, Gdańsk, Poland

Agnieszka Karkucinska-Wieckowska
Department of Pathology, The Children's Memorial Health Institute, Warsaw, Poland

Justine Lebeau
Department of Molecular Medicine, The Scripps Research Institute, La Jolla, CA, United States

Saverio Marchi
Department of Morphology Surgery and Experimental Medicine, Section of Pathology Oncology and Experimental Biology, Interdisciplinary Center for the Study of Inflammation (ICSI), Laboratory for Technologies of Advanced Therapies (LTTA), University of Ferrara, Ferrara, Italy

Maria Markaki
Institute of Molecular Biology and Biotechnology, Foundation for Research and Technology-Hellas

Pablo Mayoral
Departamento de Bioquímica y Biología Molecular, Facultad de Medicina, Instituto Universitario de Oncología (IUOPA), Universidad de Oviedo, 33006-Oviedo, Spain

Giampaolo Morciano
Department of Morphology Surgery and Experimental Medicine, Section of Pathology Oncology and Experimental Biology, Interdisciplinary Center for the Study of Inflammation (ICSI), Laboratory for Technologies of Advanced Therapies (LTTA), University of Ferrara, Ferrara, Italy; Cecilia Hospital, GVM Care & Research, 48033 Cotignola, Ravenna, Italy; Maria Pia Hospital, GVM Care & Research, Torino, Italy

Paulo J. Oliveira
CNC - Center for Neuroscience and Cell Biology, UC-Biotech, Biocant Park, University of Coimbra, Cantanhede, Portugal

Ivan Orlandi
SYSBIO Centre for Systems Biology, Milano, Italy; Dipartimento di Biotecnologie e Bioscienze, Università di Milano-Bicocca, Milano, Italy

Konstantinos Palikaras
Institute of Molecular Biology and Biotechnology, Foundation for Research and Technology-Hellas

Paulina Patalas-Krawczyk
Department of Biochemistry, Nencki Institute of Experimental Biology, Polish Academy of Sciences, Warsaw, Poland

Mariasole Perrone
Department of Morphology Surgery and Experimental Medicine, Section of Pathology Oncology and Experimental Biology, Interdisciplinary Center for the Study of Inflammation (ICSI), Laboratory for Technologies of Advanced Therapies (LTTA), University of Ferrara, Ferrara, Italy

Karolina Pierzynowska
Department of Molecular Biology, University of Gdańsk, Gdańsk, Poland

Paolo Pinton
Department of Morphology Surgery and Experimental Medicine, Section of Pathology Oncology and Experimental Biology, Interdisciplinary Center for the Study of Inflammation (ICSI), Laboratory for Technologies of Advanced Therapies (LTTA), University of Ferrara, Ferrara, Italy; Cecilia Hospital, GVM Care & Research, 48033 Cotignola, Ravenna, Italy

Piero Portincasa
Clinica Medica “A. Murri”, Dept. of Biomedical Sciences & Human Oncology, University of Bari "Aldo Moro" Medical School, Bari, Italy

Pedro M. Quirós
Departamento de Bioquímica y Biología Molecular, Facultad de Medicina, Instituto Universitario de Oncología (IUOPA), Universidad de Oviedo, 33006-Oviedo, Spain; Centro de Investigación Biomédica en Red de Cáncer (CIBERONC), Spain

T. Kelly Rainbolt
Department of Molecular Medicine, The Scripps Research Institute, La Jolla, CA, United States

Ziyu Ren
Department of Cell and Developmental Biology, Consortium for Mitochondrial Research, University College London, London, United Kingdom

Alessandro Rimessi
Department of Morphology Surgery and Experimental Medicine, Section of Pathology Oncology and Experimental Biology, Interdisciplinary Center for the Study of Inflammation (ICSI), Laboratory for Technologies of Advanced Therapies (LTTA), University of Ferrara, Ferrara, Italy

Vilma A. Sardao
CNC - Center for Neuroscience and Cell Biology, UC-Biotech, Biocant Park, University of Coimbra, Cantanhede, Portugal

Ines C.M. Simoes
Department of Biochemistry, Nencki Institute of Experimental Biology, Polish Academy of Sciences, Warsaw, Poland

Giulia Stamerra
SYSBIO Centre for Systems Biology, Milano, Italy; Dipartimento di Biotecnologie e Bioscienze, Università di Milano-Bicocca, Milano, Italy

Gyorgy Szabadkai
Department of Cell and Developmental Biology, Consortium for Mitochondrial Research, University College London, London, United Kingdom; The Francis Crick Institute, London, United Kingdom; Department of Biomedical Sciences, University of Padua, Padua, Italy

Jędrzej Szymański
Department of Biochemistry, Nencki Institute of Experimental Biology, Polish Academy of Sciences, Warsaw, Poland

Nektarios Tavernarakis
Institute of Molecular Biology and Biotechnology, Foundation for Research and Technology-Hellas; Department of Basic Sciences, Medical School, University of Crete, Heraklion, Greece

Marina Vai
SYSBIO Centre for Systems Biology, Milano, Italy; Dipartimento di Biotecnologie e Bioscienze, Università di Milano-Bicocca, Milano, Italy

David Q. Wang
Department of Medicine, Division of Gastroenterology and Liver Diseases, Marion Bessin Liver Research Center, Albert Einstein College of Medicine, Bronx, NY, United States

Grzegorz Węgrzyn
Department of Molecular Biology, University of Gdańsk, Gdańsk, Poland

Mariusz R. Wieckowski
Department of Biochemistry, Nencki Institute of Experimental Biology, Polish Academy of Sciences, Warsaw, Poland

R. Luke Wiseman
Department of Molecular Medicine, The Scripps Research Institute, La Jolla, CA, United States

Barbara Zavan
Cecilia Hospital, GVM Care & Research, 48033 Cotignola, Ravenna, Italy; Department of Biomedical Sciences, University of Padua, Padua, Italy

Hans Zischka
Institute of Molecular Toxicology and Pharmacology, Helmholtz Center Munich, German Research Center for Environmental Health, Neuherberg, Germany; Institute of Toxicology and Environmental Hygiene, Technical University Munich, Munich, Germany

CHAPTER ONE

Mitochondrial Metabolism and Aging in Yeast

Giacomo Baccolo*[,§], Giulia Stamerra*[,§], Coppola Damiano Pellegrino[§], Ivan Orlandi*[,§] and Marina Vai*[,§,1]
*SYSBIO Centre for Systems Biology, Milano, Italy
[§]Dipartimento di Biotecnologie e Bioscienze, Università di Milano-Bicocca, Milano, Italy
[1]Correspondence: E-mail: marina.vai@unimib.it

Contents

Abstract

Mitochondrial functionality is one of the main factors involved in cell survival, and mitochondrial dysfunctions have been identified as an aging hallmark. In particular, the insurgence of mitochondrial dysfunctions is tightly connected to mitochondrial metabolism. During aging, both mitochondrial oxidative and biosynthetic metabolisms are progressively altered, with the development of malfunctions, in turn affecting

International Review of Cell and Molecular Biology, Volume 340
ISSN 1937-6448
https://doi.org/10.1016/bs.ircmb.2018.05.001

mitochondrial functionality. In this context, the relation between mitochondrial pathways and aging is evolutionarily conserved from single-celled organisms, such as yeasts, to complex multicellular organisms, such as humans. Useful information has been provided by the yeast *Saccharomyces cerevisiae*, which is being increasingly acknowledged as a valuable model system to uncover mechanisms underlying cellular longevity in humans. On this basis, we review the impact of specific aspects of mitochondrial metabolism on aging supported by the contributions brought by numerous studies performed employing yeast. Initially, we will focus on the tricarboxylic acid cycle and oxidative phosphorylation, describing how their modulation has consequences on cellular longevity. Afterward, we will report information regarding the importance of nicotinamide adenine dinucleotide (NAD) metabolism during aging, highlighting its relation with mitochondrial functionality. The comprehension of these key points regarding mitochondrial metabolism and their physiological importance is an essential first step for the development of therapeutic interventions that point to increase life quality during aging, therefore promoting "healthy aging," as well as lifespan itself.

1. INTRODUCTION

Mitochondria are unquestionably essential organelles for eukaryotic aerobic metabolism since they provide a unique metabolic pathway that generates the energy required for all cellular functions in a utilizable form. In addition, mitochondria exhibit many other functions, including biosynthesis of intermediary metabolites, such as amino acids and nucleotides, formation of iron—sulfur clusters, and regulation of Ca^{2+} homeostasis.

By providing essential cellular functions, mitochondria support the metabolism of the whole cell and contribute to cellular and organismal homeostasis. Consequently, their impairment is associated not only with several diseases, but dysfunctional mitochondria are considered a hallmark of aging (López-Otín et al., 2013), and an altered mitochondrial metabolism characterizes many aging-related pathologies, including degenerative disorders (Lane et al., 2015). As far as aging is concerned, mitochondria have been/are intensively studied regarding, in particular, mitochondrial adenosine triphosphate (ATP) production and reactive oxygen species (ROS) leakage from the respiratory chain. Here, we also discuss other central functions of mitochondrial metabolism that are equally important, such as mitochondrial carbon metabolism and NAD homeostasis as parts of a complex mechanism, providing reducing equivalents and biosynthetic intermediates for the cell.

Since the single-celled yeast *Saccharomyces cerevisiae* is a valuable model system in the field of aging-related research, here we mainly refer to this organism to highlight the critical importance of this organelle. Two complementary models of aging are well established in *S. cerevisiae*: replicative and chronological aging that allow us to simulate the aging process of mitotically active and postmitotic quiescent mammalian cells, respectively (Longo and Kennedy, 2006; MacLean et al., 2001). In the former, replicative lifespan (RLS) defines the reproductive potential of individual yeast cells. In the latter, chronological lifespan (CLS) is measured as the time a culture of quiescent cells remains viable in a metabolically active nondividing state. Both models have contributed to the discovery of evolutionarily conserved factors associated with mitochondrial dysfunctions that proved to be involved in the aging process.

2. THE TRICARBOXYLIC ACID CYCLE

In terms of energetic metabolism, the mitochondrion is primarily a device for coupling the synthesis of ATP to the oxidative reactions of the tricarboxylic acid (TCA) cycle. This universal attribute of all mitochondria is referred to as oxidative phosphorylation (OXPHOS). Indeed, mitochondria catalyze aerobic oxidation of acetyl-CoA to CO_2 and water by means of the TCA cycle and produce energy, which is stored in the form of high-energy bonds of ATP.

Basically the TCA cycle consists of a series of both oxidative and hydrolytic reactions (Fig. 1). Upon condensation of acetyl-CoA with oxaloacetate generating citrate, the cycle starts with the reversible dehydration of citrate to cis-aconitate and the hydration of cis-aconitate to isocitrate, catalyzed by aconitase. The following steps are sequential redox reactions: oxidation of isocitrate to α-ketoglutarate; oxidation of α-ketoglutarate to succinyl-CoA and CO_2; and oxidation of succinate to fumarate. Then, fumarase catalyzes the reversible hydration of fumarate to malate, whose final oxidation to oxaloacetate completes the cycle. All the oxidative steps require cofactors in the oxidized forms of NAD or flavin adenine dinucleotide (FAD) as acceptors of reducing equivalents.

Besides this basic view of the cycle starting with acetyl-CoA as input to form citrate, TCA can be also fueled at the level of some of its intermediates. Indeed, the availability of glutamic acid leads to a "modified" cycle: glutamate is converted to α-ketoglutarate, which being a TCA intermediate is oxidized via the cycle to oxaloacetate. Afterward, most of the oxaloacetate

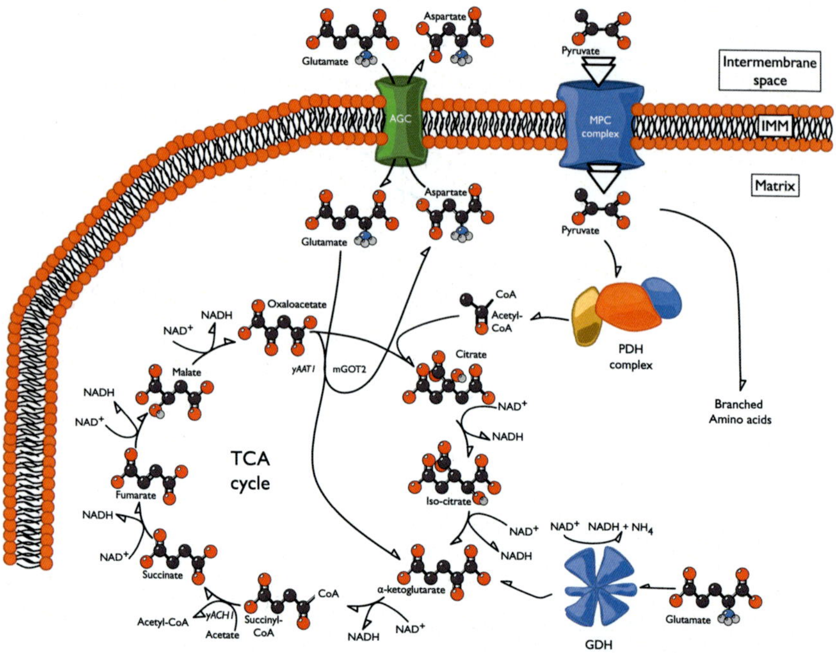

Figure 1 Scheme of the TCA cycle and its connections with the uptake of pyruvate and glutamate. *AGC*, aspartate/glutamate carrier; *GDH*, glutamate dehydrogenase complex; *IMM*, inner mitochondrial membrane; *mGOT2*, mammal glutamic-oxaloacetic transaminase 2; *MPC*, mitochondrial pyruvate carrier; *PDH*, pyruvate dehydrogenase complex; *yAat1*, yeast aspartate aminotransferase 1; *yAch1*, yeast acetyl-CoA hydrolase 1.

reacts by transamination with a second molecule of glutamate, generating aspartate and α-ketoglutarate. In such a way, the TCA cycle is connected with the anabolic biosynthesis of amino acids and vice versa (Fig. 1).

3. TRICARBOXYLIC ACID CYCLE DYSFUNCTIONS AND AGING

During chronological aging in yeast, TCA enzymatic activities have been shown to change (Samokhvalov et al., 2004). In particular, citrate synthase, α-ketoglutarate dehydrogenase, and malate dehydrogenase display a reduced activity, whilst that of succinate dehydrogenase (also known as Complex II) increases. Such an increase has been proposed to be part of a compensation mechanism induced by aged cells to maintain their respiratory competency (see Section 7.2). In this compensation mechanism is also involved the glyoxylate cycle, which leads to an increase in succinate formation (Samokhvalov et al., 2004). The glyoxylate cycle is present in yeast and operates as a shunt of the TCA cycle. During chronological aging, this cycle is operative and replenishes the TCA cycle of intermediates to keep it functioning. Briefly, both cycles share the first steps in which citrate synthase catalyzes the condensation of oxaloacetate and acetyl-CoA to form citrate and aconitase converts citrate to isocitrate. In addition, the glyoxylate cycle is completed by two unique enzymes that are isocitrate lyase (Icl1) and malate synthase. The former cleaves isocitrate to glyoxylate and succinate, and the latter converts glyoxylate and acetyl-CoA to malate. Finally, malate dehydrogenase catalyzes the oxidation of malate to oxaloacetate. This end product of the shunt can be also utilized for gluconeogenesis, a pathway that is essential for the extension of CLS (Lin et al., 2009).

In mammalian cells, TCA enzymatic activity also changes during aging, displaying a decrease that mainly concerns the NAD-dependent dehydrogenases and damage-redox sensitive enzymes such as aconitase (Ingram and Chakrabarti, 2016; Kumaran et al., 2005; Stauch et al., 2015; Yarian et al., 2006). In this context, it is well known that oxidative modifications of proteins can lead to an impairment or a complete loss of protein function. In particular, protein carbonylation is one of the most studied irreversible modifications occurring during aging (Cabiscol et al., 2014; Nyström, 2005). Both in yeast and mammalian cells, TCA enzymes, as well as the mitochondrial ATPase synthase, have been shown to be target of carbonylation (Ahmed et al., 2010; Cabiscol et al., 2014; Reverter-Branchat et al., 2004; Tamarit et al., 2012).

4. IMPAIRED MITOCHONDRIAL PYRUVATE/ACETYL-COA METABOLISM AND AGING

Glucose-derived pyruvate is one of the most important metabolic input for mitochondria where it is converted to acetyl-CoA by the pyruvate dehydrogenase (PDH) complex (Fig. 1). Pyruvate is produced at the end of glycolysis in the cytoplasm and enters the mitochondrial matrix through the mitochondrial pyruvate carrier (MPC; Bricker et al., 2012; Herzig et al., 2012). The MPC is a heteromeric complex of which Mpc1/MPC1 in yeast or mammalian cells is the major structural subunit (Bricker et al., 2012). During chronological aging, loss of Mpc1 significantly reduces CLS (Eisenberg et al., 2014; Orlandi et al., 2014). Furthermore, the *mpc1Δ* mutant accumulates pyruvate, while TCA cycle intermediates, such as citrate, succinate, and malate, are decreased (Orlandi et al., 2014).

Similarly, depletion of human MPC1 impairs pyruvate oxidation (Bricker et al., 2012), and in these cells, intracellular level of pyruvate is strongly increased, whereas citrate level is decreased (Vacanti et al., 2014).

In both yeast and human cells, a metabolic reprogramming takes place to compensate for the decrease of mitochondrial pyruvate, allowing cells to use other substrates for both respiration and anabolic purposes. In yeast, the lack of a functional MPC is associated with an increased activity of the mitochondrial malic enzyme, that by decarboxylating malate to pyruvate, it provides a mitochondrial pyruvate pool (Orlandi et al., 2014). In such a way, the TCA cycle operates in a "branched" fashion and is depleted of intermediates. In order to fulfill the needs of the TCA cycle, mutant cells increase the activities of the glyoxylate cycle, which can feed of C4 compounds the TCA cycle. In the case of human cells, which do not possess the glyoxylate cycle, in order to cope with the lack of mitochondrial pyruvate uptake, they strongly increase glutamine utilization (Vacanti et al., 2014; Yang et al., 2014). This necessarily occurs since glutamine can be deaminated to glutamate, which in turn is converted to α-ketoglutarate to replenish the TCA cycle. Then, mitochondrial isoforms of malic enzyme, ME2 and ME3, as in yeast, transform malate into pyruvate.

These mechanisms, which allow the cells to compensate for the mitochondrial pyruvate deficiency, on the other hand, render them more vulnerable. In yeast, the defect of the TCA cycle functioning, ultimately, results in an inefficient mitochondrial acetyl-CoA metabolism (Eisenberg et al., 2014; Orlandi et al., 2014). This latter consequence is associated with an increased nucleocytosolic acetyl-CoA metabolism, which in turn also affects histone acetylation, leading to a repression of genes encoding proteins required for autophagy, a process that plays a critical role for cell survival during chronological aging (Eisenberg et al., 2014). Thus, the perturbation of mitochondrial metabolism is connected to the chromatin structure.

Similarly, blocking the mitochondrial route to acetyl-CoA by deletion of *ACH1* increases nucleocytosolic acetyl-CoA metabolism that ultimately triggers histone acetylation and repression of autophagy genes (Eisenberg et al., 2014). This results in a short CLS (Eisenberg et al., 2014; Orlandi et al., 2014). The mitochondrial Ach1 displays a CoA-transferase activity: it catalyzes the CoASH transfer from succinyl-CoA (produced by the TCA cycle) to acetate producing acetyl-CoA (Fleck and Brock, 2009; Fig. 1). In the mitochondria, the reaction catalyzed by Ach1 allows, on the one hand, to save one ATP compared with the acetate activation to acetyl-CoA by the acetyl-CoA synthetase 1 and, on the other, to protect mitochondrial function from the toxic effects of acetate accumulation. In line with this the reduced CLS of the *ach1Δ* mutant is associated with an impaired mitochondrial functionality (Orlandi et al., 2012). Moreover, consistent with a role of the mitochondrial detoxifying enzyme, Ach1 has been identified to be resistant to oxidative damage (Tamarit et al., 2012).

Furthermore, *LSC1* encodes the alpha subunit of succinyl-CoA ligase, the enzyme that catalyzes the nucleotide-dependent conversion of succinyl-CoA to succinate. Interestingly, *LSC1* deletion decreases the CLS (Fabrizio et al., 2010; Powers et al., 2006). These convergent short-lived phenotypes evidence a critical role of the succinate node for the functionality of the TCA cycle.

5. MITOCHONDRIAL METABOLISM OF BRANCHED-CHAIN AMINO ACIDS AND AGING

The TCA cycle and the biosynthesis of amino acids are strongly interconnected. In addition to the α-ketoglutarate/glutamate connection, another link is found with malate, which is a precursor in the anabolic pathway of the branched-chain amino acids (BCAAs). In the reaction catalyzed by the mitochondrial malic enzyme, as mentioned previously, malate is converted in pyruvate, which is consequently used to synthesize leucine, isoleucine, and valine, collectively known as BCAAs (Fig. 1). These amino acids are synthesized in yeast by a unique pathway localized in the mitochondria, in which common enzymes catalyze similar reactions. Thus, this mitochondrial pathway contributes to the amino acids homeostasis, which is a key process involved in the long-term viability during both replicative and chronological aging (Aris et al., 2012; Kamei et al., 2014). Indeed, the levels of transcripts encoding enzymes involved in the biosynthesis of BCAAs, as well as the BCAA metabolic profile, decrease during replicative aging (Kamei et al., 2014). Furthermore, during chronological aging, not all the amino acids are equally effective. This is mirrored by the fact that the availability of specific amino acids, compared with the availability of all amino acids, has a greater importance in amino acid homeostasis. In particular, among the amino acids, leucine makes the greatest contribution to the CLS extension (Alvers et al., 2009). The importance of leucine is supported by the fact that the CLS of the *leu2* auxotroph mutant, which has a short-lived phenotype, is extended by restoring *LEU2* (Alvers et al., 2009). Intriguingly, BCAAs enhance CLS in both autophagy-deficient and autophagy-competent strains (Alvers et al., 2009). Thus, a strictly nutritional function cannot completely account for their prolongevity effect under conditions in which autophagy provides an efficient recycling of amino acids and other metabolic building blocks. In agreement, it has been proposed that isoleucine, valine, and threonine also act via a regulatory mechanism that involves the well-known general amino acid control (GAAC) pathway. This pathway regulates amino acid homeostasis in yeast (Hinnebusch, 2005) and is induced by limiting amounts of specific amino acids, such as BCAAs, with critical repercussions on longevity, while an increased level of BCAAs curtails the activity of the GAAC pathway, enhancing CLS (Alvers et al., 2009). BCAAs, by this way, prevent the activation of an anabolic process that has deep impact on the energy and metabolite resources during chronological aging.

Notably, as far as valine and threonine are concerned, it has also been reported to determine stress sensitization and reduce CLS by activating the TORC1/Sch9 pathway (Mirisola et al., 2014). This is a nutrient-sensing pathway, the activation of which promotes chronological aging (Fontana et al., 2010; Kaeberlein, 2010). The conflicting results about valine and threonine supplementations require further investigations but probably rely on different experimental conditions/yeast genetic background used (Alvers et al., 2009; Mirisola et al., 2014) and the many unresolved questions concerning the regulatory linkages between the GAAC and TOR pathways.

Amino acid homeostasis is important not only for yeast but also for humans in the maintenance of cellular integrity during aging. Since in human cells amino acid synthesis is limited to few amino acids and the availability of most of them, including BCAAs, depends on diet, a proper management of amino acids and related metabolites is essential for cellular homeostasis. However, the effects of the diet and, in particular of amino acid assumption, can vary among the different tissues, which are characterized by specific/different metabolic features. In this context, BCAAs have been shown to play an important role in the metabolic regulation of peripheral tissues, such as white adipose tissue, liver, and muscle with different outcomes according to catabolic or anabolic conditions. For example, BCAA supplementation, in particular leucine, stimulates protein synthesis in muscle and improve sarcopenia, the age-associated loss of muscle mass and function (Dillon et al., 2009; Fujita and Volpi, 2006; Pansarasa et al., 2008; Shimomura et al., 2006). On the contrary, in patients with metabolic disorders, such as obesity and insulin resistance, BCAAs exert negative effects depending on the prevalence in these conditions of catabolic signals. BCAA supplementation seems to cause an impaired amino acid oxidation, leading to the accumulation of toxic intermediates, such as branched-chain keto acids (Berg et al., 2016; Mihalik et al., 2010).

Thus given the discrepancy found in the effects of BCAA administration, the understanding of the mechanisms that regulate amino acid metabolism assumes extreme importance for the development of effective dietary approaches to contrast aging effects (see Longo et al., 2015).

6. THE OXIDATIVE PHOSPHORYLATION

The OXPHOS results from the coupling between the electron transport chain (ETC) and the F_1F_0-ATP synthase (Complex V). Under normal conditions of mitochondrial respiration, the reducing equivalents provided by the TCA cycle through redox reactions fuel the ETC (Fig. 2). The ETC consists of four enzymatic complexes (Complex I—IV) located in the inner mitochondrial membrane and two mobile electron carriers, namely coenzyme Q (ubiquinone) and cytochrome c. They allow the transfer of high-energy electrons from NADH and $FADH_2$ through redox reactions. In the ETC, electrons enter at the level of Complex I (NADH dehydrogenases in yeast) in the form of NADH and at Complex II as $FADH_2$. The last one is generated following succinate oxidation to fumarate catalyzed by succinate dehydrogenase (Complex II), which is part of both the ETC and TCA cycle. The flux of electrons along the chain terminates at the level of Complex IV (cytochrome c oxidase) in a four-electron reduction of molecular oxygen, producing two molecules of water (Fig. 2). In addition, the electron transfer is coupled to the translocation of protons into the intermembrane space. This proton pumping determines the formation of an electrochemical gradient that is dissipated by the F_1F_0-ATP synthase to drive phosphorylation of ADP to ATP (Wallace, 2009).

Respiration, and consequently the ETC, is the main source of cellular energy, making this system central and fundamental throughout the whole life of the cell and, therefore, also during the aging process.

Respiration and its dysfunctions are central in the attempt to explain the aging phenomenon. In this context, one of the most widely accepted and popular theories is the Free Radical Theory of Aging. It was proposed by Harman in 1956 (Harman, 1956) and, substantially, attributed the cause of

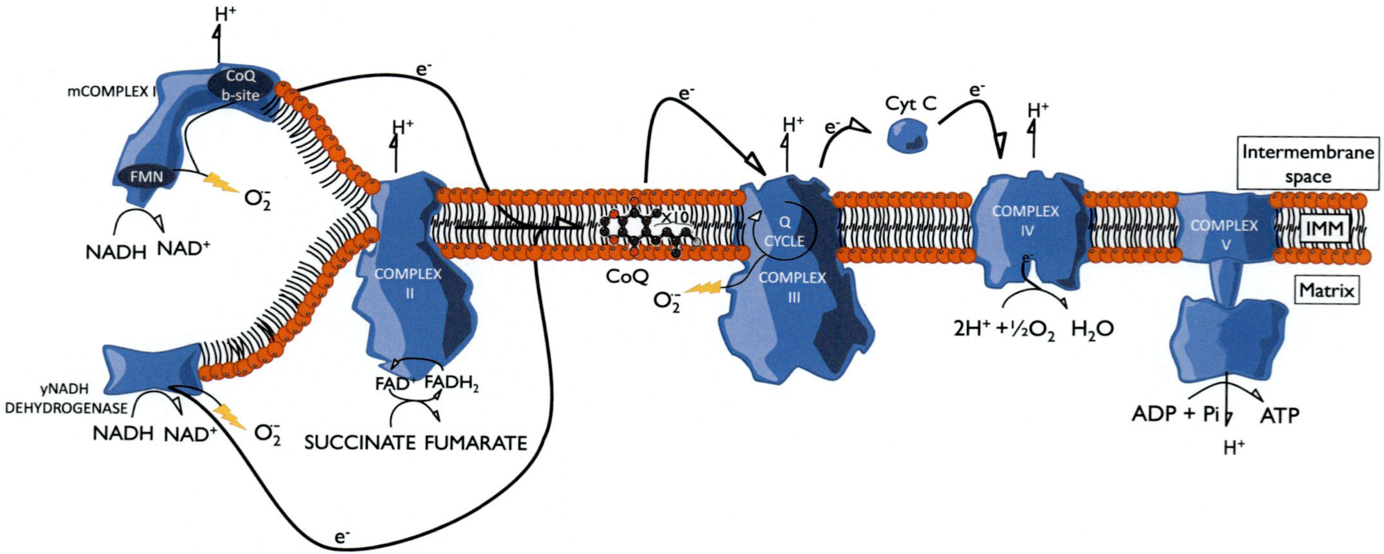

Figure 2 ***Scheme of the OXPHOS.*** Electron transfer along the electron transport chain, ATP production, and the main sites involved in ROS generation are shown. *CoQ B-site*, CoQ binding site; *CoQ*, coenzyme Q; *Cyt C*, cytochrome C; *FMN*, flavin mononucleotide; *IMM*, inner mitochondrial membrane; *mCOMPLEX I*, mammal COMPLEX I; *yNADH DEHYDROGENASE*, yeast NADH DEHYDROGENASE.

aging to damages generated by free radicals produced in redox reactions. The theory was proposed when direct evidence of ROS production by the cell was not yet available. However, Harman supposed that: "*The most likely source of* $^{\bullet}OH$ *and* HO_2 *radicals, at least in the animal cell, would be the interaction of the respiratory enzymes involved in the direct utilization of molecular oxygen*". To date, this hypothesis has been confirmed, and mitochondria are recognized as the main source of ROS within the cell. ROS are part of the free radical family, that is, atoms, molecules, and ions that have an unpaired valence electron, making these substances extremely reactive. Therefore, the term ROS does not refer to a specific molecule but to superoxide $O_2^{\bullet -}$, radical hydroxyl $\bullet OH$, hydrogen peroxide H_2O_2, and singlet oxygen $^1O_2$. Leakage of electrons from the ETC leads to the formation of the $O_2^{\bullet -}$, as oxygen can readily accept single electrons in a nonenzymatic transfer. The resulting cascade of ROS can damage not only many macromolecules but can target mitochondria with detrimental effects. ROS levels increase as a function of age, and a decline in mitochondrial functionality throughout the aging process takes place (Barros et al., 2010; Breitenbach et al., 2014).

7. RESPIRATION, ROS PRODUCTION AND AGING: A COMPLEX INTERPLAY

The specific mechanism by which each single complex in the respiratory chain is involved in the production of radical species and, thus, contributes to cell damage, and aging still remains to be completely elucidated. Here, the features of each complex will be reported in order to clarify their contribution to the aging process.

7.1 Complex I

Complex I, also known as NADH:ubiquinone oxidoreductase, transfers two electrons from NADH to ubiquinone, reducing it to ubiquinol. Complex I is the principal entry point for electrons in the ETC and plays a crucial role in ROS production (Stefanatos and Sanz, 2011) (Fig. 2). Indeed, from Complex I, electrons can leak and react with oxygen, with the resulting production of superoxide anion. Oxygen can access the electrons leaked from this complex from at least two sites: the flavin moiety and the ubiquinone-binding site (Hirst et al., 2008). Complex I dysfunctions and mutations in the genes encoding its subunits have frequently been correlated with altered ROS levels and diseases (Fato et al., 2008).

In mammals, Complex I comprises 38 subunits encoded by nuclear genes and seven encoded by the mitochondrial genome. It contains many thiol groups, which may be susceptible to oxidative damage during aging, leading to the loss of its activity. Indeed, some authors showed a strong age-dependent reduction of Complex I enzymatic activity in rat cortex mitochondria (Tatarková et al., 2011). These detrimental effects on the respiratory metabolism are dependent on an increase of oxidative stress that causes damage to protein and lipid peroxidation. Similar results were also observed in aged rats heart mitochondria (Petrosillo et al., 2009). The authors, by measuring different parameters related to mitochondrial metabolism, including basal respiration, membrane potential, ROS accumulation, and Complex I activity, found an age-dependent decline of mitochondrial performance in concert with increased levels of oxidized cardiolipin. This latter is a phospholipid required to ensure a proper Complex I activity.

In *S. cerevisiae*, Complex I is replaced by Ndi1, an NADH dehydrogenase, which does not pump protons in the inner membrane. As observed for mammals, also yeast Ndi1 has been reported to be related to the aging process. Indeed, *NDI1* overexpression in yeast determines an increase of ROS levels and $\Delta\Psi$ loss in concert with a caspase-independent apoptotic phenotype (Cui et al., 2012; Li et al., 2006). Furthermore, *NDI1*-overexpressing cells exhibit typical features of aged cells. In line with this, *NDI1* deletion determines CLS extension (Li et al., 2006). On the other hand, *NDI1* deletion in the context of replicative aging decreases RLS accompanied by elevated levels of superoxide radicals (Hacioglu et al., 2012).

7.2 Complex II

Complex II, or succinate dehydrogenase, is a tetrameric iron—sulfur flavoprotein highly conserved in mammals and yeast. It is a physical link between the ETC and the TCA cycle since it oxidizes succinate to fumarate in the TCA cycle and reduces ubiquinone to ubiquinol in the ETC.

Several studies highlight that this complex does not significantly contribute to ROS production and aging (Raha and Robinson, 2001; Robinson, 1998). However, some authors showed that by providing succinate (millimolar) to isolated mitochondria (Hansford et al., 1997), a high ROS generation occurs depending on a reverse electron transport (Scialò et al., 2017). However, in in vivo conditions, succinate concentration is not so high (micromolar) to promote this reverse electron transport, explaining why the contribution of this complex to ROS production is not significant in vivo (Hansford et al., 1997). In addition, when Complex I and III are compromised and succinate levels are low, Complex II generates high levels of H_2O_2 in rat skeletal muscle mitochondria. On the contrary, when succinate concentration is high, the levels of the aforementioned radical species reverse to physiological ones (Quinlan et al., 2012). Authors propose that at least three mechanisms may be involved:

1. High succinate levels reduce flavin semiquinone concentration, namely the species that in reaction with oxygen can generate ROS.
2. Succinate binds the complex and reduces its reactive potential.
3. Succinate occupies the access site for oxygen.

A synergistic effect of these mechanisms is not excluded.

Concerning the relationship between Complex II and the aging process, it has been demonstrated an age-dependent decline of its activity in human skin fibroblasts caused by a lower expression of genes encoding its subunits (Bowman and Birch-Machin, 2016).

Although the role of Complex II in the aging process of mammalian cells still merits a more exhaustive analysis, in yeast, there is clear evidence of its involvement in such a process. Cells deleted in *SDH1*, *SDH2*, or *SDH4* genes, encoding subunits of succinate dehydrogenase, have the shortest CLS among 33 single ETC component-deleted strains (Kwon et al., 2015). In addition, mutants for Complex II display higher levels of mitochondrial ROS than those of the other mutants (Kwon et al., 2015). The short-lived phenotype of *sdh1*, *sdh2*, and *sdh4* mutants is in line with the results of (Samokhvalov et al., 2004) showing that in chronological aging cells, an intensification of succinate formation and oxidation takes place to counteract a decrease in the activities of all NAD-dependent dehydrogenases in the TCA cycle. The oxidative metabolism of succinate necessarily involves the ETC Complex II and, probably, its enhancement is an adaptation mechanism supporting cell survival. Furthermore, the succinate formation requires the cytosolic glyoxylate shunt. During chronological aging, this shunt is triggered, providing C4 dicarboxylic acids that are required to replenish the TCA cycle with intermediates, among which is succinate. This compound is produced after cleavage of isocitrate by Icl1 and then transported into the mitochondria where is oxidized by Complex II. Notably, a high enzymatic activity of Icl1 is a property of surviving cells (Samokhvalov et al., 2004), while *ICL1* deletion significantly reduces CLS (Casatta et al., 2013). The importance of this complex is demonstrated also during replicative aging. The absence of *TCM62*, which encodes a chaperonin required for the assembly of the complex, determines a reduction in the RLS (Hacioglu et al., 2012).

7.3 Complex III

Complex III, also called cytochrome bc_1 complex, is recognized as one of the main sources of $O_2^{\bullet -}$ in the ETC. The electrons are transferred from Complex I and II to Complex III by ubiquinol. The electron transfer from ubiquinol to Complex III, known as Q-cycle, is the main cause of $O_2^{\bullet -}$ production at this ETC point (Korshunov et al., 1997; Fig. 2). The two electrons transferred to Complex III follow different fates: one is carried by cytochrome c passing through the Rieske iron–sulfur protein and cytochrome c_1. The other is carried by ubiquinone passing through the b_L and b_H heme of cytochrome b. Considering that the ubiquinone is located on the matrix side (N side), which has high levels of negative charges, the electron transfer through the b_L to b_H heme is against electric gradient. Moreover, ubiquinone requires two electrons to be completely reduced, implying that another electron will have to follow the same counter-gradient route. Therefore, the ubiquinone reduction occurs in two sequential phases and involves the formation of a highly reactive semi-reduced form $Q^{-\bullet}$. Under normal conditions, $Q^{-\bullet}$ is rapidly reduced to QH_2, preventing $O_2^{\bullet -}$ formation. On the contrary, when $\Delta\Psi$ is high, the electron transfer is slower, making $Q^{-\bullet}$ more stable, thus promoting reactivity with O_2 and leading to $O_2^{\bullet -}$ formation (Sugrue and Tatton, 2001).

It has been shown that reducing the electron flow toward Complex III strongly limits ROS production in isolated rats mitochondria, confirming the central role of this complex in radical species generation (Chen et al., 2003). Moreover, the lowering of Complex III activity in *Caenorhabditis elegans* correlates with an increase of lifespan (Dillin et al., 2002). Together, these results confirm the importance of this complex in ROS production and its impact on the aging process.

In yeast, inhibition of Complex III by treating cells with antimycin A results in a significative reduction of respiration that correlates with a strong decrease of CLS (Ocampo et al., 2012). This effect depends on the Q-cycle. Indeed, since the binding site of antimycin A is the high potential b_H heme (Fang and Beattie, 2003), this could inhibit electron transfer from b_H heme to semiquinone, favoring the accumulation of ROS. Instead, the treatment of yeast cells with sodium cyanide, a molecule that blocks the Complex IV, reduces ROS levels from Complex III, maintaining cytochrome c and ubiquinone reduced. This effect determines an increase of CLS, demonstrating that Complex III is central in ROS production and that interventions counteracting their formation promote longevity (Longo et al., 1999). In the context of replicative aging, it has been shown that the lack of *RIP1*, *CYT1*, and *QCR8* genes encoding Complex III subunits determines a strong reduction of RLS (Hacioglu et al., 2012).

7.4 Complex IV

Complex IV, or cytochrome c oxidase, is the last enzyme in the ETC. It transfers electrons from cytochrome c to molecular oxygen, reducing it to water. Differently from the other complexes, Complex IV to the best of our knowledge does not seem to be significantly involved in ROS production. However, an interesting link between the reduction of Complex IV activity and lifespan has been found (Tatarková et al., 2011). Indeed, its activity in cardiac mitochondria of 26-month-old rats significantly decreases compared with young animals, and its reduction is the highest compared with other complexes. These results outline the need for future investigations in order to clarify the connection of this complex with aging.

In *S. cerevisiae*, the deletion of some genes encoding subunits of Complex IV impairs the proton flow along the respiratory system, determining a strong reduction of CLS. This correlates with a reduction in ATP concentration (Kwon et al., 2015). Regarding replicative aging, similar results were observed, with a decrease of RLS and respiratory deficiency (Hacioglu et al., 2012).

7.5 F_1F_O ATPase

Mitochondrial ATPase, also defined as F_1F_O-ATP synthase or Complex V, is located in the inner mitochondrial membrane together with the ETC Complexes I–IV. It is a rotary enzyme that exploits the proton gradient across the inner membrane generated by the ETC to synthesize ATP. During the aging process, a decline on ETC and ATPase activities has been widely observed (Camacho-Pereira et al., 2016; Chistiakov et al., 2014; Porter et al., 2015). For example, it has been shown that ATP content and production decrease by about 50% in gastrocnemius muscle of aged rats, and this could be one of the factors promoting sarcopenia (Drew et al., 2003). Similar results were obtained in the heart of aged rats, showing a strong reduction in the RNA levels of the genes encoding the ATPase subunits. This correlates with a marked decrease in the enzymatic activity of

ATPase, with reduced ATP production and oxygen consumption (Preston et al., 2008). The role of this complex and the maintenance of adequate production of ATP are therefore crucial in the aging process. This is also supported by evidence of premature senescence in fibroblasts treated with oligomycin, an inhibitor of Complex V (Stöckl et al., 2006).

In *S. cerevisiae*, mutations in *ATP2*, encoding the β-subunit of the F_1 complex of ATPase, determine a decrease of mitochondrial membrane potential (Lai et al., 2002). In addition, daughter cells have the same RLS of old mother cells due to the inability of mother cells to retain damaged mitochondria. Furthermore, also the deletion of *ATP11*, required for the proper assembly of F_1 complex, results in a reduced RLS associated with significative levels of ROS (Hacioglu et al., 2012).

Other studies highlight that a transposon mutant (HsTnII), in which the expression of genes encoding subunits of the ETC and of Complex V is strongly downregulated, displays decreased CLS (Aerts et al., 2009). Moreover, also in this case, accelerated aging is associated with a loss of $\Delta\Psi$. In agreement with these results, it has been shown that the deletion of many genes encoding subunits of the ATPase correlated with a strong reduction in ATP levels and CLS (Kwon et al., 2015).

8. ROLE OF NAD$^+$ METABOLISM IN MITOCHONDRIAL FUNCTIONALITY

As described previously, the electrons transfer along the ETC starts from the universal cofactor NAD. Its molecular structure encompasses two covalently joined mononucleotides, namely nicotinamide mononucleotide (NMN) and adenosine monophosphate. In bioenergetic pathways, the molecule is subject to reversible reduction and oxidation cycles, from its oxidized form, NAD$^+$, to the reduced one, NADH.

Since the TCA cycle and ETC require the oxidized and the reduced form of NAD, respectively, a correct NAD$^+$/NADH balance is essential for an efficient mitochondrial metabolism and to preserve the whole-cell redox state (Stein and Imai, 2012). The correct maintenance of the NAD$^+$ pool by regulating its biosynthesis, cellular localization, and transport of its precursors plays a key role in regulating different cellular processes (Houtkooper et al., 2010).

From yeast to mammalian cells, there are different pathways for NAD$^+$ biosynthesis: the two major ones are de novo synthesis (de novo pathway) and the synthesis from its precursors (the NAD$^+$ salvage pathway;

Stein and Imai, 2012). The *de novo* pathway starts from tryptophan in the cytosol and proceeds by the kynurenine pathway, involving a total of five steps (Bogan and Brenner, 2008). The importance of this essential aromatic amino acid in NAD^+ biosynthesis emerged from studies on the positive effect given by tryptophan administration in treatment of pellagra, a disease caused by NAD^+ deficiency (Sauve, 2008). The NAD^+ salvage pathway starts from two main forms of vitamin B_3, nicotinic acid (NA) and its amidic form nicotinamide (NAM), both defined as niacin, or from the conversion of nicotinamide riboside (NR) (Houtkooper et al., 2010). In yeast, Ntd1/2 have been identified as transporters able to import NAD^+ in mitochondria (Todisco et al., 2006). For mammals, the most current model for NAD^+ synthesis in mitochondria provides that NAD^+ precursors, such as NAM and NR, are converted to NMN in cytosol from nicotinamide phosphoribosyltransferase (NAMPT) and nicotinamide riboside kinase 1/2 (NRK1/2), respectively. Then, NMN is transported into the mitochondria and converted in NAD^+ by the nicotinamide nucleotide adenyltransferase 3 (NMNAT3; Nikiforov et al., 2011). This is because the mitochondrial compartment lacks both transferase and kinase activity of NAMPT and NRK1/2 (Kitani et al., 2003; Ratajczak et al., 2016). Fig. 3 shows the most accepted model, both in mammalian and yeast cells.

9. NAD^+ METABOLISM AND AGING

Diet, exercise, and the aging process influence the NAD^+/NADH ratio, by modulating the expression of genes encoding enzymes involved in NAD^+ biosynthesis (Verdin, 2015; White and Schenk, 2012; Yang et al., 2007; Fig. 4). Calorie restriction (CR), a dietary regimen known to extend the lifespan in a wide spectrum of species (Ruetenik and Barrientos, 2015), which in most cases is obtained by reducing calorie intake of about 20%–40% compared to normal intake (Lee and Longo, 2016), increases NAD^+ levels (Yang et al., 2007). In this context, NAMPT has been identified as a nutrient-responsive gene, able to promote cell survival by boosting mitochondrial NAD^+ level in liver of rats fasted for 48 h (Yang et al., 2007). Similarly, the levels of NAMPT and NAD^+ increase in 24-h fasted mice liver (Hayashida et al., 2010), suggesting that NAD^+ levels are strongly upregulated by nutrient restriction, and this is crucial for cell survival (Yang et al., 2007).

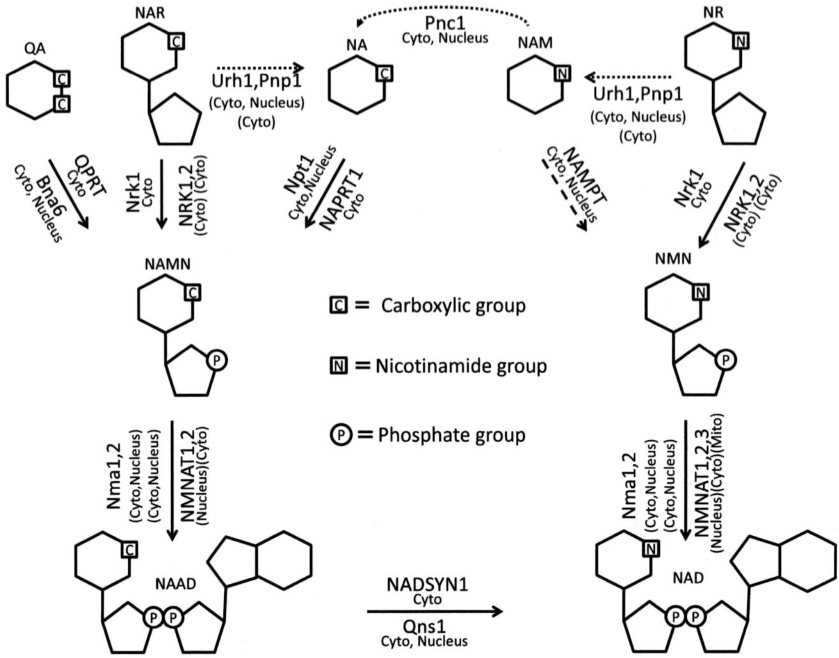

Figure 3 ***Comparative scheme of NAD$^+$ biosynthesis in mammals and yeast.*** Mammalian and yeast cells share the same precursors required for NAD$^+$ biosynthesis. Quinolinic acid (QA) derives from the kynurenine pathway. QA is converted to nicotinic acid mononucleotide (NAMN) by quinolinate phosphoribosyltransferase (QPRT) in mammals and by Bna6 in yeast. Nicotinic acid riboside (NAR) is converted by mammalian nicotinamide riboside kinase 1,2 (NRK1,2) and Nrk1 of yeast to NAMN. NAMN is also obtained from nicotinic acid (NA) by nicotinic acid phophoribosyltransferase 1 (NAPRT1) in mammal and by Npt1 in yeast. Nicotinamide mononucleotide adenylyltransferase 1,2 (NMNAT1,2) in mammals and Nma1,2 in yeast converts NAMN to nicotinic acid adenine dinucleotide (NAAD). NAD$^+$ is obtained from NAAD by nicotinamide adenine dinucleotide synthetase 1 (NADSYN1) in mammals and by Qns1 in yeast. Starting from QA, NA, and NAR, mammals and yeast share the same reactions for NAD$^+$ biosynthesis. However, yeast can also convert NAR to NA by Urh1 and Pnp1. Nicotinamide (NAM) has different metabolic routes in mammals and yeast. In mammals, NAM is converted into nicotinamide mononucleotide (NMN) by the enzyme Nicotinamide mononucleotidephophoribosyl transferase (NAMPT), which has no ortholog in yeast. Nicotinamide riboside is converted to NMN both in mammals and yeast by NRK1,2 and Nrk1, respectively. NMN is converted to NAD$^+$ by NMNAT1,2,3 in mammals and by Nma1,2 in yeast. In yeast, NR is converted to NAM by Urh1 and Pnp1. NAM can be used for NAD$^+$ biosynthesis only upon its conversion to NA by Pnc1. $\vec{x}$ indicates common reactions among mammals and yeast, $-- >$ indicates mammalian-specific reactions, and ■■■■> indicates yeast-specific reactions.

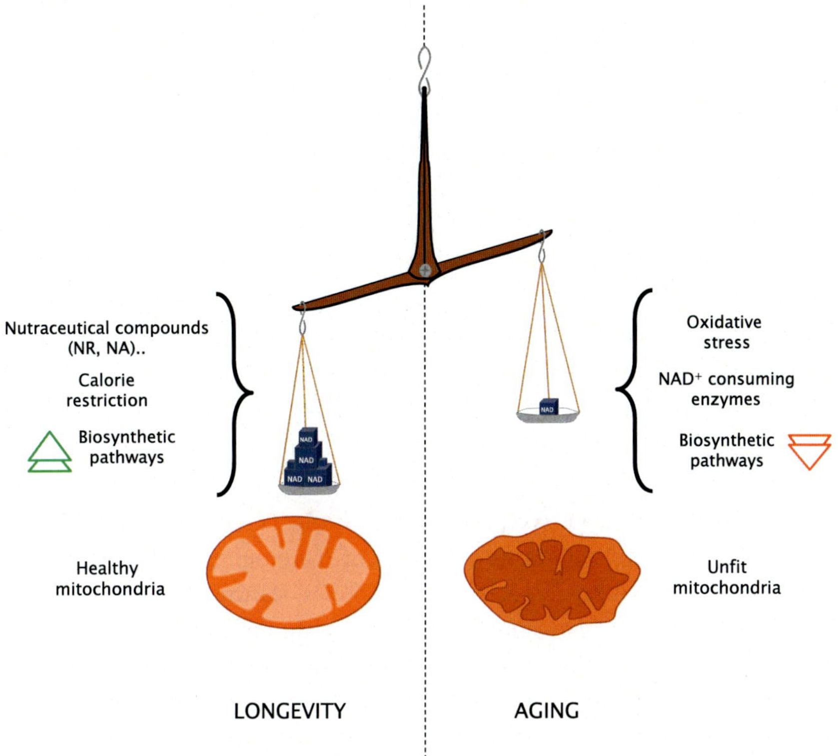

Figure 4 ***NAD$^+$ is involved in the maintenance of mitochondria functionality.*** Calorie restriction, NAD$^+$ precursors, and enzymes that synthesize NAD$^+$, all contribute to preserve NAD$^+$ levels, favoring the maintenance of mitochondria functionality (mammalian PGC1-α axis) and longevity. Oxidative stress and reduction in the activity of NAD$^+$ biosynthetic pathways unfit mitochondria and promote aging.

CR of yeast cells can be accomplished by a reduction in the glucose content in the growth media from 2% to 0.5% (Lin et al., 2002). NADH levels in these conditions decrease by about 50%, and the RLS increases significantly (Lin et al., 2002). It has been proposed that this is probably due to a metabolic shift from a predominantly fermentation-based metabolism to a respiration-based one determined by CR. This triggers a significant decrease in NADH, which is used by the ETC. Indeed, during respiratory growth, mitochondrial NADH is reoxidized by NADH dehydrogenase in the respiratory chain (Bakker et al., 2001).

However, this hypothesis is in contrast with the RLS increase observed in calorie-restricted respiratory deficient rho^0 mutants (Kaeberlein et al., 2005). Due to the absence of mitochondrial DNA, rho^0 lacks functional ETC and has an impaired TCA cycle (Jazwinski, 2005). The loss of mitochondrial membrane potential, which has a fundamental role in supporting the transport of biosynthetic intermediates outside the mitochondrion (Miceli et al., 2012; Passarella et al., 2003), results in the activation in rho^0 mutants of the retrograde response pathway, that in turn can extend RLS (Kirchman et al., 1999). The retrograde signaling is one of the most well-studied and conserved pathways involved in the cross-talk between nucleus and mitochondrion (da Cunha et al., 2015). Its main function is the upregulation of a wide range of metabolic and stress response genes in order to cope with the impairment of the TCA cycle activity. This transcriptional response resembles the one that is induced during CR (Jiang et al., 2002; Wang et al., 2010). In rho^0 cells the first three reactions of the TCA cycle are functional (Jazwinski, 2013), allowing the synthesis of α-ketoglutarate, precursor of glutamate that supplies nitrogen in biosyntheses, starting from citrate. The latter comes from the activation of the glyoxylate cycle, which synthetizes citrate from oxaloacetate and acetyl-CoA, but without any release of CO_2, preserving the two carbons of acetate, thus allowing the first three reactions of the TCA to proceed. This metabolic adaptation to the deficiency of biosynthetic intermediates is decisive for the increase of RLS in rho^0 cells (Jazwinski, 2013). Studies performed by Jiang et al. led to the identification of the target responsible for RLS extension after retrograde signaling activation (Jiang et al., 2016). Indeed, among the different candidates causing RLS lengthening, only the deletion of *PHO84*, which encodes a transporter of inorganic phosphate, suppressed the long-lived phenotype of rho^0 strains. Regarding chronological aging, the lack of mtDNA that causes respiratory-deficiency results in CLS shortening (Ocampo et al., 2012).

The aging process also influences NAD^+ levels in an opposite way compared with CR; indeed, aging is associated with different pathophysiological impacts, including the reduction of NAMPT-mediated NAD^+ biosynthesis (Imai and Yoshino, 2013). It is known that tumor necrosis factor-α and oxidative stress, which are decisive for the inflammatory process, reduce NAMPT and NAD^+ levels in primary hepatocytes (Yoshino et al., 2011). Considering that both the inflammatory cytokine production and the increase in oxidative stress significantly contribute to chronic age-associated inflammation (Singh and Newman, 2011), the development of chronic inflammation during aging may be the reason for the reduction of

NAMPT levels. However, other studies have shown that NAMPT levels and its enzymatic activity are reduced in human vascular smooth muscle cells during aging without inflammation (Van Der Veer et al., 2007), suggesting that chronic inflammation is certainly not the only cause of NAMPT reduction. Similarly, a reduction of both protein and messenger RNA (mRNA) levels of NAMPT has been found in mesenchymal stem cells (MSCs) obtained from aged rats (Ma et al., 2017). In both cases, the overexpression of NAMPT gene is sufficient to attenuate cellular senescence (Fig. 4). As mentioned previously, in mammals, NAMPT converts NAM to NMN in the cytosol; then NMN is transported in mitochondria where it is used for NAD^+ biosynthesis by NMNAT3. The activity of this enzyme is crucial to preserve the NAD^+ pool in mitochondria. NMNAT3 expression, as well as that of NAMPT, are susceptible to aging. A reduction of both NMNAT3 and mitochondrial NAD^+ levels has been also observed in aged human MSCs (Son et al., 2016); NMNAT3 overexpression in these cells is sufficient to delay replicative senescence. A proposed mechanism by which NAMPT activity might reverse the aged phenotype and increase lifespan relies on the activation of SIRT1, a member of the Sirtuin family endowed with an NAD^+-dependent deacetylase activity (Prolla and Denu, 2014).

Some studies suggest that NAMPT is the equivalent in terms of function of yeast Pnc1 (Yang et al., 2006), given that both these enzymes catalyze the first and rate-limiting step of NAD^+ biosynthesis from NAM (Revollo et al., 2004; Rongvaux et al., 2002) and that their expression increases in nutrient-restricted condition (Mei and Brenner, 2014; Yang et al., 2007). Similarly to NAMPT, RLS extension by CR requires *PNC1* (Anderson et al., 2003). The authors provide evidence that this is linked to an increase of NAD^+ levels that promotes Silent Information Regulator 2 (Sir2) activity. Sir2 is the ortholog of mammalian SIRT1 and founding member of the Sirtuin family. In the context of yeast replicative aging, *SIR2* is a prolongevity gene (Longo et al., 2012). Since *PNC1* functions in the NAD^+ salvage pathway, converting NAM in NA, it could promote Sir2 activation by increasing the availability of its co-substrate, NAD^+, and lowering that of its physiological inhibitor, NAM.

In addition to the biosynthetic pathway, the malate–aspartate shuttle also contributes to maintaining the mitochondrial physiological NAD^+ pool. It has been shown that the mitochondrial components of the malate–aspartate NADH shuttle are longevity factors required for lifespan extension in calorie-restricted yeast cells (Easlon et al., 2008). Some studies highlight that the lifespan increase observed in yeast in nutrient-restricted condition also depends on Npt1, an enzyme that catalyzes the conversion of NA to NAMN, contributing to the maintenance of NAD^+ levels (Lin et al., 2004). However, how nutritional conditions and aging affect the complex dynamics of NAMPT/*PNC1*/NAD^+ network, defined by some authors as the "NAD World" (Imai, 2010), still remains unclear. Understanding this mechanism is a fundamental investigation to develop approaches able to mitigate the detrimental effects of the aging process. Indeed, if NAMPT-mediated NAD^+ biosynthesis is compromised during aging, a way to preserve NAD^+ levels could be to provide other key intermediates such as NMN and NR. For example, Belenky et al. showed that NR supplementation to yeast cells increases RLS, through an increase of NAD^+ levels mediated by Nrk1, enzyme that converts NR to NMN (Belenky et al., 2007). For this reason, nutraceutical interventions could represent a valid alternative to pharmacological treatments to improve the aging process (Fig. 4).

10. NAD^+ IS A KEY MODULATOR OF PATHWAYS INVOLVED IN THE AGING PROCESS

NAD^+ is not only used in redox reactions but also as a co-substrate for some NAD^+-consuming enzymes in different cellular pathways (Imai and Guarente, 2014). Among these, we include Sirtuins, poly-ADP-ribose polymerase 1 (PARP-1), and CD38/CD157 ectoenzymes, which cleave the glycosidic bond between NAM moiety and ADP-ribose moiety of NAD^+.

Sirtuins are a family of NAD^+-dependent protein deacetylases. Sir2 of *S. cerevisiae* is the founding member of this family and regulates silencing at the mating-type loci, telomeres, and ribosomal DNA loci (Kaeberlein et al., 1999). In mammals, there are seven Sirtuin isoforms (SIRT1-7), which share a conserved NAD^+-binding and catalytic core domain with Sir2 (Li and Kazgan, 2011). The link between Sirtuins and aging was first observed in yeast. Indeed, Sir2 is required for RLS extension (Kaeberlein et al., 1999). In mice, at least two Sirtuins are involved in lifespan extension, SIRT6 and SIRT1. Ubiquitous SIRT6 overexpression determines an increase of lifespan, while the lack of SIRT6 results in accelerated aging, referred also as progeroid phenotype (Kanfi et al., 2012; Mostoslavsky et al., 2006). SIRT1 is the most well-studied of the mammalian Sirtuins. It is involved in epigenetic regulation, mitochondrial functionality, and aging (Yuan et al., 2016). Particularly, it has been proposed that SIRT1 may delay aging by the deacetylation of different substrates, including peroxisome proliferator-activated receptor-gamma coactivator 1 alpha (PGC-1α), which is the principal regulator of mitochondrial biogenesis (NAD^+/SIRT1/PGC-1α axis). Furthermore, SIRT1 decreases during aging, leading to a decrease of mitochondrial biogenesis in concert with an impaired mitochondrial turnover (Yuan et al., 2016). In the field of aging research, the modulation of SIRT1 activity represents a promising strategy to promote beneficial health effects and delay aging (Blum et al., 2011).

In *S. cerevisiae*, whereas Sir2 promotes RLS extension, most evidence so far supports a proaging role in CLS (Casatta et al., 2013; Fabrizio et al., 2005; Longo et al., 2012). Furthermore, when postdiauxic cells are transferred from expired medium to water (a severe CR condition), CLS significantly increases, and lack of Sir2 further exacerbates this effect (Casatta et al., 2013; Fabrizio et al., 2005). A further extension of CLS is also observed when *sir2*Δ mutant is combined with long-lived mutants that reduce the activity of two major nutrient-sensing pathways, such as TORC1-Sch9 and Ras-PKA ones. Since nutrient signaling attenuation determines an increase of CLS by ultimately activating endogenous defense mechanisms (Wei et al., 2008), it is plausible to assume that Sir2 also interferes with the same targets. Consistent with this, lack of Sir2 confers resistance to oxidative and heat stresses (Fabrizio et al., 2005).

Phosphoenolpyruvate carboxykinase (Pck1), enzyme catalyzing the rate-limiting step of gluconeogenesis, represents the cytosolic target of Sir2. Briefly, Sir2 is the enzyme responsible for the deacetylation and subsequent inactivation of Pck1 (Lin et al., 2009). During chronological aging, the lack of Sir2 deacetylase activity results in a major metabolic rearrangement toward gluconeogenesis, allowing an increase of trehalose stores essential for CLS (Casatta et al., 2013; Ocampo et al., 2012). Notably, chronological aging *sir2*Δ cells display other features that are useful for long-term survival of nondividing cells. Indeed, mutant cells have a high respiratory capacity and reduced levels of nonphosphorylating respiration in concert with a low $O_2^{\bullet -}$ content. Finally, they preserve functional mitochondria (Orlandi et al., 2017a, 2017b). NAM is an endogenous noncompetitive inhibitor of Sir2 activity; cells grown in the presence of NAM have a reduced RLS (Bitterman et al., 2002). Consistent with this, the overexpression of *PNC1*, which depletes NAM levels, increases RLS (Anderson et al., 2003). During chronological aging, NAM supplementation phenocopies the lack of Sir2. It determines the same anabolic and respiratory changes that lead to an increase of protective factors and low harmful $O_2^{\bullet -}$ levels, which ensure a longer CLS (Orlandi et al., 2017a).

PARP-1 is a multifunctional enzyme found in most Eukaryotes except for yeast (Hassa et al., 2006). It catalyzes the polymerization of ADP-ribose units of NAD^+, binding linear and branched poly ADP-ribose (PAR) polymers to itself, as well as to target proteins, following genotoxic stress (Luo and Lee Kraus, 2012). Indeed, PARP-1 is involved in the DNA damage response: it stabilizes the DNA replication fork by modifying its chromatin structure. Deficiency in PARP-1 accelerates aging in mice (Piskunova et al., 2008). In $PARP1^{-/-}$ mice, a reduction of both mean and maximum lifespan of 13.3% and 16.4%, respectively, was observed. These results are consistent with the idea that a proper DNA repair and longevity are positively correlated during aging. Other studies highlight that PARP-1 inhibition brings about an increase of oxidative metabolism, specifically an increase of O_2 consumption, via activation of SIRT1 (Bai et al., 2011). Since PARP-1 is a major cellular NAD^+ consumer, its activation depletes NAD^+ levels, which are important for Sirtuin activity. This aspect emphasizes that PARP-1 and SIRT1 compete for NAD^+ and entails that a proper balance of their enzymatic activity is crucial to regulate mitochondria maintenance and aging.

CD38 was identified in 1980 by studies focused on the identification of surface molecules involved in the immune system recognition (Reinherz et al., 1980). It is considered the homologous of CD157 because both belong to a gene family encoding enzymes that mediate cell—cell interaction (Quarona et al., 2013). Since they are located on the cellular surface, are classified as ectoenzymes able to modulate different cellular signals (from extracellular to intracellular environment and vice versa). To the best of our knowledge, CD38/CD157 are not directly involved in the aging process but are among the master regulators of intracellular NAD^+ levels in mammalian cells, influencing Sirtuin activity (Chini, 2009). As mentioned in the previous paragraphs, NAD^+ levels decrease during aging, promoting different age-related diseases. Several theories have been proposed to explain why NAD^+ levels are negatively affected during aging. However, to date, the specific contribution of metabolic pathways that modulate NAD^+ levels during the aging process still remains to be completely defined. Camacho-Pereira et al. suggest that CD38 dictates an age-dependent NAD^+ decline and mitochondria dysfunction through a SIRT3-mediated mechanism (Camacho-Pereira et al., 2016). The protein and mRNA levels of CD38 increase in several aged murine tissues, including liver, white adipose tissue, spleen, and skeletal muscle, leading to a strong reduction of NAD^+ levels. Concomitantly, a decrease of mitochondrial performance, in terms of respiration coupled to ATP synthesis, and a reduction of NAD^+/NADH ratio in mitochondria take place. These effects on oxidative mitochondrial metabolism depend on the deacetylase activity of SIRT3, which is localized in mitochondria. This age-dependent CD38 increase is also present in humans. Indeed, as observed for aged mice, also in aged human adipose tissue, CD38 mRNA levels increase of about 2.5 times compared with younger subjects (Camacho-Pereira et al., 2016).

11. CONCLUSIONS

All the aspects discussed previously indicate that mitochondrial metabolism plays a critical role in the maintenance of mitochondria functionality during aging, and its alterations lead to mitochondrial dysfunctions. These intensify progressively during aging and have detrimental consequences on cellular lifespan. On the one hand, the TCA cycle appears to function as a hub for mitochondrial metabolism. Indeed, many pathways converge at the level of this cycle, where signals are integrated in a dynamic and plastic, yet robust network that supports respiration and biosynthesis of BCAAs. As a consequence, an alteration in the TCA cycle can affect mitochondrial metabolism from different sides, with always a common negative effect on longevity. On the other hand, the complexes of ETC are the largest producers of ROS in the cell. Damage caused by ROS is frequently cited as one of the main causes of aging. In addition, the theory of the vicious circle proposes that ROS produced by respiration affect mitochondrial performance leading to more ROS and mitochondrial damage. Therefore, it is essential that the ETC remains efficient and respiration coupled in order to avoid the emergence of this vicious circle and ensure healthy aging. In addition, NAD metabolism is highlighted as a relevant connection between nutrients and respiration, capable of modulating the activity of Sirtuins.

Taking all together, mitochondrial metabolism progressively changes in time, with repercussions that extend to the whole cell causing a gradual decline of cellular functions, which define the onset of organismal weakening and the advancement of aging. The data reported in this review offer the rational basis to develop new promising strategies to prevent and control age-related disorders in humans.

ACKNOWLEDGMENTS

This work was supported by grant from CARIPLO Foundation 2015-0641 to M.V. G. S. and G. B. were supported by fellowships from SYSBIONET, Italian roadmap of ESFRI. The authors apologize for omission of relevant works and citations.

REFERENCES

Aerts, A.M., Zabrocki, P., Govaert, G., Mathys, J., Carmona-Gutierrez, D., Madeo, F., et al., 2009. Mitochondrial dysfunction leads to reduced chronological lifespan and increased apoptosis in yeast. FEBS Lett. 583, 113–117.

Ahmed, E.K., Rogowska-Wrzesinska, A., Roepstorff, P., Bulteau, A.L., Friguet, B., 2010. Protein modification and replicative senescence of WI-38 human embryonic fibroblasts. Aging Cell 9, 252–272.

Alvers, A.L., Fishwick, L.K., Wood, M.S., Hu, D., Chung, H.S., Dunn Jr., W.A., Aris, J.P., 2009. Autophagy and amino acid homeostasis are required for chronological longevity in *Saccharomyces cerevisiae*. Aging Cell 8, 353–369.

Anderson, R.M., Bitterman, K.J., Wood, J.G., Medvedik, O., Sinclair, D.A., 2003. Nicotinamide and *PNC*1 govern lifespan extension by calorie restriction in *Saccharomyces cerevisiae*. Nature 423, 181–185.

Aris, J.P., Fishwick, L.K., Marraffini, M.L., Seo, A.Y., Leeuwenburgh, C., Dunn Jr., W.A., 2012. Amino acid homeostasis and chronological longevity in *Saccharomyces cerevisiae*. Subcell. Biochem. 57, 161–186.

Bai, P., Cantó, C., Oudart, H., Brunyánszki, A., Cen, Y., Thomas, C., et al., 2011. PARP-1 inhibition increases mitochondrial metabolism through SIRT1 activation. Cell Metab. 13, 461–468.

Bakker, B.M., Overkamp, K.M., Van Maris, A.J.A., Kötter, P., Luttik, M.A.H., Van Dijken, J.P., Pronk, J.T., 2001. Stoichiometry and compartmentation of NADH metabolism in *Saccharomyces cerevisiae*. FEMS Microbiol. Rev. 25, 15–37.

Barros, M.H., da Cunha, F.M., Oliveira, G.A., Tahara, E.B., Kowaltowski, A.J., 2010. Yeast as a model to study mitochondrial mechanisms in ageing. Mech. Ageing Dev. 131, 494–502.

Belenky, P., Racette, F.G., Bogan, K.L., McClure, J.M., Smith, J.S., Brenner, C., 2007. Nicotinamide riboside promotes Sir2 silencing and extends lifespan via Nrk and Urh1/Pnp1/Meu1 pathways to NAD^+. Cell 129, 473–484.

Berg, S.M., Beck-Nielsen, H., Færgeman, N.J., Gaster, M., 2016. Carnitine acetyltransferase: a new player in skeletal muscle insulin resistance? Biochem. Biophys. Rep. 9, 47–50.

Bitterman, K.J., Anderson, R.M., Cohen, H.Y., Latorre-Esteves, M., Sinclair, D.A., 2002. Inhibition of silencing and accelerated aging by nicotinamide, a putative negative regulator of yeast Sir2 and human SIRT1. J. Biol. Chem. 277, 45099–45107.

Blum, C.A., Ellis, J.L., Loh, C., Ng, P.Y., Perni, R.B., Stein, R.L., 2011. SIRT1 modulation as a novel approach to the treatment of diseases of aging. J. Med. Chem. 54, 417–432.

Bogan, K.L., Brenner, C., 2008. Nicotinic acid, nicotinamide, and nicotinamide riboside: a molecular evaluation of NAD^+ precursor vitamins in human nutrition. Annu. Rev. Nutr. 28, 115–130.

Bowman, A., Birch-Machin, M.A., 2016. Age-dependent decrease of mitochondrial complex II activity in human skin fibroblasts. J. Invest. Dermatol. 136, 912–919.

Breitenbach, M., Rinnerthaler, M., Hartl, J., Stincone, A., Vowinckel, J., Breitenbach-Koller, H., Ralser, M., 2014. Mitochondria in ageing: there is metabolism beyond the ROS. FEMS Yeast Res. 14, 198–212.

Bricker, D.K., Taylor, E.B., Schell, J.C., Orsak, T., Boutron, A., Chen, Y.C., et al., 2012. A mitochondrial pyruvate carrier required for pyruvate uptake in yeast, *Drosophila*, and humans. Science 337, 96–100.
Cabiscol, E., Tamarit, J., Ros, J., 2014. Protein carbonylation: proteomics, specificity and relevance to aging. Mass Spectrom. Rev. 33, 21–48.
Camacho-Pereira, J., Tarragó, M.G., Chini, C.C.S., Nin, V., Escande, C., Warner, G.M., et al., 2016. CD38 dictates age-related NAD decline and mitochondrial dysfunction through an SIRT3-dependent mechanism. Cell Metab. 23, 1127–1139.
Casatta, N., Porro, A., Orlandi, I., Brambilla, L., Vai, M., 2013. Lack of Sir2 increases acetate consumption and decreases extracellular pro-aging factors. Biochim. Biophys. Acta Mol. Cell Res. 1833, 593–601.
Chen, Q., Vazquez, E.J., Moghaddas, S., Hoppel, C.L., Lesnefsky, E.J., 2003. Production of reactive oxygen species by mitochondria: central role of complex III. J. Biol. Chem. 278, 36027–36031.
Chini, E.N., 2009. CD38 as a regulator of cellular NAD: a novel potential pharmacological target for metabolic conditions. Curr. Pharm. Des. 15, 57–63.
Chistiakov, D.A., Sobenin, I.A., Revin, V.V., Orekhov, A.N., Bobryshev, Y.V., 2014. Mitochondrial aging and age-related dysfunction of mitochondria. BioMed Res. Int. 2014, 238463.
Cui, Y., Zhao, S., Wu, Z., Dai, P., Zhou, B., 2012. Mitochondrial release of the NADH dehydrogenase Ndi1 induces apoptosis in yeast. Mol. Biol. Cell 23, 4373–4382.
Da Cunha, F.M., Torelli, N.Q., Kowaltowski, A.J., 2015. Mitochondrial retrograde signaling: triggers, pathways, and outcomes. Oxid. Med. Cell. Longev. 2015, 482582.
Dillin, A., Hsu, A.L., Arantes-Oliveira, N., Lehrer-Graiwer, J., Hsin, H., Fraser, A.G., et al., 2002. Rates of behaviour and aging specified by mitochondrial function during development. Science 298, 2398–2401.
Dillon, E.L., Sheffield-Moore, M., Paddon-Jones, D., Gilkison, C., Sanford, A.P., Casperson, S.L., et al., 2009. Amino acid supplementation increases lean body mass, basal muscle protein synthesis, and insulin-like growth factor-I expression in older women. J. Clin. Endocr. Metab. 94, 1630–1637.
Drew, B., Phaneuf, S., Dirks, A., Selman, C., Gredilla, R., Lezza, A., et al., 2003. Effects of aging and caloric restriction on mitochondrial energy production in gastrocnemius muscle and heart. Am. J. Physiol. Regul. Integr. Comp. Physiol. 284, R474–R480.
Easlon, E., Tsang, F., Skinner, C., Wang, C., Lin, S., 2008. The malate-aspartate NADH shuttle components are novel metabolic longevity regulators required for calorie restriction-mediated life span extension in yeast. Genes Dev. 22, 931–944.
Eisenberg, T., Schroeder, S., Andryushkova, A., Pendl, T., Küttner, V., Bhukel, A., et al., 2014. Nucleocytosolic depletion of the energy metabolite acetyl-coenzyme a stimulates autophagy and prolongs lifespan. Cell Metab. 19, 431–444.
Fabrizio, P., Gattazzo, C., Battistella, L., Wei, M., Cheng, C., McGrew, K., Longo, V.D., 2005. Sir2 blocks extreme life-span extension. Cell 123, 655–667.
Fabrizio, P., Hoon, S., Shamalnasab, M., Galbani, A., Wei, M., Giaever, G., et al., 2010. Genome-wide screen in *Saccharomyces cerevisiae* identifies vacuolar protein sorting, autophagy, biosynthetic, and tRNA methylation genes involved in life span regulation. PLoS Genet. 6, e1001024.
Fang, J., Beattie, D.S., 2003. External alternative NADH dehydrogenase of *Saccharomyces cerevisiae*: a potential source of superoxide. Free Radic. Biol. Med. 34, 478–488.
Fato, R., Bergamini, C., Leoni, S., Lenaz, G., 2008. Mitochondrial production of reactive oxygen species: role of complex I and quinone analogues. Biofactors 32, 31–39.
Fleck, C.B., Brock, M., 2009. Re-characterisation of *Saccharomyces cerevisiae* Ach1p: fungal CoA-transferases are involved in acetic acid detoxification. Fungal. Genet. Biol. 46, 473–485.

Fontana, L., Partridge, L., Longo, V.D., 2010. Extending healthy life span—from yeast to humans. Science 328, 321–326.
Fujita, S., Volpi, E., 2006. Amino acids and muscle loss with aging. J. Nutr. 136, 277S–280S.
Hacioglu, E., Demir, A.B., Koc, A., 2012. Identification of respiratory chain gene mutations that shorten replicative life span in yeast. Exp. Gerontol. 47, 149–153.
Hansford, R.G., Hogue, B.A., Mildaziene, V., 1997. Dependence of H_2O_2formation by rat heart mitochondria on substrate availability and donor age. J. Bioenerg. Biomembr. 29, 89–95.
Harman, D., 1956. Aging: a theory based on free radical and radiation chemistry. J. Gerontol. 11, 298–300.
Hassa, P.O., Haenni, S.S., Elser, M., Hottiger, M.O., 2006. Nuclear ADP-ribosylation reactions in mammalian cells: where are we today and where are we going? Microbiol. Mol. Biol. Rev. 70, 789–829.
Hayashida, S., Arimoto, A., Kuramoto, Y., Kozako, T., Honda, S.I., Shimeno, H., Soeda, S., 2010. Fasting promotes the expression of SIRT1, an NAD^+-dependent protein deacetylase, via activation of PPARα in mice. Mol. Cell. Biochem. 339, 285–292.
Herzig, S., Raemy, E., Montessuit, S., Veuthey, J.L., Zamboni, N., Westermann, B., et al., 2012. Identification and functional expression of the mitochondrial pyruvate carrier. Science 337, 93–96.
Hinnebusch, A.G., 2005. Translational regulation of *GCN4* and the general amino acid control of yeast. Annu. Rev. Microbiol. 59, 407–450.
Hirst, J., King, M.S., Pryde, K.R., 2008. The production of reactive oxygen species by complex I. Biochem. Soc. Trans. 36, 976–980.
Houtkooper, R.H., Cantó, C., Wanders, R.J., Auwerx, J., 2010. The secret life of NAD^+: an old metabolite controlling new metabolic signaling pathways. Endocr. Rev. 31, 194–223.
Imai, S., 2010. "Clocks" in the NAD World: NAD as a metabolic oscillator for the regulation of metabolism and aging. Biochim. Biophys. Acta Proteins Proteomics 1804, 1584–1590.
Imai, S., Guarente, L., 2014. NAD^+ and sirtuins in aging and disease. Trends Cell Biol. 24, 464–471.
Imai, S., Yoshino, J., 2013. The importance of NAMPT/NAD/SIRT1 in the systemic regulation of metabolism and ageing. Diabetes Obes. Metab. 15, 26–33.
Ingram, T., Chakrabarti, L., 2016. Proteomic profiling of mitochondria: what does it tell us about the ageing brain? Aging (Albany NY) 8. Article ID: 3161.
Jazwinski, S.M., 2005. Rtg2 protein: at the nexus of yeast longevity and aging. FEMS Yeast Res. 5, 1253–1259.
Jazwinski, S.M., 2013. The retrograde response: when mitochondrial quality control is not enough. Biochim. Biophys. Acta Mol. Cell Res. 1833, 400–409.
Jiang, J.C., Stumpferl, S.W., Tiwari, A., Qin, Q., Rodriguez-Quiñones, J.F., Jazwinski, S.M., 2016. Identification of the target of the retrograde response that mediates replicative lifespan extension in *Saccharomyces cerevisiae*. Genetics 204, 659–673.
Jiang, J.C., Wawryn, J., Kumara, H.S., Jazwinski, S.M., 2002. Distinct roles of processes modulated by histone deacetylases Rpd3p, Hda1p, and Sir2p in life extension by caloric restriction in yeast. Exp. Gerontol. 37, 1023–1030.
Kaeberlein, M., 2010. Lessons on longevity from budding yeast. Nature 464, 513–519.
Kaeberlein, M., Hu, D., Kerr, E.O., Tsuchiya, M., Westman, E.A., Dang, N., Fields, S., Kennedy, B.K., 2005. Increased life span due to calorie restriction in respiratory-deficient yeast. PLoS Genet. 1, e69.
Kaeberlein, M., McVey, M., Guarente, L., 1999. The SIR2/3/4 complex and SIR2 alone promote longevity in *Saccharomyces cerevisiae* by two different mechanisms. Genes Dev. 13, 2570–2580.

Kamei, Y., Tamada, Y., Nakayama, Y., Fukusaki, E., Mukai, Y., 2014. Changes in transcription and metabolism during the early stage of replicative cellular senescence in budding yeast. J. Biol. Chem. 289, 32081–32093.

Kanfi, Y., Naiman, S., Amir, G., Peshti, V., Zinman, G., Nahum, L., et al., 2012. The sirtuin SIRT6 regulates lifespan in male mice. Nature 483, 218–221.

Kirchman, P.A., Kim, S., Lai, C.Y., Jazwinski, S.M., 1999. Interorganelle signaling is a determinant of longevity in *Saccharomyces cerevisiae*. Genetics 152, 179–190.

Kitani, T., Okuno, S., Fujisawa, H., 2003. Growth phase-dependent changes in the subcellular localization of pre-B-cell colony-enhancing factor. FEBS Lett. 544, 74–78.

Korshunov, S.S., Skulachev, V.P., Starkov, A.A., 1997. High protonic potential actuates a mechanism of production of reactive oxygen species in mitochondria. FEBS Lett. 416, 15–18.

Kumaran, S., Subathra, M., Balu, M., Panneerselvam, C., 2005. Supplementation of L-carnitine improves mitochondrial enzymes in heart and skeletal muscle of aged rats. Exp. Aging Res. 31, 55–67.

Kwon, Y., Choi, K., Cho, C., Lee, C., 2015. Mitochondrial efficiency-dependent viability of *Saccharomyces cerevisiae* mutants carrying individual electron transport chain component deletions. Mol. Cells 38, 1054–1063.

Lai, C.Y., Jaruga, E., Borghouts, C., Jazwinski, S.M., 2002. A mutation in the *ATP2* gene abrogates the age asymmetry between mother and daughter cells of the yeast *Saccharomyces cerevisiae*. Genetics 162, 73–87.

Lane, R.K., Hilsabeck, T., Rea, S.L., 2015. The role of mitochondrial dysfunction in age-related diseases. Biochim. Biophys. Acta Bioenerg. 1847, 1387–1400.

Lee, C., Longo, V., 2016. Dietary restriction with and without caloric restriction for healthy aging. F1000Res. 5, 1–7.

Li, W., Sun, L., Liang, Q., Wang, J., Mo, W., Zhou, B., 2006. Yeast AMID homologue Ndi1p displays respiration-restricted apoptotic activity and is involved in chronological aging. Mol. Biol. Cell 17, 1802–1811.

Li, X., Kazgan, N., 2011. Mammalian sirtuins and energy metabolism. Int. J. Biol. Sci. 7, 575–587.

Lin, S.J., Ford, E., Haigis, M., Liszt, G., Guarente, L., 2004. Calorie restriction extends yeast life span by lowering the level of NADH. Genes Dev. 18, 12–16.

Lin, S.J., Kaeberlein, M., Andalis, A.A., Sturtz, L.A., Defossez, P.A., Culotta, V.C., et al., 2002. Calorie restriction extends *Saccharomyces cerevisiae* lifespan by increasing respiration. Nature 418, 344–348.

Lin, Y.Y., Lu, J.Y., Zhang, J., Walter, W., Dang, W., Wan, J., et al., 2009. Protein acetylation microarray reveals that Nua4 controls key metabolic target regulating gluconeogenesis. Cell 136, 1073–1084.

Longo, V.D., Kennedy, B.K., 2006. Sirtuins in aging and age-related disease. Cell 126, 257–268.

Longo, V.D., Antebi, A., Bartke, A., Barzilai, N., Brown-Borg, H.M., Caruso, C., Ingram, D., 2015. Interventions to slow aging in humans: are we ready? Aging Cell 14, 497–510.

Longo, V.D., Liou, L.L., Valentine, J.S., Gralla, E.B., 1999. Mitochondrial superoxide decreases yeast survival in stationary phase. Arch. Biochem. Biophys. 365, 131–142.

Longo, V.D., Shadel, G.S., Kaeberlein, M., Kennedy, B., 2012. Replicative and chronological aging in *Saccharomyces cerevisiae*. Cell Metab. 16, 18–31.

López-Otín, C., Blasco, M.A., Partridge, L., Serrano, M., Kroemer, G., 2013. The hallmarks of aging. Cell 153, 1194–1217.

Luo, X., Lee Kraus, W., 2012. On par with PARP: cellular stress signaling through poly(-ADP-ribose) and PARP-1. Genes Dev. 26, 417–432.

Ma, C., Pi, C., Yang, Y., Lin, L., Shi, Y., Li, Y., et al., 2017. Nampt expression decreases age-related senescence in rat bone marrow mesenchymal stem cells by targeting Sirt1. PLoS One 12, 1–18.

MacLean, M., Harris, N., Piper, P.W., 2001. Chronological lifespan of stationary phase yeast cells; a model for investigating the factors that might influence the ageing of postmitotic tissues in higher organisms. Yeast 18, 499–509.

Mei, S.C., Brenner, C., 2014. Quantification of protein copy number in yeast: the NAD^+ metabolome. PLoS One 9, 1–10.

Miceli, M.V., Jiang, J.C., Tiwari, A., Rodriguez-Quiñones, J.F., Jazwinski, S.M., 2012. Loss of mitochondrial membrane potential triggers the retrograde response extending yeast replicative lifespan. Front. Genet. 2, e102.

Mihalik, S.J., Goodpaster, B.H., Kelley, D.E., et al., 2010. Increased levels of plasma acylcarnitines in obesity and type 2 diabetes and identification of a marker of glucolipotoxicity. Obesity 18, 1695–1700.

Mirisola, M.G., Taormina, G., Fabrizio, P., Wei, M., Hu, J., Longo, V.D., 2014. Serine-and threonine/valine-dependent activation of PDK and Tor orthologs converge on Sch9 to promote aging. PLoS Genet. 10, e1004113.

Mostoslavsky, R., Chua, K.F., Lombard, D.B., Pang, W.W., Fischer, M.R., Gellon, L., et al., 2006. Genomic instability and aging-like phenotype in the absence of mammalian SIRT6. Cell 124, 315–329.

Nikiforov, A., Dölle, C., Niere, M., Ziegler, M., 2011. Pathways and subcellular compartmentation of NAD biosynthesis in human cells: from entry of extracellular precursors to mitochondrial NAD generation. J. Biol. Chem. 286, 21767–21778.

Nyström, T., 2005. Role of oxidative carbonylation in protein quality control and senescence. EMBO J. 24, 1311–1317.

Ocampo, A., Liu, J., Schroeder, E.A., Shadel, G.S., Barrientos, A., 2012. Mitochondrial respiratory thresholds regulate yeast chronological life span and its extension by caloric restriction. Cell Metab. 16, 55–67.

Orlandi, I., Casatta, N., Vai, M., 2012. Lack of Ach1 CoA-transferase triggers apoptosis and decreases chronological lifespan in yeast. Front. Oncol. 2. Article ID: 67.

Orlandi, I., Pellegrino Coppola, D., Vai, M., 2014. Rewiring yeast acetate metabolism through *MPC1* loss of function leads to mitochondrial damage and decreases chronological lifespan. Microbial. Cell 1, 393–405.

Orlandi, I., Pellegrino Coppola, D., Strippoli, M., Ronzulli, R., Vai, M., 2017a. Nicotinamide supplementation phenocopies *SIR2* inactivation by modulating carbon metabolism and respiration during yeast chronological aging. Mech. Ageing Dev. 161, 277–287.

Orlandi, I., Stamerra, G., Strippoli, M., Vai, M., 2017b. During yeast chronological aging resveratrol supplementation results in a short-lived phenotype Sir2-dependent. Redox Biol. 12, 745–754.

Pansarasa, O., Flati, V., Corsetti, G., Brocca, L., Pasini, E., D'Antona, G., 2008. Oral amino acid supplementation counteracts age-induced sarcopenia in elderly rats. Am. J. Cardiol. 101, 35E–41E.

Passarella, S., Atlante, A., Valenti, D., de Bari, L., 2003. The role of mitochondrial transport in energy metabolism. Mitochondrion 2, 319–343.

Petrosillo, G., Matera, M., Moro, N., Ruggiero, F.M., Paradies, G., 2009. Mitochondrial complex I dysfunction in rat heart with aging: critical role of reactive oxygen species and cardiolipin. Free Radic. Biol. Med. 46, 88–94.

Piskunova, T.S., Yurova, M.N., Ovsyannikov, A.I., Semenchenko, A.V., Zabezhinski, M.A., Popovich, I.G., et al., 2008. Deficiency in poly(ADP-ribose) polymerase-1 (PARP-1) accelerates aging and spontaneous carcinogenesis in mice. Curr. Gerontol. Geriatr. Res. 2008. Article ID: 754190.

Porter, C., Hurren, N.M., Cotter, M.V., Bhattarai, N., Reidy, P.T., Dillon, E.L., et al., 2015. Mitochondrial respiratory capacity and coupling control decline with age in human skeletal muscle. Am. J. Physiol. Endocrinol. Metab. 309, E224–E232.

Powers 3rd, R.W., Kaeberlein, M., Caldwell, S.D., Kennedy, B.K., Fields, S., 2006. Extension of chronological life span in yeast by decreased TOR pathway signaling. Genes Dev. 20, 174–184.

Preston, C.C., Oberlin, A.S., Holmuhamedov, E.L., Gupta, A., Sagar, S., Syed, R.H.K., et al., 2008. Aging-induced alterations in gene transcripts and functional activity of mitochondrial oxidative phosphorylation complexes in the heart. Mech. Ageing Dev. 129, 304–312.

Prolla, T.A., Denu, J.M., 2014. NAD^+ deficiency in age-related mitochondrial dysfunction. Cell Metab. 19, 178–180.

Quarona, V., Zaccarello, G., Chillemi, A., Brunetti, E., Singh, V.K., Ferrero, E., et al., 2013. CD38 and CD157: a long journey from activation markers to multifunctional molecules. Cytom. Part B Clin. Cytom. 84, 207–217.

Quinlan, C.L., Orr, A.L., Perevoshchikova, I.V., Treberg, J.R., Ackrell, B.A., Brand, M.D., 2012. Mitochondrial complex II can generate reactive oxygen species at high rates in both the forward and reverse reactions. J. Biol. Chem. 287, 27255–27264.

Raha, S., Robinson, B.H., 2001. Mitochondria, oxygen free radicals, and apoptosis. Am. J. Med. Genet. Semin. Med. Genet. 106, 62–70.

Ratajczak, J., Joffraud, M., Trammell, S.A.J., Ras, R., Canela, N., Boutant, M., et al., 2016. NRK1 controls nicotinamide mononucleotide and nicotinamide riboside metabolism in mammalian cells. Nat. Commun. 7. Article ID: 13103.

Reinherz, E.L., Kung, P.C., Goldstein, G., Levey, R.H., Schlossman, S.F., 1980. Discrete stages of human intrathymic differentiation: analysis of normal thymocytes and leukemic lymphoblasts of T-cell lineage. Proc. Natl. Acad. Sci. U. S. A. 77, 1588–1592.

Reverter-Branchat, G., Cabiscol, E., Tamarit, J., Ros, J., 2004. Oxidative damage to specific proteins in replicative and chronological-aged *Saccharomyces cerevisiae* common targets and prevention by calorie restriction. J. Biol. Chem. 279, 31983–31989.

Revollo, J.R., Grimm, A.A., Imai, S.I., 2004. The NAD biosynthesis pathway mediated by nicotinamide phosphoribosyltransferase regulates Sir2 activity in mammalian cells. J. Biol. Chem. 279, 50754–50763.

Robinson, B.H., 1998. Human complex I deficiency: clinical spectrum and involvement of oxygen free radicals in the pathogenicity of the defect. Biochim. Biophys. Acta Bioenerg. 1364, 271–286.

Rongvaux, A., She, R.J., Mulks, M.H., Gigot, D., Urbain, J., Leo, O., Andris, F., 2002. Pre-B-cell colony-enhancing factor, whose expression is up-regulated in activated lymphocytes, is a nicotinamide phosphoribosyltransferase, a cytosolic enzyme involved in NAF biosynthesis. Eur. J. Immunol. 32, 3225–3234.

Ruetenik, A., Barrientos, A., 2015. Dietary restriction, mitochondrial function and aging: from yeast to humans. Biochim. Biophys. Acta Bioenerg. 1847, 1434–1447.

Samokhvalov, V., Ignatov, V., Kondrashova, M., 2004. Inhibition of Krebs cycle and activation of glyoxylate cycle in the course of chronological aging of *Saccharomyces cerevisiae*. Compensatory role of succinate oxidation. Biochimie 86, 39–46.

Sauve, A.A., 2008. NAD^+ and vitamin B_3: from metabolism to therapies. J. Pharmacol. Exp. Ther. 324, 883–893.

Scialò, F., Fernández-Ayala, D.J., Sanz, A., 2017. Role of mitochondrial reverse electron transport in ROS signaling: potential roles in health and disease. Front. Physiol. 8. Article ID: 428.

Shimomura, Y., Yamamoto, Y., Bajotto, G., Sato, J., Murakami, T., Shimomura, N., et al., 2006. Nutraceutical effects of branched-chainamino acids on skeletal muscle. J. Nutr. 136, 529S–532S.

Singh, T., Newman, A., 2011. Inflammatory markers in population studies of ageing. Ageing Res. Rev. 10, 2–22.

Son, M.J., Kwon, Y., Son, T., Cho, Y.S., 2016. Restoration of mitochondrial NAD^+levels delays stem cell senescence and facilitates reprogramming of aged somatic cells. Stem Cell. 34, 2840–2851.

Stauch, K.L., Purnell, P.R., Villeneuve, L.M., Fox, H.S., 2015. Proteomic analysis and functional characterization of mouse brain mitochondria during aging reveal alterations in energy metabolism. Proteomics 15, 1574–1586.

Stefanatos, R., Sanz, A., 2011. Mitochondrial complex I: a central regulator of the aging process. Cell Cycle 10, 1528–1532.

Stein, L.R., Imai, S.I., 2012. The dynamic regulation of NAD metabolism in mitochondria. Trends Endocrinol. Metab. 23, 420–428.

Stöckl, P., Hütter, E., Zwerschke, W., Jansen-Dürr, P., 2006. Sustained inhibition of oxidative phosphorylation impairs cell proliferation and induces premature senescence in human fibroblasts. Exp. Gerontol. 41, 674–682.

Sugrue, M.M., Tatton, W.G., 2001. Mitochondrial membrane potential in aging cells. Biol. Signals Recept. 10, 176–188.

Tamarit, J., de Hoogh, A., Obis, E., Alsina, D., Cabiscol, E., Ros, J., 2012. Analysis of oxidative stress-induced protein carbonylation using fluorescent hydrazides. J. Proteomics 75, 3778–3788.

Tatarková, Z., Kuka, S., Račay, P., Lehotský, J., Dobrota, D., Mištuna, D., Kaplán, P., 2011. Effects of aging on activities of mitochondrial electron transport chain complexes and oxidative damage in rat heart. Physiol. Res. 60, 281–289.

Todisco, S., Agrimi, G., Castegna, A., Palmieri, F., 2006. Identification of the mitochondrial NAD^+ transporter in *Saccharomyces cerevisiae*. J. Biol. Chem. 281, 1524–1531.

Vacanti, N.M., Divakaruni, A.S., Green, C.R., Parker, S.J., Henry, R.R., Ciaraldi, T.P., et al., 2014. Regulation of substrate utilization by the mitochondrial pyruvate Carrier. Mol. Cell 56, 425–435.

Van Der Veer, E., Ho, C., O'Neil, C., Barbosa, N., Scott, R., Cregan, S.P., Pickering, J.G., 2007. Extension of human cell lifespan by nicotinamide phosphoribosyltransferase. J. Biol. Chem. 282, 10841–10845.

Verdin, E., 2015. NAD^+ in aging, metabolism, and neurodegeneration. Science 350, 1208–1213.

Wallace, D.C., 2009. Mitochondria, bioenergetics, and the epigenome in eukaryotic and human evolution. Cold Spring Harb. Symp. Quant. Biol. 74, 383–393.

Wang, J., Jiang, J.C., Jazwinski, S.M., 2010. Gene regulatory changes in yeast during life extension by nutrient limitation. Exp. Gerontol. 45, 621–631.

Wei, M., Fabrizio, P., Hu, J., Ge, H., Cheng, C., Li, L., Longo, V.D., 2008. Life span extension by calorie restriction depends on Rim15 and transcription factors downstream of Ras/PKA, Tor, and Sch9. PLoS Genet. 4, 0139–0149.

White, A.T., Schenk, S., 2012. NAD^+/NADH and skeletal muscle mitochondrial adaptations to exercise. Am. J. Physiol. Endocrinol. Metab. 303, E308–E321.

Yang, C., Ko, B., Hensley, C.T., Jiang, L., Wasti, A.T., Kim, J., et al., 2014. Glutamine oxidation maintains the TCA cycle and cell survival during impaired mitochondrial pyruvate transport. Mol. Cell 56, 414–424.

Yang, H., Lavu, S., Sinclair, D.A., 2006. Nampt/PBEF/Visfatin: a regulator of mammalian health and longevity? Exp. Gerontol. 41, 718–726.

Yang, H., Yang, T., Baur, J.A., Perez, E., Matsui, T., Carmona, J.J., Lamming, D.W.W., et al., 2007. Nutrient-sensitive mitochondrial NAD^+ levels dictate cell survival. Cell 130, 1095–1107.

Yarian, C.S., Toroser, D., Sohal, R.S., 2006. Aconitase is the main functional target of aging in the citric acid cycle of kidney mitochondria from mice. Mech. Ageing Dev. 127, 79–84.

Yoshino, J., Mills, K.F., Yoon, M.J., Imai, S.I., 2011. Nicotinamide mononucleotide, a key NAD^+ intermediate, treats the pathophysiology of diet- and age-induced diabetes in mice. Cell Metab. 14, 528–536.

Yuan, Y., Cruzat, V.F., Newshome, P., Cheng, J., Chen, Y., Lu, Y., 2016. Regulation of SIRT1 in aging: roles in mitochondrial function and biogenesis. Mech. Ageing Dev. 155, 10–21.

CHAPTER TWO

Mitohormesis, an Antiaging Paradigm

Clea Bárcena*, Pablo Mayoral* and Pedro M. Quirós*,§,1
*Departamento de Bioquímica y Biología Molecular, Facultad de Medicina, Instituto Universitario de Oncología (IUOPA), Universidad de Oviedo, 33006-Oviedo, Spain
§Centro de Investigación Biomédica en Red de Cáncer (CIBERONC), Spain
[1]Corresponding author: E-mail: pmquiros@gmail.com

Contents

Abstract

Mitohormesis is a term used to define a biological response where the induction of a reduced amount of mitochondrial stress leads to an increment in health and viability within a cell, tissue, or organism. The mitochondrial stress response activated by a potentially damaging stimulus requires a coordinated dialogue with the cellular nucleus, known as mitonuclear communication. This interplay induced by the hormetic response in mitochondria relies in a variety of signals among which the most relevant ones are reactive oxygen species (ROS), mitochondrial metabolites, proteotoxic signals, the mitochondria–cytosol stress response, and the release of mitokines. The activation of the mitohormetic response increases lifespan in different animal models, from worms to mammals. Further, mitohormesis also enhances healthspan, particularly improving metabolism and immune system. Although multiple mediators and stress signals have been proposed to activate this protective mechanism, beneficial outcomes of mitohormesis are most probably due to an increase in mitochondrial ROS.

International Review of Cell and Molecular Biology, Volume 340
ISSN 1937-6448
https://doi.org/10.1016/bs.ircmb.2018.05.002

Activation of other protective stress mechanisms as mitochondrial unfolded protein response or the increase in the expression of mitokines are also associated with the positive benefits exerted by mitohormesis. Herein, we review the different mitohormetic signals and pathways described from worms to mammals and their effects on health and survival. The identification and description of pathways and molecules implicated in the beneficial effects of mitohormesis will help understand the complex balance between death and survival in the face of mitochondrial damage and will allow to open a novel area of therapies aimed at improving health in humans.

1. INTRODUCTION

In the XVI century, the Swiss physician Paracelsus, known as the father of the toxicology, stated in his *Third Defense* the concept that "only the dose makes the poison" (from Latin: "*sola dosis facit venenum*") (Paracelsus, 1564). This idea was used to defend that all chemical compounds can be toxic depending on the dose and not on the chemical properties themselves. Three centuries later the term "hormesis" was coined to define compounds or substances that show a "biphasic dose response," i.e., a low-dose stimulation and high-dose inhibition (Fig. 1). Nowadays, in Biology, the idea of hormesis, or the concept of hormetic responses, is

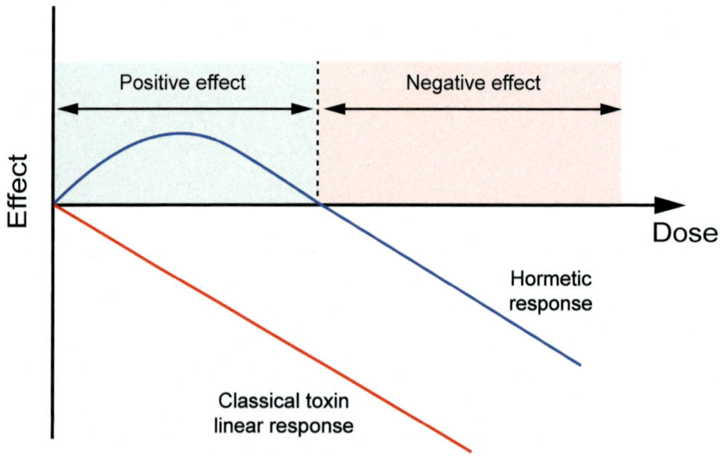

Figure 1 ***Hormetic response.*** With the classical view of toxin response, an increasing amount of toxin is expected to show a linear response between the effect (e.g., toxicity or mortality) and the dose (*red line*). On the contrary, hormesis is characterized by a response in which a low dose of a stress-inducing compound leads to positive effects, as an increment in viability, whereas high doses induce negative effects, such as cell damage or mortality (*blue line*).

well defined to describe the situation where a low dose of stress stimuli can activate an adaptive response in cells and organisms to maintain homeostasis, promoting health and even longevity, whereas higher doses become harmful. The concept of hormesis was then expanded and studied based on the stimuli. The concept of "mitohormesis" appeared in 2006 as a theory to define the hormetic response that promotes health and vitality induced by sublethal mitochondrial stress (Tapia, 2006). This theoretical hypothesis was proposed to unify a series of hormetic responses carried out by mitochondrial reactive oxygen species (mtROS), such as those induced by caloric restriction (CR), intermittent fasting, exercise, and dietary phytonutrients. A year later, in 2007, the group led by Michael Ristow provided experimental support to this theory by showing that glucose restriction extended lifespan of worms by increasing reactive oxygen species (ROS) and mitochondrial respiration (Schulz et al., 2007). This ROS-mediated hormetic response was shown to activate catalase activity and, therefore, increase oxidative stress response. Importantly, the role of mtROS was described as essential because treatment with antioxidants abolished the lifespan extension (Schulz et al., 2007). This study, together with previous evidences on ROS signaling, began to question the widespread use of antioxidants as supplements or as treatment against diverse pathological conditions, including type 2 diabetes (Kawagishi and Finkel, 2014; Ristow, 2014).

The study of hormetic responses has gained considerable relevance since the discovery that mtROS and other mitochondrial stress pathways can promote lifespan in a hormetic-dependent manner. Thus, several groups have focused their research on the pursuit of mitohormesis inducers to improve healthspan and lifespan, especially by using lower organisms, such as *Saccharomyces cerevisiae*, *Caenorhabditis elegans*, and *Drosophila melanogaster*, where it was observed that some mitochondrial mutants or the treatment with mitochondrial toxins induced an unexpected increase in lifespan. In this review, we describe recent advances in mitohormesis, beginning with a description of the major signaling pathways governing this response, and finally discussing experimental evidence in different model organisms supporting the potential of mitohormesis as a prohealth and prolongevity strategy.

2. MITONUCLEAR COMMUNICATION

Mitochondria are highly dynamic organelles derived from alphaproteobacteria that were engulfed by the precursor of the eukaryotic cell. During evolution within the eukaryotic cell there was an endosymbiosis process that triggered the loss in the mitochondrial DNA (mtDNA) of genes encoding mitochondrial proteins. Thus, among more than 1000 proteins present in mammalian mitochondria, only 13 are encoded in the mtDNA, which correspond to subunits of the oxidative phosphorylation (OXPHOS) system; mtDNA also encodes 22 transfer RNAs (tRNAs) and 2 ribosomal RNAs (rRNAs). The remaining mitochondrial proteins are encoded in the nucleus, and, therefore, they have to be translated in cytosol and transported and folded into mitochondria. This complex system makes indispensable coordination between both mitochondrial and nuclear genomes, involving proteins exerting a wide variety of functions, from ribosomes and cytosolic chaperones to mitochondrial transporters, chaperones, and proteases. Different forms of mitonuclear communication have been described to regulate mitochondrial function depending on extracellular, cytosolic, or mitochondrial signals (Quiros et al., 2016). Thus, the nucleus can regulate mitochondrial function by means of anterograde communication, integrating extracellular signals such as cold or exercise, and promoting mitochondrial biogenesis through the expression of mitochondrial genes (Scarpulla et al., 2012). Conversely, a retrograde response from mitochondria can activate the expression of nuclear genes to respond to stress conditions generated in this organelle, such as OXPHOS dysfunction, defects in mtDNA, or loss of membrane potential, which can promote a decrease in ATP production, an increase in ROS, and the release of calcium from mitochondria (Arnould et al., 2015; Guha and Avadhani, 2013). Some mitochondrial insults can also activate a bidirectional response—mitonuclear feedback—which activates the expression of nuclear-encoded mitochondrial proteins aimed at resolving the stress (Quiros et al., 2016; Topf et al., 2016); these proteins are mainly proteases and chaperones that increase the mitochondrial proteostasis (Quiros et al., 2015; Voos, 2013). Finally, mitochondrial stress can also activate a mitochondrial-cytosolic response, mainly affecting translational mechanisms, or an extracellular response mediated by mitokines—signals or molecules secreted on mitochondrial stress—which activate a response in other cells in a paracrine or endocrine manner (Lee, 2015; Quiros et al., 2016).

3. MITOHORMESIS SIGNALING

The different ways of mitonuclear communication, especially those activated under stress conditions, are aimed at resolving the stress at mitochondrial, cellular, or tissue level. However, all these signals and forms of

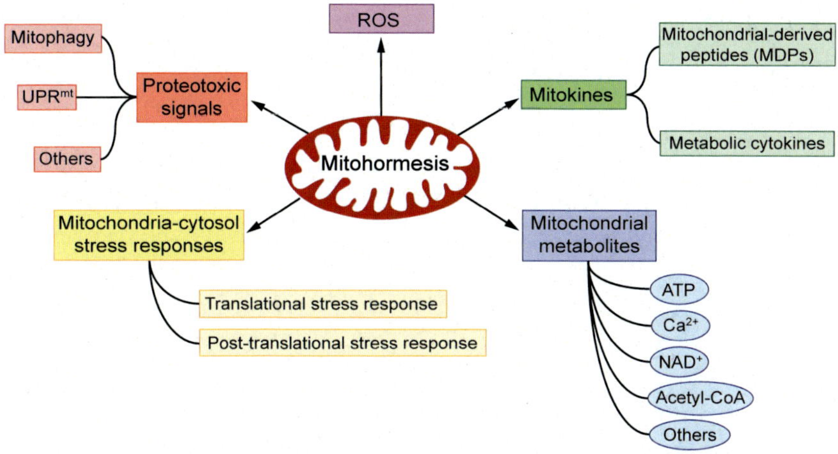

Figure 2 ***Mitohormesis signaling pathways.*** **Mitohormesis involves different cellular mechanisms and requires a meticulous coordination between the nucleus and mitochondria. Hence, the mitohormetic response depends on the activation of a variety of signaling pathways, including changes in levels of reactive oxygen species (ROS) and mitochondrial metabolites, secretion of mitokines, generation of proteotoxic signals, and activation of mitochondria–cytosol stress responses. Therefore, depending on the nature of the mitochondrial insult, one or more of these signaling pathways will be triggered to activate mitohormesis.**

mitonuclear communication can also mediate a mitohormetic response. Among all, the most important signals or processes are ROS, mitochondrial metabolites, proteotoxic signals, the mitochondria–cytosol stress response, and the release of mitokines (Fig. 2).

3.1 Reactive Oxygen Species

ROS are one of the major signals produced in mitochondria to communicate with the nucleus (Quiros et al., 2016), and the main signal in mitohormesis regulation (Ristow and Schmeisser, 2014; Yun and Finkel, 2014). Seen as detrimental and proposed as one of the causes of aging in the past (Harman, 1956), nowadays we know that ROS are essentially signaling molecules with a key role in the regulation of cellular function (Chandel, 2015; Holmstrom and Finkel, 2014; Shadel and Horvath, 2015). ROS are generated in mitochondria as end products of OXPHOS (Murphy, 2009) and, when kept under physiological levels, serve as regulatory molecules; however, increased ROS levels as a result of mitochondrial dysfunction have been implicated in the aging process and in multiple pathologies, including metabolic and cardiovascular diseases, neurodegenerative disorders, and cancer (Kim et al., 2015; Lopez-Otin et al., 2013; Moris et al., 2017; Sabharwal and Schumacker, 2014).

ROS levels are increased during exercise and CR, as well as by using CR mimetics or other compounds that alter the energetic status (Powers and Hogan, 2016; Schulz et al., 2007; Zini et al., 1999). In addition, multiple mitochondrial insults and stress stimuli, such as hypoxia or mitochondrial toxins, also increase ROS levels (Kulisz et al., 2002; Yang and Hekimi, 2010a). The rise in ROS levels activates, through a potential feedback mechanism, a hormetic response that induces host-antioxidant defense pathways, including antioxidant enzymes as catalase and superoxide dismutase, that ultimately reduce ROS levels (Zarse et al., 2012). This hormetic response has a positive effect in maintaining homeostasis and participates in multiple physiological functions, including the release of insulin and growth factors and the maintenance of muscle function, vascular tone regulation, and immune response, among others (Ristow and Schmeisser, 2014).

3.2 Mitochondrial Ions and Metabolites

Mitochondria also signal through the release of ions such as calcium, or by regulating the levels of intermediate metabolites, such as ATP, NAD^{+}/NADH, or acetyl-CoA. Mitochondria act as sensors and regulators of calcium ions (Ca^{2+}), being the calcium signaling involved in cellular metabolism and survival (Rizzuto et al., 2012). Alterations in Ca^{2+} in different cellular compartments act as signals in mitonuclear communication. Thus, increase in cytosolic Ca^{2+}—as occurs during exercise or by inhibition of mitochondrial calcium uptake—can activate mitochondrial biogenesis, whereas the release of Ca^{2+} from mitochondria—due to loss of membrane potential—can activate a retrograde response in the nucleus (Quiros et al., 2016).

Mitochondria produce ATP through OXPHOS, and changes in ATP levels can also influence hormetic responses. Decrease in ATP synthesis, as occurring during exercise or on mitochondrial dysfunction, increases the ratio AMP/ATP and activates the AMP-activated protein kinase (AMPK)—a serine/threonine kinase that acts as a master regulator of cellular metabolism (Herzig and Shaw, 2017). AMPK phosphorylates specific enzymes and growth control proteins, including regulators of lipid and glucose metabolism and autophagy proteins, to promote oxidation and to inhibit anabolism, thereby decreasing ATP consumption.

AMPK can also promote mitophagy (Egan et al., 2011) and mitochondrial fission (Toyama et al., 2016), contributing to the maintenance of mitochondrial homeostasis.

Due to their role as the central regulator of cellular metabolism by generating ATP and participating in the intermediate metabolism, mitochondria can also regulate the levels of other metabolites that can signal to nucleus and, therefore, could participate in mitohormetic responses. Thus, for example, during the oxidation of acetyl CoA to CO_2, nicotinamide adenine dinucleotide (NAD^+) and flavin adenine dinucleotide (FAD) are reduced to NADH and $FADH_2$, respectively. Therefore, levels and ratios of both cofactors are primarily maintained by mitochondria through the TCA cycle and OXPHOS function. Importantly, in addition to its role as cofactor of multiple glycolytic and TCA enzymes, NAD^+ is also an essential cosubstrate of several enzymes, including sirtuins, poly (ADP-ribose) polymerases, and histone acetyltransferases (Katsyuba and Auwerx, 2017). The main role of NAD^+ in regulating mitochondrial function is through the activation of SIRT1—an NAD^+-dependent deacetylase that positively regulates the function of peroxisome proliferator—activated receptor gamma coactivator 1-alpha (PGC-1α). The axis SIRT1—PGC-1α activates mitochondrial energy metabolism and biogenesis, increasing in turn healthspan and lifespan (Fang et al., 2017; Imai and Guarente, 2016). Also, together with acetyl CoA, mitochondria regulate levels of other TCA intermediates such as α-ketoglutarate, succinate, and fumarate. Both metabolic intermediates and cofactors participate in mitochondrial signaling, having a key role in epigenetic regulation. The function of these metabolites and their effect as signaling molecules is reviewed elsewhere (Bohovych and Khalimonchuk, 2016; Matilainen et al., 2017a).

3.3 Proteotoxic Signals

Loss of mitochondrial proteostasis leads to the accumulation of damaged or misfolded proteins in this organelle, which triggers the activation of a stress response to reactivate proteostasis. The main and most studied mitochondrial stress response is the mitochondrial unfolded protein response (UPR^{mt}) (Shpilka and Haynes, 2017). The UPR^{mt} was first described in mammalian cells as a stress response mainly mediated by proteases and chaperones (Zhao et al., 2002); however, the UPR^{mt} signaling has been mostly characterized in *C. elegans*. In worms, the UPR^{mt} is triggered by mitochondrial proteotoxic stresses induced either by RNA interference (RNAi) against some nuclear-encoded mitochondrial genes or by the treatment of toxins that affect mitochondrial translation or electron transport chain (ETC) function (Durieux et al., 2011; Houtkooper et al., 2013; Yoneda et al., 2004). The UPR^{mt} signaling in worms is regulated by the transcription factor atfs-1, which has two targeting sequences—mitochondrial and nuclear. In nonstress conditions, atfs-1 is translocated to mitochondria where it is rapidly degraded by the Lon protease (Nargund et al., 2012). Under particular stress conditions, when unfolded and/or unassembled mitochondrial proteins accumulate in the matrix, the UPR^{mt} is activated. These damaged proteins are digested into peptides by the mitochondrial matrix protease CLPP-1 (Haynes et al., 2007), orthologue of CLPP in mammals, and the resulting peptides are transported to the cytoplasm by the HAF-1 transporter (Haynes et al., 2010). By an unknown mechanism, the accumulation of these peptides in the cytosol limits the atfs-1 import to mitochondria, and, therefore, atfs-1 is translocated to the nucleus. In the nucleus, ATFS-1, together with the transcriptional coregulators DVE-1 and UBL-5, activates the expression of UPR^{mt} genes (Benedetti et al., 2006; Haynes et al., 2007). These UPR^{mt} targets include nuclear genes encoding mitochondrial proteases and chaperones, mitochondrial transporters, and antioxidant enzymes, as well as cytosolic proteins involved in mitochondrial fission, glycolysis, and detoxification (Nargund et al., 2015). The activation of this broad spectrum of genes increases mitochondrial proteostasis to resolve the stress conditions and promote the restoration of cellular metabolic function.

Although several reports have described the in vivo activation of UPR^{mt} in mammals, especially associated with the increased expression of mitochondrial proteases and chaperones in muscle and heart (Dogan et al., 2014; Khan et al., 2014, 2017; Song et al., 2015, 2017), the exact mechanisms of activation and signaling involved in it are still unknown. The first reports of UPR^{mt} activation in mammalian cells described observations made in cells lacking mtDNA (Martinus et al., 1996) or expressing a mutant form of the mitochondrial ornithine transcarbamylase (Δ-OTC), which cannot fold properly and accumulates in the mitochondrial matrix (Zhao et al., 2002). Although the use of Δ-OTC has been proved to promote other mitochondrial stress responses such as mitophagy and antioxidant signaling (Burman et al., 2017; Jin and Youle, 2013; Papa and Germain, 2014), the activation of the classical UPR^{mt} pathway defined by the activation of mitochondrial proteases (CLPP and YME1L1) and chaperones (HSP60, HSP10, and mtHSP70) has not been replicated in vitro neither with Δ-OTC nor with other mitochondrial toxins (Michel et al., 2015; Quiros et al., 2017).

Instead, the activation of the integrated stress response (ISR) has been clearly demonstrated (see 3.4.1 Translational stress response). Therefore, there are some caveats regarding the definition, regulation, and identity of the proteins involved in UPR^{mt} activation in mammals. In this regard, although ATF5 has been proposed as the functional orthologue of atfs-1 in mammals (Fiorese et al., 2016), other works have suggested that only some particular mitochondrial proteases and chaperones, including mtHSP70, TRAP1, AFG3L2, and LONP1, would be responsible for the UPR^{mt} activation in mammals (Seiferling et al., 2016). This hypothesis is based on some observations in which the mitochondrial protease CLPP, essential for UPR^{mt} activation in worms, is not required to activate UPR^{mt} in mammals (Seiferling et al., 2016). Moreover, Δ-OTC is degraded in vitro by the Lon protease and not by CLPP (Bezawork-Geleta et al., 2015). Therefore, the regulation of UPR^{mt} in mammals seems to be different from that in worms. Also, because mitochondrial stress can also activate both mitophagy and the ISR, the details of the convergence of all of these stress pathways are still unknown and will require specific works in mammals.

Besides UPR^{mt}, activation of other proteostatic stress responses has also been described under particular mitochondrial stress conditions. For instance, it has been reported that accumulation of proteins in the intermembrane space (IMS) leads to a different form of unfolded stress response mediated by the oestrogen receptor (ER) in the breast cancer cell line MCF-7. This response involves the mitochondrial transcription factor NRF1, the IMS mitochondrial protease HTRA2, and the proteasome (Papa and Germain, 2011). Similarly, another proteostatic response has been described in the context of steroidogenic regulation in adrenal cortex and gonads, involving the steroidogenic acute regulatory protein (StAR)—a protein that participates in steroid biosynthesis—facilitating cholesterol mobilization into mitochondria (Miller, 2017). Accumulation of StAR in the mitochondrial matrix is counteracted by the StAR overload response, which promotes the expression of the mitochondrial proteases LONP1, AFG3L2, and SPG7, which proteolytically degrade StAR (Bahat et al., 2014, 2015). Finally, a specific stress response mediated by chaperones has also been described in cultured cells. This stress response activates the expression of mitochondrial and cytosolic chaperones in response to heat involving the mitochondrial single-stranded DNA-binding protein 1, the heat shock factor 1, and the chromatin-remodeling factor BRG1 (Tan et al., 2015).

3.4 Mitochondria–Cytosol Stress Responses

Mitochondrial stress can also activate different cytosolic stress responses that maintain cellular proteostasis by regulating translation and protein degradation, acting at translational or post-translational levels (D'Amico et al., 2017; Topf et al., 2016).

3.4.1 Translational Stress Response

The ISR is a general stress response aimed at modulating protein synthesis. The ISR is activated by multiple stress conditions, including ER stress, amino acid starvation, low hem levels, double-stranded RNA (Pakos-Zebrucka et al., 2016), and mitochondrial stress (Quiros et al., 2017). These multiple stress conditions activate cellular kinases that phosphorylate the translation initiation factor eIF2α; this phosphorylation inhibits the canonical protein synthesis machinery favoring the translation of some stress genes, such as the transcription factors ATF4 and ATF5, which contain in their 5′ UTRs alternative open reading frames (ORFs) (Pakos-Zebrucka et al., 2016). Under nonstress conditions, these genes are translated using ORFs that produce prematurely interrupted nonfunctional proteins; however, under stress conditions, when eIF2α is phosphorylated, its availability to the translation machinery is reduced, allowing the ribosome to bypass the first ORF and access to alternative ORFs, synthesizing the functional proteins. It has been described that mitochondrial stress activates the expression of ATF4 in mammalian cells (Bao et al., 2016; Munch and Harper, 2016; Quiros et al., 2017) and that this stress response is mediated though the activation of the ISR, because chemical inhibition of eIF2α phosphorylation abolishes its activation (Quiros et al., 2017). It has also been reported that the activation of the mitochondrial-induced ISR is regulated by mTOR signaling in muscle (Khan et al., 2017); however, how mTOR complex regulates the expression of ATF4 is not well understood. In all conditions, activation of the ISR on mitochondrial stress is aimed at decreasing mitochondrial activity. This is achieved by inhibiting mitochondrial translation and promoting the remodeling of some cellular metabolic pathways such as one-carbon metabolism and the expression of metabolic cytokines (see 3.5 Mitokines) (Bao et al., 2016; Khan et al., 2017; Munch and Harper, 2016; Nikkanen et al., 2016; Quiros et al., 2017). Notably, this stress response has also been described in worms, acting in parallel to the UPR^{mt} activation and mediated though the eIF2α kinase, GCN-2 (Baker et al., 2012).

Additionally, another stress response, activated by alterations in fatty acid and ceramide metabolism on mitochondrial stress has been described in worms and defined as mitochondria to cytosol stress response (MCSR) (Kim et al., 2016). Changes in lipid balance act as a signal-activating MCSR to promote a cytosolic heat shock response that improve cellular proteostasis, reducing proteotoxic aggregation and decreasing mitochondrial damage. In worms, this response is mediated by cardiolipin and involves the transcription factors DVE-1 and HSF-1. Notably, this stress response is also conserved in mammalian cells and links mitochondrial proteostasis with cytosolic chaperons through changes in lipid homeostasis (Kim et al., 2016).

3.4.2 Post-translational Stress Responses

Beside translational responses, mitochondrial stress can also activate post-translational proteostatic repair mechanisms. These responses, mainly described in lower organisms as yeast and worms, involve the activation of proteostatic pathways in cytosol as the ubiquitin–proteasome system and mitophagy. In yeast, it has been described that defective protein import activates two similar protective responses defined as unfolded protein response activated by mistargeting of proteins (UPR^{am}) or mitochondrial precursor overaccumulation stress (mPOS) (Wang and Chen, 2015; Wrobel et al., 2015). These stress responses observed on mitochondrial dysfunction can be activated either by defects in the mitochondrial import machinery MIA and TIM23 pathways, by alterations in the mitochondrial membrane potential—i.e., by using the uncoupler FCCP—or by the overexpression of mitochondrial precursor proteins. As a consequence, mitochondrial precursor proteins are accumulated in the cytosol, promoting the inhibition of protein synthesis and activating the proteasome (Wang and Chen, 2015; Wrobel et al., 2015). Increased proteasomal activity is modulated by its assembly, which depends on the quantity of mitochondrial precursors in the cytosol, and seems to be independent of the transcription of its subunits. The mechanism proposed for protein-synthesis inhibition postulates that after mPOS there is an accumulation of negative regulators of protein synthesis, which under normal conditions are degraded by the proteasome but which, under stress conditions, cannot be degraded and therefore inhibit protein synthesis (Wang and Chen, 2015). Notably, although this response has not been identified in higher eukaryotes, some particular mitochondrial stress conditions also promote proteasome activation in mammals (Papa and Germain, 2011); however, it has also been demonstrated that several mitochondrial insults decrease proteasomal activity, mainly due to its inhibition by high ROS levels (Quiros et al., 2017). Moreover, it has been described that the ISR attenuates the TIM23-dependent mitochondrial import by reducing the levels of the Tim17A subunit, in a mechanism dependent on the mitochondrial protease YME1L (Rainbolt et al., 2013). Accordingly, decrease in Tim17A levels reduces TIM23-dependent protein import, promoting mitochondrial proteostasis in *C. elegans* and mammalian cells (Rainbolt et al., 2013).

Additionally, a different stress response system mediated by the proteasome has been described in yeast, in which outer membrane (OM) proteins are extracted from the mitochondrial OM through the highly conserved AAA-ATPase Cdc48; this mechanism was defined as mitochondria-associated protein degradation process, for its homology to endoplasmic reticulum–associated protein degradation (Heo et al., 2010). Besides the stress responses, the proteasome also participates in mitochondrial homeostasis by removing defective proteins in the OM and by regulating mitochondrial dynamics and mitophagy (Bragoszewski et al., 2017). Of note, the mitochondrial dynamics and mitophagy—two quality control mechanisms essential for mitochondrial and cellular function under both normal and stress conditions (Galluzzi et al., 2017; Sebastian et al., 2017; Wai and Langer, 2016)—can be also activated in different stress responses, participating in mitohormetic signaling.

3.5 Mitokines

Mitochondrial stress can also signal through extracellular communication promoting a cell nonautonomous response. This signaling involves the production of mitokines, defined as diffusible molecules released from a cell or a tissue in response to mitochondrial stress that transmit beneficial effects to other tissues. The existence of mitokines was proposed by Dillin et al., who observed activation of UPR^{mt} in *C. elegans* in a different tissue of the one where the stress was produced (Durieux et al., 2011). This stress response was therefore acting through cell nonautonomous signaling and was necessary to enhance longevity (Durieux et al., 2011). A similar response was also described in flies, where the knockdown of a subunit of the ETC complex I in muscle promoted a mitohormetic response, acting in the whole organism through a systemic effect mediated by insulin signaling via an IGF-binding protein (Owusu-Ansah et al., 2013). Since then, different factors have been identified acting as mitokines. Based on the origin, mitokines can be differentiated between mitochondrial-derived peptides

(MDP) and metabolic cytokines. MDPs are mtDNA-encoded peptides generated by short ORFs located in the sequences that encode 12S and 16S rRNA in mtDNA. Hitherto, several MDPs have been identified, including humanin (Hashimoto et al., 2001), six small humanin-like peptides (SHLPs) (Cobb et al., 2016), and mitochondrial ORF of the 12S rRNA-c (MOTS-c) (Lee et al., 2015). These MDPs promote multiple systemic beneficial effects, including neuroprotection and enhancement of metabolism; at the cellular level, it has been described that they can protect against apoptosis, increasing cell viability (Cobb et al., 2016; Kim et al., 2017; Lee et al., 2013, 2016). Additionally, some cytokines can also be considered as mitokines, because they are expressed on different mitochondrial insults and promote an improvement of metabolism. Among them, the best studied are fibroblast growth factor 21 (FGF21) and growth differentiation factor 15 (GDF15), whose expression has been found amplified in different mitochondrial diseases (Fujita et al., 2015; Suomalainen et al., 2011; Tyynismaa et al., 2010). These cytokines have a pleiotropic function, but in particular they are associated with an improvement of glucose and lipid metabolism and, therefore, with a regulation of energetic metabolism (Breit et al., 2017; Fisher and Maratos-Flier, 2016).

4. MITOHORMESIS AND LIFESPAN

Mitochondrial function declines during aging, contributing to cellular senescence, inflammation and stem cell exhaustion (Lopez-Otin et al., 2013; Sun et al., 2016). However, and in concordance with the concept of hormesis, mild mitochondrial damage can be beneficial to cells and extend organismal lifespan, whereas a severe stress can cause an irreparable damage. Indeed, there is ample evidence that mitochondrial dysfunction can promote longevity in a hormetic-dependent fashion in different model organisms (Fig. 3). These evidences were mostly obtained in *C. elegans*, although this phenomenon also occurs in other organisms as yeast, flies, and even mammals. The initial idea linking mitochondrial dysfunction with longevity was described in worm models where different RNAi against nuclear-encoded mitochondrial genes—targeting mainly components of the ETC—increased lifespan (Dillin et al., 2002; Feng et al., 2001; Lee et al., 2003; Yang and Hekimi, 2010a, b). These studies pointed to the inhibition of mitochondrial respiration as one of the most common mechanisms to increase longevity in worms. Importantly, to obtain the increment in

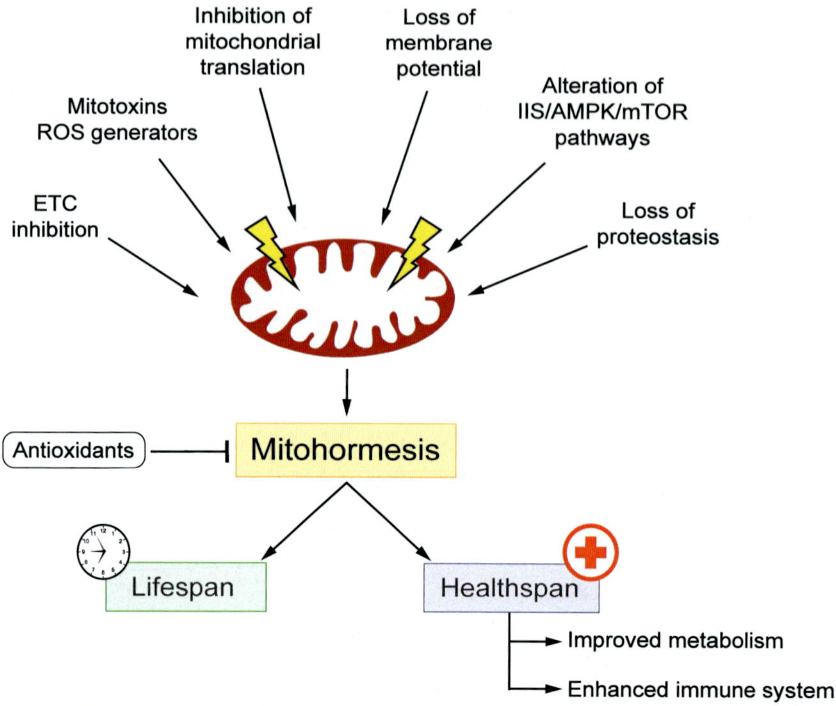

Figure 3 ***Mitohormesis promotes healthspan and lifespan.*** Mitochondrial function can be disturbed by a variety of alterations, such as inhibition of electron transport chain (ETC) and mitochondrial translation, accumulation of mitochondrial toxins (mitotoxins) and reactive oxygen species (ROS) generators, loss of mitochondrial membrane potential, alteration of IIS/AMPK/mTOR pathways, or loss of proteostasis, among others. These perturbations may elicit mitohormetic responses that increase lifespan and/or healthspan, mainly improving metabolism and enhancing the immune system. Importantly, antioxidants can inhibit multiple mitohormetic adaptations, highlighting the crucial role of ROS in mitochondrial stress signaling.

longevity, knockdown of ETC components should occur during the developmental stage (L3/L4 larval stage of development), being the effect highly sensitive to the degree of mitochondrial knockdown (Rea et al., 2007). Also, the positive effect on survival is not observed for all ETC components, as certain mutants can show a reduced longevity (Hartman et al., 2001; Ishii et al., 1998).

One of the proposed mechanisms to explain the increment in longevity is the activation of UPR^{mt} (Durieux et al., 2011). Several reports support the role of UPR^{mt} in the modulation of the aging process, as it is activated in several long-live worm mutants; however, its activation may not be sufficient to extend lifespan, as there are several inconsistencies in this regard. For example, UPR^{mt} also occurs in worm models with a reduction of lifespan (Bennett et al., 2014; Durieux et al., 2011; Ren et al., 2015; Tauffenberger et al., 2016). Further, deletion of the transcription factor atfs-1, required for UPR^{mt} induction, does not prevent lifespan extension in conditions of UPR^{mt} activation (Bennett et al., 2014; Cai et al., 2017; Labbadia et al., 2017; Ren et al., 2015). And finally, its constitutive activation by gain-of-function mutations in atfs-1 fails to extend lifespan (Bennett et al., 2014; Cai et al., 2017; Rauthan et al., 2013). Therefore, although the activation of the UPR^{mt} and its function as a cryoprotective stress response is well accepted (Shpilka and Haynes, 2017), its role as lifespan inductor is doubtful and should be revised (Bennett and Kaeberlein, 2014; Ren et al., 2015).

Besides UPR^{mt} and atfs-1, other putative mechanisms and transcription factors have been proposed to mediate the mitohormetic response that promotes longevity in worms. This includes the hypoxic response factor HIF-1 (Lee et al., 2010), the homeobox protein CEH-23 (Walter et al., 2011), the transcription factor TAF-4 (Khan et al., 2013), the chromatin remodeling factor ISW-1 (Matilainen et al., 2017b), or the heat shock factor HSF-1 (Labbadia et al., 2017), among others. Notably, overexpression of HIF-1, CEH-23, ISW-1, or HSF-1 increases longevity independently of mitochondrial stress, confirming their role enhancing lifespan (Hsu et al., 2003; Matilainen et al., 2017b; Walter et al., 2011; Zhang et al., 2009). Moreover, a translational response mediated by GCN-2 was also proposed to cooperate with UPR^{mt} during mitochondrial dysfunction (Baker et al., 2012); however, it has been reported that its activation is not required for lifespan extension in several long life–dependent mitochondrial mutants (Ren et al., 2015).

Lifespan extension induced by mitochondrial dysfunction—mainly by ETC inhibition—has also been observed in other organisms, as *Drosophila* (Borch Jensen et al., 2017; Copeland et al., 2009; Owusu-Ansah et al., 2013) and the short-lived killifish, *Nothobranchius furzeri* (Baumgart et al., 2016). In *Drosophila*, RNAi against genes encoding all OXPHOS components, except complex II, also increases longevity (Copeland et al., 2009). Notably, this lifespan extension in *Drosophila* requires RNAi treatment throughout life (Copeland et al., 2009); however, it can also be achieved by RNAi-dependent targeting of some complex I genes during adulthood (Copeland et al., 2009) or even by RNAi treatment in single tissues (Copeland et al., 2009; Owusu-Ansah et al., 2013). The key role of the mitochondrial complex I in lifespan regulation was also characterized in a longitudinal study using *N. furzeri*, where it was identified as a hub in a module of genes whose expression levels negatively correlated with lifespan (Baumgart et al., 2016). Accordingly, partial complex I inhibition with rotenone extended lifespan in *N. furzeri* (Baumgart et al., 2016).

In addition to ETC components, during the last 15 years RNAi knockdown against multiple mitochondrial genes, including genes related to coenzyme Q biosynthesis, mitochondrial proteostasis, TCA cycle, mitochondrial translation and mitochondrial transporters, was also associated with increased longevity in *C. elegans* (Curran and Ruvkun, 2007; Houtkooper et al., 2013; Lee et al., 2003; Stepanyan et al., 2006). This also applies for several toxins and proteotoxic stressors that affect mitochondrial function, such as paraquat, ethidium bromide, and doxycycline, among others (Houtkooper et al., 2013; Yang and Hekimi, 2010a). Notably, in some cases, the activation of UPR^{mt} has also been proposed as the mechanism by which longevity is enhanced (Houtkooper et al., 2013); however, the same caveats as in the knockdown of ETC components should be taken in account (Bennett and Kaeberlein, 2014).

One of the most plausible mechanism by which mitohormesis enhances longevity is through the increment of mtROS, which promotes a protective stress response mediated by the expression of several known transcription factors associated with longevity, such as FOXO, AMPK, TOR, p38, HIF1A, or NFE2L2 (also known as NRF2). Importantly, it has been described that the protective and prolongevity effects of ROS observed in worms are dependent on mitochondrial and not on cytosolic superoxide, because elevated mitochondrial ROS increases lifespan, whereas elevated cytosolic ROS decreases lifespan (Schaar et al., 2015). Multiple examples describe the activation of these signaling pathways regulating longevity in *C. elegans*. For example, treatment with D-glucosamine (GlcN)—a dietary supplement that potentially promotes cartilage health in humans—extends lifespan by impairing glucose metabolism and activating AAK-2/AMPK and mitochondrial biogenesis (Weimer et al., 2014). The positive effect of GlcN requires increasing mtROS levels, which activate the expression of the amino acid transporter aat-1, in an SKN-1/NRF2-dependent manner. GlcN also extends lifespan of aged C57BL/6 mice by promoting mitochondrial biogenesis and increasing amino-acid catabolism (Weimer et al., 2014). Similarly, treatment with epigallocatechin-3-gallate—a green tea polyphenol consumed as a dietary supplement—increases healthy lifespan in *C. elegans* by stimulating ROS production. Lifespan extension, which is achieved when administered throughout or only at early-to-mid adulthood, depend on AAK-2/AMPK, SIR-2.1/SIRT1, and DAF-16/FOXO (Xiong et al., 2017). Further, glutathione (GSH) depletion induced by treatment with diethyl maleate—a carbonyl compound that conjugates to GSH and other thiols—increases lifespan in *C. elegans* in a hormetic-dependent manner through DAF-16/FOXO and SKN-1/NRF-2 (Urban et al., 2017). Alterations in TCA cycle or treatment with some TCA intermediates, such as malate, fumarate, or oxaloacetate, also promote enhanced longevity in worms, involving DAF-16/FOXO and AAK-2/AMPK (Edwards et al., 2013; Williams et al., 2009). α-Ketoglutarate, as well as the oncometabolite 2-hydroxyglutarate, also extends lifespan by inhibiting ATPase complex V and TOR signaling (Chin et al., 2014; Fu et al., 2015). Accordingly, knockdown of ATP synthase subunit d in *Drosophila* also extends lifespan in a TOR and mitogen-activated protein kinase (MAPK)—dependent manner (Sun et al., 2014). In *Drosophila*, increasing ROS production through complex I also extends lifespan (Scialo et al., 2016). The increase in ROS is generated by reverse electron transport, which also improves mitochondrial function and extends lifespan in a model of Parkinson's disease (Scialo et al., 2016). Mild mitochondrial stress induced by the expression of ΔOTC throughout larval development also extends lifespan in *Drosophila*, involving the phosphoglycerate mutase 5—dependent activation of Foxo (Borch Jensen et al., 2017).

In addition to interfering with mitochondrial function, there are other ways to activate a mitohormetic response and influence longevity. For example, inhibition of transaldolase—an enzyme of the cytosolic pentose phosphate pathway—impairs mitochondrial respiration and promotes lifespan extension in *C. elegans* by inducing a starvation-like response (Bennett et al., 2017). This lifespan extension is mediated by a hormetic response signal that involves p38 and JNK, and the transcription factor HLH-30/TFEB. This transcription factor promotes autophagy and increases the expression of flavin-containing monooxygenase 2 (fmo-2) (Bennett et al., 2017), a prolongevity xenobiotic detoxification enzyme that is activated by dietary restriction and the hypoxic response (Leiser et al., 2015). Also, treatment with the glycolytic inhibitor lonidamine extends lifespan in *C. elegans* by promoting ROS formation in a p38-mediated pathway (Schmeisser et al., 2011). Low-dose arsenite also promotes stress resistance and extends lifespan by a transient increase in ROS levels, which requires the transcription factors DAF-16/FOXO and SKN-1/NRF2 (Schmeisser et al., 2013b). Importantly, and according to the definition of hormesis, higher concentrations of arsenite reduce longevity (Schmeisser et al., 2013b). Similarly, treatment with juglone, a ROS generator, also extends lifespan in a hormetic way, depending on DAF-16/FOXO and SIR-2.1 (Heidler et al., 2010).

Some classical interventions known to extend lifespan have also been associated to the mitohormetic response both in yeast and in worm models. Thus, reduced insulin and IGF-1 signaling (IIS) and reduced TOR pathway, activation of sirtuins, and CR rely, at least in part, on this adaptive response of mitochondria to exert their beneficial effects (Mesquita et al., 2010; Pan et al., 2011; Robida-Stubbs et al., 2012; Schmeisser et al., 2013a; Schulz et al., 2007; Zarse et al., 2012), presumably in a mtROS-dependent way. It has been reported that impaired IIS reduces glucose uptake and activates an AMP kinase–driven metabolic shift toward oxidative proline metabolism, inducing expression of mitochondrial proline oxidase, both in worms and mammalian cells (Zarse et al., 2012). This metabolic shift increases mtROS and promotes an adaptive response through PMK-1/p38 and SKN-1/NRF-2, which ultimately activates the expression of the antioxidant enzymes SOD and catalase, promoting stress resistance and extending lifespan in worms (Zarse et al., 2012). Notably, and consistently with the mitohormetic theory, treatment with antioxidants reduces the lifespan extension induced by inhibition of IIS. In agreement with these facts, inhibition of mitochondrial L-proline catabolism impairs the lifespan extension induced by reduced IIS whereas L-proline supplementation extends *C. elegans* lifespan (Zarse et al., 2012). On the other hand, it has been described that the induction of several cytoprotection-regulatory genes from stress pathways such as UPR^{er}, UPR^{mt}, ROS response, or xenobiotic detoxification are essential for the lifespan extension conferred by this classical longevity pathways—decreased IIS, disruption of mitochondrial function, or CR (Shore et al., 2012).

Additionally, the ROS-dependent mitohormetic response can also be coupled to other stress mechanisms, such as the activation of mitophagy or UPR^{mt}, potentially benefiting from them and cooperating in the response to stress and in lifespan extension. In particular, this cooperation has been described for the inhibition of ubiquinone biosynthesis, a conserved mechanism implicated in the control of lifespan from worms to mammals. Decreased COQ7 levels—a mitochondrial enzyme involved in ubiquinone biosynthesis (also known as MCLK1 in mammals and CLK-1 in worms)—extend lifespan in both worms and mice (Lakowski and Hekimi, 1996; Liu et al., 2005). This lifespan extension, which is independent of the IIS (Stepanyan et al., 2006), acts synergistically through mtROS and UPR^{mt} signaling and is mediated by the dual location—nuclear and mitochondrial—of COQ7/CLK-1 (Monaghan et al., 2015). Similarly, defective coenzyme Q biosynthesis extends lifespan in flies although the mechanism is not clear (Liu et al., 2011). Additionally, activation of mitophagy also cooperates with the mitohormetic response to promote longevity. On mitochondrial stress, concomitant activation of SKN-1 increases mitophagy and promotes mitochondrial biogenesis by inducing the expression of DCT-1—a putative orthologue to the mammalian mitophagy receptors NIX/BNIP3L and BNIP3—and several mitochondrial genes (Palikaras et al., 2015). This finding suggests that, after mitochondrial stress, a mitohormetic response coordinating mitophagy and mitochondrial biogenesis is activated to promote the rejuvenation of the mitochondrial pool. Notably, DCT-1 is also activated by DAF-16 linking impaired IIS to mitophagy activation (Palikaras et al., 2015). Also, partial deficiency of frataxin—a nuclear-encoded mitochondrial protein involved in the biogenesis of Fe–S cluster–containing proteins and ETC components—extends *C. elegans* lifespan partly by activating mitophagy through an adaptive iron starvation response (Schiavi et al., 2015). Notably, severe frataxin deficiency impairs respiration and decreases lifespan in *C. elegans* (Zarse et al., 2007), whereas in humans it causes Friedreich's ataxia, a life-threatening neurodegenerative disorder (Campuzano et al., 1996).

Interestingly, and in addition to a reduction in mitochondrial respiration, worm-longevity pathways also converge in a decreased mitochondrial membrane potential. This occurs not only for mitochondrial mutant worms but also for classical genes, such as daf-2 (IGF1R) or age-1(PI3K) (Lemire et al., 2009), and, accordingly, treatment with the mitochondrial uncoupler carbonyl cyanide 3-chlorophenylhydrazone increases lifespan in worms (Lemire et al., 2009). This effect has also been observed in mice, as mild mitochondrial uncoupling generated by low doses of the protonophore 2,4-dinitrophenol increases longevity by affecting energy metabolism and redox balance (Caldeira da Silva et al., 2008).

Mitochondrial dysfunction, mainly characterized by ETC deficiency, has also been linked to longevity in mice. Deficiency in the complex IV assembly factor Surf1, linked to Leigh syndrome in humans, promotes an increase in mouse survival (Dell'agnello et al., 2007). Surf1 deficiency in mice has a milder effect than in humans and activates a hormetic response promoting mitochondrial biogenesis and antioxidant response (Pulliam et al., 2014). However, a mouse model with mild ETC impairment through a loss-of-function mutation in a complex III subunit has sex-specific effects on lifespan and aging, decreasing survival only in males (Hughes and Hekimi, 2011). Aged females, however, live longer than controls, and, besides the lifespan differences between the sexes, mortality rate in mutant animals was lower than in controls in both sexes (Hughes and Hekimi, 2011). Interestingly, the mitochondrial–nuclear DNA mismatch that occurs in a conplastic mouse model—mice having a nuclear genome derived from one strain and mitochondrial genome from another—results in healthspan and lifespan extension (Latorre-Pellicer et al., 2016). This mitohormetic response is triggered by enhanced mitochondrial proteostasis and ROS generation, and, importantly, ROS buffering altered some mitochondrial parameters, indicating that ROS are essential for this response (Latorre-Pellicer et al., 2016). Notably, analysis of a different conplastic mouse model carrying a nuclear mutation in nucleotide transhydrogenase—an enzyme that participates in cytosolic antioxidant capacity increasing the production of hydrogen peroxide—do not show any alterations in healthspan and lifespan (Hirose et al., 2016), demonstrating the relevance of ROS in the mitohormetic response.

Cell nonautonomous signals also have the capability to exert prolongevity effects and increase lifespan in several model organisms, possibly by activating a systemic hormetic response that coordinates different stress responses. The first report described results obtained in *D. melanogaster*, where knockdown of two complex I genes solely in neurons was sufficient to extend lifespan (Copeland et al., 2009). A similar response was then observed in *C. elegans*, where neuron-specific knockdown of *cco-1* activated UPR^{mt} not only in neurons but also in the intestine, exerting a beneficial effect in the whole organism and enhancing survival (Durieux et al., 2011). In sight of this, the above-introduced theory suggesting the existence of mitokines was proposed. These mitokines would therefore be released on mitochondrial dysfunction in a particular tissue, exerting their function in the whole organism and resulting in increased survival (Durieux et al., 2011). Recently, it has been reported that the positive effect of cell nonautonomous UPR^{mt} activation requires a neural subcircuit of sensory neurons and one interneuron and that it is regulated by the neuropeptide FLP-2 (Shao et al., 2016). Similarly, cell nonautonomous mitohormesis responses have been described in muscle in different animal models. In *D. melanogaster*, a mild muscle mitochondrial distress also triggered a protective signal that preserved mitochondrial function, impeding the age-dependent muscle deterioration and increasing lifespan. In this case, this stress signal is composed of a muscle-specific activation of UPR^{mt} and redox response, and a cell nonautonomous response mediated by ImpL2/IGFBP7, which systemically antagonizes IIS and facilitates mitophagy (Owusu-Ansah et al., 2013). Finally, some reports also suggest that mitokines can increase survival in mice because it has been described that overexpression of nuclear-encoded mitokines FGF21 and GDF15 extends both healthspan and lifespan, mainly by altering the IIS (Wang et al., 2014; Zhang et al., 2012).

5. MITOHORMESIS AND HEALTHSPAN

Besides the positive effects of mitohormesis in lifespan, its activation can also have beneficial outcomes enhancing healthspan and improving metabolism (Fig. 3). Although the activation of mitohormesis has been reported to protect from different stress conditions in multiple model organisms, such as yeast and worms, we will focus mainly on healthspan improvements observed in mammals.

One of the most studied effectors that mediate the health benefits of mitochondrial hormesis is probably the increase in mtROS. As previously described, the mitohormetic effect activated in a conplastic mouse model not only enhances longevity but also promotes healthspan by improving insulin signaling, protecting against high-fat diet (HFD)—induced obesity, and ameliorating aging parameters including telomere shortening and mitochondrial dysfunction (Latorre-Pellicer et al., 2016). This also occurs in aged mice treated with GlcN, in which, in addition to increasing longevity, this treatment also improved glucose metabolism and enhanced amino-acid catabolism (Weimer et al., 2014). mtROS exert positive effects in metabolism by activating multiple signaling pathways. For example, mtROS participate in the regulation and adaptation of fuel utilization in mitochondria, which is essential to adjust metabolism on fasting conditions, hypoxia and reoxygenation, and T cell activation (Acin-Perez et al., 2014). Mitohormesis also drives positive effects in exercise by inducing transcriptional adaptive-responses though mitonuclear signaling pathways (Merry and Ristow, 2016a). mtROS are the primary candidate to signal mitohormesis during exercise, promoting mitochondrial biogenesis though a NFE2L2/NRF-1—dependent pathway (Merry and Ristow, 2016b). Indeed, it has been described that deficiency in the antioxidative hepatokine selenoprotein P (SeP) increases endurance after exercise training by enhancing mtROS production. Increase in mtROS activates the AMPK/PGC-1α signaling in muscle, and its decrease by treatment with NAC reduces endurance capacity of SeP-deficient mice (Misu et al., 2017). Importantly, the key role of mitohormesis during exercise has also been described in humans, where it has been observed that treatment with antioxidants prevents health-promoting effects of physical exercise (Ristow et al., 2009).

Mitohormesis also activates protective pathways in cells, enhancing cellular healthspan. For example, keratinocytes deficient in rictor—an adaptor protein essential for mechanistic target of rapamycin complex 2 activity—show increased lifespan, protection from senescence, and enhanced tolerance to cellular stressors. These beneficial effects are mediated by a mitohormetic response that relies on mtROS and promotes a metabolic rewiring toward increased glutaminolysis (Tassone et al., 2017). In cultured human neuroblastoma SH-SY5Y cells, mild stress caused by treatment with molecular hydrogen (H2), a gas that ameliorates oxidative stress-induced acute injuries in the brain, activate a mitohormetic response against oxidative stress in cells mediated by NRF2 pathway (Murakami et al., 2017b).

ROS are also essential regulators of the innate immune response against intracellular bacteria. mtROS production contributes to mouse macrophage bactericidal activity by coupling Toll-like receptors signaling to mitochondria (West et al., 2011). The increase in mtROS is essential for this response because reducing mtROS levels in macrophages results in defective bacterial killing (Break et al., 2012; West et al., 2011). ROS impact the regulation of macrophage signaling and polarization, which implies that the maintenance of a proper mitochondrial function is essential for macrophage activation and role as antibacterial host defense (Mehta et al., 2017). In this regard, it has been described that mtROS-dependent mitohormesis activation protects the intestine from inflammation by inducing macrophage polarization to functional M2 antiinflammatory phenotype (Formentini et al., 2017). The increase in mtROS production is mediated through partial inhibition of the mitochondrial ATP synthase by upregulating the ATPase inhibitor factor 1, a key factor that can regulate mitohormetic responses through ROS, ATP, and TOR pathways (Esparza-Molto et al., 2017). Importantly, and in concordance to the mitohormetic theory, the use of mitochondria-targeted antioxidants prevents the antiinflammatory positive effects of mtROS (Formentini et al., 2017). ROS also enhanced immunity of the long-lived *Mclk1*-heterozygous mice via HIF-1α, showing a strong immune response against *Salmonella* infection, and also displaying partial resistance to fibrosis, cancer, and the development of aging-related disorders (Wang et al., 2010, 2012). Besides the protective effects of mtROS, other forms of mitochondrial stress also enhance immunity. For example, mild mtDNA stress provoked by TFAM deficiency engages cytosolic antiviral signaling to promote the expression of a subset of interferon-stimulated genes. This response is mediated by the activation of cGAS–STING–IRF3 pathway, which potentiates type I interferon responses and confers broad viral resistance (West et al., 2015). Notably, mitochondrial stress appears to be essential for the antiviral response because herpes virus infection also triggers mtDNA stress and reduces TFAM levels, which is required to reach a complete antiviral response (West et al., 2015). Mitochondrial function also regulates T cells, participating in their activation, memory response, and exhaustion (Desdin-Mico et al., 2017; Mehta et al., 2017). Many of these functions are regulated by mtROS, and, consequently, their reduction leads to failure in T cell activation (Sena et al., 2013).

Mitochondrial stress and immune response activation are closely related probably due to the bacterial origin of mitochondria. Certain mitochondrial stress conditions can induce the release of signals from mitochondria, which can be interpreted by the host (the cell) as bacterial infection, triggering an immune response. Fragments of mtDNA released by ruptures in the inner mitochondrial membrane can be one of these signals, as well as mitochondrial toxins such as inhibitors of mitochondrial translation and ETC, which can also affect bacterial function. In this regard, it has been demonstrated that *C. elegans* respond to mitochondrial dysfunction by activating mitochondrial stress and detoxification response pathways together with an innate immune response (Liu et al., 2014). Additionally, bacterial infection causes mitochondrial dysfunction and, interestingly, some bacterial species from *Pseudomonas* genera can also inhibit the stress response induced by mitochondrial toxins (Liu et al., 2014). Particularly, infection with *P. aeruginosa* causes mitochondrial dysfunction and activates the UPR^{mt} in worms, promoting an effective immune response (Pellegrino et al., 2014). These observations in worm models suggest a close relationship between bacterial infection and mitochondrial stress, being both mitochondrial stress and immune responses activated by each other.

In addition to mtROS, many of the positive outcomes of mitohormesis described in mammals are associated with the expression of metabolic cytokines or mitokines. MDPs, including humanin, MOTS-c, and SHLPs, exert beneficial healthspan-improving effects, ranging from neuroprotection to enhanced metabolism. For example, humanin and its analogues have multiple healthspan-improving roles in mice, among which we can find amelioration of obesity (Gong et al., 2015), protection against myocardial ischemia-reperfusion injury (Muzumdar et al., 2010), or improved memory (Murakami et al., 2017a). MOTS-c regulates insulin sensitivity and metabolic homeostasis, acting in skeletal muscle. Mice treated with MOTS-c are protected against obesity and insulin resistance (Lee et al., 2015), have increased survival and decreased bacteria loads after infection (Zhai et al., 2017), and are resistant to ovariectomy-induced bone loss (Ming et al., 2016). Moreover, MDPs SHLP2 and SHLP3 improve mitochondrial metabolism in vitro and enhance 3T3-L1 preadipocyte differentiation. Also, intracerebral-infused SHLP2 increases glucose uptake and suppresses hepatic glucose production, suggesting that it can act as an insulin sensitizer (Cobb et al., 2016). Although several works support the beneficial effect of MDPs at improving healthspan and metabolism (Kim et al., 2017), their putative role as mitohormetic signals is still unknown. Further studies will be required to explore the conditions under which those MDPs are activated and if they follow or are triggered by an hormetic response.

Other mechanism proposed for mitohormesis activation of a pro-health response is through the secretion of the metabolic cytokines FGF21 and GDF15. The role of FGF21 was linked to mitochondrial dysfunction in humans, being proposed as a marker of mitochondrial myopathy (Suomalainen et al., 2011; Tyynismaa et al., 2010). The observation that mice with mitochondrial myopathy showed, in addition to high levels of Fgf21, widespread positive effects on lipid metabolism such as decreased adipocyte size, low liver fat content, and resistance to high-fat diet suggested that the role of Fgf21 could have important implications in metabolism via mitochondrial dysfunction (Tyynismaa et al., 2010). These positive effects were demonstrated in mice with skeletal muscle– and liver-specific deficiency in autophagy (Kim et al., 2013). Tissue-specific deletion of *Atg7*—an essential gene in autophagy—induced a mitochondrial dysfunction that promoted the expression of Fgf21 through Atf4, the master regulator of the ISR (Pakos-Zebrucka et al., 2016) and implicated in mitochondrial stress in mammals (Quiros et al., 2017). Fgf21 expression protected these mutant mice from obesity and insulin resistance, showing an increase in fatty acid oxidation and browning of WAT (Kim et al., 2013). Because ETC inhibitors also induced Fgf21 in an Atf4-dependent manner (Kim et al., 2013), this observation highlighted the link between mitochondrial dysfunction and Fgf21 expression, promoting a pro-health mitohormetic response. Several examples had already showed metabolic beneficial effects on mild mitochondrial dysfunction, as observed in mouse models with muscle-specific knockdown of mitochondrial transcription factor A (*Tfam*) or with muscle- and liver-specific ablation of the apoptosis inducing factor (*Aif*) (Pospisilik et al., 2007; Wredenberg et al., 2006). These tissue-specific mutant mice showed an increase in insulin sensitivity and protection against obesity. Notably, these metabolic improvements were associated with insulin target tissues, such as liver and skeletal muscle. Recently, other multiple studies have demonstrated a protective role of FGF21 in metabolism after mitochondrial dysfunction, highlighting the in vivo role of mitohormesis in enhancing healthspan. For example, it has been reported that in the young preprogeroid polymerase gamma mutator mice mitochondrial mutations induce a starvation-like metabolic state (Wall et al., 2015). This results in a metabolic imbalance characterized by defects in BAT thermogenesis; however, at the same time, these mice show a complete resistance to HFD-induced obesity, increasing mitochondrial biogenesis in BAT and thermogenesis in a Fgf21-dependent manner (Wall et al., 2015). Disruption of the mitochondrial fission protein dynamin-related protein 1 in liver promotes a decrease in fat mass, protecting against HFD-induced obesity. This positive effect is mediated by ATF4, which promotes the expression of FGF21 to increase energy expenditure (Wang et al., 2015). Partial deletion of the mitochondrial fusion protein optic atrophy 1 (OPA1) in skeletal muscle also promotes a mitohormetic response mediated by the secretion of FGF21 from muscle that prevents obesity and insulin resistance. Muscle-specific OPA1 hypomorphism in young mice results in nonlethal progressive mitochondrial dysfunction and loss of muscle mass; however, these mutant mice are resistant to age- and diet-induced obesity and insulin resistance by a mechanism that involves activation of UPR^{er} and secretion of FGF21 from skeletal muscle (Pereira et al., 2017). Contrary to this study, another group has reported that muscle-specific Opa1 deletion is lethal in newborns, whereas acute muscle-specific deletion of Opa1 in adult mice induces a precocious senescence phenotype and premature death (Tezze et al., 2017). Mechanistically, they also found activation of ER stress and increase in FGF21 secretion, which induces a catabolic program of muscle loss and systemic aging. Importantly, pharmacological inhibition of ER stress or muscle-specific deletion of Fgf21 compensates the Opa1 deficiency, restoring a normal metabolic state, and preventing muscle atrophy and premature death (Tezze et al., 2017). This apparent contradiction can be explained by the hormetic nature of this stress response mediated by FGF21; in the first study, muscle-specific Opa1 deletion generates nonlethal mice (Pereira et al., 2017), whereas in the second study Opa1 deletion is lethal (Tezze et al., 2017). The use of different systems to decrease Opa1 levels in muscle seems to induce a different degree of mitochondrial dysfunction, which would explain the hormetic response in both cases. FGF21, as well as GDF15, were also found upregulated in muscle from mice deficient in adenine nucleotide translocator isoform 1 (ANT1)—a mitochondrial inner membrane translocator that exchanges mitochondrial ATP for cytosolic ADP (Morrow et al., 2017). Mice deficient in ANT1, which is predominantly expressed in heart, muscle, and brain, develop mitochondrial myopathy associated with hyperproliferation of this organelle in skeletal muscle (Graham et al., 1997). However, these mice also show insulin hypersensitivity, improved glucose tolerance, and resistance to HFD-induced obesity (Morrow et al., 2017).

GDF15 has also been proposed as other putative mitokine activated on mitochondrial stress. Increased levels of GDF15 were identified in human patients with mitochondrial dysfunction. As for FGF21, GDF15 has therefore being suggested as a marker of mitochondrial respiratory chain deficiency (Fujita et al., 2015; Yatsuga et al., 2015). Increased expression of FGF21 and GDF15 has been considered as one of the stress responses activated by mitochondrial dysfunction in patients with adult-onset mitochondrial myopathy (Khan et al., 2017). FGF21 and GDF15 were also upregulated in mice deficient in CR6-interacting factor 1 (CRIF1), a mitoribosomal protein (Chung et al., 2017; Kang et al., 2017). Adipose tissue–specific deletion of Crif1 in mice causes upregulation of FGF21 and angiopoietin-like 6, a cytokine also upregulated on inhibition of OXPHOS in cultured adipocytes (Kang et al., 2017). Skeletal muscle–specific deficiency of Crif1 protects against obesity and increases insulin sensitivity by boosting the expression of GDF15, which enhanced systemic energy homeostasis (Chung et al., 2017).

Notably, although both FGF21 and GDF15 can be considered as good biomarkers of mitochondrial myopathy compared to other myopathies, a recent meta-analysis has evidenced that GDF15 is also increased in a wide range of nonmitochondrial conditions. Therefore, FGF21 should be preferentially used as a marker for mitochondrial myopathy in humans (Lehtonen et al., 2016). Nevertheless, the role of these mitokines is not exempt from controversy. The role of FGF21 has been questioned, being reported that the beneficial outcome after mitochondrial stress in skeletal muscle could be independent of the endogenous levels of FGF21 (Ost et al., 2016). Also, it has been reported that deficiency of mitochondrial fatty acid oxidation stimulates FGF21 secretion, increasing glucose uptake into muscles in a paracrine manner but without contributing to fat mass decrease (Vandanmagsar et al., 2016). Therefore, despite the vast number of studies relating FGF21 secretion after mitochondrial stress and its beneficial outcome (Salminen et al., 2017), its role as a systemic mitokine with endocrine function still needs to be clarified. Similarly, as GDF15 has been found associated with multiple pathological conditions, it has been speculated that it can work as a general stress cytokine (Fujita et al., 2016). Further studies will be required to elucidate the role of these mitokines in mitohormesis.

6. CONCLUDING REMARKS AND FUTURE OUTLOOK

Mitochondrial dysfunction is one of the hallmarks of aging (Lopez-Otin et al., 2013), and its role during the aging process, as player and trigger, has been widely studied during many years (Sun et al., 2016). However, the identification of a mitochondrial hormetic response, defined as mitohormesis, has modified the initial ideas about the negative role of mitochondrial stress during the aging process (Yun and Finkel, 2014). Activation of mitohormesis signaling after mild mitochondrial or cellular stress promotes an enhancement of healthspan and/or lifespan in multiple model organisms (Fig. 4). This mechanism, not characterized nor understood during many years, is behind the first results published more than 15 years ago linking mitochondrial dysfunction with increased longevity (Dillin et al., 2002; Feng et al., 2001; Lee et al., 2003). Nowadays, although many details are still not well understood, we have a more precise idea about the signaling pathways and molecular mechanisms that are behind this antiaging paradigm. This finding, together with the identification of different forms of mitonuclear communication that regulate cellular homeostasis (Quiros et al., 2016), also participating in mitohormesis signaling, has centered the study of mitochondria as a hub regulator of healthspan and lifespan.

After observing all the positive effects of mitohormesis on lifespan and healthspan, an important question arises: which are the mediators that promote these beneficial outcomes? As described in this review, multiple downstream signaling pathways have been suggested to regulate the effects observed at the cellular and organismal levels. These include all the classical pathways associated with longevity, embracing those related to nuclear-sensing routes as IIS, AMPK, and TOR signaling, as well as pathways that promote mitochondrial biogenesis and antioxidant responses

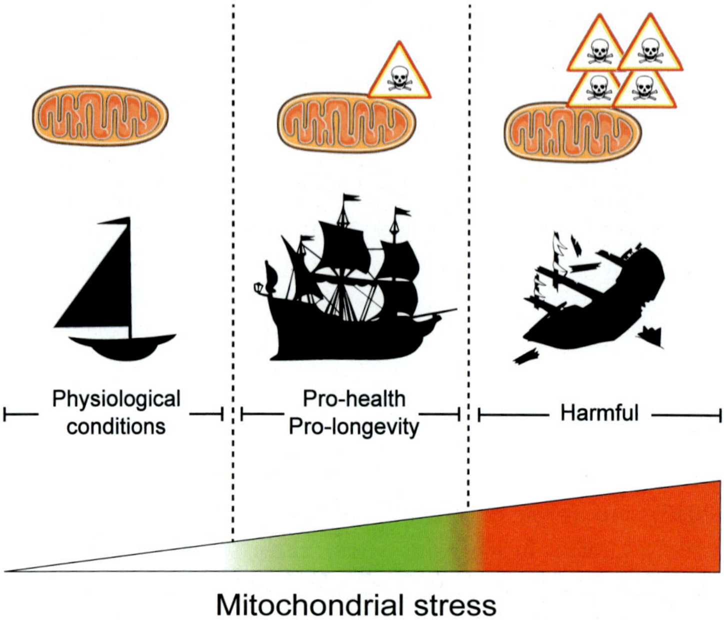

Figure 4 ***Biological responses to mitochondrial stress.*** **According to mitohormesis, low levels of mitochondrial stress induce adaptive responses that increase healthspan and/or lifespan. On the other hand, an overexposure to mitochondrial stress can impair cellular homeostasis and lead to cellular damage, triggering cell death. This biphasic response could evoke the metaphor of the benefits of moderate wind for navigation and the consequences of a gale in the open sea.**

(Lopez-Otin et al., 2016; Riera et al., 2016). These multiple effectors can be also found associated with the activation of proteostatic pathways such as the UPR^{mt} or mitophagy, which collaborate in enhancing both healthspan and lifespan. This pleiotropy in the downstream signaling pathways is also observed in the upstream events that trigger the activation of hormetic signaling in mitochondria. Multiple stressors that affect mitochondrial function have also been described to activate mitohormesis. In some cases, these stress stimuli can trigger contradictory phenotypes, as increase and decrease in mitochondrial respiration. However, even in opposing conditions, diverse stressors can activate the same pathway by different mechanisms. For example, a decline in mitochondrial respiration, induced by ETC dysfunction, diminishes ATP production, which activates AMPK (Martinez-Reyes et al., 2016); notwithstanding, AMPK activation can also occur independently of the decrease in ATP, being activated by mtROS (Emerling et al., 2009). Importantly, both conditions lead to an enhancement of lifespan (Curtis et al., 2006; Hwang et al., 2014). Therefore, the positive effects promoted by mitohormesis cannot be explained by a single mechanism, but through a complex system that includes multiple factors and sometimes different pathways that converge on the same effect. Importantly, this intricacy increases as the organisms become more complex, where there may be pleiotropic outcomes in different tissues, even deleterious effects in some and beneficial in others.

Among all signals described to activate mitohormesis, the most vastly documented so far is the increase in mtROS. Multiple stimuli of stress, compounds, and physiological situations promote this increment, which subsequently activates transcriptional programs that promote resistance to stress. An increase in ROS production can lead to cellular damage and is associated with multiple pathological and age-related conditions, such as cancer, diabetes, cardiovascular diseases, and neurodegenerative disorders. However, exposure to low levels of ROS activates protective mechanisms that increase lifespan and healthspan in multiple organisms, and is also associated with health benefits in humans (Ristow and Schmeisser, 2014). The identification of ROS as mitohormetic inducer contradicted the widely extended idea of using antioxidants as antiaging therapy. While the use of antioxidants in different pathological conditions can be positive (Gioscia-Ryan et al., 2014; Rehman et al., 2016), scientific evidences suggest that their use in healthy conditions should be avoided. Indeed, the identification of ROS as signaling molecules and regulators of multiple functions such as immunity (Sena and Chandel, 2012; Shadel and Horvath, 2015), also suggests that their elimination in nonpathological conditions can bring more deleterious than beneficial effects. Similarly, as ROS are key mediators of stem cell function (Chandel et al., 2016), being essential in differentiation and cell reprogramming (Hamanaka et al., 2013; Morimoto et al., 2013; Owusu-Ansah and Banerjee, 2009; Skamagki et al., 2017), the manipulation of their levels with the use of antioxidants could bring undesirable effects.

The identification of the mitohormesis signaling pathways is an interesting field of study focused on the development of future therapies to improve quality of life in humans. However, this raises another question: what are the signals that elicit a mitohormesis response? Although ROS are the most extensively studied signals and are very well documented, basing on the literature, the central role of mitochondria in the cellular metabolism, and the different ways of mitonuclear communication, we speculate that additional signaling molecules may be acting. These other molecules that can act together or independently of ROS could be of a varied chemical nature, such as ions, metabolites, proteins, nucleic acids, lipids, or even vesicles. These signals would be released on mitochondrial stress activating different stress response pathways, increasing cellular and organismal homeostasis. Importantly, in addition to the internal activation of cellular stress responses, it has been reported that these signals could act in an autocrine, paracrine, and even endocrine manner, in the latter case being defined as mitokines. Importantly, the identification of mitokines with a mitochondrial origin, the MDPs, opens a new and interesting field of study because these molecules could work as signals and effectors. Comparably, we also speculate that other mitochondrial molecules, such as mtDNA fragments, formyl peptides—small peptides formylated in the amino-terminal that are produced in mitochondria and bacteria—or some specific lipids, may also have a role triggering and mediating particular stress responses with key functions in immunity and metabolism. Further studies will be required to uncover the specific function of these molecules and their roles in mitohormesis signaling.

It has been proved that mitohormesis promotes healthspan and lifespan in multiple model organisms, existing also evidences of positive effects in humans. Although numerous pathways, factors, and molecules have been identified as signals and mediators of mitohormesis, more studies are needed to map completely the complex landscape of activation and regulation of this antiaging paradigm. This, together with the pursuit of mitokines, opens an optimistic future for the development of therapies to improve and promote health in humans, keeping in mind that, as Paracelso said, everything is a matter of dose.

REFERENCES

Acin-Perez, R., Carrascoso, I., Baixauli, F., Roche-Molina, M., Latorre-Pellicer, A., Fernandez-Silva, P., Mittelbrunn, M., Sanchez-Madrid, F., Perez-Martos, A., Lowell, C.A., Manfredi, G., Enriquez, J.A., 2014. ROS-triggered phosphorylation of complex II by Fgr kinase regulates cellular adaptation to fuel use. Cell Metab. 19, 1020–1033.

Arnould, T., Michel, S., Renard, P., 2015. Mitochondria retrograde signaling and the UPR mt: where are we in mammals? Int. J. Mol. Sci. 16, 18224–18251.

Bahat, A., Perlberg, S., Melamed-Book, N., Isaac, S., Eden, A., Lauria, I., Langer, T., Orly, J., 2015. Transcriptional activation of LON Gene by a new form of mitochondrial stress: a role for the nuclear respiratory factor 2 in StAR overload response (SOR). Mol. Cell. Endocrinol. 408, 62–72.

Bahat, A., Perlberg, S., Melamed-Book, N., Lauria, I., Langer, T., Orly, J., 2014. StAR enhances transcription of genes encoding the mitochondrial proteases involved in its own degradation. Mol. Endocrinol. 28, 208–224.

Baker, B.M., Nargund, A.M., Sun, T., Haynes, C.M., 2012. Protective coupling of mitochondrial function and protein synthesis via the eIF2alpha kinase GCN-2. PLoS Genet. 8, e1002760.

Bao, X.R., Ong, S.E., Goldberger, O., Peng, J., Sharma, R., Thompson, D.A., Vafai, S.B., Cox, A.G., Marutani, E., Ichinose, F., Goessling, W., Regev, A., Carr, S.A., Clish, C.B., Mootha, V.K., 2016. Mitochondrial dysfunction remodels one-carbon metabolism in human cells. Elife 5.

Baumgart, M., Priebe, S., Groth, M., Hartmann, N., Menzel, U., Pandolfini, L., Koch, P., Felder, M., Ristow, M., Englert, C., Guthke, R., Platzer, M., Cellerino, A., 2016. Longitudinal RNA-Seq analysis of vertebrate aging identifies mitochondrial complex I as a small-molecule-sensitive modifier of lifespan. Cell Syst. 2, 122–132.

Benedetti, C., Haynes, C.M., Yang, Y., Harding, H.P., Ron, D., 2006. Ubiquitin-like protein 5 positively regulates chaperone gene expression in the mitochondrial unfolded protein response. Genetics 174, 229–239.

Bennett, C.F., Kaeberlein, M., 2014. The mitochondrial unfolded protein response and increased longevity: cause, consequence, or correlation? Exp. Gerontol. 56, 142–146.

Bennett, C.F., Kwon, J.J., Chen, C., Russell, J., Acosta, K., Burnaevskiy, N., Crane, M.M., Bitto, A., Vander Wende, H., Simko, M., Pineda, V., Rossner, R., Wasko, B.M., Choi, H., Chen, S., Park, S., Jafari, G., Sands, B., Perez Olsen, C., Mendenhall, A.R., Morgan, P.G., Kaeberlein, M., 2017. Transaldolase inhibition impairs mitochondrial respiration and induces a starvation-like longevity response in *Caenorhabditis elegans*. PLoS Genet. 13, e1006695.

Bennett, C.F., Vander Wende, H., Simko, M., Klum, S., Barfield, S., Choi, H., Pineda, V.V., Kaeberlein, M., 2014. Activation of the mitochondrial unfolded protein response does not predict longevity in *Caenorhabditis elegans*. Nat. Commun. 5, 3483.

Bezawork-Geleta, A., Brodie, E.J., Dougan, D.A., Truscott, K.N., 2015. LON is the master protease that protects against protein aggregation in human mitochondria through direct degradation of misfolded proteins. Sci. Rep. 5, 17397.

Bohovych, I., Khalimonchuk, O., 2016. Sending out an SOS: mitochondria as a signaling hub. Front Cell Dev. Biol. 4, 109.

Borch Jensen, M., Qi, Y., Riley, R., Rabkina, L., Jasper, H., 2017. PGAM5 promotes lasting FoxO activation after developmental mitochondrial stress and extends lifespan in Drosophila. Elife 6.

Bragoszewski, P., Turek, M., Chacinska, A., 2017. Control of mitochondrial biogenesis and function by the ubiquitin-proteasome system. Open Biol. 7.

Break, T.J., Jun, S., Indramohan, M., Carr, K.D., Sieve, A.N., Dory, L., Berg, R.E., 2012. Extracellular superoxide dismutase inhibits innate immune responses and clearance of an intracellular bacterial infection. J. Immunol. 188, 3342–3350.

Breit, S.N., Tsai, V.W., Brown, D.A., 2017. Targeting obesity and cachexia: identification of the GFRAL receptor-MIC-1/GDF15 pathway. Trends Mol. Med. 23, 1065–1067.

Burman, J.L., Pickles, S., Wang, C., Sekine, S., Vargas, J.N.S., Zhang, Z., Youle, A.M., Nezich, C.L., Wu, X., Hammer, J.A., Youle, R.J., 2017. Mitochondrial fission facilitates the selective mitophagy of protein aggregates. J. Cell Biol. 216, 3231–3247.

Cai, H., Rasulova, M., Vandemeulebroucke, L., Meagher, L., Vlaeminck, C., Dhondt, I., Braeckman, B.P., 2017. Life-span extension by axenic dietary restriction is independent of the mitochondrial unfolded protein response and mitohormesis in *Caenorhabditis elegans*. J. Gerontol. A Biol. Sci. Med. Sci. 72, 1311–1318.

Caldeira da Silva, C.C., Cerqueira, F.M., Barbosa, L.F., Medeiros, M.H., Kowaltowski, A.J., 2008. Mild mitochondrial uncoupling in mice affects energy metabolism, redox balance and longevity. Aging Cell 7, 552–560.

Campuzano, V., Montermini, L., Molto, M.D., Pianese, L., Cossee, M., Cavalcanti, F., Monros, E., Rodius, F., Duclos, F., Monticelli, A., Zara, F., Canizares, J., Koutnikova, H., Bidichandani, S.I., Gellera, C., Brice, A., Trouillas, P., De Michele, G., Filla, A., De Frutos, R., Palau, F., Patel, P.I., Di Donato, S., Mandel, J.L., Cocozza, S., Koenig, M., Pandolfo, M., 1996. Friedreich's ataxia: autosomal recessive disease caused by an intronic GAA triplet repeat expansion. Science 271, 1423–1427.

Chandel, N.S., 2015. Evolution of mitochondria as signaling organelles. Cell Metab. 22, 204–206.

Chandel, N.S., Jasper, H., Ho, T.T., Passegue, E., 2016. Metabolic regulation of stem cell function in tissue homeostasis and organismal ageing. Nat. Cell Biol. 18, 823–832.

Chin, R.M., Fu, X., Pai, M.Y., Vergnes, L., Hwang, H., Deng, G., Diep, S., Lomenick, B., Meli, V.S., Monsalve, G.C., Hu, E., Whelan, S.A., Wang, J.X., Jung, G., Solis, G.M., Fazlollahi, F., Kaweeteerawat, C., Quach, A., Nili, M., Krall, A.S., Godwin, H.A., Chang, H.R., Faull, K.F., Guo, F., Jiang, M., Trauger, S.A., Saghatelian, A., Braas, D., Christofk, H.R., Clarke, C.F., Teitell, M.A., Petrascheck, M., Reue, K., Jung, M.E., Frand, A.R., Huang, J., 2014. The metabolite alpha-ketoglutarate extends lifespan by inhibiting ATP synthase and TOR. Nature 510, 397–401.

Chung, H.K., Ryu, D., Kim, K.S., Chang, J.Y., Kim, Y.K., Yi, H.S., Kang, S.G., Choi, M.J., Lee, S.E., Jung, S.B., Ryu, M.J., Kim, S.J., Kweon, G.R., Kim, H., Hwang, J.H., Lee, C.H., Lee, S.J., Wall, C.E., Downes, M., Evans, R.M., Auwerx, J., Shong, M., 2017. Growth differentiation factor 15 is a myomitokine governing systemic energy homeostasis. J. Cell Biol. 216, 149–165.

Cobb, L.J., Lee, C., Xiao, J., Yen, K., Wong, R.G., Nakamura, H.K., Mehta, H.H., Gao, Q., Ashur, C., Huffman, D.M., Wan, J., Muzumdar, R., Barzilai, N., Cohen, P., 2016. Naturally occurring mitochondrial-derived peptides are age-dependent regulators of apoptosis, insulin sensitivity, and inflammatory markers. Aging (Albany NY) 8, 796–809.

Copeland, J.M., Cho, J., Lo Jr., T., Hur, J.H., Bahadorani, S., Arabyan, T., Rabie, J., Soh, J., Walker, D.W., 2009. Extension of Drosophila life span by RNAi of the mitochondrial respiratory chain. Curr. Biol. 19, 1591–1598.

Curran, S.P., Ruvkun, G., 2007. Lifespan regulation by evolutionarily conserved genes essential for viability. PLoS Genet. 3, e56.

Curtis, R., O'Connor, G., DiStefano, P.S., 2006. Aging networks in *Caenorhabditis elegans*: AMP-activated protein kinase (aak-2) links multiple aging and metabolism pathways. Aging Cell 5, 119–126.

D'Amico, D., Sorrentino, V., Auwerx, J., 2017. Cytosolic proteostasis networks of the mitochondrial stress response. Trends Biochem. Sci. 42, 712–725.

Dell'agnello, C., Leo, S., Agostino, A., Szabadkai, G., Tiveron, C., Zulian, A., Prelle, A., Roubertoux, P., Rizzuto, R., Zeviani, M., 2007. Increased longevity and refractoriness to Ca(2+)-dependent neurodegeneration in Surf1 knockout mice. Hum. Mol. Genet. 16, 431–444.

Desdin-Mico, G., Soto-Heredero, G., Mittelbrunn, M., 2017. Mitochondrial activity in T cells. Mitochondrion 41, 51–57.

Dillin, A., Hsu, A.L., Arantes-Oliveira, N., Lehrer-Graiwer, J., Hsin, H., Fraser, A.G., Kamath, R.S., Ahringer, J., Kenyon, C., 2002. Rates of behavior and aging specified by mitochondrial function during development. Science 298, 2398–2401.

Dogan, S.A., Pujol, C., Maiti, P., Kukat, A., Wang, S., Hermans, S., Senft, K., Wibom, R., Rugarli, E.I., Trifunovic, A., 2014. Tissue-specific loss of DARS2 activates stress responses independently of respiratory chain deficiency in the heart. Cell Metab. 19, 458–469.

Durieux, J., Wolff, S., Dillin, A., 2011. The cell-non-autonomous nature of electron transport chain-mediated longevity. Cell 144, 79–91.

Edwards, C.B., Copes, N., Brito, A.G., Canfield, J., Bradshaw, P.C., 2013. Malate and fumarate extend lifespan in *Caenorhabditis elegans*. PLoS One 8, e58345.
Egan, D.F., Shackelford, D.B., Mihaylova, M.M., Gelino, S., Kohnz, R.A., Mair, W., Vasquez, D.S., Joshi, A., Gwinn, D.M., Taylor, R., Asara, J.M., Fitzpatrick, J., Dillin, A., Viollet, B., Kundu, M., Hansen, M., Shaw, R.J., 2011. Phosphorylation of ULK1 (hATG1) by AMP-activated protein kinase connects energy sensing to mitophagy. Science 331, 456–461.
Emerling, B.M., Weinberg, F., Snyder, C., Burgess, Z., Mutlu, G.M., Viollet, B., Budinger, G.R., Chandel, N.S., 2009. Hypoxic activation of AMPK is dependent on mitochondrial ROS but independent of an increase in AMP/ATP ratio. Free Radic. Biol. Med. 46, 1386–1391.
Esparza-Molto, P.B., Nuevo-Tapioles, C., Cuezva, J.M., 2017. Regulation of the H+-ATP synthase by IF1: a role in mitohormesis. Cell. Mol. Life Sci. 74, 2151–2166.
Fang, E.F., Lautrup, S., Hou, Y., Demarest, T.G., Croteau, D.L., Mattson, M.P., Bohr, V.A., 2017. NAD+ in aging: molecular mechanisms and translational implications. Trends Mol. Med. 23, 899–916.
Feng, J., Bussiere, F., Hekimi, S., 2001. Mitochondrial electron transport is a key determinant of life span in *Caenorhabditis elegans*. Dev. Cell 1, 633–644.
Fiorese, C.J., Schulz, A.M., Lin, Y.F., Rosin, N., Pellegrino, M.W., Haynes, C.M., 2016. The transcription factor ATF5 mediates a mammalian mitochondrial UPR. Curr. Biol. 26, 2037–2043.
Fisher, F.M., Maratos-Flier, E., 2016. Understanding the physiology of FGF21. Annu. Rev. Physiol. 78, 223–241.
Formentini, L., Santacatterina, F., Nunez de Arenas, C., Stamatakis, K., Lopez-Martinez, D., Logan, A., Fresno, M., Smits, R., Murphy, M.P., Cuezva, J.M., 2017. Mitochondrial ROS production protects the intestine from inflammation through functional M2 macrophage polarization. Cell Rep. 19, 1202–1213.
Fu, X., Chin, R.M., Vergnes, L., Hwang, H., Deng, G., Xing, Y., Pai, M.Y., Li, S., Ta, L., Fazlollahi, F., Chen, C., Prins, R.M., Teitell, M.A., Nathanson, D.A., Lai, A., Faull, K.F., Jiang, M., Clarke, S.G., Cloughesy, T.F., Graeber, T.G., Braas, D., Christofk, H.R., Jung, M.E., Reue, K., Huang, J., 2015. 2-Hydroxyglutarate inhibits ATP synthase and mTOR signaling. Cell Metab. 22, 508–515.
Fujita, Y., Ito, M., Kojima, T., Yatsuga, S., Koga, Y., Tanaka, M., 2015. GDF15 is a novel biomarker to evaluate efficacy of pyruvate therapy for mitochondrial diseases. Mitochondrion 20, 34–42.
Fujita, Y., Taniguchi, Y., Shinkai, S., Tanaka, M., Ito, M., 2016. Secreted growth differentiation factor 15 as a potential biomarker for mitochondrial dysfunctions in aging and age-related disorders. Geriatr. Gerontol. Int. 16 (Suppl. 1), 17–29.
Galluzzi, L., Baehrecke, E.H., Ballabio, A., Boya, P., Bravo-San Pedro, J.M., Cecconi, F., Choi, A.M., Chu, C.T., Codogno, P., Colombo, M.I., Cuervo, A.M., Debnath, J., Deretic, V., Dikic, I., Eskelinen, E.L., Fimia, G.M., Fulda, S., Gewirtz, D.A., Green, D.R., Hansen, M., Harper, J.W., Jaattela, M., Johansen, T., Juhasz, G., Kimmelman, A.C., Kraft, C., Ktistakis, N.T., Kumar, S., Levine, B., Lopez-Otin, C., Madeo, F., Martens, S., Martinez, J., Melendez, A., Mizushima, N., Munz, C., Murphy, L.O., Penninger, J.M., Piacentini, M., Reggiori, F., Rubinsztein, D.C., Ryan, K.M., Santambrogio, L., Scorrano, L., Simon, A.K., Simon, H.U., Simonsen, A., Tavernarakis, N., Tooze, S.A., Yoshimori, T., Yuan, J., Yue, Z., Zhong, Q., Kroemer, G., 2017. Molecular definitions of autophagy and related processes. EMBO J. 36, 1811–1836.
Gioscia-Ryan, R.A., LaRocca, T.J., Sindler, A.L., Zigler, M.C., Murphy, M.P., Seals, D.R., 2014. Mitochondria-targeted antioxidant (MitoQ) ameliorates age-related arterial endothelial dysfunction in mice. J. Physiol. 592, 2549–2561.

Gong, Z., Su, K., Cui, L., Tas, E., Zhang, T., Dong, H.H., Yakar, S., Muzumdar, R.H., 2015. Central effects of human in on hepatic triglyceride secretion. Am. J. Physiol. Endocrinol. Metab. 309, E283–E292.

Graham, B.H., Waymire, K.G., Cottrell, B., Trounce, I.A., MacGregor, G.R., Wallace, D.C., 1997. A mouse model for mitochondrial myopathy and cardiomyopathy resulting from a deficiency in the heart/muscle isoform of the adenine nucleotide translocator. Nat. Genet. 16, 226–234.

Guha, M., Avadhani, N.G., 2013. Mitochondrial retrograde signaling at the crossroads of tumor bioenergetics, genetics and epigenetics. Mitochondrion 13, 577–591.

Hamanaka, R.B., Glasauer, A., Hoover, P., Yang, S., Blatt, H., Mullen, A.R., Getsios, S., Gottardi, C.J., DeBerardinis, R.J., Lavker, R.M., Chandel, N.S., 2013. Mitochondrial reactive oxygen species promote epidermal differentiation and hair follicle development. Sci. Signal. 6, ra8.

Harman, D., 1956. Aging: a theory based on free radical and radiation chemistry. J. Gerontol. 11, 298–300.

Hartman, P.S., Ishii, N., Kayser, E.B., Morgan, P.G., Sedensky, M.M., 2001. Mitochondrial mutations differentially affect aging, mutability and anesthetic sensitivity in *Caenorhabditis elegans*. Mech. Ageing Dev. 122, 1187–1201.

Hashimoto, Y., Niikura, T., Tajima, H., Yasukawa, T., Sudo, H., Ito, Y., Kita, Y., Kawasumi, M., Kouyama, K., Doyu, M., Sobue, G., Koide, T., Tsuji, S., Lang, J., Kurokawa, K., Nishimoto, I., 2001. A rescue factor abolishing neuronal cell death by a wide spectrum of familial Alzheimer's disease genes and Abeta. Proc. Natl. Acad. Sci. U. S. A 98, 6336–6341.

Haynes, C.M., Petrova, K., Benedetti, C., Yang, Y., Ron, D., 2007. ClpP mediates activation of a mitochondrial unfolded protein response in *C. elegans*. Dev. Cell 13, 467–480.

Haynes, C.M., Yang, Y., Blais, S.P., Neubert, T.A., Ron, D., 2010. The matrix peptide exporter HAF-1 signals a mitochondrial UPR by activating the transcription factor ZC376.7 in *C. elegans*. Mol. Cell 37, 529–540.

Heidler, T., Hartwig, K., Daniel, H., Wenzel, U., 2010. *Caenorhabditis elegans* lifespan extension caused by treatment with an orally active ROS-generator is dependent on DAF-16 and SIR-2.1. Biogerontology 11, 183–195.

Heo, J.M., Livnat-Levanon, N., Taylor, E.B., Jones, K.T., Dephoure, N., Ring, J., Xie, J., Brodsky, J.L., Madeo, F., Gygi, S.P., Ashrafi, K., Glickman, M.H., Rutter, J., 2010. A stress-responsive system for mitochondrial protein degradation. Mol. Cell 40, 465–480.

Herzig, S., Shaw, R.J., 2017. AMPK: guardian of metabolism and mitochondrial homeostasis. Nat. Rev. Mol. Cell Biol. 19, 121–135.

Hirose, M., Schilf, P., Gupta, Y., Wright, M.N., Wright, M.N., Johren, O., Wagner, A.E., Sina, C., Ziegler, A., Ristow, M., Ibrahim, S.M., 2016. Lifespan effects of mitochondrial mutations. Nature 540, E13–E14.

Holmstrom, K.M., Finkel, T., 2014. Cellular mechanisms and physiological consequences of redox-dependent signalling. Nat. Rev. Mol. Cell Biol. 15, 411–421.

Houtkooper, R.H., Mouchiroud, L., Ryu, D., Moullan, N., Katsyuba, E., Knott, G., Williams, R.W., Auwerx, J., 2013. Mitonuclear protein imbalance as a conserved longevity mechanism. Nature 497, 451–457.

Hsu, A.L., Murphy, C.T., Kenyon, C., 2003. Regulation of aging and age-related disease by DAF-16 and heat-shock factor. Science 300, 1142–1145.

Hughes, B.G., Hekimi, S., 2011. A mild impairment of mitochondrial electron transport has sex-specific effects on lifespan and aging in mice. PLoS One 6, e26116.

Hwang, A.B., Ryu, E.A., Artan, M., Chang, H.W., Kabir, M.H., Nam, H.J., Lee, D., Yang, J.S., Kim, S., Mair, W.B., Lee, C., Lee, S.S., Lee, S.J., 2014. Feedback regulation via AMPK and HIF-1 mediates ROS-dependent longevity in *Caenorhabditis elegans*. Proc. Natl. Acad. Sci. U. S. A 111, E4458–E4467.

Imai, S.I., Guarente, L., 2016. It takes two to tango: NAD+ and sirtuins in aging/longevity control. NPJ Aging Mech. Dis. 2, 16017.
Ishii, N., Fujii, M., Hartman, P.S., Tsuda, M., Yasuda, K., Senoo-Matsuda, N., Yanase, S., Ayusawa, D., Suzuki, K., 1998. A mutation in succinate dehydrogenase cytochrome b causes oxidative stress and ageing in nematodes. Nature 394, 694–697.
Jin, S.M., Youle, R.J., 2013. The accumulation of misfolded proteins in the mitochondrial matrix is sensed by PINK1 to induce PARK2/Parkin-mediated mitophagy of polarized mitochondria. Autophagy 9, 1750–1757.
Kang, S.G., Yi, H.S., Choi, M.J., Ryu, M.J., Jung, S., Chung, H.K., Chang, J.Y., Kim, Y.K., Lee, S.E., Kim, H.W., Choi, H., Kim, D.S., Lee, J.H., Kim, K.S., Kim, H.J., Lee, C.H., Oike, Y., Shong, M., 2017. ANGPTL6 expression is coupled with mitochondrial OXPHOS function to regulate adipose FGF21. J. Endocrinol. 233, 105–118.
Katsyuba, E., Auwerx, J., 2017. Modulating NAD+ metabolism, from bench to bedside. EMBO J. 36, 2670–2683.
Kawagishi, H., Finkel, T., 2014. Unraveling the truth about antioxidants: ROS and disease: finding the right balance. Nat. Med. 20, 711–713.
Khan, M.H., Ligon, M., Hussey, L.R., Hufnal, B., Farber 2nd, R., Munkacsy, E., Rodriguez, A., Dillow, A., Kahlig, E., Rea, S.L., 2013. TAF-4 is required for the life extension of isp-1, clk-1 and tpk-1 Mit mutants. Aging (Albany NY) 5, 741–758.
Khan, N.A., Auranen, M., Paetau, I., Pirinen, E., Euro, L., Forsstrom, S., Pasila, L., Velagapudi, V., Carroll, C.J., Auwerx, J., Suomalainen, A., 2014. Effective treatment of mitochondrial myopathy by nicotinamide riboside, a vitamin B3. EMBO Mol. Med. 6, 721–731.
Khan, N.A., Nikkanen, J., Yatsuga, S., Jackson, C., Wang, L., Pradhan, S., Kivela, R., Pessia, A., Velagapudi, V., Suomalainen, A., 2017. mTORC1 regulates mitochondrial integrated stress response and mitochondrial myopathy progression. Cell Metab 26, 419–428 e415.
Kim, G.H., Kim, J.E., Rhie, S.J., Yoon, S., 2015. The role of oxidative stress in neurodegenerative diseases. Exp. Neurobiol. 24, 325–340.
Kim, H.E., Grant, A.R., Simic, M.S., Kohnz, R.A., Nomura, D.K., Durieux, J., Riera, C.E., Sanchez, M., Kapernick, E., Wolff, S., Dillin, A., 2016. Lipid biosynthesis coordinates a mitochondrial-to-cytosolic stress response. Cell 166, 1539–1552.
Kim, K.H., Jeong, Y.T., Oh, H., Kim, S.H., Cho, J.M., Kim, Y.N., Kim, S.S., Kim do, H., Hur, K.Y., Kim, H.K., Ko, T., Han, J., Kim, H.L., Kim, J., Back, S.H., Komatsu, M., Chen, H., Chan, D.C., Konishi, M., Itoh, N., Choi, C.S., Lee, M.S., 2013. Autophagy deficiency leads to protection from obesity and insulin resistance by inducing Fgf21 as a mitokine. Nat. Med. 19, 83–92.
Kim, S.J., Xiao, J., Wan, J., Cohen, P., Yen, K., 2017. Mitochondrially derived peptides as novel regulators of metabolism. J. Physiol. 595, 6613–6621.
Kulisz, A., Chen, N., Chandel, N.S., Shao, Z., Schumacker, P.T., 2002. Mitochondrial ROS initiate phosphorylation of p38 MAP kinase during hypoxia in cardiomyocytes. Am. J. Physiol. Lung Cell Mol. Physiol. 282, L1324–L1329.
Labbadia, J., Brielmann, R.M., Neto, M.F., Lin, Y.F., Haynes, C.M., Morimoto, R.I., 2017. Mitochondrial stress restores the heat shock response and prevents proteostasis collapse during aging. Cell Rep. 21, 1481–1494.
Lakowski, B., Hekimi, S., 1996. Determination of life-span in *Caenorhabditis elegans* by four clock genes. Science 272, 1010–1013.
Latorre-Pellicer, A., Moreno-Loshuertos, R., Lechuga-Vieco, A.V., Sanchez-Cabo, F., Torroja, C., Acin-Perez, R., Calvo, E., Aix, E., Gonzalez-Guerra, A., Logan, A., Bernad-Miana, M.L., Romanos, E., Cruz, R., Cogliati, S., Sobrino, B., Carracedo, A., Perez-Martos, A., Fernandez-Silva, P., Ruiz-Cabello, J., Murphy, M.P., Flores, I., Vazquez, J., Enriquez, J.A., 2016. Mitochondrial and nuclear DNA matching shapes metabolism and healthy ageing. Nature 535, 561–565.

Lee, C., Kim, K.H., Cohen, P., 2016. MOTS-c: a novel mitochondrial-derived peptide regulating muscle and fat metabolism. Free Radic. Biol. Med. 100, 182–187.

Lee, C., Yen, K., Cohen, P., 2013. Humanin: a harbinger of mitochondrial-derived peptides? Trends Endocrinol. Metab. 24, 222–228.

Lee, C., Zeng, J., Drew, B.G., Sallam, T., Martin-Montalvo, A., Wan, J., Kim, S.J., Mehta, H., Hevener, A.L., de Cabo, R., Cohen, P., 2015. The mitochondrial-derived peptide MOTS-c promotes metabolic homeostasis and reduces obesity and insulin resistance. Cell Metab. 21, 443–454.

Lee, M.S., 2015. Effect of mitochondrial stress on systemic metabolism. Ann. N. Y. Acad. Sci. 1350, 61–65.

Lee, S.J., Hwang, A.B., Kenyon, C., 2010. Inhibition of respiration extends *C. elegans* life span via reactive oxygen species that increase HIF-1 activity. Curr. Biol. 20, 2131–2136.

Lee, S.S., Lee, R.Y., Fraser, A.G., Kamath, R.S., Ahringer, J., Ruvkun, G., 2003. A systematic RNAi screen identifies a critical role for mitochondria in *C. elegans* longevity. Nat. Genet. 33, 40–48.

Lehtonen, J.M., Forsstrom, S., Bottani, E., Viscomi, C., Baris, O.R., Isoniemi, H., Hockerstedt, K., Osterlund, P., Hurme, M., Jylhava, J., Leppa, S., Markkula, R., Helio, T., Mombelli, G., Uusimaa, J., Laaksonen, R., Laaksovirta, H., Auranen, M., Zeviani, M., Smeitink, J., Wiesner, R.J., Nakada, K., Isohanni, P., Suomalainen, A., 2016. FGF21 is a biomarker for mitochondrial translation and mtDNA maintenance disorders. Neurology 87, 2290–2299.

Leiser, S.F., Miller, H., Rossner, R., Fletcher, M., Leonard, A., Primitivo, M., Rintala, N., Ramos, F.J., Miller, D.L., Kaeberlein, M., 2015. Cell nonautonomous activation of flavin-containing monooxygenase promotes longevity and health span. Science 350, 1375–1378.

Lemire, B.D., Behrendt, M., DeCorby, A., Gaskova, D., 2009. *C. elegans* longevity pathways converge to decrease mitochondrial membrane potential. Mech. Ageing Dev. 130, 461–465.

Liu, J., Wu, Q., He, D., Ma, T., Du, L., Dui, W., Guo, X., Jiao, R., 2011. Drosophila sbo regulates lifespan through its function in the synthesis of coenzyme Q in vivo. J. Genet. Genomics 38, 225–234.

Liu, X., Jiang, N., Hughes, B., Bigras, E., Shoubridge, E., Hekimi, S., 2005. Evolutionary conservation of the clk-1-dependent mechanism of longevity: loss of mclk1 increases cellular fitness and lifespan in mice. Genes Dev. 19, 2424–2434.

Liu, Y., Samuel, B.S., Breen, P.C., Ruvkun, G., 2014. *Caenorhabditis elegans* pathways that surveil and defend mitochondria. Nature 508, 406–410.

Lopez-Otin, C., Blasco, M.A., Partridge, L., Serrano, M., Kroemer, G., 2013. The hallmarks of aging. Cell 153, 1194–1217.

Lopez-Otin, C., Galluzzi, L., Freije, J.M., Madeo, F., Kroemer, G., 2016. Metabolic control of longevity. Cell 166, 802–821.

Martinez-Reyes, I., Diebold, L.P., Kong, H., Schieber, M., Huang, H., Hensley, C.T., Mehta, M.M., Wang, T., Santos, J.H., Woychik, R., Dufour, E., Spelbrink, J.N., Weinberg, S.E., Zhao, Y., DeBerardinis, R.J., Chandel, N.S., 2016. TCA cycle and mitochondrial membrane potential are necessary for diverse biological functions. Mol. Cell 61, 199–209.

Martinus, R.D., Garth, G.P., Webster, T.L., Cartwright, P., Naylor, D.J., Hoj, P.B., Hoogenraad, N.J., 1996. Selective induction of mitochondrial chaperones in response to loss of the mitochondrial genome. Eur. J. Biochem. 240, 98–103.

Matilainen, O., Quiros, P.M., Auwerx, J., 2017a. Mitochondria and epigenetics - crosstalk in homeostasis and stress. Trends Cell Biol. 27, 453–463.

Matilainen, O., Sleiman, M.S.B., Quiros, P.M., Garcia, S., Auwerx, J., 2017b. The chromatin remodeling factor ISW-1 integrates organismal responses against nuclear and mitochondrial stress. Nat. Commun. 8, 1818.

Mehta, M.M., Weinberg, S.E., Chandel, N.S., 2017. Mitochondrial control of immunity: beyond ATP. Nat. Rev. Immunol. 17, 608–620.

Merry, T.L., Ristow, M., 2016a. Mitohormesis in exercise training. Free Radic. Biol. Med. 98, 123–130.

Merry, T.L., Ristow, M., 2016b. Nuclear factor erythroid-derived 2-like 2 (NFE2L2, Nrf2) mediates exercise-induced mitochondrial biogenesis and the anti-oxidant response in mice. J. Physiol. 594, 5195–5207.

Mesquita, A., Weinberger, M., Silva, A., Sampaio-Marques, B., Almeida, B., Leao, C., Costa, V., Rodrigues, F., Burhans, W.C., Ludovico, P., 2010. Caloric restriction or catalase inactivation extends yeast chronological lifespan by inducing H_2O_2 and superoxide dismutase activity. Proc. Natl. Acad. Sci. U. S. A 107, 15123–15128.

Michel, S., Canonne, M., Arnould, T., Renard, P., 2015. Inhibition of mitochondrial genome expression triggers the activation of CHOP-10 by a cell signaling dependent on the integrated stress response but not the mitochondrial unfolded protein response. Mitochondrion 21, 58–68.

Miller, W.L., 2017. Steroidogenesis: unanswered questions. Trends Endocrinol. Metab. 28, 771–793.

Ming, W., Lu, G., Xin, S., Huanyu, L., Yinghao, J., Xiaoying, L., Chengming, X., Banjun, R., Li, W., Zifan, L., 2016. Mitochondria related peptide MOTS-c suppresses ovariectomy-induced bone loss via AMPK activation. Biochem. Biophys. Res. Commun. 476, 412–419.

Misu, H., Takayama, H., Saito, Y., Mita, Y., Kikuchi, A., Ishii, K.A., Chikamoto, K., Kanamori, T., Tajima, N., Lan, F., Takeshita, Y., Honda, M., Tanaka, M., Kato, S., Matsuyama, N., Yoshioka, Y., Iwayama, K., Tokuyama, K., Akazawa, N., Maeda, S., Takekoshi, K., Matsugo, S., Noguchi, N., Kaneko, S., Takamura, T., 2017. Deficiency of the hepatokine selenoprotein P increases responsiveness to exercise in mice through upregulation of reactive oxygen species and AMP-activated protein kinase in muscle. Nat. Med. 23, 508–516.

Monaghan, R.M., Barnes, R.G., Fisher, K., Andreou, T., Rooney, N., Poulin, G.B., Whitmarsh, A.J., 2015. A nuclear role for the respiratory enzyme CLK-1 in regulating mitochondrial stress responses and longevity. Nat. Cell Biol. 17, 782–792.

Morimoto, H., Iwata, K., Ogonuki, N., Inoue, K., Atsuo, O., Kanatsu-Shinohara, M., Morimoto, T., Yabe-Nishimura, C., Shinohara, T., 2013. ROS are required for mouse spermatogonial stem cell self-renewal. Cell Stem Cell 12, 774–786.

Moris, D., Spartalis, M., Spartalis, E., Karachaliou, G.S., Karaolanis, G.I., Tsourouflis, G., Tsilimigras, D.I., Tzatzaki, E., Theocharis, S., 2017. The role of reactive oxygen species in the pathophysiology of cardiovascular diseases and the clinical significance of myocardial redox. Ann. Transl. Med. 5, 326.

Morrow, R.M., Picard, M., Derbeneva, O., Leipzig, J., McManus, M.J., Gouspillou, G., Barbat-Artigas, S., Dos Santos, C., Hepple, R.T., Murdock, D.G., Wallace, D.C., 2017. Mitochondrial energy deficiency leads to hyperproliferation of skeletal muscle mitochondria and enhanced insulin sensitivity. Proc. Natl. Acad. Sci. U. S. A 114, 2705–2710.

Munch, C., Harper, J.W., 2016. Mitochondrial unfolded protein response controls matrix pre-RNA processing and translation. Nature 534, 710–713.

Murakami, M., Nagahama, M., Maruyama, T., Niikura, T., 2017a. Humanin ameliorates diazepam-induced memory deficit in mice. Neuropeptides 62, 65–70.

Murakami, Y., Ito, M., Ohsawa, I., 2017b. Molecular hydrogen protects against oxidative stress-induced SH-SY5Y neuroblastoma cell death through the process of mitohormesis. PLoS One 12, e0176992.

Murphy, M.P., 2009. How mitochondria produce reactive oxygen species. Biochem. J. 417, 1–13.

Muzumdar, R.H., Huffman, D.M., Calvert, J.W., Jha, S., Weinberg, Y., Cui, L., Nemkal, A., Atzmon, G., Klein, L., Gundewar, S., Ji, S.Y., Lavu, M., Predmore, B.L., Lefer, D.J., 2010. Acute humanin therapy attenuates myocardial ischemia and reperfusion injury in mice. Arterioscler. Thromb. Vasc. Biol. 30, 1940–1948.

Nargund, A.M., Fiorese, C.J., Pellegrino, M.W., Deng, P., Haynes, C.M., 2015. Mitochondrial and nuclear accumulation of the transcription factor ATFS-1 promotes OXPHOS recovery during the UPR(mt). Mol. Cell 58, 123–133.

Nargund, A.M., Pellegrino, M.W., Fiorese, C.J., Baker, B.M., Haynes, C.M., 2012. Mitochondrial import efficiency of ATFS-1 regulates mitochondrial UPR activation. Science 337, 587–590.

Nikkanen, J., Forsstrom, S., Euro, L., Paetau, I., Kohnz, R.A., Wang, L., Chilov, D., Viinamaki, J., Roivainen, A., Marjamaki, P., Liljenback, H., Ahola, S., Buzkova, J., Terzioglu, M., Khan, N.A., Pirnes-Karhu, S., Paetau, A., Lonnqvist, T., Sajantila, A., Isohanni, P., Tyynismaa, H., Nomura, D.K., Battersby, B.J., Velagapudi, V., Carroll, C.J., Suomalainen, A., 2016. Mitochondrial DNA replication defects disturb cellular dNTP pools and remodel one-carbon metabolism. Cell Metab. 23, 635–648.

Ost, M., Coleman, V., Voigt, A., van Schothorst, E.M., Keipert, S., van der Stelt, I., Ringel, S., Graja, A., Ambrosi, T., Kipp, A.P., Jastroch, M., Schulz, T.J., Keijer, J., Klaus, S., 2016. Muscle mitochondrial stress adaptation operates independently of endogenous FGF21 action. Mol. Metab. 5, 79–90.

Owusu-Ansah, E., Banerjee, U., 2009. Reactive oxygen species prime Drosophila haematopoietic progenitors for differentiation. Nature 461, 537–541.

Owusu-Ansah, E., Song, W., Perrimon, N., 2013. Muscle mitohormesis promotes longevity via systemic repression of insulin signaling. Cell 155, 699–712.

Pakos-Zebrucka, K., Koryga, I., Mnich, K., Ljujic, M., Samali, A., Gorman, A.M., 2016. The integrated stress response. EMBO Rep. 17, 1374–1395.

Palikaras, K., Lionaki, E., Tavernarakis, N., 2015. Coordination of mitophagy and mitochondrial biogenesis during ageing in *C. elegans*. Nature 521, 525–528.

Pan, Y., Schroeder, E.A., Ocampo, A., Barrientos, A., Shadel, G.S., 2011. Regulation of yeast chronological life span by TORC1 via adaptive mitochondrial ROS signaling. Cell Metab 13, 668–678.

Papa, L., Germain, D., 2011. Estrogen receptor mediates a distinct mitochondrial unfolded protein response. J. Cell Sci. 124, 1396–1402.

Papa, L., Germain, D., 2014. SirT3 regulates the mitochondrial unfolded protein response. Mol. Cell Biol. 34, 699–710.

Paracelsus, 1564. Labyrintus und Irrgang der vermeinten Artzet: Item, Siben defensiones, oder Schirmreden. Petrus Perna, Basel, Switzerland.

Pellegrino, M.W., Nargund, A.M., Kirienko, N.V., Gillis, R., Fiorese, C.J., Haynes, C.M., 2014. Mitochondrial UPR-regulated innate immunity provides resistance to pathogen infection. Nature 516, 414–417.

Pereira, R.O., Tadinada, S.M., Zasadny, F.M., Oliveira, K.J., Pires, K.M.P., Olvera, A., Jeffers, J., Souvenir, R., McGlauflin, R., Seei, A., Funari, T., Sesaki, H., Potthoff, M.J., Adams, C.M., Anderson, E.J., Abel, E.D., 2017. OPA1 deficiency promotes secretion of FGF21 from muscle that prevents obesity and insulin resistance. EMBO J. 36, 2126–2145.

Pospisilik, J.A., Knauf, C., Joza, N., Benit, P., Orthofer, M., Cani, P.D., Ebersberger, I., Nakashima, T., Sarao, R., Neely, G., Esterbauer, H., Kozlov, A., Kahn, C.R., Kroemer, G., Rustin, P., Burcelin, R., Penninger, J.M., 2007. Targeted deletion of AIF decreases mitochondrial oxidative phosphorylation and protects from obesity and diabetes. Cell 131, 476–491.

Powers, S.K., Hogan, M.C., 2016. Exercise and oxidative stress. J. Physiol. 594, 5079–5080.

Pulliam, D.A., Deepa, S.S., Liu, Y., Hill, S., Lin, A.L., Bhattacharya, A., Shi, Y., Sloane, L., Viscomi, C., Zeviani, M., Van Remmen, H., 2014. Complex IV-deficient Surf1(-/-) mice initiate mitochondrial stress responses. Biochem. J. 462, 359–371.

Quiros, P.M., Langer, T., Lopez-Otin, C., 2015. New roles for mitochondrial proteases in health, ageing and disease. Nat. Rev. Mol. Cell Biol. 16, 345–359.

Quiros, P.M., Mottis, A., Auwerx, J., 2016. Mitonuclear communication in homeostasis and stress. Nat. Rev. Mol. Cell Biol. 17, 213–226.

Quiros, P.M., Prado, M.A., Zamboni, N., D'Amico, D., Williams, R.W., Finley, D., Gygi, S.P., Auwerx, J., 2017. Multi-omics analysis identifies ATF4 as a key regulator of the mitochondrial stress response in mammals. J. Cell Biol. 216, 2027–2045.

Rainbolt, T.K., Atanassova, N., Genereux, J.C., Wiseman, R.L., 2013. Stress-regulated translational attenuation adapts mitochondrial protein import through Tim17A degradation. Cell Metab. 18, 908–919.

Rauthan, M., Ranji, P., Aguilera Pradenas, N., Pitot, C., Pilon, M., 2013. The mitochondrial unfolded protein response activator ATFS-1 protects cells from inhibition of the mevalonate pathway. Proc. Natl. Acad. Sci. U. S. A 110, 5981–5986.

Rea, S.L., Ventura, N., Johnson, T.E., 2007. Relationship between mitochondrial electron transport chain dysfunction, development, and life extension in *Caenorhabditis elegans*. PLoS Biol. 5, e259.

Rehman, H., Liu, Q., Krishnasamy, Y., Shi, Z., Ramshesh, V.K., Haque, K., Schnellmann, R.G., Murphy, M.P., Lemasters, J.J., Rockey, D.C., Zhong, Z., 2016. The mitochondria-targeted antioxidant MitoQ attenuates liver fibrosis in mice. Int. J. Physiol. Pathophysiol. Pharmacol. 8, 14–27.

Ren, Y., Chen, S., Ma, M., Yao, X., Sun, D., Li, B., Lu, J., 2015. The activation of protein homeostasis protective mechanisms perhaps is not responsible for lifespan extension caused by deficiencies of mitochondrial proteins in *C. elegans*. Exp. Gerontol. 65, 53–57.

Riera, C.E., Merkwirth, C., De Magalhaes Filho, C.D., Dillin, A., 2016. Signaling networks determining life span. Annu. Rev. Biochem. 85, 35–64.

Ristow, M., 2014. Unraveling the truth about antioxidants: mitohormesis explains ROS-induced health benefits. Nat. Med. 20, 709–711.

Ristow, M., Schmeisser, K., 2014. Mitohormesis: promoting health and lifespan by increased levels of reactive oxygen species (ROS). Dose Response 12, 288–341.

Ristow, M., Zarse, K., Oberbach, A., Kloting, N., Birringer, M., Kiehntopf, M., Stumvoll, M., Kahn, C.R., Bluher, M., 2009. Antioxidants prevent health-promoting effects of physical exercise in humans. Proc. Natl. Acad. Sci. U. S. A 106, 8665–8670.

Rizzuto, R., De Stefani, D., Raffaello, A., Mammucari, C., 2012. Mitochondria as sensors and regulators of calcium signalling. Nat. Rev. Mol. Cell Biol. 13, 566–578.

Robida-Stubbs, S., Glover-Cutter, K., Lamming, D.W., Mizunuma, M., Narasimhan, S.D., Neumann-Haefelin, E., Sabatini, D.M., Blackwell, T.K., 2012. TOR signaling and rapamycin influence longevity by regulating SKN-1/Nrf and DAF-16/FoxO. Cell Metab. 15, 713–724.

Sabharwal, S.S., Schumacker, P.T., 2014. Mitochondrial ROS in cancer: initiators, amplifiers or an Achilles' heel? Nat. Rev. Cancer 14, 709–721.

Salminen, A., Kaarniranta, K., Kauppinen, A., 2017. Regulation of longevity by FGF21: interaction between energy metabolism and stress responses. Ageing Res. Rev. 37, 79–93.
Scarpulla, R.C., Vega, R.B., Kelly, D.P., 2012. Transcriptional integration of mitochondrial biogenesis. Trends Endocrinol. Metab. 23, 459–466.
Schaar, C.E., Dues, D.J., Spielbauer, K.K., Machiela, E., Cooper, J.F., Senchuk, M., Hekimi, S., Van Raamsdonk, J.M., 2015. Mitochondrial and cytoplasmic ROS have opposing effects on lifespan. PLoS Genet. 11, e1004972.
Schiavi, A., Maglioni, S., Palikaras, K., Shaik, A., Strappazzon, F., Brinkmann, V., Torgovnick, A., Castelein, N., De Henau, S., Braeckman, B.P., Cecconi, F., Tavernarakis, N., Ventura, N., 2015. Iron-starvation-induced mitophagy mediates lifespan extension upon mitochondrial stress in *C. elegans*. Curr. Biol. 25, 1810–1822.
Schmeisser, K., Mansfeld, J., Kuhlow, D., Weimer, S., Priebe, S., Heiland, I., Birringer, M., Groth, M., Segref, A., Kanfi, Y., Price, N.L., Schmeisser, S., Schuster, S., Pfeiffer, A.F., Guthke, R., Platzer, M., Hoppe, T., Cohen, H.Y., Zarse, K., Sinclair, D.A., Ristow, M., 2013a. Role of sirtuins in lifespan regulation is linked to methylation of nicotinamide. Nat. Chem. Biol. 9, 693–700.
Schmeisser, S., Schmeisser, K., Weimer, S., Groth, M., Priebe, S., Fazius, E., Kuhlow, D., Pick, D., Einax, J.W., Guthke, R., Platzer, M., Zarse, K., Ristow, M., 2013b. Mitochondrial hormesis links low-dose arsenite exposure to lifespan extension. Aging Cell 12, 508–517.
Schmeisser, S., Zarse, K., Ristow, M., 2011. Lonidamine extends lifespan of adult *Caenorhabditis elegans* by increasing the formation of mitochondrial reactive oxygen species. Horm. Metab. Res. 43, 687–692.
Schulz, T.J., Zarse, K., Voigt, A., Urban, N., Birringer, M., Ristow, M., 2007. Glucose restriction extends *Caenorhabditis elegans* life span by inducing mitochondrial respiration and increasing oxidative stress. Cell Metab. 6, 280–293.
Scialo, F., Sriram, A., Fernandez-Ayala, D., Gubina, N., Lohmus, M., Nelson, G., Logan, A., Cooper, H.M., Navas, P., Enriquez, J.A., Murphy, M.P., Sanz, A., 2016. Mitochondrial ROS produced via reverse electron transport extend animal lifespan. Cell Metab. 23, 725–734.
Sebastian, D., Palacin, M., Zorzano, A., 2017. Mitochondrial dynamics: coupling mitochondrial fitness with healthy aging. Trends Mol. Med. 23, 201–215.
Seiferling, D., Szczepanowska, K., Becker, C., Senft, K., Hermans, S., Maiti, P., Konig, T., Kukat, A., Trifunovic, A., 2016. Loss of CLPP alleviates mitochondrial cardiomyopathy without affecting the mammalian UPRmt. EMBO Rep. 17, 953–964.
Sena, L.A., Chandel, N.S., 2012. Physiological roles of mitochondrial reactive oxygen species. Mol. Cell 48, 158–167.
Sena, L.A., Li, S., Jairaman, A., Prakriya, M., Ezponda, T., Hildeman, D.A., Wang, C.R., Schumacker, P.T., Licht, J.D., Perlman, H., Bryce, P.J., Chandel, N.S., 2013. Mitochondria are required for antigen-specific T cell activation through reactive oxygen species signaling. Immunity 38, 225–236.
Shadel, G.S., Horvath, T.L., 2015. Mitochondrial ROS signaling in organismal homeostasis. Cell 163, 560–569.
Shao, L.W., Niu, R., Liu, Y., 2016. Neuropeptide signals cell non-autonomous mitochondrial unfolded protein response. Cell Res. 26, 1182–1196.
Shore, D.E., Carr, C.E., Ruvkun, G., 2012. Induction of cytoprotective pathways is central to the extension of lifespan conferred by multiple longevity pathways. PLoS Genet. 8, e1002792.
Shpilka, T., Haynes, C.M., 2018. The mitochondrial UPR: mechanisms, physiological functions and implications in ageing. Nat. Rev. Mol. Cell Biol. 19, 109–120.

Skamagki, M., Correia, C., Yeung, P., Baslan, T., Beck, S., Zhang, C., Ross, C.A., Dang, L., Liu, Z., Giunta, S., Chang, T.P., Wang, J., Ananthanarayanan, A., Bohndorf, M., Bosbach, B., Adjaye, J., Funabiki, H., Kim, J., Lowe, S., Collins, J.J., Lu, C.W., Li, H., Zhao, R., Kim, K., 2017. ZSCAN10 expression corrects the genomic instability of iPSCs from aged donors. Nat. Cell Biol. 19, 1037–1048.

Song, M., Franco, A., Fleischer, J.A., Zhang, L., Dorn 2nd, G.W., 2017. Abrogating mitochondrial dynamics in mouse hearts accelerates mitochondrial senescence. Cell Metab. 26, 872–883.

Song, M., Mihara, K., Chen, Y., Scorrano, L., Dorn 2nd, G.W., 2015. Mitochondrial fission and fusion factors reciprocally orchestrate mitophagic culling in mouse hearts and cultured fibroblasts. Cell Metab. 21, 273–285.

Stepanyan, Z., Hughes, B., Cliche, D.O., Camp, D., Hekimi, S., 2006. Genetic and molecular characterization of CLK-1/mCLK1, a conserved determinant of the rate of aging. Exp. Gerontol. 41, 940–951.

Sun, N., Youle, R.J., Finkel, T., 2016. The mitochondrial basis of aging. Mol. Cell 61, 654–666.

Sun, X., Wheeler, C.T., Yolitz, J., Laslo, M., Alberico, T., Sun, Y., Song, Q., Zou, S., 2014. A mitochondrial ATP synthase subunit interacts with TOR signaling to modulate protein homeostasis and lifespan in Drosophila. Cell Rep. 8, 1781–1792.

Suomalainen, A., Elo, J.M., Pietilainen, K.H., Hakonen, A.H., Sevastianova, K., Korpela, M., Isohanni, P., Marjavaara, S.K., Tyni, T., Kiuru-Enari, S., Pihko, H., Darin, N., Ounap, K., Kluijtmans, L.A., Paetau, A., Buzkova, J., Bindoff, L.A., Annunen-Rasila, J., Uusimaa, J., Rissanen, A., Yki-Jarvinen, H., Hirano, M., Tulinius, M., Smeitink, J., Tyynismaa, H., 2011. FGF-21 as a biomarker for muscle-manifesting mitochondrial respiratory chain deficiencies: a diagnostic study. Lancet Neurol. 10, 806–818.

Tan, K., Fujimoto, M., Takii, R., Takaki, E., Hayashida, N., Nakai, A., 2015. Mitochondrial SSBP1 protects cells from proteotoxic stresses by potentiating stress-induced HSF1 transcriptional activity. Nat. Commun. 6, 6580.

Tapia, P.C., 2006. Sublethal mitochondrial stress with an attendant stoichiometric augmentation of reactive oxygen species may precipitate many of the beneficial alterations in cellular physiology produced by caloric restriction, intermittent fasting, exercise and dietary phytonutrients: "Mitohormesis" for health and vitality. Med. Hypotheses 66, 832–843.

Tassone, B., Saoncella, S., Neri, F., Ala, U., Brusa, D., Magnuson, M.A., Provero, P., Oliviero, S., Riganti, C., Calautti, E., 2017. Rictor/mTORC2 deficiency enhances keratinocyte stress tolerance via mitohormesis. Cell Death Differ. 24, 731–746.

Tauffenberger, A., Vaccaro, A., Parker, J.A., 2016. Fragile lifespan expansion by dietary mitohormesis in *C. elegans*. Aging (Albany NY) 8, 50–61.

Tezze, C., Romanello, V., Desbats, M.A., Fadini, G.P., Albiero, M., Favaro, G., Ciciliot, S., Soriano, M.E., Morbidoni, V., Cerqua, C., Loefler, S., Kern, H., Franceschi, C., Salvioli, S., Conte, M., Blaauw, B., Zampieri, S., Salviati, L., Scorrano, L., Sandri, M., 2017. Age-associated loss of OPA1 in muscle impacts muscle mass, metabolic homeostasis, systemic inflammation, and epithelial senescence. Cell Metab. 25, 1374–1389 e1376.

Topf, U., Wrobel, L., Chacinska, A., 2016. Chatty mitochondria: keeping balance in cellular protein homeostasis. Trends Cell Biol. 26, 577–586.

Toyama, E.Q., Herzig, S., Courchet, J., Lewis Jr., T.L., Loson, O.C., Hellberg, K., Young, N.P., Chen, H., Polleux, F., Chan, D.C., Shaw, R.J., 2016. Metabolism. AMP-activated protein kinase mediates mitochondrial fission in response to energy stress. Science 351, 275–281.

Tyynismaa, H., Carroll, C.J., Raimundo, N., Ahola-Erkkila, S., Wenz, T., Ruhanen, H., Guse, K., Hemminki, A., Peltola-Mjosund, K.E., Tulkki, V., Oresic, M., Moraes, C.T., Pietilainen, K., Hovatta, I., Suomalainen, A., 2010. Mitochondrial myopathy induces a starvation-like response. Hum. Mol. Genet. 19, 3948–3958.

Urban, N., Tsitsipatis, D., Hausig, F., Kreuzer, K., Erler, K., Stein, V., Ristow, M., Steinbrenner, H., Klotz, L.O., 2017. Non-linear impact of glutathione depletion on *C. elegans* life span and stress resistance. Redox Biol. 11, 502–515.

Vandanmagsar, B., Warfel, J.D., Wicks, S.E., Ghosh, S., Salbaum, J.M., Burk, D., Dubuisson, O.S., Mendoza, T.M., Zhang, J., Noland, R.C., Mynatt, R.L., 2016. Impaired mitochondrial fat oxidation induces FGF21 in muscle. Cell Rep. 15, 1686–1699.

Voos, W., 2013. Chaperone-protease networks in mitochondrial protein homeostasis. Biochim. Biophys. Acta 1833, 388–399.

Wai, T., Langer, T., 2016. Mitochondrial dynamics and metabolic regulation. Trends Endocrinol. Metab. 27, 105–117.

Wall, C.E., Whyte, J., Suh, J.M., Fan, W., Collins, B., Liddle, C., Yu, R.T., Atkins, A.R., Naviaux, J.C., Li, K., Bright, A.T., Alaynick, W.A., Downes, M., Naviaux, R.K., Evans, R.M., 2015. High-fat diet and FGF21 cooperatively promote aerobic thermogenesis in mtDNA mutator mice. Proc. Natl. Acad. Sci. U. S. A 112, 8714–8719.

Walter, L., Baruah, A., Chang, H.W., Pace, H.M., Lee, S.S., 2011. The homeobox protein CEH-23 mediates prolonged longevity in response to impaired mitochondrial electron transport chain in *C. elegans*. PLoS Biol. 9, e1001084.

Wang, D., Malo, D., Hekimi, S., 2010. Elevated mitochondrial reactive oxygen species generation affects the immune response via hypoxia-inducible factor-1alpha in long-lived Mclk1+/- mouse mutants. J. Immunol. 184, 582–590.

Wang, D., Wang, Y., Argyriou, C., Carriere, A., Malo, D., Hekimi, S., 2012. An enhanced immune response of Mclk1(+)/(-) mutant mice is associated with partial protection from fibrosis, cancer and the development of biomarkers of aging. PLoS One 7, e49606.

Wang, L., Ishihara, T., Ibayashi, Y., Tatsushima, K., Setoyama, D., Hanada, Y., Takeichi, Y., Sakamoto, S., Yokota, S., Mihara, K., Kang, D., Ishihara, N., Takayanagi, R., Nomura, M., 2015. Disruption of mitochondrial fission in the liver protects mice from diet-induced obesity and metabolic deterioration. Diabetologia 58, 2371–2380.

Wang, X., Chen, X.J., 2015. A cytosolic network suppressing mitochondria-mediated proteostatic stress and cell death. Nature 524, 481–484.

Wang, X., Chrysovergis, K., Kosak, J., Kissling, G., Streicker, M., Moser, G., Li, R., Eling, T.E., 2014. hNAG-1 increases lifespan by regulating energy metabolism and insulin/IGF-1/mTOR signaling. Aging (Albany NY) 6, 690–704.

Weimer, S., Priebs, J., Kuhlow, D., Groth, M., Priebe, S., Mansfeld, J., Merry, T.L., Dubuis, S., Laube, B., Pfeiffer, A.F., Schulz, T.J., Guthke, R., Platzer, M., Zamboni, N., Zarse, K., Ristow, M., 2014. D-Glucosamine supplementation extends life span of nematodes and of ageing mice. Nat. Commun. 5, 3563.

West, A.P., Brodsky, I.E., Rahner, C., Woo, D.K., Erdjument-Bromage, H., Tempst, P., Walsh, M.C., Choi, Y., Shadel, G.S., Ghosh, S., 2011. TLR signalling augments macrophage bactericidal activity through mitochondrial ROS. Nature 472, 476–480.

West, A.P., Khoury-Hanold, W., Staron, M., Tal, M.C., Pineda, C.M., Lang, S.M., Bestwick, M., Duguay, B.A., Raimundo, N., MacDuff, D.A., Kaech, S.M., Smiley, J.R., Means, R.E., Iwasaki, A., Shadel, G.S., 2015. Mitochondrial DNA stress primes the antiviral innate immune response. Nature 520, 553–557.

Williams, D.S., Cash, A., Hamadani, L., Diemer, T., 2009. Oxaloacetate supplementation increases lifespan in *Caenorhabditis elegans* through an AMPK/FOXO-dependent pathway. Aging Cell 8, 765–768.

Wredenberg, A., Freyer, C., Sandstrom, M.E., Katz, A., Wibom, R., Westerblad, H., Larsson, N.G., 2006. Respiratory chain dysfunction in skeletal muscle does not cause insulin resistance. Biochem. Biophys. Res. Commun. 350, 202–207.

Wrobel, L., Topf, U., Bragoszewski, P., Wiese, S., Sztolsztener, M.E., Oeljeklaus, S., Varabyova, A., Lirski, M., Chroscicki, P., Mroczek, S., Januszewicz, E., Dziembowski, A., Koblowska, M., Warscheid, B., Chacinska, A., 2015. Mistargeted mitochondrial proteins activate a proteostatic response in the cytosol. Nature 524, 485–488.

Xiong, L.G., Chen, Y.J., Tong, J.W., Gong, Y.S., Huang, J.A., Liu, Z.H., 2017. Epigallocatechin-3-gallate promotes healthy lifespan through mitohormesis during early-to-mid adulthood in *Caenorhabditis elegans*. Redox Biol. 14, 305–315.

Yang, W., Hekimi, S., 2010a. A mitochondrial superoxide signal triggers increased longevity in *Caenorhabditis elegans*. PLoS Biol. 8, e1000556.

Yang, W., Hekimi, S., 2010b. Two modes of mitochondrial dysfunction lead independently to lifespan extension in *Caenorhabditis elegans*. Aging Cell 9, 433–447.

Yatsuga, S., Fujita, Y., Ishii, A., Fukumoto, Y., Arahata, H., Kakuma, T., Kojima, T., Ito, M., Tanaka, M., Saiki, R., Koga, Y., 2015. Growth differentiation factor 15 as a useful biomarker for mitochondrial disorders. Ann. Neurol. 78, 814–823.

Yoneda, T., Benedetti, C., Urano, F., Clark, S.G., Harding, H.P., Ron, D., 2004. Compartment-specific perturbation of protein handling activates genes encoding mitochondrial chaperones. J. Cell Sci. 117, 4055–4066.

Yun, J., Finkel, T., 2014. Mitohormesis. Cell Metab. 19, 757–766.

Zarse, K., Schmeisser, S., Groth, M., Priebe, S., Beuster, G., Kuhlow, D., Guthke, R., Platzer, M., Kahn, C.R., Ristow, M., 2012. Impaired insulin/IGF1 signaling extends life span by promoting mitochondrial L-proline catabolism to induce a transient ROS signal. Cell Metab. 15, 451–465.

Zarse, K., Schulz, T.J., Birringer, M., Ristow, M., 2007. Impaired respiration is positively correlated with decreased life span in *Caenorhabditis elegans* models of Friedreich Ataxia. FASEB J. 21, 1271–1275.

Zhai, D., Ye, Z., Jiang, Y., Xu, C., Ruan, B., Yang, Y., Lei, X., Xiang, A., Lu, H., Zhu, Z., Yan, Z., Wei, D., Li, Q., Wang, L., Lu, Z., 2017. MOTS-c peptide increases survival and decreases bacterial load in mice infected with MRSA. Mol. Immunol. 92, 151–160.

Zhang, Y., Shao, Z., Zhai, Z., Shen, C., Powell-Coffman, J.A., 2009. The HIF-1 hypoxia-inducible factor modulates lifespan in *C. elegans*. PLoS One 4, e6348.

Zhang, Y., Xie, Y., Berglund, E.D., Coate, K.C., He, T.T., Katafuchi, T., Xiao, G., Potthoff, M.J., Wei, W., Wan, Y., Yu, R.T., Evans, R.M., Kliewer, S.A., Mangelsdorf, D.J., 2012. The starvation hormone, fibroblast growth factor-21, extends lifespan in mice. Elife 1, e00065.

Zhao, Q., Wang, J., Levichkin, I.V., Stasinopoulos, S., Ryan, M.T., Hoogenraad, N.J., 2002. A mitochondrial specific stress response in mammalian cells. EMBO J. 21, 4411–4419.

Zini, R., Morin, C., Bertelli, A., Bertelli, A.A., Tillement, J.P., 1999. Effects of resveratrol on the rat brain respiratory chain. Drugs Exp. Clin. Res. 25, 87–97.

CHAPTER THREE

Coordinating Mitochondrial Biology Through the Stress-Responsive Regulation of Mitochondrial Proteases

Justine Lebeau, T. Kelly Rainbolt and R. Luke Wiseman[1]
Department of Molecular Medicine, The Scripps Research Institute, La Jolla, CA, United States
[1]Corresponding author: Email: wiseman@scripps.edu

Contents

International Review of Cell and Molecular Biology, Volume 340
ISSN 1937-6448
https://doi.org/10.1016/bs.ircmb.2018.05.003

Abstract

Proteases are localized throughout mitochondria and function as critical regulators of all aspects of mitochondrial biology. As such, the activities of these proteases are sensitively regulated through transcriptional and post-translational mechanisms to adapt mitochondrial function to specific cellular demands. Here, we discuss the stress-responsive mechanisms responsible for regulating mitochondrial protease activity and the implications of this regulation on mitochondrial function. Furthermore, we describe how imbalances in the activity or regulation of mitochondrial proteases induced by genetic, environmental, or aging-related factors influence mitochondria in the context of disease. Understanding the molecular mechanisms by which cells regulate mitochondrial function through alterations in protease activity provide insights into the contributions of these proteases in pathologic mitochondrial dysfunction and reveals new therapeutic opportunities to ameliorate this dysfunction in the context of diverse classes of human disease.

1. INTRODUCTION

Mitochondria are responsible for essential metabolic and signaling activities within eukaryotic cells (Fig. 1). These include energy production, phospholipid metabolism, and the regulation of apoptotic signaling pathways, all of which require the maintenance of the mitochondrial proteome, also referred to as mitochondrial proteostasis, for their efficient activity. Imbalances in mitochondrial proteostasis induced by genetic, environmental, or aging-related insults are implicated in the pathologic mitochondrial dysfunction associated with diverse diseases including cancer, cardiovascular disease, and many neurodegenerative diseases. A key factor in preventing this pathologic dysfunction is the regulation of mitochondrial proteostasis through the activity of proteases localized throughout mitochondria. These proteases act through a variety of mechanisms to regulate mitochondrial function in response to a given insult and prevent stress-induced dysregulation of mitochondrial proteostasis. The central importance of mitochondrial proteases provides an opportunity for cells to sensitively adapt mitochondrial function to specific cellular demands through the stress-responsive regulation of proteolytic activity. Here, we discuss the mechanisms by which cells regulate mitochondrial proteolytic activity in

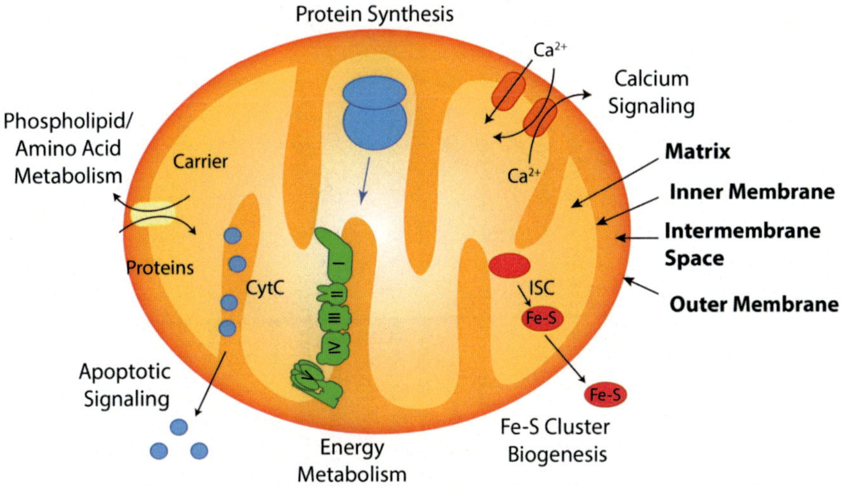

Figure 1 Illustration showing select biologic activities localized to mitochondria. The four compartments of mitochondria are shown in bold text.

response to stress. Furthermore, we describe the functional importance of stress-responsive protease regulation and how imbalances in this regulation can promote pathologic mitochondrial dysfunction implicated in diverse diseases.

1.1 Organization and Activity of the Mitochondrial Proteolytic Network

Mitochondria are defined by a double membrane architecture resulting in four intramitochondrial compartments: the outer membrane (OM), the intermembrane space (IMS), the inner membrane (IM), and the mitochondrial matrix (Fig. 1). Except for those localized to the OM, mitochondrial proteins are not accessible to the ubiquitin proteasome system (UPS) for degradation. Instead, these mitochondrial proteins are regulated by proteases localized to the IMS, IM, and mitochondrial matrix that promote the maintenance of mitochondrial proteostasis and regulate specific aspects of mitochondrial function (Table 1). Mitochondrial proteases can be generally classified into three categories based on their predominant role in regulating mitochondrial proteostasis and function (Baker et al., 2011; Quiros et al., 2015). Mitochondrial processing peptidases function in the biogenesis and maturation of the mitochondrial proteome. Quality control proteases degrade damaged mitochondrial proteins and regulate specific aspects of mitochondrial biology. Finally, oligopeptidases degrade peptides derived

Table 1 List of Mitochondrial Proteases and Peptidases Discussed in This Review

Protease/Peptidase	Catalytic Activity	Localization	Category	Yeast Homolog
USP15	Cys	OM	Deubiquitinylase	Ubp12
USP30	Cys	OM	Deubiquitinylase	Ubp16
MEP	Metallo	IMS	Oligopeptidase	Prd1
MIP	Metallo	Matrix	Processing peptidase	Oct1
PMPCB	Metallo	Matrix	Processing peptidase	Mas1
PREP (PITRM1)	Metallo	Matrix	Oligopeptidase	Mop112
OSGEPL1	Metallo	Matrix	Unknown	Qri7
METAP1D (MAP1D)	Metallo	Matrix	Processing peptidase	Map1
XPNPEP3	Metallo	Matrix	Processing peptidase	Icp55
YME1L	Metallo	IM	Quality Control Protease	Yme1
PARAPLEGIN (SPG7)	Metallo	IM	Quality Control Protease	Yta10
AFG3L2	Metallo	IM	Quality Control Protease	Yta12
OMA1	Metallo	IM	Quality Control Protease	Oma1
ATP23	Metallo	IMS	Quality Control and processing	Atp23
HTRA2	Ser	IMS	Quality Control Protease	
LACTB	Ser	IMS	Unknown	
CLPP	Ser	Matrix	Quality Control Protease	
LONP	Ser	Matrix	Quality Control Protease	Pim1/Lon
IMMP2L	Ser	IM	Processing peptidase	Imp2
IMMP1L	Ser	IM	Processing peptidase	Imp1
PARL	Ser	IM	Quality Control Protease	Pcp1

from mitochondrial processing peptidases (MPPs) and quality control proteases into amino acids. Proteases in these categories are diverse in their catalytic activity, ATP dependence, and method of regulation, which reflect their unique functions in coordinating mitochondrial biology (Baker et al., 2011; Quiros et al., 2015). Below we discuss these three classes of mitochondrial proteases and their generalized roles in regulating mitochondrial proteostasis and function.

1.1.1 Processing Peptidases Facilitate Establishment of the Mitochondrial Proteome

The majority of mitochondrial-localized proteins (>99%) are encoded by the nuclear genome and translated on cytosolic ribosomes. These mitochondrial proteins are directed to the translocase of the outer membrane (TOM) complex by targeting sequences localized at the N-terminus or internally within the polypeptide sequence (Chacinska et al., 2009; Harbauer et al., 2014; Schmidt et al., 2010). TOM facilitates translocation of these polypeptides across the OM to the IMS where they are further targeted to additional translocase and localization machinery that directs these proteins to specific mitochondrial subcompartments (e.g., the translocase of the inner membrane 23 [TIM23] translocates proteins into the IM and across the IM to the mitochondrial matrix) (Chacinska et al., 2009; Harbauer et al., 2014; Schmidt et al., 2010). Once localized, these polypeptides are often proteolytically processed by MPPs to release the mature polypeptide.

The matrix-localized metalloprotease MPP is a heterodimeric complex comprising the catalytic MPPβ subunit (*PMPCB*) and the noncatalytic MPPα subunit (*PMPCA*) (Kleiber et al., 1990; Ou et al., 1989; Yang et al., 1988). MPP is responsible for removing the N-terminal—targeting sequences from mitochondrial proteins targeted to the mitochondrial matrix (Gakh et al., 2002). The removal of this N-terminal—targeting sequence releases the mature polypeptide to engage matrix chaperones and folding enzymes that facilitate folding to its native conformation.

MPP-dependent processing of N-terminal—targeting sequences can also be the first step in the maturation of mitochondrial proteins that require multiple proteolytic steps for maturation and/or localization. Mitochondrial intermediate peptidase removes an N-terminal octapeptide sequence from selected polypeptides that is revealed following MPP-dependent processing of the mitochondrial-targeting sequence (Vogtle et al., 2011). The removal of this octapeptide is important as it replaces a destabilizing N-terminal amino acid with a stabilizing N-terminal amino acid. This substitution can prevent degradation of the immature polypeptide through the N-end rule degradation mechanism of the mitochondrial matrix (Vogtle et al., 2011). Similarly, the mitochondrial x-prolyl aminopeptidase XPNPEP3 can cleave a single destabilizing N-terminal amino acid from MPP-processed peptides to reveal a stable N-terminal amino acid (Vogtle et al., 2009). The METAP1D aminopeptidase (Methionyl Aminopeptidase Type 1D, also referred to as MAP1D) also regulates N-terminal amino acids of mitochondrial-encoded proteins such as CoxIII by removing the N-terminal methionine as part of an N-terminal methionine excision process (Hu et al., 2007; Serero et al., 2003; Walker et al., 2009). Alternatively, inner membrane protease subunit 1 and inner membrane protease subunit 2 remove hydrophobic sorting signals from MPP-processed proteins, releasing the mature polypeptide to the IMS (Nunnari et al., 1993). Thus, through the activities discussed above, MPPs have critical roles in regulating the maturation, stability, and localization of the mitochondrial proteome.

1.1.2 Mitochondrial Quality Control Proteases Regulate the Integrity and Function of the Mitochondrial Proteome

In contrast to mitochondrial processing peptidases, mitochondrial quality control proteases primarily function to remove damaged or misfolded proteins localized to intramitochondrial environments. Four AAA + ATP-dependent proteases—the *i*-AAA protease, the *m*-AAA protease, LON, and ClpXP—are predominantly responsible for performing quality control functions within mitochondria (Fig. 2). These proteases all function through a similar biochemical mechanism that appears to be conserved from bacteria to mammals (Puchades et al., 2017; Sauer and Baker, 2011). Damaged and/or aberrantly folded proteins are directed to the AAA + domains of these proteases where they are unfolded through an ATP-dependent process (Puchades et al., 2017; Sauer and Baker, 2011). The unfolded proteins are then translocated to a protected proteolytic core for degradation. Within mitochondria, AAA+ proteases localize to specific intramitochondrial environments such as the IM or the mitochondrial matrix. The localization of these proteases to specific compartments is a primary determinant in dictating their substrate specificity and biologic function.

The *i*-AAA and *m*-AAA proteases are both mitochondrial IM proteases with active sites oriented toward different sides of the membrane. The *i*-AAA protease is a homooligomer comprising six subunits of the zinc metalloprotease YME1L. The active site of the *i*-AAA protease, or more

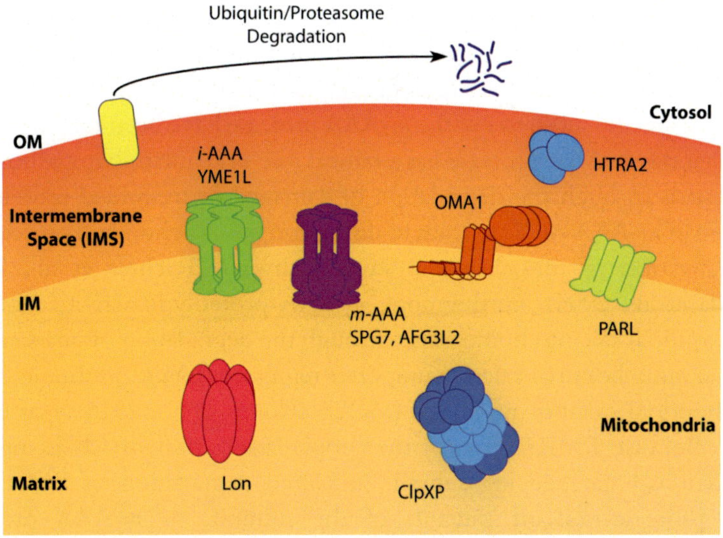

Figure 2 The organization and localization of mitochondrial quality control proteases within mitochondria.

simply YME1L, is oriented toward the IMS, making this the only ATP-dependent protease accessing this mitochondrial subcompartment (Fig. 2). YME1L predominantly functions to degrade damaged, unassembled, or misfolded proteins localized to the IM or IMS. This includes many established substrates such as complex I subunits (Stiburek et al., 2012), the core TIM23 subunit Tim17A (Rainbolt et al., 2013), small Tim proteins (Baker et al., 2012b; Spiller et al., 2015), and the phosphatidic acid transporter PRELID1 (Potting et al., 2013). As such, YME1L influences many aspects of mitochondrial biology, including electron transport chain activity, mitochondrial protein import, and phospholipid metabolism. YME1L also functions to regulate mitochondrial morphology through the processing of the inner membrane GTPase optic atrophy protein 1 (OPA1) in a process that is discussed in more detail below (Griparic et al., 2007; Ishihara et al., 2006; Song et al., 2007). Through these (and other) activities, YME1L has a key role in dictating mitochondrial biology both during normal physiology and in response to pathologic insults.

In contrast to YME1L, the active site of the *m*-AAA protease is oriented toward the mitochondrial matrix side of the IM (Fig. 2). The *m*-AAA protease exists in two different oligomeric conformations composed of homooligomers of the zinc metalloprotease AFG3L2 or heterooligomers of AFG3L2 and the zinc metalloprotease paraplegin (Koppen et al., 2007).

Interestingly, the relative populations of AFG3L2 and paraplegin differ across mammalian tissues, suggesting that homooligomeric and heterooligomeric *m*-AAA proteases have distinct functions (Koppen et al., 2007). However, few specific substrates of the *m*-AAA protease have been identified. It is clear that the *m*-AAA protease has a critical role in regulating mitochondrial proteostasis through the degradation of damaged or misfolded proteins, as deletion of *m*-AAA subunits leads to disruption in mitochondrial proteostasis and activation of mitochondrial stress pathways (Richter et al., 2015; Yoneda et al., 2004). Furthermore, *m*-AAA proteolytic activity regulates mitochondrial calcium homeostasis through the degradation of unassembled essential mitochondrial calcium uniporter regulator (EMRE) subunits of the mitochondrial calcium uniporter (MCU) (Konig et al., 2016; Tsai et al., 2017). Because EMRE ensures the proper association of MCU protein with MICU1 gatekeeping subunit, impaired degradation of EMRE by either protease-resistant mutants of the subunit, or *m*-AAA protease knockout lead to calcium overload in mitochondria and sensitization to mitochondrial permeability transition pore (MPTP) opening (Konig et al., 2016; Tsai et al., 2017). The *m*-AAA protease is also responsible for the maturation of the mitochondrial ribosomal subunit MRPL32, required for mitochondrial ribosome assembly and mitochondrial protein translation (Almajan et al., 2012; Nolden et al., 2005).

Apart from the *m*-AAA, the soluble AAA+ proteases LON and CLPXP also regulate proteostasis within the mitochondrial matrix. LON is a serine protease with orthologs found in organisms from bacteria to humans (Sauer and Baker, 2011). LON assembles as a homohexamer with each subunit containing both an AAA+ ATPase domain and a protected proteolytic core (Sauer and Baker, 2011). LON has a critical role in regulating mitochondrial matrix proteostasis through the degradation of oxidatively damaged and/or misfolded proteins such as aconitase (Bota and Davies, 2002, 2016; Bota et al., 2002). Furthermore, LON regulates many other aspects of mitochondrial function such as electron transport chain (ETC) activity, steroid synthesis, heme biosynthesis, and mitochondrial transcription through the degradation of specific proteins such as the complex IV subunit COX4-1, steroid acute regulatory protein, and 5-aminolevulinic acid synthase, respectively (Fukuda et al., 2007; Granot et al., 2007; Matsushima et al., 2010; Quiros et al., 2014; Tian et al., 2011). LON has also been shown to bind mtDNA in a process linked to the degradation of mitochondrial transcription factor A (TFAM) (Liu et al., 2004; Lu et al., 2007, 2013).

The importance of LON for mitochondrial function is highlighted by the fact that mice lacking *Lon* are not viable (Quiros et al., 2014).

The CLPXP protease is a heterooligomeric protease composed of two different proteins: the AAA+ ATPase (CLPX) and the serine protease (CLPP). Like LON, the CLPXP protease can function as a general quality control protease in the mitochondrial matrix through the degradation of misfolded or nonnative protein conformations. The CLPXP protease has an important role in regulating mammalian mitochondrial proteostasis. Deletion of CLPP in mice leads to increased expression of mitochondrial proteostasis factors and mitochondrial dysfunction, likely reflecting an imbalance in mitochondrial proteome integrity (Deepa et al., 2015; Gispert et al., 2013). Additionally, CLPP is involved in the regulation of mitochondrial proteostasis through the mitochondrial unfolded protein response (UPR^{mt}) in *C. elegans* (Haynes et al., 2007), although similar roles for mammalian UPR^{mt} signaling have not been identified (Seiferling et al., 2016). Interestingly, apart from its role in proteolytic degradation, CLPX has been shown to work independent of CLPP in chaperoning the folding of specific proteins such as ALA synthase, indicating that this protein has a nonproteolytic role in regulating mitochondrial proteostasis (Kardon et al., 2015). Thus, while the substrate specificity of CLPXP remains to be established, it is clear that this protease has an important role in regulating mitochondrial proteostasis and function.

Apart from AAA+ ATP-dependent proteases, mitochondrial proteostasis is also regulated by a network of ATP-independent quality control proteases. Unlike the AAA+ proteases, these proteases generally have more specific proteolytic functions associated with aspects of mitochondrial biology such as the regulation of apoptotic signaling and mitochondrial dynamics. The presenilin-associated rhomboid–like (PARL) protease is an ATP-independent IM protease that can assemble into a proteolytic complex consisting of the AAA+ protease YME1L and the cardiolipin-binding protein stomatin-like protein 2 (SLP2) (Wai et al., 2016). This complex regulates PARL-dependent functions including the cleavage of the PTEN-induced putative kinase 1 (PINK1) involved in mitophagy (Jin et al., 2010) and the Ser/Thr protein phosphatase phosphoglycerate mutase family member 5 (PGAM5) linked to mitochondrial dynamics, respiration, and cell survival (Sekine et al., 2012). HTRA2 is a highly conserved trimeric serine protease that localizes to the IMS (Clausen et al., 2011). Deletion of *HTRA2* leads to accumulation of unfolded mitochondrial proteins suggesting that this protein has an important role in degrading misfolded, nonnative, or damaged proteins within the IMS (Moisoi et al., 2009). HTRA2 activity has also been implicated in other mitochondrial functions, including mtDNA regulation, mitophagy, and apoptotic signaling (Chao et al., 2008; Cilenti et al., 2014; Kang et al., 2013; Suzuki et al., 2001). Finally, OMA1 is a stress-activated ATP-independent IM zinc metalloprotease with an active site oriented to the IMS (Baker et al., 2014; Kaser et al., 2003). OMA1 is normally maintained in a quiescent state that suppresses its protease activity (Anand et al., 2014; Head et al., 2009; Rainbolt et al., 2016; Zhang et al., 2014). In response to specific insults that induce membrane depolarization such as oxidative stress, the OMA1 protease is activated resulting in the regulated processing of the dynamin-like GTPase OPA1 involved in dictating mitochondrial morphology during stress (discussed in detail below).

1.1.3 Oligopeptidases Degrade Polypeptides Within Mitochondria

Processing peptidases and quality control proteases both generate peptide fragments within mitochondria. The build-up of these peptides could disrupt mitochondrial proteostasis through mechanisms such as the feedback inhibition of mitochondrial proteases or binding to mitochondrial chaperones. Interestingly, mitochondrial-derived peptides can also be effluxed from mitochondria and serve as stress signals involved in the activation of stress-responsive signaling pathways (Arnold et al., 2006; Haynes et al., 2010; Young et al., 2001). Thus, controlling the levels of mitochondrial peptides is an important parameter in dictating mitochondrial proteostasis. This is primarily achieved by the activity of mitochondrial-localized oligopeptidases.

The most prominent mitochondrial oligopeptidase in the mitochondrial matrix is the metalloprotease PITRM1, also known as PreP. PreP degrades N-terminal—targeting sequences produced by MPP-dependent processing of precursor proteins (Alikhani et al., 2011a; Chow et al., 2009; Stahl et al., 2002). This degradation is important to prevent feedback inhibition of MPP, which can lead to mitochondrial proteotoxic stress through the accumulation of precursor proteins within the matrix. PreP has also been implicated in the degradation of mitochondrial-localized amyloid β peptides associated with Alzheimer's disease (Brunetti et al., 2015; Falkevall et al., 2006; Mossmann et al., 2014), indicating that this peptidase has a more global role in regulating peptide levels within mitochondria. PreP has also been suggested to facilitate the maturation of frataxin, a protein involved in iron–sulfur biogenesis within mitochondria (Nabhan et al., 2015). Other peptidases including neurolysin (also known as oligopeptidase M or MEP) also function to degrade peptides within the IMS (Chow et al., 2009; Serizawa et al., 1995). These peptidases serve critical roles in regulating the levels of peptides localized to the matrix and IMS, preventing imbalances in mitochondrial proteostasis and stress-signaling.

1.2 Mitochondrial Proteases Are Key Regulators of Organellar Quality Control

Mitochondria are structurally heterogeneous organelles that demonstrate distinct morphologies in different mammalian tissues (Chan, 2012; Shutt and McBride, 2013; Wai and Langer, 2016). Furthermore, mitochondrial morphology is dynamically regulated to adapt mitochondrial function in response to environmental or metabolic challenges. These different morphologies can be accessed through the regulation of biologic pathways involved in mitochondrial fusion and fission. The capacity to adapt mitochondrial morphology through alterations in the relative activities of mitochondrial fusion and fission processes provides a unique opportunity to match mitochondrial function to tissue-specific or environmental challenges. For example, elongated mitochondria, which result from increased fusion or decreased fission, are associated with enhanced ETC activity and desensitization to apoptotic insult (Chan, 2012; Shutt and McBride, 2013; Wai and Langer, 2016). However, fragmented mitochondria, which result from increased fission or decreased fusion, show reduced ETC activity and are sensitized to stress-induced apoptosis (Chan, 2012; Shutt and McBride, 2013; Wai and Langer, 2016) but can promote mitochondrial proteostasis through the sequestration and degradation of damaged mitochondria by mitophagy (Youle and Narendra, 2011). Thus, the capacity to dynamically regulate mitochondrial morphology provides another level of quality control that cells utilize to sensitively regulate both mitochondrial proteostasis and function.

Mitochondrial dynamics are regulated by the activities of dynamin-like GTPases localized to the OM and IM (Chan, 2012; Shutt and McBride, 2013; van der Bliek et al., 2013; Wai and Langer, 2016). Mitochondrial fission is largely dictated by the dynamin-like GTPase dynamin-related protein 1 (DRP1). DRP1 is mainly located in the cytosol under basal condition and is recruited to the mitochondrial OM by receptors such as mitochondrial fission factor (MFF) and FIS1 for the fission process (Loson et al., 2013). This recruitment facilitates DRP1 oligomerization into ring structures that promote mitochondrial fission through the constriction of the outer membrane. Conversely, mitochondrial fusion is regulated by the outer membrane GTPases mitofusin 1 (MFN1) and mitofusin 2 (MFN2) and the inner membrane GTPase OPA1. In this process, mitofusins form complexes through homotypic or heterotypic interactions that drive OM fusion through a GTP-dependent process. Long isoforms of OPA1 (l-OPA1) similarly utilize GTP to drive IM fusion events. Not surprisingly, the activities of these GTPases are highly regulated through a variety of posttranslational mechanisms, including phosphorylation, ubiquitination, sumoylation, and proteolytic processing, providing a mechanism to sensitively regulate mitochondrial morphology in response to specific cues (Chan, 2012; Shutt and McBride, 2013; van der Bliek et al., 2013; Wai and Langer, 2016). Here, we specifically focus on the role of mitochondrial proteases in the regulation of mitochondrial dynamics. For a more comprehensive review of other posttranslational mechanisms involved in the regulation of mitochondrial dynamics see (Chan, 2012; Shutt and McBride, 2013; van der Bliek et al., 2013; Wai and Langer, 2016).

1.2.1 Mitochondrial Morphology Is Regulated by the Activity of Mitochondrial Proteases

Mitochondrial fusion/fission factors localized to the OM are regulated by the UPS. E3 ubiquitin ligases such as PARKIN and MARCH5 reside on the OM and can ubiquitinate OM membrane proteins (Covill-Cooke et al., 2017). Interestingly, ubiquitination of MFN1/2 by one or both of these E3 ligases leads to the degradation of these OM fusion factors (Park and Cho, 2012; Tanaka et al., 2010). UPS-dependent degradation of MFN1/2 attenuates mitochondrial fusion and in turn promotes mitochondrial fission (Yue et al., 2014). Interestingly, the OM deubiquitinase USP15 and USP30 suppress MFN1/2 ubiquitination, inhibiting proteasomal degradation and thus promoting mitochondrial fusion (Cunningham et al., 2015; Liang et al., 2015; Yue et al., 2014). Thus, the relative activities of PARKIN and MARCH5 versus USP15/30 regulate MFN1/2 stability, directly influencing mitochondrial morphology. DRP1 can also be ubiquitinated by MARCH5, although this modification does not appear to be associated with DRP1 degradation, but instead could be involved in promoting DRP1-dependent fission (Karbowski et al., 2007; Nakamura et al., 2006).

By contrast, the predominant IM dynamin-like GTPase OPA1 is not accessible to the UPS. Instead, OPA1 is regulated through a highly

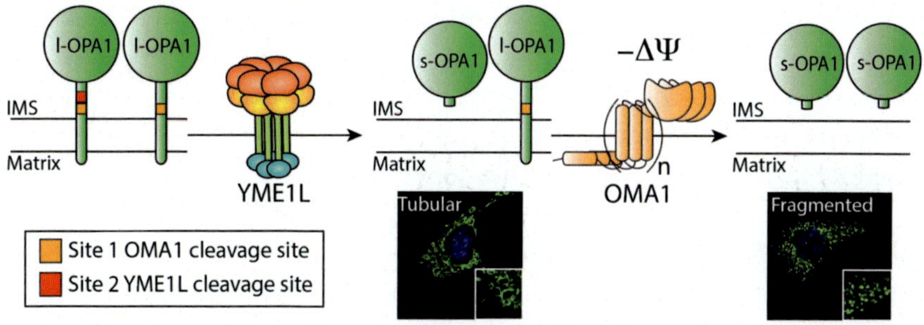

Figure 3 Illustration showing the functional role for YME1L and OMA1 in regulating mitochondrial morphology through the regulated processing of the inner membrane GTPAse OPA1. YME1L constitutively cleaves a subset of OPA1 containing the Site 2 YME1L cleavage site to establish a distribution of long-OPA1 (l-OPA1) and short-OPA1 (s-OPA1) required for maintaining mitochondrial tubular morphology. Upon depolarization of mitochondria, OMA1 is activated resulting in the cleavage of all OPA1 at the Site 1 OMA1 cleavage site. This OMA1-dependent processing of OPA1 to s-OPA1 promoting mitochondrial fragmentation.

coordinated mechanism involving the IM proteases YME1L and OMA1 (Fig. 3). OPA1 is expressed as eight distinct mRNA transcripts that predominantly differ in the presence or absence of the S5b exon (Ishihara et al., 2006; Song et al., 2007). All of these transcripts contain an OMA1 proteolytic cleavage site referred to as Site 1. However, only a population of these transcripts contains the Site 2 cleavage site encoded by the S5b exon, which is selectively cleaved by YME1L. OPA1 exists as long (l-OPA1) and short (s-OPA1) isoforms that distinctly influence mitochondrial fusion and fission (Anand et al., 2014; Ishihara et al., 2006). l-OPA1 isoforms are full-length, membrane-integrated OPA1 proteins that are generally associated with mitochondrial fusion. YME1L and OMA1 differentially cleave l-OPA1 isoforms to produce s-OPA1 (Anand et al., 2014). YME1L constitutively cleaves the subset of l-OPA1 containing the Site 2 cleavage site to produce a distinct isoform of s-OPA1. This YME1L-dependent OPA1 processing is important for establishing the distribution of l-OPA1 and s-OPA1 involved in promoting and maintaining tubular mitochondrial morphology (Anand et al., 2014; Griparic et al., 2007; Mishra et al., 2014; Song et al., 2007). Alternatively, the stress-activated OMA1 protease cleaves all l-OPA1 isoforms at the Site 1 cleavage site to produce s-OPA1 isoforms, which promotes mitochondrial fragmentation (Anand et al., 2014; Mishra et al., 2014; Song et al., 2007). The recovery of tubular mitochondrial morphology from this fragmented state requires the suppression of OMA1 protease activity and de novo synthesis of l-OPA1 (Anand et al., 2014; Mishra et al., 2014; Rainbolt et al., 2016; Song et al., 2007). As such, the balance of YME1L- and OMA1-dependent OPA1 processing is a key determinant in dictating mitochondrial morphology, and thus mitochondrial function, in response to acute insults (discussed in detail below).

1.2.2 Proteolytic Control of Mitophagy

The regulation of mitochondrial dynamics also facilitates the degradation of damaged mitochondria through mitophagy. In this process, mitochondria are sequestered in autophagosomes that are then directed to the lysosome for degradation. Mitophagy can function to nonselectively degrade mitochondria in response to stress insults such as starvation that induces a global cellular autophagic response (Youle and Narendra, 2011). However, mitophagy also provides a mechanism to selectively remove damaged mitochondria in response to stresses such as mitochondrial depolarization or the accumulation of misfolded mitochondrial proteins. The ability to selectively degrade damaged mitochondria allows cells to maintain the global integrity of the mitochondrial population, preventing potential imbalances in mitochondrial function that could challenge cellular viability.

The mitochondrial proteolytic network has a central role in regulating mitophagy of damaged mitochondria in response to stress. Damaged mitochondria are predominantly identified through a mechanism involving PINK1 and the E3 ligase PARKIN (Matsuda et al., 2010). Under normal conditions, PINK1 is imported into mitochondria and incorporated into the IM as a transmembrane protein. In the IM, PINK1 is processed by MPP and then cleaved by the mitochondrial protease PARL, releasing PINK1 to the cytosol where it can be degraded by the proteasome (Greene et al., 2012; Thomas et al., 2014). PINK1 degradation can also be regulated by other mitochondrial proteases, including AFG3L2 and LON (Greene et al., 2012; Thomas et al., 2014). This degradation of PINK1 functions to suppress its activity on the mitochondrial OM (Yamano and Youle, 2013). Conversely, mitochondrial insults that induce depolarization impair PINK1 import into the mitochondrial IM, inhibiting its degradation and facilitating PINK1 incorporation into the OM (Okatsu et al., 2013). Localization of PINK1 to the OM recruits the E3 ligase PARKIN to mitochondria, which promotes fragmentation through the ubiquitination of MFN1/2 and degradation of other OM proteins through the UPS (Chan et al., 2011; Ziviani et al., 2010). PARKIN recruitment and subsequent ubiquitination of OM proteins facilitate formation of the autophagasome required for lysosomal-dependent degradation of mitochondria through autophagy (Harper et al., 2018; Koyano et al., 2014; Okatsu et al., 2013, 2015). Apart from the PINK1/PARKIN pathway, mitophagy can be initiated through other mechanisms involving mitochondrial receptors such as NIX, BNIP3, and FUNDC1 that interact with the microtubule-associated protein LC3 directly to induce autophagasome formation (Hamacher-Brady and Brady, 2016; Novak and Dikic, 2011; Novak et al., 2010; Sandoval et al., 2008; Wu et al., 2014).

1.3 Mitochondrial Proteases Link Mitochondrial Proteostasis to Apoptotic Signaling

Mitochondrial integrity is a key regulator of intrinsic apoptotic signaling pathways. In this process, mitochondrial outer membrane permeabilization (MOMP) leads to the release of proapoptotic factors, including cytochrome *c*, Smac, and AIF, from mitochondria to the cytosol. The release of these factors induces assembly of the apoptosome allowing the activation of initiator and executioner caspases that mediate the apoptotic program (Tait and Green, 2013). MOMP is regulated by the relative activities of several proapoptotic (e.g., Bad, Bax, Bak) and prosurvival (Bcl-2 and Bcl-xl) Bcl-2 family members located on the OM. Many factors can influence intrinsic apoptotic signaling through the expression of Bcl-2 family members or altering the activity of downstream signaling mechanisms (Gillies and Kuwana, 2014). However, it is clear that mitochondrial proteases are contributors to the activation and regulation of apoptosis induced by intrinsic signals.

The IMS quality control serine protease HTRA2 has been implicated in diverse regulatory roles involved in apoptotic signaling. HTRA2 is expressed as a transmembrane protein that can be activated through proteolytic processing by PARL and/or an autocatalytic mechanism (Chao et al., 2008; Seong et al., 2004). The release of activated HTRA2 to the cytosol promotes apoptosis through the degradation of antiapoptotic factors such as X-linked inhibitor of apoptosis (Hegde et al., 2002; Suzuki et al., 2001). However, PARL-dependent HTRA2 activation has been proposed to suppress accumulation of proapoptotic BCL-2 proteins on the OM, suggesting an antiapoptotic role for HTRA2 (Chao et al., 2008). These results highlight that HTRA2 appears to be mitochondrial protease that is a key regulator of apoptotic signaling pathways.

Apoptotic signaling also involves both mitochondrial fragmentation and disruption of cristae, which is thought to facilitate the release of cytochrome *c* during apoptosis (Cipolat et al., 2006; Frezza et al., 2006; Varanita et al., 2015). This suggests that mitochondrial proteases such as YME1L and OMA1 involved in regulating mitochondrial morphology through OPA1 processing could influence apoptotic signaling because OPA1 maintains cristae organization (Olichon et al., 2003). *OMA1* deletion inhibits stress-induced OPA1 processing required for mitochondrial fragmentation and disruption of cristae integrity necessary for cytochrome *c* release (Anand et al., 2014; Head et al., 2009; Olichon et al., 2003). As such, *OMA1*-deficient cells show a reduced sensitivity to apoptotic insult (Anand et al., 2014; Head et al., 2009; Xiao et al., 2014). Alternatively, *YME1L*-deficient cells show increased mitochondrial fragmentation and disrupted cristae morphology, resulting in increased sensitivity to apoptotic stimuli (Anand et al., 2014; Rainbolt et al., 2015; Stiburek et al., 2012). Interestingly, *OMA1* deletion in *YME1L*-deficient cells rescues these phenotypes, indicating that the increased sensitivity to apoptotic stimuli in *YME1L*-deficient cells largely results from imbalances in OPA1 processing (Anand et al., 2014). PARL has also been implicated in the regulation of apoptotic signaling through OPA1 processing (Pellegrini and Scorrano, 2007) and other mechanisms involved in regulating cristae morphology and cytochrome *c* release (Cipolat et al., 2006; Frezza et al., 2006). In addition, PARL processes the proapoptotic SMAC/Diablo to release this proapoptotic factor to the cytosol (Saita et al., 2017).

Other mitochondrial proteases have also been implicated in the regulation of apoptotic signaling, although direct roles for these proteases in the regulation of apoptotic signaling mechanisms remain to be established. LON has been suggested to influence apoptosis through its association with the mitochondrial chaperones HSP60 and mtHSP70 (HSPA9); however, the dependence of apoptosis on LON protease activity is currently undefined (Kao et al., 2015). YME1L can also influence apoptotic signaling through the degradation of PRELID1, which is involved in cardiolipin synthesis and subsequently cytochrome *c* release (Potting et al., 2013). While the specific contributions for these other proteases in intrinsic apoptotic signaling induced by specific stresses remain to be further defined; it is clear that the maintenance of mitochondrial proteolytic systems is important to regulate apoptotic signaling pathways in response to diverse pathologic insults.

1.4 Stress-Responsive Regulation of the Mitochondrial Proteolytic Network

Mitochondrial proteostasis and function are challenged by diverse insults including environmental toxins, metabolic stress, and aging. All of these insults can increase the accumulation of misfolded or aggregated proteins within mitochondria and lead to the pathologic disruption of mitochondrial function. To protect mitochondria in response to these insults, cells evolved a network of stress-responsive signaling mechanisms that regulate the composition and activity of mitochondrial proteostasis pathways. Mitochondrial proteases and their activities are key determinants in this stress-responsive regulation. The capacity to influence mitochondrial proteostasis, organellar morphology, and apoptotic signaling through the regulation of proteolytic activity provides a mechanism to adapt mitochondrial function to the cellular or metabolic demands imposed by a specific type of stress or insult (Quiros et al., 2015; Shpilka and Haynes, 2018). Here, we describe the transcriptional and translational stress-responsive signaling mechanisms by which cells adapt mitochondrial proteostasis and function to specific cell stresses.

1.4.1 Mitochondrial Unfolded Protein Response-Dependent Regulation of Mitochondrial Proteostasis

The preeminent stress-responsive signaling pathway responsible for regulating mitochondrial proteostasis is the mitochondrial UPR^{mt} (Shpilka and Haynes, 2018). The UPR^{mt} is activated in response to imbalances in mitochondrial matrix proteostasis (e.g., the accumulation of misfolded or unassembled mitochondrial proteins). Many different types of cellular insults can activate the UPR^{mt} in eukaryotes, including deletion of mitochondrial quality control proteases (Benedetti et al., 2006; Yoneda et al., 2004), overexpression of a misfolding-prone mitochondrial matrix protein (e.g., ΔOTC) (Aldridge et al., 2007; Zhao et al., 2002), inhibition of mitochondrial translation (Houtkooper et al., 2013), and alterations in mitochondrial cofactor levels (Baker et al., 2012a; Durieux et al., 2011). Activation of the UPR^{mt} functions to adapt mitochondrial proteostasis through the transcriptional upregulation of mitochondrial proteostasis factors involved in protein import, folding, and proteolysis. Through this upregulation, stress-induced imbalances in mitochondrial proteostasis are alleviated, and mitochondrial function is adapted to match demands associated with the specific insult.

The signaling mechanism of the UPR^{mt} has predominantly been defined in *C. elegans*, where it has been shown that mitochondrial proteases play a key role in regulating UPR^{mt} signaling. The worm UPR^{mt} can be activated by several genetic manipulations that increase protein misfolding within the mitochondrial matrix (e.g., RNAi-depletion of the *m*-AAA subunit *spg-7*) (Benedetti et al., 2006). This increase in protein misfolding is initially detected through the increased degradation of mitochondrial proteins by the CLPXP quality control protease (Haynes et al., 2007). Peptides derived from the CLPXP-dependent proteolysis are exported across the IM to the IMS through the HAF-1 ABC peptide transporter (Haynes et al., 2010). In the IMS, these peptides function to dictate the localization of the UPR^{mt}-associated transcription factor ATFS-1. In the absence of stress, ATFS-1 is targeted to mitochondria by an N-terminal—targeting sequence and imported across the OM and IM into the mitochondrial matrix (Nargund et al., 2012). In the matrix, ATFS-1 is degraded by the LON protease, suppressing its transcriptional activity. However, the increased efflux of peptides through HAF-1 induced by mitochondrial stress attenuates ATFS-1 import into mitochondria through an undefined mechanism (Nargund et al., 2012). This reduced import of ATFS-1 allows this transcription factor to be targeted to the nucleus by a nuclear localization sequence located on the C-terminus of the ATFS-1 protein. In the nucleus, ATFS-1 coordinates with other nuclear transcriptional regulators such as DVE-1 and UBL-5 to induce expression of UPR^{mt} target genes, which includes many mitochondrial proteases (Benedetti et al., 2006; Haynes et al., 2007; Nargund et al., 2012).

Mammals also possess a UPR^{mt}, although the signaling mechanism(s) responsible for the mammalian UPR^{mt} remains less well defined (Fig. 4). The mammalian UPR^{mt} is activated by stress insults that disrupt proteostasis within mitochondria such as overexpression of misfolding-prone proteins (Aldridge et al., 2007; Horibe and Hoogenraad, 2007; Zhao et al., 2002), inhibition of mitochondrial translation (Houtkooper et al., 2013), and the modulation of NAD+ levels (Mouchiroud et al., 2013). Interestingly, signaling through the mammalian UPR^{mt} appears to be different than that observed in *C. elegans*. Recently, the mammalian homolog of ATFS-1, ATF5, was identified and shown to be regulated through mechanism similar to that shown for ATFS-1 and involving stress-dependent localization between mitochondria and the nucleus (Fiorese et al., 2016). However, the

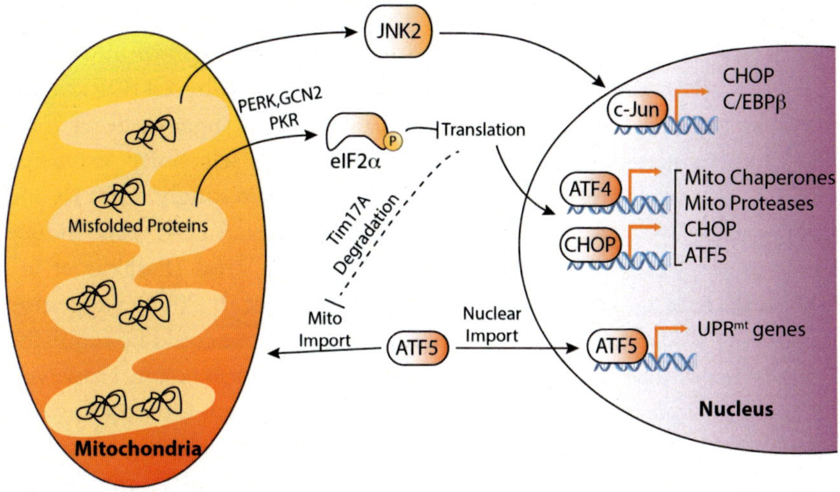

Figure 4 Illustration showing the mechanism(s) predicted to be involved in mammalian UPR^{mt} signaling. In the absence of stress, ATF5 transcriptional activity is suppressed through its targeting to the mitochondrial matrix. In response to proteotoxic stress within mitochondria, ATF5 import is attenuated, allowing its localization to the nucleus and subsequent activation of UPR^{mt}-associated genes. Mitochondrial proteotoxic stress can potentially influence UPR^{mt} signaling through multiple other mechanisms including activation of JNK2 and eIF2α phosphorylation, as discussed in the text.

potential dependence of this regulated import on peptide-efflux through an HAF-1—dependent channel has not been defined. Furthermore, CLPP is not required for activating the UPR^{mt} in the mammalian heart, suggesting that mammalian UPR^{mt} signaling proceeds through a CLPP-independent mechanism at least in this tissue (Seiferling et al., 2016).

In contrast, mammalian UPR^{mt} signaling has been proposed to involve many other stress-responsive signaling mechanisms. The JNK2 signaling cascade can be activated by mitochondrial stress and lead to downstream activation of the transcription factor c-Jun (Horibe and Hoogenraad, 2007). c-Jun in turn binds to the AP-1 sites in the promoters of the stress-responsive transcription factors C/EBP homologous protein (CHOP) and C/EBPβ inducing their expression (Horibe and Hoogenraad, 2007) (Fig. 4). These transcription factors form homodimers and/or heterodimers that bind to CHOP elements found in the promoters of many mitochondrial proteostasis genes (Zhao et al., 2002). However, it is unlikely that these transcription factors function alone to induce mammalian UPR^{mt} signaling, as they are activated in response to many different types of stress (see below), making it difficult to convey selectivity of a UPR^{mt} transcriptional response. Consistent with this, identification of mitochondrial UPR elements (MUREs), which are not bound by CHOP, in the promoter region of mitochondrial proteostasis genes suggests other transcriptional regulators are involved in dictating the selectivity associated with UPR^{mt} activation (Aldridge et al., 2007). Interestingly, CHOP and other stress-responsive transcription factors (e.g., ATF4) can induce ATF5 in response to other types of proteotoxic stress (Fusakio et al., 2016; Kilberg et al., 2012; Teske et al., 2013; Zhou et al., 2008), suggesting that an ATF5-selective UPR^{mt} transcriptional response could be regulated downstream of these other stress-responsive transcription factor. However, this transcriptional relationship remains to be established. Other pathways, including the AKT signaling pathway and estrogen receptor signaling, are activated through a reactive oxygen species (ROS)-mediated mechanism in response to proteotoxic insults localized to the IMS in a process referred to the IMS UPR^{mt} (Papa and Germain, 2011). This suggests that different signaling mechanisms may be utilized to adapt mammalian mitochondrial proteostasis to specific types of proteotoxic insults that challenge distinct mitochondrial subcompartments.

The lack of specific signaling factors involved in the mammalian UPR^{mt} has challenged the ability to explicitly define the transcriptional program induced by this stress-responsive pathway. However, the transcriptional response for the worm UPR^{mt} has been defined by identifying genes regulated by the UPR^{mt} transcription factor ATFS-1 (Nargund et al., 2012). This analysis showed that the worm UPR^{mt} sensitively adapts the composition and activity of mitochondrial proteostasis pathways through the selective upregulation of specific import factors (e.g., Tim-17 and Tim-23), chaperones (e.g., HSP60), and proteases (e.g., YME-1) (Nargund et al., 2012). However, key mitochondrial proteostasis factors such as LON and CLPP were not induced by ATFS-1 in response to mitochondrial stress. This selective remodeling of mitochondrial proteostasis pathways may provide a mechanism to support adaption of mitochondrial function in response to mitochondrial proteotoxic stress. For example, ATFS-1 also regulates expression of glycolytic genes suggesting that UPR^{mt} activation promotes a metabolic shift for ATP production from respiration to glycolysis (Nargund et al., 2012). ATFS-1 also induces expression of transcriptional programs involved in ROS detoxification and innate immunity (Nargund et al., 2012). Selective adaptation of mitochondrial proteostasis pathways could provide a mechanism to facilitate alterations in mitochondrial functions to support the increased activities of these other programs. For example, the upregulation of YME-1 could increase YME-1—dependent degradation of damaged or unassembled respiratory chain subunits, preventing the accumulation of ROS-damaged ETC subunits and promoting an energetic shift to glycolysis.

It is currently unclear whether the mammalian UPR^{mt} also induces selective remodeling of mitochondrial proteostasis pathways. Initial experiments using a transcriptional reporter approach where mitochondrial stress was induced by overexpression of a misfolding-prone matrix protein showed that selective mitochondrial proteostasis factors, including the import subunit Tim17 A, the mitochondrial chaperone HSP60, and the mitochondrial proteases MPP, YME1L, AFG3L2, and CLPP, were induced by the UPR^{mt}, whereas other proteostasis factors such as mitochondrial HSP70 and LON were not (Aldridge et al., 2007; Zhao et al., 2002). However, more recent studies using quantitative polymerase chain reaction (qPCR) show that *mtHSP70* and *LON* are both induced by UPR^{mt}-activating insults in mammalian cells (Fiorese et al., 2016). These results suggest that, unlike the worm UPR^{mt}, mammalian UPR^{mt} signaling may not promote the same type of selective remodeling of mitochondrial proteostasis pathways. However, more experiments are required to further define the scope of mitochondrial proteostasis remodeling induced by the mammalian UPR^{mt}.

While it is becoming increasingly clear that the UPR^{mt} is involved in regulating many aspects of mitochondria and organismal health (e.g., energy metabolism, longevity, and innate immunity), the functional importance of UPR^{mt}-dependent adaptation of mitochondrial proteostasis pathways for mitochondrial function remain to be established. In mammals, establishing this type of relationship is hindered by our poor understanding of the signaling mechanism(s) involved in the mammalian UPR^{mt} activation. As new, specific components of this pathway are found, the transcriptional profile of the mammalian UPR^{mt} can be better established, revealing the scope of the transcriptional program involved in regulating mitochondrial and global cellular physiology in response to mitochondrial proteotoxic insults. This will in turn provide new opportunities to define the specific role of UPR^{mt}-dependent proteostasis adaptation in adapting mitochondrial function to environmental or metabolic demands induced by a specific stress insult.

1.4.2 The Integrated Stress Response Coordinates Mitochondrial Proteostasis in Response to Diverse Insults

Apart from the UPR^{mt}, the Integrated Stress Response (ISR) has also been identified as a key stress-responsive signaling mechanism responsible for regulating mitochondrial proteostasis and function (Pakos-Zebrucka et al., 2016; Quiros et al., 2016). The ISR is a multifaceted stress-responsive signaling pathway regulated by a network of stress-responsive kinases, including protein kinase R (PKR), general control nonderepressible 2 (GCN2), heme-regulated inhibitor (HRI), and the PKR-like ER kinase (PERK) (Fig. 5) (Pakos-Zebrucka et al., 2016). These kinases are activated in response to diverse cellular insults including viral infection, amino acid deprivation, oxidative stress, and endoplasmic reticulum (ER) stress. Upon activation, these kinases phosphorylate the S51 residue of the eukaryotic initiation factor 2α subunit (eIF2α). Phosphorylation of eIF2α inhibits the activity of the eIF2B GTP exchange factor involved in translation initiation, resulting in a transient attenuation in new protein synthesis. This attenuation in synthesis functions to promote global cellular proteostasis by reducing the load of newly synthesized proteins, freeing proteostasis factors to prevent pathologic disruptions in cellular proteome integrity.

Apart from reducing translation, eIF2α phosphorylation also activates stress-responsive transcription factors such as activating transcription factor

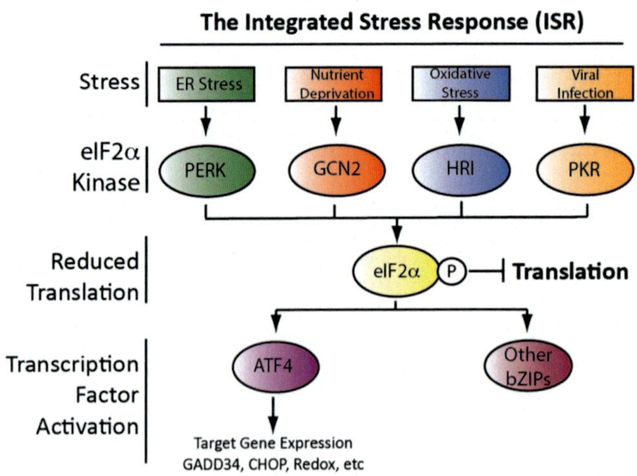

Figure 5 Illustration showing the signaling mechanisms associated with the integrated stress response. Four different kinases activated by distinct insults initiate ISR signaling through the phosphorylation of eIF2α. This leads to both translation attenuation and activation of ISR-associated transcription factors such as ATF4.

4 (ATF4) (Pakos-Zebrucka et al., 2016; Quiros et al., 2016; Young and Wek, 2016). ATF4 is selectively translated in response to eIF2α phosphorylation through a mechanism involving uORFs in its 5′ UTR (Young and Wek, 2016). In response to stresses that increase eIF2α phosphorylation, ATF4 is translated allowing ATF4-dependent expression of cellular proteostasis factors involved in redox homeostasis, amino acid biosynthesis, and ER proteostasis maintenance (Han et al., 2013; Harding et al., 2003). ATF4 also induces the expression of additional stress-responsive transcription factors such as CHOP—the same transcription factor involved in UPR^{mt} signaling (Harding et al., 2000). As part of the ISR, CHOP induces expression of additional proteostasis factors including GADD34—a regulatory phosphatase subunit that interacts with protein phosphatase 1 to dephosphorylate eIF2α in an ISR negative feedback pathway that restores translational integrity following acute insult (Ma and Hendershot, 2003; Novoa et al., 2001). CHOP activation also promotes proapoptotic signaling in response to severe or sustained toxic insults through complex mechanisms involving the expression of proapoptotic factors and/or increasing cellular oxidative stress (Han et al., 2013; Hetz and Papa, 2018; Sano and Reed, 2013). Thus, ISR-dependent activation of stress-responsive transcription factors such a CHOP and ATF4 serves a critical role in dictating both prosurvival adaptation of cellular proteostasis and activation of proapoptotic signaling pathways in response to varying types and extents of cellular stress.

Mitochondrial stress has been shown to activate the ISR through multiple mechanisms. In *C. elegans*, mitochondrial stress induced by RNAi-depletion of the *m*-AAA subunit *spg-7* leads to increased activation of the ISR kinase GCN2 (Baker et al., 2012a). The activation of the ISR in response to this stress is predicted to function in a parallel mechanism to the UPR^{mt} to regulate mitochondrial proteostasis and function in response to mitochondrial proteotoxic insults. Similarly, mitochondrial stress in mammals activates eIF2α kinases such as GCN2, PKR, or HRI (Martinez-Reyes et al., 2012; Michel et al., 2015; Rath et al., 2012; Viader et al., 2013), indicating mitochondrial dysfunction can activate multiple ISR kinases in metazoans. Not surprisingly considering the link between mitochondrial stress and the ISR, ATF4 is also activated in response to diverse mitochondrial insults, including membrane depolarization, disruption of protein import, imbalances in OXPHOS protein stability, and inhibition of mitochondrial translation in an eIF2α phosphorylation-dependent manner (Quiros et al., 2017). Interestingly, knocking down individually the different ISR kinases failed to inhibit ATF4 activation under these conditions, likely reflecting the capacity for mitochondrial stress to signal through multiple eIF2α kinases. CHOP activity is also activated in response to mitochondrial stress through an ISR-associated mechanism (Michel et al., 2015).

The above results suggest that the ISR has an important role in regulating mitochondrial proteostasis and function in response to diverse pathologic insults that activate ISR-associated kinases. Consistent with this, ISR activation transcriptionally regulates many aspects of mitochondrial proteostasis factors. The PERK-dependent activation of ATF4 and CHOP in response to ER stress, a condition induced by accumulation of misfolded proteins within the ER lumen, increases expression of mitochondrial chaperones (e.g., mtHSP70) and the quality control protease LON (Han et al., 2013; Harding et al., 2003; Hori et al., 2002; Lebeau et al., 2018). The increased expression of *LON* could reflect a mechanism to increase mitochondrial proteostastic capacity to degrade misfolded, nonnative, or oxidatively damaged mitochondrial proteins that accumulate in response to this insult. Many insults that activate the ISR lead to increased production of ROS that could damage mitochondrial proteome. LON preferentially degrades oxidatively modified proteins (Bota and Davies, 2002), suggesting that its increased expression could serve to prevent the potentially toxic accumulation of damaged proteins during these stresses. Consistent with this, overexpression of *LON* protects cells from mitochondrial dysfunction induced by ER stress—a potent activator of the ISR kinase PERK (Hori et al., 2002). However, overexpression of a proteolytically inactive LON can also alleviate ER stress-induced mitochondrial dysfunction, suggesting LON could also promote mitochondrial function through nondegradative mechanisms in response to ER stress (Hori et al., 2002).

The yeast mtHSP70 has been shown to cooperate with the yeast LON homolog PIM1 to degrade misfolded proteins in the mitochondrial matrix (Savel'ev et al., 1998; Wagner et al., 1994). This suggests that the coordinated regulation of LON and mtHSP70 by the ISR may provide a mechanism to selectively facilitate LON-dependent degradation of damaged mitochondrial proteins through mechanisms such as preventing stress-induced aggregation of misfolded proteins and/or altering the substrate specificity of LON (Bender et al., 2011; Savel'ev et al., 1998; Wagner et al., 1994). Interestingly, the subset of mitochondrial stress-genes regulated by ATF4 during mitochondrial stress is predicted to be distinct from those regulated by the UPR^{mt} (Quiros et al., 2017). This suggests that the ISR and UPR^{mt} could serve as complementary mechanisms to regulate mitochondria in response to stress, mirroring the relationship between these pathways observed in worm. However, the shared involvement of many ISR and UPR^{mt} signaling factors, including ATF4, CHOP, and ATF5, suggests that these two pathways could also be highly integrated to regulate mitochondria in response to distinct types of proteotoxic insults (Fig. 4). Thus, further experiments are necessary to define the transcriptional relationship between ISR and UPR^{mt} signaling in mammals.

Apart from transcriptional signaling, translational attenuation induced by the ISR also leads to remodeling of mitochondrial protein import pathways through the selective degradation of the core TIM23 subunit Tim17A. In mammals, the IM TIM23 protein import complex exists in two distinct subunit compositions defined by the presence of a Tim17A or Tim17B subunit. Currently, functional differences between these two Tim17 homologs are poorly defined, although some experiments suggest that they differ in their ability to import nontraditional mitochondrial substrates (Sinha et al., 2014). In response to ISR-dependent translational attenuation, Tim17A, but not Tim17B, is rapidly degraded by the YME1L quality control protease (Rainbolt et al., 2013). YME1L-dependent Tim17A degradation decreases the population of import competent TIM23 complexes and thus reduces mitochondrial protein import into the matrix. Interestingly, depletion of *Tim17A* increases stress resistance to diverse types of stresses such as As(III) and paraquat, indicating that ISR-dependent Tim17A degradation is a protective mechanism to suppress import during acute insult (Rainbolt et al., 2013). This protection could be afforded by reducing the population of newly imported proteins entering mitochondria, which decreases the folding load in the mitochondrial matrix and frees mitochondrial chaperones and proteases to protect the conformational integrity of the established mitochondrial proteome. Alternatively, Tim17A degradation could function as a mechanism to regulate the import of specific stress-responsive transcription factors such as ATF5, facilitating their activation (Fig. 4). While RNAi depletion of the *Tim17A* homolog *tim-17* in *C. elegans* does activate the UPR^{mt} (Rainbolt et al., 2013), further experiments are necessary to confirm this potential relationship in mammalian cells. Interestingly, the posttranslational regulation of Tim17A induced downstream of ISR-regulated translational attenuation suggests that the stability of other mitochondrial proteins could similarly be regulated through this mechanism, although other examples have not yet been identified to date.

Consistent with the above, translational attenuation induced by ER stress-dependent activation of the ISR kinase PERK also influences mitochondrial morphology during ER stress. PERK-regulated translational attenuation promotes stress-induced mitochondrial hyperfusion (SIMH) (Tondera et al., 2009) in response to ER stress through a mechanism requiring the cardiolipin-binding protein SLP2 (Lebeau et al., 2018). ER stress-induced SIMH is attenuated in cells deficient in the inner membrane protease YME1L, indicating that this protease regulates both SIMH and TIM17 A degradation in response to PERK-regulated translational attenuation. These results show that the PERK—YME1L signaling axis is a critical stress-signaling mechanism responsible for regulating mitochondria in response to ER stress. While the mechanism and functional implications of PERK-regulated SIMH remains to be fully established, SLP2-dependent elongation appears to protect mitochondrial metabolic capacity and prevent premature fragmentation in response to acute ER stress (Lebeau et al., 2018). Although it is currently unclear whether other ISR-activating insults similarly promote SIMH downstream of eIF2α-phosphorylation—dependent translation attenuation, these results show that PERK-dependent ISR activation integrates transcriptional and translational signaling to coordinate mitochondrial proteostasis and morphology, revealing a global role for this pathway in regulating mitochondria in response to ER insults (Lebeau et al., 2018; Rainbolt et al., 2014).

The above results show that ISR signaling is a key determinant in dictating protective remodeling of mitochondrial proteostasis and function in response to acute cellular insults. However, critical questions remain about the specific role for the ISR in regulating mitochondria. For example, what is the functional interplay between the mammalian ISR and UPR^{mt}? Do other stress pathways integrate their signaling to modulate ISR-dependent regulation of mitochondria in response to distinct insults? What biochemical pathways are dependent on ISR-dependent regulation of mitochondrial proteostasis? As we and others continue to define the impact of ISR activation on mitochondria, these (and other) questions will be answered, further revealing ISR signaling as a core stress-responsive signaling pathway important for regulating mitochondria.

1.4.3 LON Is Also Transcriptionally Regulated by Other Stress-Responsive Signaling Pathways

Apart from the ISR and UPR^{mt}, the mitochondrial protease LON is also transcriptionally regulated by other stress-responsive signaling pathways

(Bota and Davies, 2016). For example, hypoxia-inducible factor 1 (HIF-1) induces *LON* expression in response to hypoxic conditions (Fukuda et al., 2007). HIF-1 is a heterodimeric transcription factor comprising HIF-1α and HIF-1β subunits. Under normoxic conditions, HIF-1 is targeted for proteasomal degradation through a mechanism involving the hydroxylation and ubiquitination of HIF-1α. Reductions in cellular oxygen levels inhibit HIF-1α hydroxylation, stabilizing this protein and facilitating its nuclear localization. In the nucleus, HIF-1α interacts with HIF-1β to form the active HIF-1 transcription factor. Activated HIF-1 binds to hypoxia-related elements (HREs) in the promoters of genes involved in the regulation of oxygen homeostasis including angiogenesis factors and metabolic genes. *LON* also contains HREs in its promoter, allowing it to be efficiently induced by HIF-1 activation (Fukuda et al., 2007).

HIF-1—dependent expression of LON promotes mitochondrial proteostasis and function in response to hypoxic injury by suppressing the potentially toxic accumulation of oxidatively damaged mitochondrial proteins that can accumulate under these conditions (Bota and Davies, 2002; Bota et al., 2002). Interestingly, HIF-1—dependent increases in LON may also be involved in the functional adaptation of complex IV during hypoxia. In response to reductions in oxygen, the complex IV subunit COXIV-1 is degraded and replaced by the stress-resistant COXIV-2 subunit, another HIF-1 transcriptional target (Fukuda et al., 2007). COXIV-2 is more efficient at electron transfer in hypoxic cells, allowing improved respiratory chain activity under these conditions. LON may facilitate this process through mechanisms such as the degradation of COXIV-1 or incorporation of COXIV-2 into complex IV, although a specific role for LON in this process remains to be established.

LON is also transcriptionally regulated by the nuclear respiratory transcription factor 2 (NRF-2) (Bahat et al., 2015; Pinti et al., 2011). NRF-2 is a stress-responsive transcription factor responsible for regulating the expression of many antioxidant and proteostasis factors following acute oxidative stress. In the absence of stress, NRF-2 activity is repressed by the activity of the KEAP E3 ligase, which ubiquitinates NRF-2 and directs this protein to proteasomal degradation. Oxidative stress impairs KEAP-dependent ubiquitination of NRF-2 allowing nuclear localization and transcriptional activity. The LON gene contains two NRF-2—binding elements in its promoter region (Bahat et al., 2015; Pinti et al., 2011), suggesting that NRF-2 can regulate *LON* expression. NRF-2—dependent induction of LON has been shown in cell models overexpressing the mitochondrial matrix protein StAR (Granot et al., 2007). In response to the accumulation of StAR within mitochondria, cells initiate a stress response, referred to as the StAR overload response (SOR), which activates NRF-2 signaling and increases LON expression through an unknown mechanism (Bahat et al., 2015). Interestingly, the increased expression of LON through the SOR likely reflects the need for increased cellular capacity to degrade mitochondrial StAR, which has been shown to be a LON substrate. Similarly, SOR transcriptionally induces other mitochondrial proteases involved in StAR degradation, including *YME1L*, *SPG7*, and *AFG3L2* (Bahat et al., 2014, 2015), although the specific requirement for NRF-2 in the induction of these other proteases remains to be defined. NRF-2 activation has also been proposed to be induced downstream of the ISR kinases such as PERK (Cullinan et al., 2003), suggesting that NRF-2 could also influence *LON* expression during ISR-activating insults.

The capacity for cells to regulate *LON* through multiple stress-responsive signaling pathways highlights the importance of this protease in protecting mitochondrial proteostasis and function. A predominant function of LON is to degrade damaged proteins in the mitochondrial matrix before they misfold and/or aggregate into toxic conformations that disrupt mitochondrial function (Bota and Davies, 2016). However, alterations in LON activity could also serve to sensitively adapt mitochondrial functions during stress through mechanisms such as COXIV remodeling and StAR degradation. This suggests the intriguing possibility that cells can actively adapt mitochondrial function to specific environmental or metabolic demands through the specific stress-responsive regulation of the LON protease.

1.4.4 Posttranslational Regulation of Mitochondrial Proteolytic Activity

Apart from transcriptional regulation, mitochondrial proteases are also subject to posttranslational regulation. One of the best characterized examples of this is the regulation of OMA1 proteolytic activity (Fig. 6). OMA1 exists in mitochondria as an inactive protease. In response to cellular insults that activate OMA1 proteolytic activity (e.g., membrane depolarization), OMA1 induces rapid processing of the inner membrane GTPase OPA1 to promote mitochondrial fission (Anand et al., 2014; Head et al., 2009; Rainbolt et al., 2016; Zhang et al., 2014). Interestingly, the activated OMA1 protease then undergoes degradation, which serves to restrict OMA1 proteolytic activity (Baker et al., 2014; Rainbolt et al., 2016; Zhang et al., 2014). This OMA1 degradation requires both OMA1 proteolytic

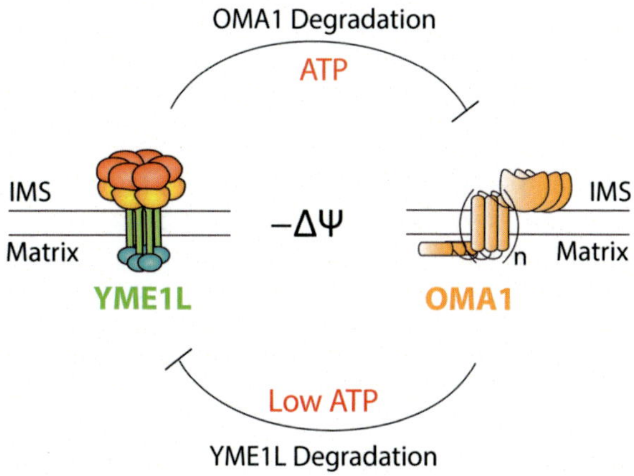

Figure 6 Illustration showing the reciprocal regulation of OMA1 and YME1L induced by distinct types of cellular insults. Acute insults that depolarize mitochondria activate OMA1 protease activity, which is then repressed through a mechanism involving ATP-dependent degradation by the protease YME1L. In response to insults that depolarize mitochondria and deplete cellular ATP (e.g., oxidative stress), YME1L is degraded, suppressing protease activity through a mechanism involving activated OMA1.

activity and the ATP-dependent activity of the inner membrane protease YME1L. The dependence on the activities of these two proteases appears to result from a two-step mechanism involving autocatalytic processing of OMA1 required for protease activation followed by ATP-dependent degradation of the active OMA1 protease by YME1L (Rainbolt et al., 2016). Through this mechanism, OMA1 activity is suppressed after an acute insult allowing effective recovery of mitochondrial morphology.

Interestingly, the inner membrane protease YME1L is also susceptible to stress-induced degradation (Rainbolt et al., 2015, 2016). Oxidative insults that reduce cellular ATP and depolarize mitochondria increase degradation of YME1L through a mechanism involving the activity of the inner membrane protease OMA1 (Fig. 6). This degradation of YME1L prevents many protective YME1L functions, including Tim17 A degradation and suppression of OMA1 proteolytic activity, increasing cellular sensitivity to apoptotic stimuli (Rainbolt et al., 2015, 2016). Degradation of YME1L also defines a reciprocal relationship between the regulation of YME1L and OMA1 proteolytic activity during stresses that induce mitochondrial damage in the presence or absence of metabolic crisis (Rainbolt et al., 2016). Mitochondrial depolarization in the absence of metabolic crisis will lead to transient activation of the OMA1 protease, which is then suppressed through ATP-dependent degradation by YME1L. However, mitochondrial depolarization in the presence of a metabolic crisis that reduces cellular ATP will stabilize active OMA1 and lead to YME1L degradation. A consequence of this regulation is the inability to recover from a mitochondrial-fragmented state. Thus, the reciprocal regulation between YME1L and OMA1 indicates that the sensitive regulation of these proteases provides a unique opportunity to dictate mitochondrial morphology in response to different types and extents of cellular stress.

Apart from degradation, proteases are also susceptible to regulation through posttranslational modifications. While few examples are known, there are some cases where protease stability and/or activity have been reported to be regulated through posttranslational modification. Phosphorylation of the inner membrane protease AFG3L2 at tyrosine 179 suppresses processing of paraplegin during coassembly of the AFG3L2/paraplegin *m*-AAA heterooligomeric complex (Almontashiri et al., 2014). Following assembly, AFG3L2 is dephosphorylated allowing mature paraplegin to be proteolytically processed by AFG3L2. Similarly, LON acetylation by sirtuin-3 has been proposed to regulate the stability and activity of this protease (Gibellini et al., 2014). Posttranslational modifications of substrates can also influence their susceptibility to proteolysis. For example, phosphorylation of the TFAM impairs binding to DNA and promotes degradation by the quality control protease LON. Similarly, phosphorylation of cytochrome *c* oxidase subunits I, IVi1, and Vb increase targeting to LON for degradation upon hypoxic/ischemic insult (Lu et al., 2013; Sepuri et al., 2017).

The capacity to sensitively influence mitochondrial proteolytic activity through diverse posttranslational mechanisms provides additional levels for the stress-responsive regulation of mitochondrial proteostasis and function. While the mechanisms by which these posttranslational modifications integrate with transcriptional aspects of protease regulation remain to be established, it is clear that the regulation of these proteases is a central mechanism by which cells adapt mitochondrial function to match the specific demands required by a given insult.

1.5 Altered Mitochondrial Protease Activity in Aging and Disease

As outlined above, the diverse mechanisms of mitochondrial protease regulation provide a platform that allows cells to adapt mitochondrial function in response to diverse cellular insults. Imbalances in this regulation induced by protease mutations, alterations in stress-responsive signaling pathways, or aging predispose individuals to mitochondrial dysfunction in etiologically diverse human disorders, including cancer, cardiovascular disease, and neurodegenerative disorders (Quiros et al., 2015; Rugarli and Langer, 2012). Below, we describe how alterations in mitochondrial protease activity could influence pathologic mitochondrial dysfunction in these types of disorders.

1.5.1 Mutations in Mitochondrial Proteases Are Genetically Linked to Diverse Neurodegenerative Disorders

Mutations in mitochondrial quality control proteases have been causatively associated with the onset and pathology of numerous neurological disorders (Quiros et al., 2015; Rugarli and Langer, 2012). Most notably, mutations in the two *m*-AAA protease subunits AFG3L2 and paraplegin are implicated in different neurodegenerative diseases that present with varying levels of severity. Heterozygous mutations in *AFG3L2* cause spinocerebellar ataxia type 28 (Di Bella et al., 2010), while homozygous mutations in the same gene cause different neurological syndromes with onset in early infancy (Eskandrani et al., 2017; Muona et al., 2015; Pierson et al., 2011). Alternatively, *SPG7* mutations cause hereditary spastic paraplegia type 7 (Brugman et al., 2008; Casari et al., 1998; Elleuch et al., 2006). While the specific mechanism(s) by which these mutations influence disease pathogenesis remain to be fully elucidated, mouse models deficient in these two *m*-AAA subunits have provided significant insights into disease etiology associated with proteolytic dysregulation. Mice lacking either *Afg3l2* or *Spg7* show neurologic defects, including disrupted mitochondrial morphology, impaired Ca^{2+} buffering, axon degeneration, and impaired axon development (Almajan et al., 2012; Ferreirinha et al., 2004; Kondadi et al., 2014; Maltecca et al., 2008, 2009, 2015)—all phenotypes also observed in patients. Loss of *AFG3L2* has also been shown to impair mitochondrial axonal transport and is associated with tau hyperphosphorylation (Kondadi et al., 2014), the latter of which can promote proteotoxic tau aggregation and neurodegeneration. These results highlight the importance of the *m*-AAA protease in neuronal mitochondrial function and indicate that genetic mutations that disrupt *m*-AAA proteolytic activity cause pathologic mitochondrial dysfunction thereby impairing diverse neurologic functions.

Mutations in other mitochondrial quality control proteases have also been implicated in neurodegenerative disease. Mutations in the IMS quality control protease HTRA2 are associated with Parkinson's disease (PD) (Bogaerts et al., 2008; Lin et al., 2011; Strauss et al., 2005); however, the strength of this association remains to be fully established (Kruger et al., 2011; Spataro et al., 2015). Regardless, a potential role for HTRA2 in PD is supported by in vivo evidence showing that *HTRA2*-deficient mice and mice expressing a mutant *HTRA2* present parkinsonian phenotypes (Jones et al., 2003; Kruger et al., 2011; Martins et al., 2004). Furthermore, *HTRA2* mutations have been suggested to impair the autophagic degradation of α-synuclein aggregates, providing a potential mechanism to explain how mutations in this gene could induce the neuromuscular phenotypes associated with PD (Li et al., 2010). *HTRA2* mutations have also been suggested to influence other neurodegenerative diseases, including Alzheimer's disease (Westerlund et al., 2011) and infantile neurodegeneration (Mandel et al., 2016), although the molecular underpinnings of these relationships have not been elucidated. Rare *PARL* mutations have also been suggested to influence PD (Heinitz et al., 2011; Shi et al., 2011; Wust et al., 2015). A potential role for PARL in PD could reflect the involvement of this protease in regulating the PINK1-Parkin mitophagy axis (Greene et al., 2012; Jin et al., 2010; Meissner et al., 2015; Yamano and Youle, 2013), which would be consistent with the link between PD and other proteins involved in this pathway (Youle and Narendra, 2011). *YME1L1* mutations have also been found to be causatively associated with mitochondriopathy in familial optic atrophy (Hartmann et al., 2016). Here, mutations in the YME1L1 mitochondrial-targeting sequence affect MPP-dependent maturation of the protease, decreasing YME1L proteolytic activity and disrupting mitochondrial morphology and function.

Apart from mitochondrial quality control proteases, mutations in MPPs are also associated with neuronal disease. *IMMPL2* mutations are linked to many diseases, including autism, attention deficit/hyperactivity disorder, and Tourette syndrome (Elia et al., 2010; Maestrini et al., 2010; Petek et al., 2001, 2007). Furthermore, *Immpl2* mutant mice show increased levels of oxidative stress, providing a potential mechanism to explain neuronal phenotypes associated with alterations in this protein (George et al., 2011; Liu et al., 2016). Altered activity of the peptidase PreP has also been suggested to influence neurodegeneration in AD. PreP degrades the amyloidogenic peptide amyloid β in mitochondria (Falkevall et al., 2006; Mossmann et al., 2014). Thus, mutations that alter the activity of this protease could impair its function and promote mitochondrial dysfunction in association with AD (Alikhani et al., 2011b; Brunetti et al., 2016; Fang et al., 2015).

Apart from neuronal disorders, mutations in mitochondrial proteases also influence mitochondrial function in association with other types of diseases. *LON* mutations that disrupt the oligomerization and/or activity of this quality control protease lead to impaired mitochondrial proteostasis and function in association with cerebral, ocular, dental, auricular, and skeletal (CODAS) syndrome (Strauss et al., 2015). *CLPP* mutations are implicated in Perrault disease, which involve ovarian failure and sensorineural hearing loss (Ahmed et al., 2015; Jenkinson et al., 2013). Interestingly, mice deficient in *Clpp* recapitulate many Perrault-associated phenotypes, suggesting that alteration in CLPP-dependent regulation of mitochondrial proteostasis is causative in this disease (Gispert et al., 2013). Finally, *XPNPEP3* mutations are associated with the autosomal recessive kidney disease nephronophthisis, likely resulting from a mechanism involving impaired ciliary function (O'Toole et al., 2010). The genetic links between mitochondrial proteases and human disease described above clearly highlights the central importance of these proteases for regulating mitochondrial function.

Mutations in the stress-responsive signaling pathways responsible for regulating mitochondrial proteolytic activity and proteostasis could also directly contribute to pathologic mitochondrial dysfunction in association with disease. For example, mutations in the ISR-associated kinase PERK have been implicated in the onset and pathogenesis of diseases including Wolcott-Rallison Syndrome (Delepine et al., 2000; Durocher et al., 2006; Julier and Nicolino, 2010). Interestingly, the pathogenesis of these diseases has been observed to include impaired mitochondrial morphology and function (Collardeau-Frachon et al., 2015; Engelmann et al., 2008; Sovik et al., 2008). While PERK has many functions involved in the global regulation of cellular proteostasis and physiology during ER stress, it is possible that impaired PERK-dependent regulation of mitochondrial proteostasis directly contributes to mitochondrial dysfunction in association with this disease. As we continue to learn more of the integrative mechanisms by which stress-responsive signaling pathways such as the ISR and UPR^{mt} regulate mitochondrial proteostasis, similar types of relationships can be established to demonstrate how genetic perturbations of these pathways can influence mitochondrial function by impairing stress-responsive regulation of proteostasis.

1.5.2 Aging Dependent Alterations in Mitochondrial Proteolytic Capacity

Imbalances in mitochondrial proteostasis and function are a pathologic hallmark of the aging process in various organisms and in diverse tissues (Lopez-Otin et al., 2013). Mitochondrial proteases and their regulation are centrally implicated in the decline of mitochondrial function associated with the aging process (Quiros et al., 2015). Notably, the stress-responsive protease LON is intricately linked with organismal aging. Aging reduces the protein levels and activity of the LON protease in tissues including skeletal muscles and liver (Bakala et al., 2003; Bota et al., 2002; Lee et al., 1999). Similarly, while LON protein levels increase in the aging heart (likely reflecting stress-responsive regulation of this protease), LON activity does not increase, indicating that the specific activity of LON decreases with age in this tissue (Delaval et al., 2004). The importance of LON in aging is evident from studies in model organisms. Overexpression of LON in *P. anserina* increases organismal lifespan (Luce and Osiewacz, 2009). Alternatively, deletion of the *S. cerevisiae* Lon homolog, Pim1, accelerates aging in yeast (Erjavec et al., 2013). These results implicate LON activity as being an important determinant in eukaryotic aging.

Other mitochondrial proteases are also implicated in aging. For example, genetic depletion of *CLPP* increases *P. anserina* strain lifespan, indicating that overactivity of this protease could contribute to age-dependent loss of mitochondrial function in this organism (Fischer et al., 2013). Similarly, *HTRA2* depletion in mice induces mtDNA loss in systemic tissues, accelerating the aging process (Kang et al., 2013). *YME1L* depletion has also been shown to induce premature aging in flies (Qi et al., 2016). Furthermore, SNPs in the *m*-AAA protease subunit *AFG3L2* are linked to improved cognitive function in elderly humans (Lopez et al., 2012). Collectively, these results indicate that mitochondrial function is tightly regulated by the balanced activity of mitochondrial proteases and that aging-related alterations in this balance can perturb mitochondrial proteostasis and promote organismal aging.

The regulation of mitochondrial proteases through stress-responsive signaling pathways can also influence mitochondrial proteostasis capacity during organismal aging. Stress-responsive signaling pathways such as the ISR and UPR^{mt} are intricately related to the aging process (Schulz and Haynes, 2015; Shpilka and Haynes, 2018). Furthermore, the activity and/or regulation of stress-responsive signaling pathways is altered during aging, which challenges the ability for old organisms to efficiently regulate mitochondrial proteostasis and function in response to acute stress (Martinez et al., 2017; Naidoo, 2009; Rainbolt et al., 2014). Thus, apart from direct alterations in protease activity, the impaired ability to regulate mitochondrial proteostasis through stress-responsive signaling can also contribute to the mitochondrial dysfunction implicated in organismal aging.

1.5.3 Indirect Consequences of Altered Mitochondrial Proteolytic Activity on Mitochondrial Proteostasis and Function

Alterations in the activity or regulation of proteases are predicted to influence mitochondrial function through the altered capacity for these proteases to perform specific mitochondrial functions (e.g., LON-dependent degradation of oxidatively damaged proteins). However, alterations in these proteases could also influence mitochondrial function through imbalances in the capacity for mitochondria to sensitively regulate proteostasis. For example, LON and CLPP are reported to be overexpressed in multiple types of cancer and mice lacking a single copy of *Lon* show reduced malignant progression in melanoma and colorectal cancer models (Quiros et al., 2014; Seo et al., 2016). These results indicate that increased activity of mitochondrial proteases can support the malignant state. Alternatively, the reduced activity of a specific protease could lead to adaptation of the entire proteolytic network through stress-responsive signaling mechanisms that could lead to dysregulation of mitochondrial function. For example, spastic paraplegia patients harboring mutations in the mitochondrial chaperone *HSP60* show lower levels of LON and CLPP, suggesting that these cells may have lowered activity of these two proteases to compensate for impaired chaperoning activity (Hansen et al., 2008). The reduction in the activity of these proteases could reduce their ability to regulate specific mitochondrial functions and/or increase cellular sensitivity to subsequent insults. Furthermore, alterations in YME1L and OMA1 proteolytic activity are implicated in neonatal hypoxic-ischemic brain injury, indicating that stress-dependent changes in the activities of these proteases are implicated in this disorder (Baburamani et al., 2015). It is important to consider to these types of indirect alterations in protease activity on global mitochondrial proteostasis, as it could reveal compensatory mechanisms involved in the regulation of mitochondrial proteostasis that could contribute to disease pathogenesis and potentially be accessed to develop new therapeutic approaches to ameliorate pathologic mitochondrial dysfunction. While it is currently difficult to define how this type of imbalance can influence mitochondria, the importance of this type of mechanism will become more apparent as we learn more of the intricate relationship between mitochondrial protease regulation and function.

2. CONCLUDING COMMENTS AND FUTURE PERSPECTIVES

Mitochondrial proteases are critical regulators of mitochondrial proteostasis. It is clear from the results discussed throughout this review that the activity and regulation of these proteases are central in adapting mitochondrial function to various types of environmental, pathologic, or aging-related insults. This suggests that imbalances in the activity of these proteases and the pathways involved in their regulation represent a constellation of mechanisms that can contribute to pathologic mitochondrial dysfunction implicated in diverse disease states. Furthermore, the results discussed above indicate that influencing mitochondrial proteases through either direct targeting of the protease or intervention in stress-responsive signaling pathways represent new potential therapeutic opportunities to correct imbalances in mitochondrial proteostasis associated with disease. As we continue to learn more about mitochondrial proteases and their regulation in the context of disease, we anticipate that new roles for these proteases will be revealed and new strategies focused on manipulating proteolytic activity will be established to intervene in diverse classes of human disease.

ACKNOWLEDGMENTS

We apologize to all of our colleagues whose work we were unable to discuss in this review. We thank Aparajita Madhavan, Cristina Puchades, Mia Shin, and Vivian Moraes for critical reading of the manuscript. We thank NIH NS095892 (RLW) and the Hewitt Foundation (JL) for funding.

REFERENCES

Ahmed, S., Jelani, M., Alrayes, N., Mohamoud, H.S., Almramhi, M.M., Anshasi, W., Ahmed, N.A., Wang, J., Nasir, J., Al-Aama, J.Y., 2015. Exome analysis identified a novel missense mutation in the CLPP gene in a consanguineous Saudi family expanding the clinical spectrum of Perrault Syndrome type-3. J. Neurol. Sci. 353, 149–154.

Aldridge, J.E., Horibe, T., Hoogenraad, N.J., 2007. Discovery of genes activated by the mitochondrial unfolded protein response (mtUPR) and cognate promoter elements. PLoS One 2, e874.

Alikhani, N., Berglund, A.K., Engmann, T., Spanning, E., Vogtle, F.N., Pavlov, P., Meisinger, C., Langer, T., Glaser, E., 2011a. Targeting capacity and conservation of PreP homologues localization in mitochondria of different species. J. Mol. Biol. 410, 400–410.

Alikhani, N., Guo, L., Yan, S., Du, H., Pinho, C.M., Chen, J.X., Glaser, E., Yan, S.S., 2011b. Decreased proteolytic activity of the mitochondrial amyloid-beta degrading enzyme, PreP peptidasome, in Alzheimer's disease brain mitochondria. J. Alzheimers Dis. 27, 75–87.

Almajan, E.R., Richter, R., Paeger, L., Martinelli, P., Barth, E., Decker, T., Larsson, N.G., Kloppenburg, P., Langer, T., Rugarli, E.I., 2012. AFG3L2 supports mitochondrial protein synthesis and Purkinje cell survival. J. Clin. Invest. 122, 4048–4058.

Almontashiri, N.A., Chen, H.H., Mailloux, R.J., Tatsuta, T., Teng, A.C., Mahmoud, A.B., Ho, T., Stewart, N.A., Rippstein, P., Harper, M.E., et al., 2014. SPG7 variant escapes phosphorylation-regulated processing by AFG3L2, elevates mitochondrial ROS, and is associated with multiple clinical phenotypes. Cell Rep. 7, 834–847.

Anand, R., Wai, T., Baker, M.J., Kladt, N., Schauss, A.C., Rugarli, E., Langer, T., 2014. The i-AAA protease YME1L and OMA1 cleave OPA1 to balance mitochondrial fusion and fission. Journal Biol. 204, 919–929.

Arnold, I., Wagner-Ecker, M., Ansorge, W., Langer, T., 2006. Evidence for a novel mitochondria-to-nucleus signalling pathway in respiring cells lacking i-AAA protease and the ABC-transporter Mdl1. Gene 367, 74–88.

Baburamani, A.A., Hurling, C., Stolp, H., Sobotka, K., Gressens, P., Hagberg, H., Thornton, C., 2015. Mitochondrial optic atrophy (OPA) 1 processing is altered in response to neonatal hypoxic-ischemic brain injury. Int. J. Mol. Sci. 16, 22509–22526.

Bahat, A., Perlberg, S., Melamed-Book, N., Isaac, S., Eden, A., Lauria, I., Langer, T., Orly, J., 2015. Transcriptional activation of LON Gene by a new form of mitochondrial stress: a role for the nuclear respiratory factor 2 in StAR overload response (SOR). Mol. Cell. Endocrinol. 408, 62–72.

Bahat, A., Perlberg, S., Melamed-Book, N., Lauria, I., Langer, T., Orly, J., 2014. StAR enhances transcription of genes encoding the mitochondrial proteases involved in its own degradation. Mol. Endocrinol. 28, 208–224.

Bakala, H., Delaval, E., Hamelin, M., Bismuth, J., Borot-Laloi, C., Corman, B., Friguet, B., 2003. Changes in rat liver mitochondria with aging. Lon protease-like reactivity and N(epsilon)-carboxymethyllysine accumulation in the matrix. Eur. J. Biochem./FEBS 270, 2295–2302.

Baker, B.M., Nargund, A.M., Sun, T., Haynes, C.M., 2012a. Protective coupling of mitochondrial function and protein synthesis via the eIF2alpha kinase GCN-2. PLoS Genetics 8, e1002760.

Baker, M.J., Lampe, P.A., Stojanovski, D., Korwitz, A., Anand, R., Tatsuta, T., Langer, T., 2014. Stress-induced OMA1 activation and autocatalytic turnover regulate OPA1-dependent mitochondrial dynamics. EMBO J. 33, 578–593.

Baker, M.J., Mooga, V.P., Guiard, B., Langer, T., Ryan, M.T., Stojanovski, D., 2012b. Impaired folding of the mitochondrial small TIM chaperones induces clearance by the i-AAA protease. J. Mol. Biol. 424, 227–239.

Baker, M.J., Tatsuta, T., Langer, T., 2011. Quality control of mitochondrial proteostasis. Cold Spring Harb. Perspect. Biol. 3.

Bender, T., Lewrenz, I., Franken, S., Baitzel, C., Voos, W., 2011. Mitochondrial enzymes are protected from stress-induced aggregation by mitochondrial chaperones and the Pim1/LON protease. Mol. Biol. Cell 22, 541–554.

Benedetti, C., Haynes, C.M., Yang, Y., Harding, H.P., Ron, D., 2006. Ubiquitin-like protein 5 positively regulates chaperone gene expression in the mitochondrial unfolded protein response. Genetics 174, 229–239.

Bogaerts, V., Nuytemans, K., Reumers, J., Pals, P., Engelborghs, S., Pickut, B., Corsmit, E., Peeters, K., Schymkowitz, J., De Deyn, P.P., et al., 2008. Genetic variability in the mitochondrial serine protease HTRA2 contributes to risk for Parkinson disease. Hum. Mutat. 29, 832–840.

Bota, D.A., Davies, K.J., 2002. Lon protease preferentially degrades oxidized mitochondrial aconitase by an ATP-stimulated mechanism. Nat. Cell Biol. 4, 674–680.

Bota, D.A., Davies, K.J., 2016. Mitochondrial Lon protease in human disease and aging: including an etiologic classification of Lon-related diseases and disorders. Free Radic. Biol. Med. 100, 188–198.

Bota, D.A., Van Remmen, H., Davies, K.J., 2002. Modulation of Lon protease activity and aconitase turnover during aging and oxidative stress. FEBS Letters 532, 103–106.

Brugman, F., Scheffer, H., Wokke, J.H., Nillesen, W.M., de Visser, M., Aronica, E., Veldink, J.H., van den Berg, L.H., 2008. Paraplegin mutations in sporadic adult-onset upper motor neuron syndromes. Neurology 71, 1500–1505.

Brunetti, D., Torsvik, J., Dallabona, C., Teixeira, P., Sztromwasser, P., Fernandez-Vizarra, E., Cerutti, R., Reyes, A., Preziuso, C., D'Amati, G., et al., 2015. Defective PITRM1 mitochondrial peptidase is associated with Abeta amyloidotic neurodegeneration. EMBO Molecular Medicine.

Brunetti, D., Torsvik, J., Dallabona, C., Teixeira, P., Sztromwasser, P., Fernandez-Vizarra, E., Cerutti, R., Reyes, A., Preziuso, C., D'Amati, G., et al., 2016. Defective PITRM1 mitochondrial peptidase is associated with Abeta amyloidotic neurodegeneration. EMBO Molecular Medicine 8, 176–190.

Casari, G., De Fusco, M., Ciarmatori, S., Zeviani, M., Mora, M., Fernandez, P., De Michele, G., Filla, A., Cocozza, S., Marconi, R., et al., 1998. Spastic paraplegia and OXPHOS impairment caused by mutations in paraplegin, a nuclear-encoded mitochondrial metalloprotease. Cell 93, 973–983.

Chacinska, A., Koehler, C.M., Milenkovic, D., Lithgow, T., Pfanner, N., 2009. Importing mitochondrial proteins: machineries and mechanisms. Cell 138, 628–644.

Chan, D.C., 2012. Fusion and fission: interlinked processes critical for mitochondrial health. Annu. Rev. Genet. 46, 265–287.

Chan, N.C., Salazar, A.M., Pham, A.H., Sweredoski, M.J., Kolawa, N.J., Graham, R.L., Hess, S., Chan, D.C., 2011. Broad activation of the ubiquitin-proteasome system by Parkin is critical for mitophagy. Hum. Mol. Genet. 20, 1726–1737.

Chao, J.R., Parganas, E., Boyd, K., Hong, C.Y., Opferman, J.T., Ihle, J.N., 2008. Hax1-mediated processing of HtrA2 by Parl allows survival of lymphocytes and neurons. Nature 452, 98–102.

Chow, K.M., Gakh, O., Payne, I.C., Juliano, M.A., Juliano, L., Isaya, G., Hersh, L.B., 2009. Mammalian pitrilysin: substrate specificity and mitochondrial targeting. Biochemistry 48, 2868–2877.

Cilenti, L., Ambivero, C.T., Ward, N., Alnemri, E.S., Germain, D., Zervos, A.S., 2014. Inactivation of Omi/HtrA2 protease leads to the deregulation of mitochondrial Mulan E3 ubiquitin ligase and increased mitophagy. Biochim. Biophys. Acta 1843, 1295–1307.

Cipolat, S., Rudka, T., Hartmann, D., Costa, V., Serneels, L., Craessaerts, K., Metzger, K., Frezza, C., Annaert, W., D'Adamio, L., et al., 2006. Mitochondrial rhomboid PARL regulates cytochrome c release during apoptosis via OPA1-dependent cristae remodeling. Cell 126, 163–175.

Clausen, T., Kaiser, M., Huber, R., Ehrmann, M., 2011. HTRA proteases: regulated proteolysis in protein quality control. Nat. Rev. Mol. Cell Biol. 12, 152–162.

Collardeau-Frachon, S., Vasiljevic, A., Jouvet, A., Bouvier, R., Senee, V., Nicolino, M., 2015. Microscopic and ultrastructural features in Wolcott-Rallison syndrome, a permanent neonatal diabetes mellitus: about two autopsy cases. Pediatr. Diabetes 16, 510–520.

Covill-Cooke, C., Howden, J.H., Birsa, N., Kittler, J.T., 2017. Ubiquitination at the mitochondria in neuronal health and disease. Neurochem. Int.

Cullinan, S.B., Zhang, D., Hannink, M., Arvisais, E., Kaufman, R.J., Diehl, J.A., 2003. Nrf2 is a direct PERK substrate and effector of PERK-dependent cell survival. Mol. Cell. Biol. 23, 7198–7209.

Cunningham, C.N., Baughman, J.M., Phu, L., Tea, J.S., Yu, C., Coons, M., Kirkpatrick, D.S., Bingol, B., Corn, J.E., 2015. USP30 and parkin homeostatically regulate atypical ubiquitin chains on mitochondria. Nat. Cell Biol. 17, 160–169.

Deepa, S.S., Bhaskaran, S., Ranjit, R., Qaisar, R., Nair, B.C., Liu, Y., Walsh, M.E., Fok, W.C., Van Remmen, H., 2015. Down-regulation of the mitochondrial matrix peptidase ClpP in muscle cells causes mitochondrial dysfunction and decreases cell proliferation. Free Radic. Biol. Med.

Delaval, E., Perichon, M., Friguet, B., 2004. Age-related impairment of mitochondrial matrix aconitase and ATP-stimulated protease in rat liver and heart. Eur. J. Biochem./FEBS 271, 4559–4564.

Delepine, M., Nicolino, M., Barrett, T., Golamaully, M., Lathrop, G.M., Julier, C., 2000. EIF2AK3, encoding translation initiation factor 2-alpha kinase 3, is mutated in patients with Wolcott-Rallison syndrome. Nat. Genet. 25, 406–409.

Di Bella, D., Lazzaro, F., Brusco, A., Plumari, M., Battaglia, G., Pastore, A., Finardi, A., Cagnoli, C., Tempia, F., Frontali, M., et al., 2010. Mutations in the mitochondrial protease gene AFG3L2 cause dominant hereditary ataxia SCA28. Nat. Genet. 42, 313–321.

Durieux, J., Wolff, S., Dillin, A., 2011. The cell-non-autonomous nature of electron transport chain-mediated longevity. Cell 144, 79–91.

Durocher, F., Faure, R., Labrie, Y., Pelletier, L., Bouchard, I., Laframboise, R., 2006. A novel mutation in the EIF2AK3 gene with variable expressivity in two patients with Wolcott-Rallison syndrome. Clin. Genet. 70, 34–38.

Elia, J., Gai, X., Xie, H.M., Perin, J.C., Geiger, E., Glessner, J.T., D'Arcy, M., deBerardinis, R., Frackelton, E., Kim, C., et al., 2010. Rare structural variants found in attention-deficit hyperactivity disorder are preferentially associated with neurodevelopmental genes. Mol. Psychiatr. 15, 637–646.

Elleuch, N., Depienne, C., Benomar, A., Hernandez, A.M., Ferrer, X., Fontaine, B., Grid, D., Tallaksen, C.M., Zemmouri, R., Stevanin, G., et al., 2006. Mutation analysis of the paraplegin gene (SPG7) in patients with hereditary spastic paraplegia. Neurology 66, 654–659.

Engelmann, G., Meyburg, J., Shahbek, N., Al-Ali, M., Hairetis, M.H., Baker, A.J., Rodenburg, R.J., Wenning, D., Flechtenmacher, C., Ellard, S., et al., 2008. Recurrent acute liver failure and mitochondriopathy in a case of Wolcott-Rallison syndrome. J. Inherit. Metab. Dis. 31, 540–546.

Erjavec, N., Bayot, A., Gareil, M., Camougrand, N., Nystrom, T., Friguet, B., Bulteau, A.L., 2013. Deletion of the mitochondrial Pim1/Lon protease in yeast results in accelerated aging and impairment of the proteasome. Free Radic. Biol. Med. 56, 9–16.

Eskandrani, A., AlHashem, A., Ali, E.S., AlShahwan, S., Tlili, K., Hundallah, K., Tabarki, B., 2017. Recessive AFG3L2 mutation causes progressive microcephaly, early onset seizures, spasticity, and basal ganglia involvement. Pediatr. Neurol. 71, 24–28.

Falkevall, A., Alikhani, N., Bhushan, S., Pavlov, P.F., Busch, K., Johnson, K.A., Eneqvist, T., Tjernberg, L., Ankarcrona, M., Glaser, E., 2006. Degradation of the amyloid beta-protein by the novel mitochondrial peptidasome, PreP. J. Biol. Chem. 281, 29096–29104.

Fang, D., Wang, Y., Zhang, Z., Du, H., Yan, S., Sun, Q., Zhong, C., Wu, L., Vangavaragu, J.R., Yan, S., et al., 2015. Increased neuronal PreP activity reduces Abeta accumulation, attenuates neuroinflammation and improves mitochondrial and synaptic function in Alzheimer disease's mouse model. Hum. Mol. Genet. 24, 5198–5210.

Ferreirinha, F., Quattrini, A., Pirozzi, M., Valsecchi, V., Dina, G., Broccoli, V., Auricchio, A., Piemonte, F., Tozzi, G., Gaeta, L., et al., 2004. Axonal degeneration in paraplegin-deficient mice is associated with abnormal mitochondria and impairment of axonal transport. J. Clin. Invest. 113, 231–242.
Fiorese, C.J., Schulz, A.M., Lin, Y.F., Rosin, N., Pellegrino, M.W., Haynes, C.M., 2016. The transcription factor ATF5 mediates a mammalian mitochondrial UPR. Curr. Biol. 26, 2037–2043.
Fischer, F., Weil, A., Hamann, A., Osiewacz, H.D., 2013. Human CLPP reverts the longevity phenotype of a fungal ClpP deletion strain. Nat. Commun. 4, 1397.
Frezza, C., Cipolat, S., Martins de Brito, O., Micaroni, M., Beznoussenko, G.V., Rudka, T., Bartoli, D., Polishuck, R.S., Danial, N.N., De Strooper, B., et al., 2006. OPA1 controls apoptotic cristae remodeling independently from mitochondrial fusion. Cell 126, 177–189.
Fukuda, R., Zhang, H., Kim, J.W., Shimoda, L., Dang, C.V., Semenza, G.L., 2007. HIF-1 regulates cytochrome oxidase subunits to optimize efficiency of respiration in hypoxic cells. Cell 129, 111–122.
Fusakio, M.E., Willy, J.A., Wang, Y., Mirek, E.T., Al Baghdadi, R.J., Adams, C.M., Anthony, T.G., Wek, R.C., 2016. Transcription factor ATF4 directs basal and stress-induced gene expression in the unfolded protein response and cholesterol metabolism in the liver. Mol. Biol. Cell 27, 1536–1551.
Gakh, O., Cavadini, P., Isaya, G., 2002. Mitochondrial processing peptidases. Biochim. Biophys. Acta 1592, 63–77.
George, S.K., Jiao, Y., Bishop, C.E., Lu, B., 2011. Mitochondrial peptidase IMMP2L mutation causes early onset of age-associated disorders and impairs adult stem cell self-renewal. Aging Cell 10, 584–594.
Gibellini, L., Pinti, M., Beretti, F., Pierri, C.L., Onofrio, A., Riccio, M., Carnevale, G., De Biasi, S., Nasi, M., Torelli, F., et al., 2014. Sirtuin 3 interacts with Lon protease and regulates its acetylation status. Mitochondrion 18, 76–81.
Gillies, L.A., Kuwana, T., 2014. Apoptosis regulation at the mitochondrial outer membrane. J. Cell. Biochem. 115, 632–640.
Gispert, S., Parganlija, D., Klinkenberg, M., Drose, S., Wittig, I., Mittelbronn, M., Grzmil, P., Koob, S., Hamann, A., Walter, M., et al., 2013. Loss of mitochondrial peptidase Clpp leads to infertility, hearing loss plus growth retardation via accumulation of CLPX, mtDNA and inflammatory factors. Hum. Mol. Genet. 22, 4871–4887.
Granot, Z., Kobiler, O., Melamed-Book, N., Eimerl, S., Bahat, A., Lu, B., Braun, S., Maurizi, M.R., Suzuki, C.K., Oppenheim, A.B., et al., 2007. Turnover of mitochondrial steroidogenic acute regulatory (StAR) protein by Lon protease: the unexpected effect of proteasome inhibitors. Mol. Endocrinol. 21, 2164–2177.
Greene, A.W., Grenier, K., Aguileta, M.A., Muise, S., Farazifard, R., Haque, M.E., McBride, H.M., Park, D.S., Fon, E.A., 2012. Mitochondrial processing peptidase regulates PINK1 processing, import and Parkin recruitment. EMBO Rep. 13, 378–385.
Griparic, L., Kanazawa, T., van der Bliek, A.M., 2007. Regulation of the mitochondrial dynamin-like protein Opa1 by proteolytic cleavage. J. Cell Biol. 178, 757–764.
Hamacher-Brady, A., Brady, N.R., 2016. Mitophagy programs: mechanisms and physiological implications of mitochondrial targeting by autophagy. Cell. Mol. Life Sci. 73, 775–795.
Han, J., Back, S.H., Hur, J., Lin, Y.H., Gildersleeve, R., Shan, J., Yuan, C.L., Krokowski, D., Wang, S., Hatzoglou, M., et al., 2013. ER-stress-induced transcriptional regulation increases protein synthesis leading to cell death. Nat. Cell Biol. 15, 481–490.

Hansen, J., Corydon, T.J., Palmfeldt, J., Durr, A., Fontaine, B., Nielsen, M.N., Christensen, J.H., Gregersen, N., Bross, P., 2008. Decreased expression of the mitochondrial matrix proteases Lon and ClpP in cells from a patient with hereditary spastic paraplegia (SPG13). Neuroscience 153, 474–482.

Harbauer, A.B., Zahedi, R.P., Sickmann, A., Pfanner, N., Meisinger, C., 2014. The protein import machinery of mitochondria-a regulatory hub in metabolism, stress, and disease. Cell Metab. 19, 357–372.

Harding, H.P., Novoa, I., Zhang, Y., Zeng, H., Wek, R., Schapira, M., Ron, D., 2000. Regulated translation initiation controls stress-induced gene expression in mammalian cells. Mol. cell 6, 1099–1108.

Harding, H.P., Zhang, Y., Zeng, H., Novoa, I., Lu, P.D., Calfon, M., Sadri, N., Yun, C., Popko, B., Paules, R., et al., 2003. An integrated stress response regulates amino acid metabolism and resistance to oxidative stress. Mol. cell 11, 619–633.

Harper, J.W., Ordureau, A., Heo, J.M., 2018. Building and decoding ubiquitin chains for mitophagy. Nat. Rev. Mol. Cell Biol. 19, 93–108.

Hartmann, B., Wai, T., Hu, H., MacVicar, T., Musante, L., Fischer-Zirnsak, B., Stenzel, W., Graf, R., van den Heuvel, L., Ropers, H.H., et al., 2016. Homozygous YME1L1 mutation causes mitochondriopathy with optic atrophy and mitochondrial network fragmentation. Elife 5.

Haynes, C.M., Petrova, K., Benedetti, C., Yang, Y., Ron, D., 2007. ClpP mediates activation of a mitochondrial unfolded protein response in *C. elegans*. Dev. Cell 13, 467–480.

Haynes, C.M., Yang, Y., Blais, S.P., Neubert, T.A., Ron, D., 2010. The matrix peptide exporter HAF-1 signals a mitochondrial UPR by activating the transcription factor ZC376.7 in *C. elegans*. Mol. cell 37, 529–540.

Head, B., Griparic, L., Amiri, M., Gandre-Babbe, S., van der Bliek, A.M., 2009. Inducible proteolytic inactivation of OPA1 mediated by the OMA1 protease in mammalian cells. J. Cell Biol. 187, 959–966.

Hegde, R., Srinivasula, S.M., Zhang, Z., Wassell, R., Mukattash, R., Cilenti, L., DuBois, G., Lazebnik, Y., Zervos, A.S., Fernandes-Alnemri, T., et al., 2002. Identification of Omi/HtrA2 as a mitochondrial apoptotic serine protease that disrupts inhibitor of apoptosis protein-caspase interaction. J. Biol. Chem. 277, 432–438.

Heinitz, S., Klein, C., Djarmati, A., 2011. The p.S77N presenilin-associated rhomboid-like protein mutation is not a frequent cause of early-onset Parkinson's disease. Mov. Disord. : Official Journal of the Movement Disorder Society 26, 2441–2442.

Hetz, C., Papa, F.R., 2018. The unfolded protein response and cell fate control. Mol. Cell 69, 169–181.

Hori, O., Ichinoda, F., Tamatani, T., Yamaguchi, A., Sato, N., Ozawa, K., Kitao, Y., Miyazaki, M., Harding, H.P., Ron, D., et al., 2002. Transmission of cell stress from endoplasmic reticulum to mitochondria: enhanced expression of Lon protease. J. Cell Biol. 157, 1151–1160.

Horibe, T., Hoogenraad, N.J., 2007. The chop gene contains an element for the positive regulation of the mitochondrial unfolded protein response. PLoS One 2, e835.

Houtkooper, R.H., Mouchiroud, L., Ryu, D., Moullan, N., Katsyuba, E., Knott, G., Williams, R.W., Auwerx, J., 2013. Mitonuclear protein imbalance as a conserved longevity mechanism. Nature 497, 451–457.

Hu, X.V., Chen, X., Han, K.C., Mildvan, A.S., Liu, J.O., 2007. Kinetic and mutational studies of the number of interacting divalent cations required by bacterial and human methionine aminopeptidases. Biochemistry 46, 12833–12843.

Ishihara, N., Fujita, Y., Oka, T., Mihara, K., 2006. Regulation of mitochondrial morphology through proteolytic cleavage of OPA1. EMBO J. 25, 2966–2977.

Jenkinson, E.M., Rehman, A.U., Walsh, T., Clayton-Smith, J., Lee, K., Morell, R.J., Drummond, M.C., Khan, S.N., Naeem, M.A., Rauf, B., et al., 2013. Perrault syndrome is caused by recessive mutations in CLPP, encoding a mitochondrial ATP-dependent chambered protease. Am. J. Hum. Genet. 92, 605–613.

Jin, S.M., Lazarou, M., Wang, C., Kane, L.A., Narendra, D.P., Youle, R.J., 2010. Mitochondrial membrane potential regulates PINK1 import and proteolytic destabilization by PARL. J. Cell Biol. 191, 933–942.

Jones, J.M., Datta, P., Srinivasula, S.M., Ji, W., Gupta, S., Zhang, Z., Davies, E., Hajnoczky, G., Saunders, T.L., Van Keuren, M.L., et al., 2003. Loss of Omi mitochondrial protease activity causes the neuromuscular disorder of mnd2 mutant mice. Nature 425, 721–727.

Julier, C., Nicolino, M., 2010. Wolcott-Rallison syndrome. Orphanet J. Rare Dis. 5, 29.

Kang, S., Louboutin, J.P., Datta, P., Landel, C.P., Martinez, D., Zervos, A.S., Strayer, D.S., Fernandes-Alnemri, T., Alnemri, E.S., 2013. Loss of HtrA2/Omi activity in non-neuronal tissues of adult mice causes premature aging. Cell Death Differ 20, 259–269.

Kao, T.Y., Chiu, Y.C., Fang, W.C., Cheng, C.W., Kuo, C.Y., Juan, H.F., Wu, S.H., Lee, A.Y., 2015. Mitochondrial Lon regulates apoptosis through the association with Hsp60-mtHsp70 complex. Cell Death Dis. 6, e1642.

Karbowski, M., Neutzner, A., Youle, R.J., 2007. The mitochondrial E3 ubiquitin ligase MARCH5 is required for Drp1 dependent mitochondrial division. J. Cell Biol. 178, 71–84.

Kardon, J.R., Yien, Y.Y., Huston, N.C., Branco, D.S., Hildick-Smith, G.J., Rhee, K.Y., Paw, B.H., Baker, T.A., 2015. Mitochondrial ClpX activates a key enzyme for heme biosynthesis and erythropoiesis. Cell 161, 858–867.

Kaser, M., Kambacheld, M., Kisters-Woike, B., Langer, T., 2003. Oma1, a novel membrane-bound metallopeptidase in mitochondria with activities overlapping with the m-AAA protease. J. Biol. Chem. 278, 46414–46423.

Kilberg, M.S., Balasubramanian, M., Fu, L., Shan, J., 2012. The transcription factor network associated with the amino acid response in mammalian cells. Adv. Nutr. 3, 295–306.

Kleiber, J., Kalousek, F., Swaroop, M., Rosenberg, L.E., 1990. The general mitochondrial matrix processing protease from rat liver: structural characterization of the catalytic subunit. Proc. Natl. Acad. Sci. U S A 87, 7978–7982.

Kondadi, A.K., Wang, S., Montagner, S., Kladt, N., Korwitz, A., Martinelli, P., Herholz, D., Baker, M.J., Schauss, A.C., Langer, T., et al., 2014. Loss of the m-AAA protease subunit AFG(3)L(2) causes mitochondrial transport defects and tau hyperphosphorylation. EMBO J. 33, 1011–1026.

Konig, T., Troder, S.E., Bakka, K., Korwitz, A., Richter-Dennerlein, R., Lampe, P.A., Patron, M., Muhlmeister, M., Guerrero-Castillo, S., Brandt, U., et al., 2016. The m-AAA protease associated with neurodegeneration limits MCU activity in mitochondria. Mol. Cell 64, 148–162.

Koppen, M., Metodiev, M.D., Casari, G., Rugarli, E.I., Langer, T., 2007. Variable and tissue-specific subunit composition of mitochondrial m-AAA protease complexes linked to hereditary spastic paraplegia. Mol. Cell. Biol. 27, 758–767.

Koyano, F., Okatsu, K., Kosako, H., Tamura, Y., Go, E., Kimura, M., Kimura, Y., Tsuchiya, H., Yoshihara, H., Hirokawa, T., et al., 2014. Ubiquitin is phosphorylated by PINK1 to activate parkin. Nature 510, 162–166.

Kruger, R., Sharma, M., Riess, O., Gasser, T., Van Broeckhoven, C., Theuns, J., Aasly, J., Annesi, G., Bentivoglio, A.R., Brice, A., et al., 2011. A large-scale genetic association study to evaluate the contribution of Omi/HtrA2 (PARK13) to Parkinson's disease. Neurobiol. Aging 32 (548), e518–e549.

Lebeau, J., Saunders, J.M., Moraes, V.W.R., Madhavan, A., Madrazo, N., Anthony, M.C., Wiseman, R.L., 2018. The PERK arm of the unfolded protein response regulates mitochondrial morphology during acute endoplasmic reticulum stress. Cell Rep. 22, 2827–2836.

Lee, C.K., Klopp, R.G., Weindruch, R., Prolla, T.A., 1999. Gene expression profile of aging and its retardation by caloric restriction. Science 285, 1390–1393.

Li, B., Hu, Q., Wang, H., Man, N., Ren, H., Wen, L., Nukina, N., Fei, E., Wang, G., 2010. Omi/HtrA2 is a positive regulator of autophagy that facilitates the degradation of mutant proteins involved in neurodegenerative diseases. Cell Death Differ. 17, 1773–1784.

Liang, J.R., Martinez, A., Lane, J.D., Mayor, U., Clague, M.J., Urbe, S., 2015. USP30 deubiquitylates mitochondrial Parkin substrates and restricts apoptotic cell death. EMBO Rep. 16, 618–627.

Lin, C.H., Chen, M.L., Chen, G.S., Tai, C.H., Wu, R.M., 2011. Novel variant Pro143Ala in HTRA2 contributes to Parkinson's disease by inducing hyperphosphorylation of HTRA2 protein in mitochondria. Hum. Genet. 130, 817–827.

Liu, C., Li, X., Lu, B., 2016. The Immp2l mutation causes age-dependent degeneration of cerebellar granule neurons prevented by antioxidant treatment. Aging Cell 15, 167–176.

Liu, T., Lu, B., Lee, I., Ondrovicova, G., Kutejova, E., Suzuki, C.K., 2004. DNA and RNA binding by the mitochondrial lon protease is regulated by nucleotide and protein substrate. J. Biol. Chem. 279, 13902–13910.

Lopez, L.M., Harris, S.E., Luciano, M., Liewald, D., Davies, G., Gow, A.J., Tenesa, A., Payton, A., Ke, X., Whalley, L.J., et al., 2012. Evolutionary conserved longevity genes and human cognitive abilities in elderly cohorts. Eur. J. Hum. Genet. 20, 341–347.

Lopez-Otin, C., Blasco, M.A., Partridge, L., Serrano, M., Kroemer, G., 2013. The hallmarks of aging. Cell 153, 1194–1217.

Loson, O.C., Song, Z., Chen, H., Chan, D.C., 2013. Fis1, Mff, MiD49, and MiD51 mediate Drp1 recruitment in mitochondrial fission. Mol. Biol. Cell 24, 659–667.

Lu, B., Lee, J., Nie, X., Li, M., Morozov, Y.I., Venkatesh, S., Bogenhagen, D.F., Temiakov, D., Suzuki, C.K., 2013. Phosphorylation of human TFAM in mitochondria impairs DNA binding and promotes degradation by the AAA+ Lon protease. Mol. Cell 49, 121–132.

Lu, B., Yadav, S., Shah, P.G., Liu, T., Tian, B., Pukszta, S., Villaluna, N., Kutejova, E., Newlon, C.S., Santos, J.H., et al., 2007. Roles for the human ATP-dependent Lon protease in mitochondrial DNA maintenance. J. Biol. Chem. 282, 17363–17374.

Luce, K., Osiewacz, H.D., 2009. Increasing organismal healthspan by enhancing mitochondrial protein quality control. Nat. Cell Biol. 11, 852–858.

Ma, Y., Hendershot, L.M., 2003. Delineation of a negative feedback regulatory loop that controls protein translation during endoplasmic reticulum stress. J. Biol. Chem. 278, 34864–34873.

Maestrini, E., Pagnamenta, A.T., Lamb, J.A., Bacchelli, E., Sykes, N.H., Sousa, I., Toma, C., Barnby, G., Butler, H., Winchester, L., et al., 2010. High-density SNP association study and copy number variation analysis of the AUTS1 and AUTS5 loci implicate the IMMP2L-DOCK4 gene region in autism susceptibility. Mol. Psychiatr. 15, 954–968.

Maltecca, F., Aghaie, A., Schroeder, D.G., Cassina, L., Taylor, B.A., Phillips, S.J., Malaguti, M., Previtali, S., Guenet, J.L., Quattrini, A., et al., 2008. The mitochondrial protease AFG3L2 is essential for axonal development. J. Neurosci. 28, 2827–2836.

Maltecca, F., Baseggio, E., Consolato, F., Mazza, D., Podini, P., Young Jr., S.M., Drago, I., Bahr, B.A., Puliti, A., Codazzi, F., et al., 2015. Purkinje neuron Ca2+ influx reduction rescues ataxia in SCA28 model. J. Clin. Invest. 125, 263–274.

Maltecca, F., Magnoni, R., Cerri, F., Cox, G.A., Quattrini, A., Casari, G., 2009. Haploinsufficiency of AFG3L2, the gene responsible for spinocerebellar ataxia type 28, causes mitochondria-mediated Purkinje cell dark degeneration. J. Neurosci. 29, 9244–9254.

Mandel, H., Saita, S., Edvardson, S., Jalas, C., Shaag, A., Goldsher, D., Vlodavsky, E., Langer, T., Elpeleg, O., 2016. Deficiency of HTRA2/Omi is associated with infantile neurodegeneration and 3-methylglutaconic aciduria. J. Med. Genet. 53, 690–696.

Martinez, G., Duran-Aniotz, C., Cabral-Miranda, F., Vivar, J.P., Hetz, C., 2017. Endoplasmic reticulum proteostasis impairment in aging. Aging Cell 16, 615–623.

Martinez-Reyes, I., Sanchez-Arago, M., Cuezva, J.M., 2012. AMPK and GCN2-ATF4 signal the repression of mitochondria in colon cancer cells. Biochem. J. 444, 249–259.

Martins, L.M., Morrison, A., Klupsch, K., Fedele, V., Moisoi, N., Teismann, P., Abuin, A., Grau, E., Geppert, M., Livi, G.P., et al., 2004. Neuroprotective role of the Reaper-related serine protease HtrA2/Omi revealed by targeted deletion in mice. Mol. Cell. Biol. 24, 9848–9862.

Matsuda, N., Sato, S., Shiba, K., Okatsu, K., Saisho, K., Gautier, C.A., Sou, Y.S., Saiki, S., Kawajiri, S., Sato, F., et al., 2010. PINK1 stabilized by mitochondrial depolarization recruits Parkin to damaged mitochondria and activates latent Parkin for mitophagy. J. Cell Biol. 189, 211–221.

Matsushima, Y., Goto, Y., Kaguni, L.S., 2010. Mitochondrial Lon protease regulates mitochondrial DNA copy number and transcription by selective degradation of mitochondrial transcription factor A (TFAM). Proc. Natl. Acad. Sci. U S A 107, 18410–18415.

Meissner, C., Lorenz, H., Hehn, B., Lemberg, M.K., 2015. Intramembrane protease PARL defines a negative regulator of PINK1- and PARK2/Parkin-dependent mitophagy. Autophagy 11, 1484–1498.

Michel, S., Canonne, M., Arnould, T., Renard, P., 2015. Inhibition of mitochondrial genome expression triggers the activation of CHOP-10 by a cell signaling dependent on the integrated stress response but not the mitochondrial unfolded protein response. Mitochondrion 21, 58–68.

Mishra, P., Carelli, V., Manfredi, G., Chan, D.C., 2014. Proteolytic cleavage of Opa1 stimulates mitochondrial inner membrane fusion and couples fusion to oxidative phosphorylation. Cell Metab. 19, 630–641.

Moisoi, N., Klupsch, K., Fedele, V., East, P., Sharma, S., Renton, A., Plun-Favreau, H., Edwards, R.E., Teismann, P., Esposti, M.D., et al., 2009. Mitochondrial dysfunction triggered by loss of HtrA2 results in the activation of a brain-specific transcriptional stress response. Cell Death Differ. 16, 449–464.

Mossmann, D., Vogtle, F.N., Taskin, A.A., Teixeira, P.F., Ring, J., Burkhart, J.M., Burger, N., Pinho, C.M., Tadic, J., Loreth, D., et al., 2014. Amyloid-beta peptide induces mitochondrial dysfunction by inhibition of preprotein maturation. Cell Metab. 20, 662–669.

Mouchiroud, L., Houtkooper, R.H., Moullan, N., Katsyuba, E., Ryu, D., Canto, C., Mottis, A., Jo, Y.S., Viswanathan, M., Schoonjans, K., et al., 2013. The NAD(+)/sirtuin pathway modulates longevity through activation of mitochondrial UPR and FOXO signaling. Cell 154, 430–441.

Muona, M., Berkovic, S.F., Dibbens, L.M., Oliver, K.L., Maljevic, S., Bayly, M.A., Joensuu, T., Canafoglia, L., Franceschetti, S., Michelucci, R., et al., 2015. A recurrent de novo mutation in KCNC1 causes progressive myoclonus epilepsy. Nat. Genet. 47, 39–46.

Nabhan, J.F., Gooch, R.L., Piatnitski Chekler, E.L., Pierce, B., Bulawa, C.E., 2015. Perturbation of cellular proteostasis networks identifies pathways that modulate precursor and intermediate but not mature levels of frataxin. Sci. Rep. 5, 18251.

Naidoo, N., 2009. ER and aging-Protein folding and the ER stress response. Ageing Res. Rev. 8, 150–159.

Nakamura, N., Kimura, Y., Tokuda, M., Honda, S., Hirose, S., 2006. MARCH-V is a novel mitofusin 2- and Drp1-binding protein able to change mitochondrial morphology. EMBO Rep. 7, 1019–1022.

Nargund, A.M., Pellegrino, M.W., Fiorese, C.J., Baker, B.M., Haynes, C.M., 2012. Mitochondrial import efficiency of ATFS-1 regulates mitochondrial UPR activation. Science 337, 587–590.

Nolden, M., Ehses, S., Koppen, M., Bernacchia, A., Rugarli, E.I., Langer, T., 2005. The m-AAA protease defective in hereditary spastic paraplegia controls ribosome assembly in mitochondria. Cell 123, 277–289.

Novak, I., Dikic, I., 2011. Autophagy receptors in developmental clearance of mitochondria. Autophagy 7, 301–303.

Novak, I., Kirkin, V., McEwan, D.G., Zhang, J., Wild, P., Rozenknop, A., Rogov, V., Lohr, F., Popovic, D., Occhipinti, A., et al., 2010. Nix is a selective autophagy receptor for mitochondrial clearance. EMBO Rep. 11, 45–51.

Novoa, I., Zeng, H., Harding, H.P., Ron, D., 2001. Feedback inhibition of the unfolded protein response by GADD34-mediated dephosphorylation of eIF2alpha. J. Cell Biol. 153, 1011–1022.

Nunnari, J., Fox, T.D., Walter, P., 1993. A mitochondrial protease with two catalytic subunits of nonoverlapping specificities. Science 262, 1997–2004.

O'Toole, J.F., Liu, Y., Davis, E.E., Westlake, C.J., Attanasio, M., Otto, E.A., Seelow, D., Nurnberg, G., Becker, C., Nuutinen, M., et al., 2010. Individuals with mutations in XPNPEP3, which encodes a mitochondrial protein, develop a nephronophthisis-like nephropathy. J. Clin. Invest. 120, 791–802.

Okatsu, K., Koyano, F., Kimura, M., Kosako, H., Saeki, Y., Tanaka, K., Matsuda, N., 2015. Phosphorylated ubiquitin chain is the genuine Parkin receptor. J. Cell Biol. 209, 111–128.

Okatsu, K., Uno, M., Koyano, F., Go, E., Kimura, M., Oka, T., Tanaka, K., Matsuda, N., 2013. A dimeric PINK1-containing complex on depolarized mitochondria stimulates Parkin recruitment. J. Biol. Chem. 288, 36372–36384.

Olichon, A., Baricault, L., Gas, N., Guillou, E., Valette, A., Belenguer, P., Lenaers, G., 2003. Loss of OPA1 perturbates the mitochondrial inner membrane structure and integrity, leading to cytochrome c release and apoptosis. J. Biol. Chem. 278, 7743–7746.

Ou, W.J., Ito, A., Okazaki, H., Omura, T., 1989. Purification and characterization of a processing protease from rat liver mitochondria. EMBO J. 8, 2605–2612.

Pakos-Zebrucka, K., Koryga, I., Mnich, K., Ljujic, M., Samali, A., Gorman, A.M., 2016. The integrated stress response. EMBO Rep. 17, 1374–1395.

Papa, L., Germain, D., 2011. Estrogen receptor mediates a distinct mitochondrial unfolded protein response. J. Cell Sci. 124, 1396–1402.

Park, Y.Y., Cho, H., 2012. Mitofusin 1 is degraded at G2/M phase through ubiquitylation by MARCH5. Mol. Biol. Cell 23.

Pellegrini, L., Scorrano, L., 2007. A cut short to death: Parl and Opa1 in the regulation of mitochondrial morphology and apoptosis. Cell Death Differ. 14, 1275–1284.

Petek, E., Schwarzbraun, T., Noor, A., Patel, M., Nakabayashi, K., Choufani, S., Windpassinger, C., Stamenkovic, M., Robertson, M.M., Aschauer, H.N., et al., 2007. Molecular and genomic studies of IMMP2L and mutation screening in autism and Tourette syndrome. Mol. Genet. Genom. : MGG (Mol. Gen. Genet.) 277, 71–81.

Petek, E., Windpassinger, C., Vincent, J.B., Cheung, J., Boright, A.P., Scherer, S.W., Kroisel, P.M., Wagner, K., 2001. Disruption of a novel gene (IMMP2L) by a breakpoint in 7q31 associated with Tourette syndrome. Am. J. Hum. Genet. 68, 848–858.

Pierson, T.M., Adams, D., Bonn, F., Martinelli, P., Cherukuri, P.F., Teer, J.K., Hansen, N.F., Cruz, P., Mullikin For The Nisc Comparative Sequencing Program, J.C., Blakesley, R.W., et al., 2011. Whole-exome sequencing identifies homozygous AFG3L2 mutations in a spastic ataxia-neuropathy syndrome linked to mitochondrial m-AAA proteases. PLoS Genetics 7, e1002325.

Pinti, M., Gibellini, L., De Biasi, S., Nasi, M., Roat, E., O'Connor, J.E., Cossarizza, A., 2011. Functional characterization of the promoter of the human Lon protease gene. Mitochondrion 11, 200–206.

Potting, C., Tatsuta, T., Konig, T., Haag, M., Wai, T., Aaltonen, M.J., Langer, T., 2013. TRIAP1/PRELI complexes prevent apoptosis by mediating intramitochondrial transport of phosphatidic acid. Cell Metab. 18, 287–295.

Puchades, C., Rampello, A.J., Shin, M., Giuliano, C.J., Wiseman, R.L., Glynn, S.E., Lander, G.C., 2017. Structure of the mitochondrial inner membrane AAA+ protease YME1 gives insight into substrate processing. Science 358.

Qi, Y., Liu, H., Daniels, M.P., Zhang, G., Xu, H., 2016. Loss of Drosophila i-AAA protease, dYME1L, causes abnormal mitochondria and apoptotic degeneration. Cell Death Differ. 23, 291–302.

Quiros, P.M., Espanol, Y., Acin-Perez, R., Rodriguez, F., Barcena, C., Watanabe, K., Calvo, E., Loureiro, M., Fernandez-Garcia, M.S., Fueyo, A., et al., 2014. ATP-dependent Lon protease controls tumor bioenergetics by reprogramming mitochondrial activity. Cell Rep. 8, 542–556.

Quiros, P.M., Langer, T., Lopez-Otin, C., 2015. New roles for mitochondrial proteases in health, ageing and disease. Nat. Rev. Mol. Cell Biol. 16, 345–359.

Quiros, P.M., Mottis, A., Auwerx, J., 2016. Mitonuclear communication in homeostasis and stress. Nat. Rev. Mol. Cell Biol. 17, 213–226.

Quiros, P.M., Prado, M.A., Zamboni, N., D'Amico, D., Williams, R.W., Finley, D., Gygi, S.P., Auwerx, J., 2017. Multi-omics analysis identifies ATF4 as a key regulator of the mitochondrial stress response in mammals. J. Cell Biol. 216, 2027–2045.

Rainbolt, T.K., Atanassova, N., Genereux, J.C., Wiseman, R.L., 2013. Stress-regulated translational attenuation adapts mitochondrial protein import through Tim17A degradation. Cell Metab. 18, 908–919.

Rainbolt, T.K., Lebeau, J., Puchades, C., Wiseman, R.L., 2016. Reciprocal degradation of YME1L and OMA1 adapts mitochondrial proteolytic activity during stress. Cell Rep. 14, 2041–2049.

Rainbolt, T.K., Saunders, J.M., Wiseman, R.L., 2014. Stress-responsive regulation of mitochondria through the ER unfolded protein response. Trends Endocrinol. Metab. 25, 528–537.

Rainbolt, T.K., Saunders, J.M., Wiseman, R.L., 2015. YME1L degradation reduces mitochondrial proteolytic capacity during oxidative stress. EMBO Rep. 16, 97–106.

Rath, E., Berger, E., Messlik, A., Nunes, T., Liu, B., Kim, S.C., Hoogenraad, N., Sans, M., Sartor, R.B., Haller, D., 2012. Induction of dsRNA-activated protein kinase links mitochondrial unfolded protein response to the pathogenesis of intestinal inflammation. Gut 61, 1269–1278.

Richter, U., Lahtinen, T., Marttinen, P., Suomi, F., Battersby, B.J., 2015. Quality control of mitochondrial protein synthesis is required for membrane integrity and cell fitness. J. Cell Biol. 211, 373–389.

Rugarli, E.I., Langer, T., 2012. Mitochondrial quality control: a matter of life and death for neurons. EMBO J. 31, 1336–1349.

Saita, S., Nolte, H., Fiedler, K.U., Kashkar, H., Venne, A.S., Zahedi, R.P., Kruger, M., Langer, T., 2017. PARL mediates Smac proteolytic maturation in mitochondria to promote apoptosis. Nat. Cell Biol. 19, 318–328.

Sandoval, H., Thiagarajan, P., Dasgupta, S.K., Schumacher, A., Prchal, J.T., Chen, M., Wang, J., 2008. Essential role for Nix in autophagic maturation of erythroid cells. Nature 454, 232–235.

Sano, R., Reed, J.C., 2013. ER stress-induced cell death mechanisms. Biochim. Biophys. Acta 1833, 3460–3470.

Sauer, R.T., Baker, T.A., 2011. AAA+ proteases: ATP-fueled machines of protein destruction. Annu. Rev. Biochem. 80, 587–612.

Savel'ev, A.S., Novikova, L.A., Kovaleva, I.E., Luzikov, V.N., Neupert, W., Langer, T., 1998. ATP-dependent proteolysis in mitochondria. m-AAA protease and PIM1 protease exert overlapping substrate specificities and cooperate with the mtHsp70 system. J. Biol. Chem. 273, 20596–20602.

Schmidt, O., Pfanner, N., Meisinger, C., 2010. Mitochondrial protein import: from proteomics to functional mechanisms. Nat. Rev. Mol. Cell Biol. 11, 655–667.

Schulz, A.M., Haynes, C.M., 2015. UPR(mt)-mediated cytoprotection and organismal aging. Biochim. Biophys. Acta 1847, 1448–1456.

Seiferling, D., Szczepanowska, K., Becker, C., Senft, K., Hermans, S., Maiti, P., Konig, T., Kukat, A., Trifunovic, A., 2016. Loss of CLPP alleviates mitochondrial cardiomyopathy without affecting the mammalian UPRmt. EMBO Rep. 17, 953–964.

Sekine, S., Kanamaru, Y., Koike, M., Nishihara, A., Okada, M., Kinoshita, H., Kamiyama, M., Maruyama, J., Uchiyama, Y., Ishihara, N., et al., 2012. Rhomboid protease PARL mediates the mitochondrial membrane potential loss-induced cleavage of PGAM5. J. Biol. Chem. 287, 34635–34645.

Seo, J.H., Rivadeneira, D.B., Caino, M.C., Chae, Y.C., Speicher, D.W., Tang, H.Y., Vaira, V., Bosari, S., Palleschi, A., Rampini, P., et al., 2016. The mitochondrial unfoldase-peptidase complex ClpXP controls bioenergetics stress and metastasis. PLoS Biol. 14, e1002507.

Seong, Y.M., Choi, J.Y., Park, H.J., Kim, K.J., Ahn, S.G., Seong, G.H., Kim, I.K., Kang, S., Rhim, H., 2004. Autocatalytic processing of HtrA2/Omi is essential for induction of caspase-dependent cell death through antagonizing XIAP. J. Biol. Chem. 279, 37588–37596.

Sepuri, N.B.V., Angireddy, R., Srinivasan, S., Guha, M., Spear, J., Lu, B., Anandatheerthavarada, H.K., Suzuki, C.K., Avadhani, N.G., 2017. Mitochondrial LON protease-dependent degradation of cytochrome c oxidase subunits under hypoxia and myocardial ischemia. Biochim. Biophys. Acta 1858, 519–528.

Serero, A., Giglione, C., Sardini, A., Martinez-Sanz, J., Meinnel, T., 2003. An unusual peptide deformylase features in the human mitochondrial N-terminal methionine excision pathway. J. Biol. Chem. 278, 52953–52963.

Serizawa, A., Dando, P.M., Barrett, A.J., 1995. Characterization of a mitochondrial metallopeptidase reveals neurolysin as a homologue of thimet oligopeptidase. J. Biol. Chem. 270, 2092–2098.

Shi, G., Lee, J.R., Grimes, D.A., Racacho, L., Ye, D., Yang, H., Ross, O.A., Farrer, M., McQuibban, G.A., Bulman, D.E., 2011. Functional alteration of PARL contributes to mitochondrial dysregulation in Parkinson's disease. Hum. Mol. Genet. 20, 1966–1974.

Shpilka, T., Haynes, C.M., 2018. The mitochondrial UPR: mechanisms, physiological functions and implications in ageing. Nat. Rev. Mol. Cell Biol. 19, 109–120.

Shutt, T.E., McBride, H.M., 2013. Staying cool in difficult times: mitochondrial dynamics, quality control and the stress response. Biochim. Biophys. Acta 1833, 417–424.

Sinha, D., Srivastava, S., Krishna, L., D'Silva, P., 2014. Unraveling the intricate organization of mammalian mitochondrial presequence translocases: existence of multiple translocases for maintenance of mitochondrial function. Mol. Cell. Biol. 34, 1757–1775.

Song, Z., Chen, H., Fiket, M., Alexander, C., Chan, D.C., 2007. OPA1 processing controls mitochondrial fusion and is regulated by mRNA splicing, membrane potential, and Yme1L. J. Cell Biol. 178, 749–755.

Sovik, O., Njolstad, P.R., Jellum, E., Molven, A., 2008. Wolcott-Rallison syndrome with 3-hydroxydicarboxylic aciduria and lethal outcome. J. Inherit. Metab. Dis. 31 (Suppl. 2), S293–S297.

Spataro, N., Calafell, F., Cervera-Carles, L., Casals, F., Pagonabarraga, J., Pascual-Sedano, B., Campolongo, A., Kulisevsky, J., Lleo, A., Navarro, A., et al., 2015. Mendelian genes for Parkinson's disease contribute to the sporadic forms of the disease. Hum. Mol. Genet. 24, 2023–2034.

Spiller, M.P., Guo, L., Wang, Q., Tran, P., Lu, H., 2015. Mitochondrial Tim9 protects Tim10 from degradation by the protease Yme1. Biosci. Rep. 35.

Stahl, A., Moberg, P., Ytterberg, J., Panfilov, O., Brockenhuus Von Lowenhielm, H., Nilsson, F., Glaser, E., 2002. Isolation and identification of a novel mitochondrial metalloprotease (PreP) that degrades targeting presequences in plants. J. Biol. Chem. 277, 41931–41939.

Stiburek, L., Cesnekova, J., Kostkova, O., Fornuskova, D., Vinsova, K., Wenchich, L., Houstek, J., Zeman, J., 2012. YME1L controls the accumulation of respiratory chain subunits and is required for apoptotic resistance, cristae morphogenesis, and cell proliferation. Mol. Biol. Cell 23, 1010–1023.

Strauss, K.A., Jinks, R.N., Puffenberger, E.G., Venkatesh, S., Singh, K., Cheng, I., Mikita, N., Thilagavathi, J., Lee, J., Sarafianos, S., et al., 2015. CODAS syndrome is associated with mutations of LONP1, encoding mitochondrial AAA+ Lon protease. Am. J. Hum. Genet. 96, 121–135.

Strauss, K.M., Martins, L.M., Plun-Favreau, H., Marx, F.P., Kautzmann, S., Berg, D., Gasser, T., Wszolek, Z., Muller, T., Bornemann, A., et al., 2005. Loss of function mutations in the gene encoding Omi/HtrA2 in Parkinson's disease. Hum. Mol. Genet. 14, 2099–2111.

Suzuki, Y., Imai, Y., Nakayama, H., Takahashi, K., Takio, K., Takahashi, R., 2001. A serine protease, HtrA2, is released from the mitochondria and interacts with XIAP, inducing cell death. Mol. Cell 8, 613–621.

Tait, S.W., Green, D.R., 2013. Mitochondrial regulation of cell death. Cold Spring Harb. Perspect. Biol. 5.

Tanaka, A., Cleland, M.M., Xu, S., Narendra, D.P., Suen, D.F., Karbowski, M., Youle, R.J., 2010. Proteasome and p97 mediate mitophagy and degradation of mitofusins induced by Parkin. J. Cell Biol. 191, 1367–1380.

Teske, B.F., Fusakio, M.E., Zhou, D., Shan, J., McClintick, J.N., Kilberg, M.S., Wek, R.C., 2013. CHOP induces activating transcription factor 5 (ATF5) to trigger apoptosis in response to perturbations in protein homeostasis. Mol. Biol. Cell 24, 2477–2490.

Thomas, R.E., Andrews, L.A., Burman, J.L., Lin, W.Y., Pallanck, L.J., 2014. PINK1-Parkin pathway activity is regulated by degradation of PINK1 in the mitochondrial matrix. PLoS Genetics 10, e1004279.

Tian, Q., Li, T., Hou, W., Zheng, J., Schrum, L.W., Bonkovsky, H.L., 2011. Lon peptidase 1 (LONP1)-dependent breakdown of mitochondrial 5-aminolevulinic acid synthase protein by heme in human liver cells. J. Biol. Chem. 286, 26424–26430.

Tondera, D., Grandemange, S., Jourdain, A., Karbowski, M., Mattenberger, Y., Herzig, S., Da Cruz, S., Clerc, P., Raschke, I., Merkwirth, C., et al., 2009. SLP-2 is required for stress-induced mitochondrial hyperfusion. EMBO J. 28, 1589–1600.

Tsai, C.W., Wu, Y., Pao, P.C., Phillips, C.B., Williams, C., Miller, C., Ranaghan, M., Tsai, M.F., 2017. Proteolytic control of the mitochondrial calcium uniporter complex. Proc. Natl. Acad. Sci. U S A 114, 4388–4393.

van der Bliek, A.M., Shen, Q., Kawajiri, S., 2013. Mechanisms of mitochondrial fission and fusion. Cold Spring Harb. Perspect. Biol. 5.
Varanita, T., Soriano, M.E., Romanello, V., Zaglia, T., Quintana-Cabrera, R., Semenzato, M., Menabo, R., Costa, V., Civiletto, G., Pesce, P., et al., 2015. The OPA1-dependent mitochondrial cristae remodeling pathway controls atrophic, apoptotic, and ischemic tissue damage. Cell Metab. 21, 834–844.
Viader, A., Sasaki, Y., Kim, S., Strickland, A., Workman, C.S., Yang, K., Gross, R.W., Milbrandt, J., 2013. Aberrant Schwann cell lipid metabolism linked to mitochondrial deficits leads to axon degeneration and neuropathy. Neuron 77, 886–898.
Vogtle, F.N., Prinz, C., Kellermann, J., Lottspeich, F., Pfanner, N., Meisinger, C., 2011. Mitochondrial protein turnover: role of the precursor intermediate peptidase Oct1 in protein stabilization. Mol. Biol. Cell 22, 2135–2143.
Vogtle, F.N., Wortelkamp, S., Zahedi, R.P., Becker, D., Leidhold, C., Gevaert, K., Kellermann, J., Voos, W., Sickmann, A., Pfanner, N., et al., 2009. Global analysis of the mitochondrial N-proteome identifies a processing peptidase critical for protein stability. Cell 139, 428–439.
Wagner, I., Arlt, H., van Dyck, L., Langer, T., Neupert, W., 1994. Molecular chaperones cooperate with PIM1 protease in the degradation of misfolded proteins in mitochondria. EMBO J. 13, 5135–5145.
Wai, T., Langer, T., 2016. Mitochondrial dynamics and metabolic regulation. Trends Endocrinol. Metab. 27, 105–117.
Wai, T., Saita, S., Nolte, H., Muller, S., Konig, T., Richter-Dennerlein, R., Sprenger, H.G., Madrenas, J., Muhlmeister, M., Brandt, U., et al., 2016. The membrane scaffold SLP2 anchors a proteolytic hub in mitochondria containing PARL and the i-AAA protease YME1L. EMBO Rep. 17, 1844–1856.
Walker, J.E., Carroll, J., Altman, M.C., Fearnley, I.M., 2009. Chapter 6 mass spectrometric characterization of the thirteen subunits of bovine respiratory complexes that are encoded in mitochondrial DNA. Meth. Enzymol. 456, 111–131.
Westerlund, M., Behbahani, H., Gellhaar, S., Forsell, C., Belin, A.C., Anvret, A., Zettergren, A., Nissbrandt, H., Lind, C., Sydow, O., et al., 2011. Altered enzymatic activity and allele frequency of OMI/HTRA2 in Alzheimer's disease. FASEB J. 25, 1345–1352.
Wu, W., Tian, W., Hu, Z., Chen, G., Huang, L., Li, W., Zhang, X., Xue, P., Zhou, C., Liu, L., et al., 2014. ULK1 translocates to mitochondria and phosphorylates FUNDC1 to regulate mitophagy. EMBO Rep. 15, 566–575.
Wust, R., Maurer, B., Hauser, K., Woitalla, D., Sharma, M., Kruger, R., 2015. Mutation analyses and association studies to assess the role of the presenilin-associated rhomboid-like gene in Parkinson's disease. Neurobiol. Aging.
Xiao, X., Hu, Y., Quiros, P.M., Wei, Q., Lopez-Otin, C., Dong, Z., 2014. OMA1 mediates OPA1 proteolysis and mitochondrial fragmentation in experimental models of ischemic kidney injury. Am. J. Physiol. Renal. Physiol. 306, F1318–F1326.
Yamano, K., Youle, R.J., 2013. PINK1 is degraded through the N-end rule pathway. Autophagy 9, 1758–1769.
Yang, M., Jensen, R.E., Yaffe, M.P., Oppliger, W., Schatz, G., 1988. Import of proteins into yeast mitochondria: the purified matrix processing protease contains two subunits which are encoded by the nuclear MAS1 and MAS2 genes. EMBO J. 7, 3857–3862.
Yoneda, T., Benedetti, C., Urano, F., Clark, S.G., Harding, H.P., Ron, D., 2004. Compartment-specific perturbation of protein handling activates genes encoding mitochondrial chaperones. J. Cell Sci. 117, 4055–4066.
Youle, R.J., Narendra, D.P., 2011. Mechanisms of mitophagy. Nat. Rev. Mol. Cell Biol. 12, 9–14.

Young, L., Leonhard, K., Tatsuta, T., Trowsdale, J., Langer, T., 2001. Role of the ABC transporter Mdl1 in peptide export from mitochondria. Science 291, 2135–2138.

Young, S.K., Wek, R.C., 2016. Upstream open reading frames differentially regulate gene-specific translation in the integrated stress response. J. Biol. Chem. 291, 16927–16935.

Yue, W., Chen, Z., Liu, H., Yan, C., Chen, M., Feng, D., Yan, C., Wu, H., Du, L., Wang, Y., et al., 2014. A small natural molecule promotes mitochondrial fusion through inhibition of the deubiquitinase USP30. Cell Res. 24, 482–496.

Zhang, K., Li, H., Song, Z., 2014. Membrane depolarization activates the mitochondrial protease OMA1 by stimulating self-cleavage. EMBO Rep. 15, 576–585.

Zhao, Q., Wang, J., Levichkin, I.V., Stasinopoulos, S., Ryan, M.T., Hoogenraad, N.J., 2002. A mitochondrial specific stress response in mammalian cells. EMBO J. 21, 4411–4419.

Zhou, D., Palam, L.R., Jiang, L., Narasimhan, J., Staschke, K.A., Wek, R.C., 2008. Phosphorylation of eIF2 directs ATF5 translational control in response to diverse stress conditions. J. Biol. Chem. 283, 7064–7073.

Ziviani, E., Tao, R.N., Whitworth, A.J., 2010. Drosophila parkin requires PINK1 for mitochondrial translocation and ubiquitinates mitofusin. Proc. Natl. Acad. Sci. U S A 107, 5018–5023.

CHAPTER FOUR

Mitochondrial Dynamics: Shaping Metabolic Adaptation

Carles Cantó*,§

*Nestle Institute of Health Sciences NIHS, Lausanne, Switzerland
§School of Life Sciences, Ecole Polytechnique Fédérale de Lausanne (EPFL), Lausanne, Switzerland
Email: carlos.cantoalvarez@rd.nestle.com

Contents

Abstract

Despite their classic bean-shaped depiction, mitochondria have very different aspects in each cell type. From long filamentous structures to punctuated small round organelles. These shapes can dynamically change in response to nutrients and in situations of metabolic disease. However, why do mitochondria adapt different shapes and how is this determined? In this review, we will aim to understand different visions on how

International Review of Cell and Molecular Biology, Volume 340
ISSN 1937-6448
https://doi.org/10.1016/bs.ircmb.2018.05.004

metabolic cues influence mitochondrial shape and vice-versa. This response can be dramatically different between tissues and cells, as illustrated by a large array of genetically engineered mouse models reported to date. We will use these models to understand the role of different mitochondrial dynamics—related proteins and processes.

1. INTRODUCTION

More than a century ago, Dr. Carl Benda, a German microbiologist, was one of the first people to experience the fascination of visualizing mitochondria under the microscope. In one of his studies on spermatogenesis, he described: "They lie within of a portion of the protoplasmic threads, sometimes aggregated to particular bodies, and are identical to at least a portion of the already known cell microsomes, but differ from Altmann's and Ehrlich's granulations. I would like to suggest tentatively reserving for them a special status as 'mitochondria', which I will justify in further work" (Benda, 1898). The term mitochondria was coined from the Greek "mitos" (thread) and chondros (grain), given the different morphological features presented by these organelles.

The observation that mitochondria were morphologically diverse was overshadowed by the discoveries of mitochondrial bioenergetic functions. Nonetheless, experiments in yeast during the late 1960s revealed that the genomes from different mitochondria could complement and recombine (Thomas and Wilkie, 1968), providing initial evidence that mitochondria might not act as individual organelles. A decade later, detailed microscopic studies in yeast demonstrated how astonishing configurational changes occurred during meiosis, where individual small mitochondrial organelles could become giant elongate mitochondria via fusion mechanisms (Miyakawa et al., 1984). This ability of mitochondria to fuse and divide was later confirmed in mammalian cells and tissues (Wakabayashi and Green, 1977; Kimberg and Loeb, 1972). Thus the question was how, when, and why do mitochondria undergo fusion or fission processes?

Initial findings indicated a role for mitochondrial dynamics in the mitochondrial life cycle. The encounters between different mitochondria allow them to exchange their components and to set aside through fission processes the damaged parts and enzymes in a selective manner (Liesa and Shirihai, 2013). This reorganization allows the selective maintenance of a functional pool of mitochondria. Hence, it is not surprising that the ablation of fission-related proteins leads to the formation of abnormally long and dysfunctional mitochondria, as they fail to divide mitochondria and remove the damaged parts.

During the past decade, however, it has become clear that mitochondrial dynamics serve a purpose in not only quality control but also the adaptation to cellular bioenergetics needs (Liesa and Shirihai, 2013). Mitochondrial architecture can acutely and chronically change in response to nutrient availability. Most importantly, changes in mitochondrial connectivity impact the development of multiple metabolic complications. Here we will first explore the different components influencing mitochondrial dynamics. Then we will wonder how nutrient availability influences mitochondrial dynamics and vice versa. Finally, we will overview what different animal models have taught us on how mitochondrial dynamic proteins influence dietary and age-related diseases.

2. THE PERFECT ELEMENTS: THE MITOCHONDRIAL FUSION AND FISSION MACHINERY

2.1 Mitochondrial Fusion

The fusion of two mitochondria can be divided into two separable events. First, the outer mitochondrial membrane (OMM) fusion takes place. Then a similar process between the inner mitochondrial membrane (IMM) takes place (Liesa et al., 2009). While generally coordinated, these two processes can be differentiated and segregated, as different families of GTPase proteins account for them (Malka et al., 2005). The mitofusin enzymes control OMM fusion, whereas the dynamin-related GTPase optic atrophy 1 (OPA1) mediates IMM fusion. The activity of the mitofusin enzymes can be regulated by external stimuli, as mitofusins display large cytoplasmic domains. In contrast, IMM through OPA1 seems fine-tuned to the activity of the oxidative phosphorylation (OXPHOS) flux (Mishra et al., 2014). To date, there is still a significant gap in our knowledge on the elements that initiate fusion/fission processes at a particular region of the mitochondrial membrane. Similarly, we are still far from grasping a full picture of the gatekeeping elements for the coordination of OMM and IMM fusion.

2.1.1 Mitofusins

It was not until 1997 that Hales and Fuller identified that the *Drosophila* fuzzy onions (*fzo*) gene encoded a protein-mediating mitochondrial fusion (Hales and Fuller, 1997). This protein was predicted to be a mitochondrial GTPase, and mutation of the key residues for the GTPase activity abolished the ability of the protein to fuse mitochondria (Hales and Fuller, 1997). A few years later, two human homologs were identified and named mitofusin 1 and mitofusin 2 (*Mfn1* and *Mfn2*) (Santel and Fuller, 2001). The high similarity between *Mfn1* and *Mfn2* (around 63% in humans (Santel and Fuller, 2001)) and the fact that many lower eukaryotes, including yeast and worms, display only one mitofusin homolog suggest that *Mfn1* and *Mfn2* appeared as a duplication of a single gene during the expansion of chordates. Nevertheless, their functions have diverged during evolution and their role in mitochondrial fusion is only partially redundant.

The human Mfn1 and Mfn2 proteins expand for 741 and 751 residues, respectively. The classical view of Mfn proteins indicate that, while anchored on the OMM, they expose both N- and C-terminal parts to the cytosol. The cytosolic N-terminal part of the protein (~350 amino acids) contains the five domains that coordinately confer the protein GTPase activity of these proteins (Rojo et al., 2002; Santel et al., 2003). This is closely followed by a coiled-coil domain that precedes the transmembrane domain that anchors the protein on the OMM. This transmembrane domain loops to render a second cytoplasmic coiled-coil domain at the very C terminal of the protein. Of note, this C-terminal region determines the mitochondrial localization of the mitofusin proteins (Santel et al., 2003; Rojo et al., 2002). This view has been challenged by experiments demonstrating that the C-terminal part of Mfn enzymes might reside on the mitochondrial intermembrane space (Mattie et al., 2018).

The coiled-coil domains are critical for the tethering of two mitochondria in order to trigger the fusion process (Chen et al., 2003; Koshiba et al., 2004). Two mitofusins, one from each mitochondrion, will interact with each other in a homotypic (e.g., Mfn1-Mfn1) or heterotypic (i.e., Mfn1-Mfn2) fashion (Koshiba et al., 2004). In this sense, Mfn1 displays a more robust GTPase activity compared with Mfn2 (Ishihara et al., 2004). Both mitofusins can be found in most tissues, albeit their abundance in different tissues is not always proportional. For example, while Mfn2 finds its highest levels in brown adipose tissue (BAT) and brain, Mfn1 levels are relatively low in these tissues compared with heart or skeletal muscle (Boutant et al., 2017).

Mfn1 is highly expressed in most mitochondria-rich tissues such as the heart, liver, and muscles (Boutant et al., 2017). Testifying for the key role of Mfn1 in the mitochondrial fusion process, all models of genetic ablation of Mfn1 display a largely fragmented mitochondrial network

(Chen et al., 2003; Kulkarni et al., 2016; Papanicolaou et al., 2012). This indicates that Mfn2, whose expression is often increased in Mfn1-null cells and tissues (Kulkarni et al., 2016), cannot compensate the defective fusion rates triggered by the lack of Mfn1. However, this might only be true for functions related to mitochondrial fusion. For example, the single deletion of *Mfn1* or *Mfn2* does not dramatically impair the mitochondrial respiratory capacity or mitochondrial DNA (mtDNA) content, whereas the double deletion does (Chen et al., 2003). Mfn1 expression can be modulated in response to multiple stimuli that, in general, enhance mitochondrial biogenesis, such as exercise in skeletal muscles (Cartoni et al., 2005; Ding et al., 2010). However, it has been found that the Mfn1 GTPase activity can also be modulated at the posttranscriptional level. For example, Mfn1 is an acetylated protein whose deacetylation by HDAC6 potentiates its GTPase activity (Lee et al., 2014). Similarly the Scorrano laboratory has demonstrated how Mfn1 fusion activity can also be regulated by phosphorylation (Pyakurel et al., 2015). In particular, the extracellular-regulated protein kinase was identified to phosphorylate Mfn1 at T562 immediately before the transmembrane domain. Phosphorylation of T562 impairs oligomerization of Mfn1, hence impeding mitochondrial tethering and fusion (Pyakurel et al., 2015).

Despite the multiple structural similarities to Mfn1, Mfn2 seems to play very different biological roles. *Mfn2* null mouse embryonic fibroblasts (MEFs) show only mild defects on mitochondrial fusion (Chen et al., 2003). This can be, in part, due to the fact that the GTP hydrolysis activity of Mfn2 is much lower than that of Mfn1. In exchange, Mfn2 seems critical for the tethering and functional relationship between the mitochondria and the endoplasmic reticulum (ER) (De Brito and Scorrano, 2008). Mfn2 has been shown to be present in the ER mitochondrial associated membranes, influencing lipid and Ca^{2+} transfer (De Brito and Scorrano, 2008). While the exact role of Mfn2 in the ER—mitochondria interaction is still a matter of debate (Naon et al., 2016; Filadi et al., 2015; Cosson et al., 2012), Mfn2 deletion has been consistently associated with an increase in ER stress in most, if not all, cells and tissues tested to date (De Brito and Scorrano, 2008; Munoz et al., 2013; Schneeberger et al., 2013; Sebastian et al., 2012; Boutant et al., 2017). In contrast, Mfn1 deficiency does not lead to ER stress (Kulkarni et al., 2016). Our laboratory has also identified Mfn2 as a key facilitator of the interaction between mitochondria and the lipid droplet (LD) in the BAT (Boutant et al., 2017). These Mfn2 roles beyond mitochondrial fusion might explain why Mfn2, but not Mfn1, has been linked to multiple diseases (for review see (Liesa et al., 2009)). Mutations in the *Mfn2* gene are causal to the type 2A form of the Charcot-Marie-Tooth (CMT2A) disease, one of the most common hereditary neuropathies (Zuchner et al., 2004). CMT2A is clinically characterized by weakness, muscle atrophy, and sensory loss due to chronic axonal degeneration and reduced motor nerve conduction speed (Zuchner et al., 2004). In addition, alterations in the *Mfn2* gene have also been related to atherosclerosis (Guo et al., 2007), hypertension (Wang et al., 2011b; Ryan et al., 2013), and diabetes (Bach et al., 2003, 2005).

2.1.2 Optic Atrophy 1

OPA1 is a 93-kDa protein that can be found either anchored in the IMM or as cleaved soluble forms in the intermembrane space (Alexander et al., 2000; Akepati et al., 2008; Olichon et al., 2002; Delettre et al., 2000). As mitofusins, OPA1 is a GTPase enzyme and contains coiled-coil domains that allow the formation of homotypic interactions with OPA1 molecules from adjacent mitochondria during the fusion process (Akepati et al., 2008; Alexander et al., 2000; Delettre et al., 2000). Mutations in *Opa1* lead to autosomal dominant optic nerve atrophy (Alexander et al., 2000; Delettre et al., 2000). This disease is characterized by an early onset of a moderate to severe loss of vision. Up to 117 mutations of the *Opa1* gene have been related to this disease, causing premature truncation of the protein or impairing its GTPase activity (Delettre et al., 2000; Alexander et al., 2000).

The biology of OPA1 is highly regulated at the posttranscriptional and posttranslational stages. In humans, there are at least eight messenger RNA (mRNA) isoforms formed by alternative splicing (Delettre et al., 2001; Olichon et al., 2007). The roles of these different forms are still a matter of study. At the posttranslational level, OPA1 can also be cleaved into different protein products (see Macvicar and Langer, 2016 for review). In most eukaryotic cells, OPA1 can be detected in regular Western blots as a five-band pattern (Macvicar and Langer, 2016). The membrane-bound long (L)-OPA1 form contains two protease cleavage sites, S1 and S2, that allow the generation of short (S)-OPA1 forms (Ishihara et al., 2006). These two cleavage sites are the target of two proteases residing in the IMM: the zinc metalloprotease overlapping with m-AAA protease (OMA1), which acts at S1 (Kaser et al., 2003), and the ATP-dependent protease yeast mtDNA escape 1-like (YME1L), which acts at S2 (Song et al., 2007). The double genetic deletion of *Yme1l* and *Oma1* completely blunts L-OPA1 proteolytic processing (Anand et al., 2014; Wai et al., 2015). Deletion of *Oma1* abolishes OPA1 cleavage at S1, resulting in an elongated mitochondrial network (Anand et al., 2014; Quiros et al., 2012). In contrast, the loss of *Yme1l* impairs OPA1 cleavage at S2 and leads to mitochondrial fragmentation (Griparic et al., 2007; Song et al., 2007; Stiburek et al., 2012). Interestingly, experiments using photoactivable probes demonstrated that YME1L-deficient mitochondria do not show any defect in their fusion capacity, but their fission rates were dramatically increased (Anand et al., 2014; Ruan et al., 2013). The dramatic accumulation of short OPA1 forms and mitochondrial fragmentation in MEFs lacking *Yme1l* could be prevented by the deletion of *Oma1* (Anand et al., 2014). This supports a model in which long OPA1 forms sustain mitochondrial fusion, whereas cleaved short OPA1 forms lead to mitochondrial fission. In agreement the overexpression of OPA1 forms lacking an S1 cleavage site leads to mitochondrial hyperfusion (Duvezin-Caubet et al., 2006). Oppositely, the overexpression of S-OPA1 leads to mitochondrial fragmentation (Anand et al., 2014). Nevertheless, the contribution of the different OPA1 forms to mitochondrial dynamics is still unclear.

OMA1-mediated processing of OPA1 at S1 is triggered by numerous conditions that render mitochondria dysfunctional, such as heat shock, the dissipation of $\Delta\psi m$ (mitochondrial membrane potential [MMP]) with the uncoupler carbonyl cyanide *m*-chlorophenyl hydrazine or upon inhibition of the F_1F_0-ATP synthase with oligomycin (Ehses et al., 2009; Head et al., 2009). The precise mechanism by which such a varied constellation of stimuli impact on OMA1 activity is currently an active field of research. In contrast, it is not clear to what extent YME1L is a regulated enzyme (Griparic et al., 2007; Song et al., 2007). Interestingly, a paper from the Chan laboratory indicates that the activity of YME1L, but not that of OMA1, can be triggered by increased OXPHOS activity, leading to OPA1 processing at S2 and increased mitochondrial fusion (Mishra et al., 2014). Oppositely, increased glycolytic metabolism leads to mitochondrial fragmentation (Mishra et al., 2014). The mechanistic insights on how YME1L is regulated by OXPHOS activity are not well understood. Nevertheless, YME1L has been reported to interact with subunits of complex V (Stiburek et al., 2012), raising the possibility that it may directly sense ATP generation by the respiratory chain.

2.2 Mitochondrial Fission

Mitochondrial fission plays an important role in mitochondrial proliferation, mitochondrial distribution during cell division, and in the removal of damaged mitochondria through mitophagy (Otera et al., 2013). Mitochondrial fission is driven by the constricting action of the GTPase enzyme dynamin-related protein 1 (Drp1) (Otera et al., 2013). Drp1 lacks membrane-binding domains and therefore requires docking proteins at the OMM (Chan, 2012). Several proteins participating in the mitochondrial recruitment of Drp1 have been identified, including the mitochondrial fission 1 (Fis1), mitochondrial fission factor (MFF), mitochondrial dynamic protein of 49 kDa (MiD49), and mitochondrial dynamic protein of 51 kDa (MiD51, also known as mitochondrial elongation factor 1).

2.2.1 Dynamin-related protein 1

Drp1, encoded by the *Dnm1l* gene, is a ubiquitously expressed protein that was initially related to the constriction of nascent endocytic vesicles (Praefcke and Mcmahon, 2004). However, mitochondrial architecture was dramatically altered in cell models with impaired Drp1 activity (Pitts et al., 1999; Smirnova et al., 1998). In particular, mitochondria were perinuclear and displayed abnormally elongated shapes, mainly because Drp1 is one of the key active players in mitochondrial fission. Drp1 monomers can assemble in the OMM, where they oligomerize and form constricting filaments around the mitochondria (Bui and Shaw, 2013).

Drp1 contains most of its domains accounting for the GTPase activity of the protein on the first half of its sequence (Smirnova et al., 1998). There is a large middle domain and an assembly domain at the C-terminal part. These last two sections will participate in the regulation of the GTPase activity and oligomerization of Drp1 protein (Otera et al., 2013). Drp1 does not contain transmembrane domains and it is predominantly found in the cytosol (Smirnova et al., 1998). However, part of the Drp1 pool is bound to mitochondria, where it is recruited via adaptor proteins.

The GTPase activity and localization of Drp1 is influenced by posttranslational modifications (for review see Otera et al. (2013)). Two critical phosphorylation sites have been widely described. A decade ago the Blackstone and Strack laboratories identified a phosphorylation site that is conserved in all metazoan Drp1 orthologs (Chang and Blackstone, 2007; Cribbs and Strack, 2007). Mouse Drp1 S643 (S637 in human Drp1) is phosphorylated by the cyclic AMP (cAMP)-dependent protein kinase (PKA) (Chang and Blackstone, 2007; Cribbs and Strack, 2007), while calcineurin acts as its main phosphatase (Cereghetti et al., 2008; Cribbs and Strack, 2007). Hence, this phosphorylation event can link Drp1 function with multiple hormonal inputs and calcium levels in the cell. Drp1 phosphorylation at S643 has been shown to decrease Drp1 oligomerization and prompt the sequestration of the protein to the cytosol (Cereghetti et al., 2008). In addition to S643, Taguchi et al. (2007) identified S622 as a phosphorylation site that critically affected mitochondrial fission rates. This residue is phosphorylated by cyclin-dependent kinase 1/cyclin B, a key mitotic kinase. Contrary to S643, phosphorylation at S622 (S616 in humans) induces Drp1 translocation to the mitochondria, triggering mitochondrial fragmentation (Taguchi et al., 2007). This has been proposed to be an important mechanism ensuring an equilibrated distribution of mitochondria between the daughter cells during mitosis.

2.2.2 Dynamin-Related Protein 1 Adaptor Proteins

The first mammalian adaptor of Drp1 described was Fis1. It was identified as the homolog of the sole Drp1 adaptor in yeasts, Fis1p (James et al., 2003; Stojanovski et al., 2004; Yoon et al., 2003). Fis1 is a small 17-kDa protein that resides on the OMM. Interestingly, its insertion in the OMM seems to be critically influenced by lipid composition (Kemper et al., 2008). Interestingly, Drp1 can also directly interact with different mitochondrial phospholipids (Adachi et al., 2016). For example, Drp1 interacts with phosphatidic acid in the OMM, and this protein—lipid interaction suppresses the activation of GTP hydrolysis that is required for Drp1-driven mitochondrial constriction (Adachi et al., 2016). Oppositely, cardiolipin binding is required for Drp1-mediated fission (Bustillo-Zabalbeitia et al., 2014). These observations indicate that the lipid composition of the OMM might be key to determine the susceptibility to fission events.

Fis1 was rapidly considered the mammalian limiting factor for mitochondrial fission. However, Fis1 requires additional molecular adaptors, Mdv1 or Caf4, in order to interact with Drp1 in yeasts (Zhao et al., 2013). Mammals have no orthologs for these additional adaptors. Evidences thereafter have provided mixed support for the role of Fis1 in mitochondrial fission, as in some cell models reduced Fis1 activity does not impair Drp1 mitochondrial recruitment and fission (Otera et al., 2010). Instead, the van der Bliek and Mihara laboratories provided strong evidence for the MFF as a key protein to recruit Drp1 at the OMM (Gandre-Babbe and Van Der Bliek, 2008; Otera et al., 2010). Knockdown of MFF impairs Drp1 recruitment to the mitochondria, prompting mitochondrial elongation (Otera et al., 2010; Gandre-Babbe and Van Der Bliek, 2008). MiD49 and MiD51 have also been identified as possible receptors for Drp1 on the mitochondrial surface (Palmer et al., 2011). However, there is conflicting evidence concerning their mechanism of action because, paradoxically, both the knockdown and overexpression of either MiD49 or MiD51 lead to mitochondrial elongation (Palmer et al., 2011; Zhao et al., 2011).

In order to bring light into the role of this constellation of proteins mediating Drp1 recruitment to the OMM, the Chan laboratory performed systematic analyses using models for the genetic deletion or knockdown of these proteins (Loson et al., 2013). *Fis1*-null MEFs suggest a minor role for Fis1 in Drp1 recruitment and mitochondrial fusion/fission rates (Loson et al., 2013). In contrast, MFF deletion has a major impact on mitochondrial dynamics, leading to abnormally elongated mitochondria, in line with a key role in the recruitment of Drp1 (Loson et al., 2013). Additionally, these proteins might participate not only in the docking of Drp1 but also in its oligomerization at the OMM (Loson et al., 2013). Based on short interfering RNA studies, Mid49 and MiD51 also proved to be implicated in the recruitment of Drp1 (Loson et al., 2013). The requirement for these different adaptors in differentiated tissues or in response to different stimuli has not yet been investigated in depth. In addition, it has to be noted that the actions of Drp1, MFF, and Fis1 are not restricted to the mitochondria. Indeed, these players are also mediators, for example, of peroxisome fission (Schrader et al., 2016). This is critical when interpreting the outcomes of gain-of-function or loss-of-function strategies for mitochondrial fission—related proteins.

3. PROTEINS PASSING BY: MITOCHONDRIAL DYNAMICS AND METABOLIC CONTROL

3.1 The Impact of Mitochondrial Dynamics on Mitochondrial Function

3.1.1 Mitochondrial Fusion and Respiratory Capacity

One of the initial hypothesis around the existence of long mitochondrial filaments is that they could transmit the MMP to better coordinate the mitochondrial output across the cell (Amchenkova et al., 1988; Skulachev, 2001). This could be particularly important for large cells, such as muscle fibers and neurons. Experiments from the Balaban Lab demonstrate the existence of a mitochondrial reticulum providing a conductive pathway for energy distribution in mouse skeletal muscle cells (Glancy et al., 2015). Nevertheless, whether this model applies to all cells and whether it relies solely in the fusion of mitochondria are not year clear. Indeed, experiments in COS-7 and INS-1 cells using a mitochondria-targeted photoactivable green fluorescent protein suggest that mitochondrial morphology does not necessarily dictate mitochondrial matrix continuity and that morphological and functional networks can be disentangled (Twig et al., 2006). Mitochondrial functional networks, at the level of MMP distribution, were often altered in the absence of changes in mitochondrial morphology (Twig et al., 2006). Therefore the mitochondrial functional reticulum probably has several layers of complexity, involving not only the mitochondrial fusion/fission machinery but also other elements within mitochondria and in the mitochondrial environment (e.g., ER connectivity).

Early attempts to evaluate the impact of mitochondrial dynamics on mitochondrial function were based on classical gain-of-function or loss-of-function approaches through overexpression or genetic knockdown/ablation, respectively. For more than a decade, these experiments were largely based on cell models, but this has now been extended to mouse tissues derived from conditional knockout (KO) models. In general, the large overexpression of mitofusins leads to mitochondrial perinuclear aggregation and bioenergetic impairment (Pich et al., 2005; Rojo et al., 2002; Santel and Fuller, 2001; Huang et al., 2007; Santel et al., 2003). In MEF cells the single deletion of *Mfn1* does not lead to reductions in respiratory capacity (Chen et al., 2005). Instead, *Mfn1* KO MEF cells display increased respiratory capacity and enhanced respiratory coupling, despite their limited mitochondrial fusion capacity (Kulkarni et al., 2016; Papanicolaou et al., 2012). Importantly, the simple reintroduction of *Mfn1* in *Mfn1*$^{-/-}$ MEFs is enough to normalize respiratory profiles (Kulkarni et al., 2016). This increased respiratory capacity can also be seen in primary hepatocytes, liver homogenates, and isolated mitochondria from *Mfn1*-deficient livers and is tightly related to an increased abundance of mitochondria with higher Complex I activity (Kulkarni et al., 2016).

In contrast to Mfn1, the reduction of Mfn2 levels in differentiated tissues has generally led to decreased respiratory capacity (Sebastian et al., 2012, 2016; Schneeberger et al., 2013; Debattisti et al., 2014; Boutant et al., 2017). However, the impact of *Mfn2* deletion on the overall respiratory capacity can dramatically change between cell types. *Mfn2* KO MEFs display no major reductions in respiratory capacity when cultured in regular high-glucose media (Chen et al., 2005; Boutant et al., 2017; Munoz et al., 2013). Nevertheless, when cells are grown in galactose, which forces oxidative metabolism for ATP production, respiratory defects become more prominent (Mourier et al., 2015). The defective mitochondrial respiration in

Mfn2 KO MEFs was associated with a deficiency of coenzyme Q9 and Q10 (Mourier et al., 2015). Accordingly, *Mfn2*-deficient cardiomyocytes are characterized by defects in terpenoid biosynthetic pathways, severe coenzyme Q deficiency, and impaired respiratory chain capacity (Mourier et al., 2015). The reduction of *Mfn2* levels via antisense oligonucleotides or microRNA in myotubes is also enough to decrease MMP and impair respiration (Pich et al., 2005). In mouse *tibialis anterior* muscle devoid of *Mfn2*, state 3 and state 4 respiration from Complex I were not significantly affected (Sebastian et al., 2012). Nevertheless, they showed an impaired respiratory control ratio and increased reactive oxygen species (ROS) production (Sebastian et al., 2012), which were accentuated during mice aging (Sebastian et al., 2016). The detrimental effects of *Mfn2* deletion on mitochondrial respiratory capacity were more pronounced in liver cells, with prominent impairments in the activity of Complexes I + III (Sebastian et al., 2012). Dramatic defects in Complex I activity were also at the root of the mitochondrial respiratory deficits in neurons or BAT lacking the *Mfn2* gene (Schneeberger et al., 2013; Boutant et al., 2017). Interestingly, Mfn2 has been reported to directly interact with Complex I in cardiac muscles (Segales et al., 2013), which could potentially explain why Mfn2 deficiency leads to defective Complex I function in most, if not all, tissues examined to date. However, a report shows the impact of Mfn2 in the BAT respiratory function might be gender specific (Mahdaviani et al., 2017). This observation suggests that the impact of Mfn2 on respiratory function is unlikely to simply rely on a direct structural relationship with the electron transport chain.

The large divergence in the impact of *Mfn2* deficiency on mitochondrial respiration might be related to the basal mitochondrial respiratory rates and turnover in each tissue. In this sense, Mfn2 has been reported as a key component in the mitophagy process. When the mitochondrial phosphatase and tensin homolog (PTEN)-induced putative kinase protein 1 (PINK1) is activated upon mitochondrial damage, it phosphorylates Mfn2 at T111 and S442 (Chen and Dorn, 2013). This phosphorylation acts as a docking element for the recruitment of the cytosolic ubiquitin ligase Parkin, promoting Parkin-mediated ubiquitination of the mitochondria (Chen and Dorn, 2013). In agreement the ablation of *Mfn2* in cardiac myocytes blunted mitophagy after trifluoromethoxy carbonylcyanide phenylhydrazone (FCCP)-induced mitochondrial damage (Chen and Dorn, 2013). This could explain the accumulation of defective mitochondria in *Mfn2*-deficient mouse tissues. Whether the effects of *Mfn2* deletion in different tissues is solely linked to compromised mitophagy is, however, unclear.

Interestingly, the simultaneous deletion of both mitofusin enzymes dramatically compromises respiratory capacity (Chen et al., 2005). This has been associated to the fact that residual fusion events still occur in the single mitofusin KO cells, while no fusion is observed in the double KOs. Later, it was also characterized that the complete loss of mitofusin activity could lead to severe mtDNA depletion, at least in skeletal muscles where mitofusin deficiency leads to a rapid accumulation of point mutations and deletion in mtDNA (Chen et al., 2010).

Alterations in OPA1 also lead to dramatic effects on mitochondrial metabolism. MEFs lacking OPA1 manifest a major loss of basal and 2,4-dinitrophenol-stimulated O_2 consumption (Chen et al., 2005). Experiments in HeLa cells and MEFs demonstrated that OPA1 deficiency leads to the dissipation of the MMP and perturbations in the cristae structures (Olichon et al., 2003; Frezza et al., 2006). Hence, in addition to its role in mitochondrial fusion, OPA1 is also required for the proper arrangement of the IMM and cristae integrity. Cristae morphology determines the assembly and stability of respiratory complexes and supercomplexes, and therefore optimal respiratory function (Cogliati et al., 2013). Not surprisingly, OPA1-deficient mice die early during the embryonic development (Davies et al., 2007; Alavi et al., 2007). In this sense, it is important to note that OPA1 requires Mfn1, and vice versa, to consolidate fusion events (Cipolat et al., 2004). The fact that Mfn1 KO cells display enhanced respiration illustrates that fusion and respiratory regulation can be mechanistically uncoupled. Interestingly, studies have tried to evaluate the impact of *Opa1* overexpression in mice. Uncontrolled *Opa1* overexpression is toxic (Cipolat et al., 2004), complicating the generation of suitable in vivo models. To circumvent these difficulties, the Scorrano laboratory has reported a knockin mouse model where *Opa1*, under the control of the ubiquitous human β-actin promoter, was integrated into a permissive X chromosome region without altering the endogenous gene expression (Cogliati et al., 2013). This led to a mild OPA1 overexpression that is compatible with life. Controlled OPA1 overexpression triggered a slight mitochondrial elongation and increased respiratory capacity (Varanita et al., 2015). In addition, OPA1 could alleviate many of the dramatic symptoms of two independent mouse models of mitochondrial myopathy (Civiletto et al., 2015). Strikingly, at least in the Sv129 background, OPA1 overexpression led to a higher incidence of spontaneous cancers, thus shortening life span (Varanita et al., 2015). As discussed by the authors, this increase in cancer prevalence might contribute to explain why animals do not naturally display constitutively high OPA1 levels.

3.1.2 Mitochondrial Fission and Respiratory Capacity

Mitochondrial fission is required to ensure the proper removal of damaged mitochondrial components. Therefore, it is not surprising that defective mitochondrial fission leads to even worse effects than defective mitochondrial fusion. In most, if not all, cell lines tested to date, the knockdown of Drp1 leads to different degrees of mitochondrial respiratory function impairments (Benard et al., 2007; Twig et al., 2008). In line with the embryonic lethality upon deletion of mitochondrial fusion proteins, *Dnm1l*-null embryos die at E11.5 and are smaller, suggesting that Drp1 participates during early developmental stages of cell growth, proliferation, and differentiation (Ishihara et al., 2009; Wakabayashi et al., 2009). Given the high expression of Drp1 in the brain, one of the first conditional *Dnm1l* KOs generated was that where *Dnm1l* was deleted in the cerebellum using the Cre recombinase driven by the En1 promoter (Wakabayashi et al., 2009). However, these mice died within 24 h of birth, exhibiting impaired cerebellar development (Wakabayashi et al., 2009). Similarly, when *Dnm1l* was deleted in a broad region of the brain using the Nestin-Cre mouse model, mice died shortly after birth as a result of brain hypoplasia with apoptosis (Ishihara et al., 2009). This was in line with an abnormal presence of aggregated mitochondria, which probably failed to distribute properly during cell division processes (Ishihara et al., 2009). A liver-specific KO (LKO) for *Dnm1l* has been developed (Wang et al., 2015). The deletion of *Dnm1l* specifically in the liver (Drp1-LKO) does not lead to any major phenotype when mice are fed a regular diet (Wang et al., 2015). However, upon high-fat feeding, electron microscopic analysis revealed that the mitochondria of Drp1-LKO mice were markedly swollen, but not elongated (Wang et al., 2015). In addition, the ER in hepatocytes from Drp1-LKO mice were irregular, with disrupted contacts with the mitochondria (Wang et al., 2015). Despite being larger, the total number of mitochondria in hepatocytes from Drp1-LKO mice was significantly reduced, suggesting exacerbated fusion (Wang et al., 2015). However, other than these morphometric aspects, no major differences were observed at the level of the expression/activity of mitochondrial respiratory chain complexes or intracellular ATP levels.

Notably, a synthetic compound called Mdivi-1 was identified in a screen as an inhibitor of Drp1 GTPase activity (Cassidy-Stone et al., 2008). This compound has been widely used to blunt Drp1 action. However, the specificity of Mdivi-1 has been challenged, as it might promote its actions primarily by acting as a mild Complex I inhibitor (Bordt et al., 2017). Indeed, it is often observed that mitochondria fuse in situations of mild chronic energy deficit. In such cases, AMPK initially might drive mitochondrial fission (Toyama et al., 2016) while transcriptionally activating programs related to mitochondrial biogenesis and fusion (Cantó et al., 2009). A small chronic increase in basal AMPK activation or nutrient scarcity has been related to mitochondrial hyperfusion (Cantó et al., 2009; Canto et al., 2010; Garcia-Roves et al., 2008). Based on that, and while looking for alternative inhibitors, the use of Mdivi-1 is highly discouraged or at least requires strict controls (e.g., genetic controls) in order to certify the origin of its impact in mitochondrial biology.

3.2 Aligning Mitochondrial Architecture and Metabolic Needs

Almost 15 years ago, the Zorzano laboratory reported one of the pioneering observations linking mitochondrial dynamics with cellular metabolism. Using a differential mRNA screen, the Mfn2 gene turned out to be one of the most significantly downregulated in skeletal muscles of obese diabetic rats (Bach et al., 2003). In line with this, samples from obese rodents and humans displayed a fragmented mitochondrial network (Bach et al., 2003). This observation has been solidly confirmed, leading to the hypothesis that nutrient overload in the absence of a matching energy requirement leads to mitochondrial fission (Liesa and Shirihai, 2013). Pioneering work from the Shirihai Lab demonstrated that the mitochondrial network of INS-1 cells became largely fragmented when exposed for 4 h to a lipid-loaded medium (0.4 mmol/L of palmitate) (Molina et al., 2009). Very similar observations have been reported in MEF cells and AML12 hepatocytes, where palmitate loading led to a dramatic fragmentation of the mitochondrial network within minutes (Kulkarni et al., 2016). Glucose overload has also been shown to promote mitochondrial fission (Yu et al., 2006, 2011; Theurey et al., 2016), even though this effect might be cell type dependent. Notably, while fission did not occur in INS-1 in response to high glucose (20 mmol/L), the simultaneous loading of high glucose and high lipids led to even more pronounced fission effects than the single loading with lipids (Molina et al., 2009).

The interrelation between mitochondrial morphology and tissue metabolic properties has been studied in the mouse skeletal muscle. Mitochondrial morphology and connectivity correlated with fiber type in the mouse skeletal muscle (Mishra et al., 2015). Oxidative muscles displayed a higher fusion activity. When mice were submitted to exercise training for 4 weeks, the glycolytic muscle gained oxidative capacity and mitochondrial architecture was characterized by a highly fused network (Mishra et al., 2015). When the isolated glycolytic skeletal muscle was incubated in acetoacetate-containing media to enhance oxidative metabolism, there was not only a higher O_2 consumption than in glucose-containing media but also it prompted a more tubular mitochondrial network (Mishra et al., 2015). In contrast, the treatment with oligomycin, which inhibits ATP synthase, resulted in a completely fragmented mitochondrial network (Mishra et al., 2015). Therefore these observations indicate that mitochondrial morphology can dynamically respond to extracellular and intracellular metabolic cues in vivo.

So how does nutrient availability modulate mitochondrial architecture? And why should it be modulated? A landmark manuscript from the Scorrano laboratory brought some additional light on those aspects. When fibroblasts, muscle cells, or hepatocytes were submitted to glucose starvation, their mitochondrial network elongated (Gomes et al., 2011). This elongation was triggered by an intracellular increase in cAMP levels, which led to Drp1 phosphorylation at S643 (Gomes et al., 2011). Accordingly, treatment with forskolin was enough to elongate mitochondria. Surprisingly, rapamycin treatment was also enough to trigger mitochondrial elongation (Gomes et al., 2011), even if the connection between the mTOR pathway and cAMP signaling was not further explored. The fusion of mitochondria spared them from autophagy and allowed the cell to sustain energy production (Gomes et al., 2011). Mitochondrial fusion, additionally, had an intrinsic effect in the mitochondrial bioenergetics, as it increased the dimerization and activity of the ATP synthase complex (Gomes et al., 2011). This allows cells under stress to optimize energy production from nutrients in order to maintain energy levels even on extreme nutrient-scarce conditions. Furthermore, it was shown that if fusion is blocked, mitochondria do not optimize their ATP production upon starvation, prompting cell death (Gomes et al., 2011).

The enhanced oligomerization and optimized ATP production upon mitochondrial fusion might also explain why mitochondria undergo the opposite path (i.e., fission) upon lipid loading or nutrient excess. It has been proposed that mitochondrial fragmentation is a physiological response that increases mitochondrial uncoupling capacity in brown adipocytes (Wikstrom et al., 2014). Hence, the BAT thermogenic capacity is determined not only by lipolytic rates or uncoupling protein 1 (UCP1) expression but also by its ability to fire mitochondrial fission upon adrenergic stimulation. Norepinephrine stimulation triggered mitochondrial fission through Drp1 activation (Wikstrom et al., 2014). The treatment of brown adipocytes with orlistat, a lipolysis inhibitor, did not impair mitochondrial fission in response to adrenergic stimulation, but it completely prevented the increase in energy expenditure, indicating that mitochondrial fission as such is not enough to increase energy expenditure (Wikstrom et al., 2014). Accordingly, models of impaired mitochondrial fusion, which are characterized by a fissioned mitochondrial network, do not show any change in basal energy expenditure, albeit they display higher respiratory capacity when stimulated with oxidative substrates (Kulkarni et al., 2016). In a similar line, adipocytes where *Mfn2* has been knocked down display larger energy expenditure than control cells when treated with oleate and palmitate (Wikstrom et al., 2014). It has been proposed that changes in mitochondrial dynamics in the adipocyte leading to a higher fissioned state might facilitate the access of fatty acids to UCP1, hence facilitating its activation. In this sense, one could speculate that small mitochondria might be more easily driven to the LD or might allow maximizing the surface contact for fatty acids. If this were true, the failure to fission mitochondria upon cold stimulation would compromise the BAT thermogenic function. Accordingly, norepinephrine-induced O_2 consumption in brown adipocytes is impaired by more than 50% when infected with adenoviruses encoding a dominant negative form of Drp1 display (Wikstrom et al., 2014). Mitochondrial depolarization further contributes to mitochondrial fission by triggering OPA1 processing and impeding its pro-fusion activity (Wikstrom et al., 2014). This ensures the long-term thermogenic function of brown adipocytes. This initial cleavage is mediated by OMA1. Accordingly, the deletion of *Oma1* in mice leads to severely impaired thermogenic function (Quiros et al., 2012).

In a twist of the tale, norepinephrine-induced Drp1-mediated fission was achieved via phosphorylation of S643 (Wikstrom et al., 2014). This point is not entirely consistent with the original findings indicating that the phosphorylation of S643 leads to decreased Drp1-depedent fission rates (Cereghetti et al., 2008; Chang and Blackstone, 2007; Cribbs and Strack, 2007). However, other reports have also related the phosphorylation of

S643 to enhanced mitochondrial fission rates (Han et al., 2008; Wang et al., 2012). This apparent discrepancy might stem from the fact that most works reporting that S643 phosphorylation impedes fission are based on the use of the brain-specific isoform of Drp1, which is 37 amino acid longer than the ubiquitous one. In contrast, most reports indicating that Drp1 phosphorylation at S643 can increase fission were performed using the other, shorter, isoforms. Therefore the confusion from the literature might stem from the fact that the effects of S643 phosphorylation might be isoform specific. Another alternative is that the phosphorylation of S643 might not act as an on/off switch for Drp1, but rather as an accelerator of either process depending on other signaling cues or additional Drp1 modifications.

Drp1 recruitment to the mitochondria can also be influenced through modifications in Drp1 docking proteins. A beautiful model illustrating this point is the physiological response to acute energy stress. For example, upon ATP depletion, the AMP-activated protein kinase directly phosphorylates MFF on S155 and S172 (Ducommun et al., 2015; Toyama et al., 2016). This phosphorylation promotes the recruitment of Drp1 to the mitochondrial surface and engages the fission process (Toyama et al., 2016). In parallel to triggering the fission of mitochondria, AMPK promotes autophagy through the phosphorylation of the ULK1 protein (Toyama et al., 2016; Egan et al., 2011). Hence, upon energy stress, mitochondrial fission is coordinated with autophagy processes in order to ensure the efficient removal of dysfunctional mitochondria. This serves a twofold purpose: first, to generate energy and second, to ensure the optimal function of the mitochondrial pool in order to maximize energy production efficiency.

Mfn2 can also be phosphorylated in response to stress stimuli. In particular, JNK can phosphorylate Mfn2 at S27, prompting the recruitment of E3 ubiquitin ligases and the consequent ubiquitination and degradation of Mfn2 (Leboucher et al., 2012). This would provide an explanation on why decreases in Mfn2 levels have been prominently linked to obesity-related metabolic complications, as JNK activity is higher in these situations (Hirosumi et al., 2002). Interestingly, it has been reported that JNK-mediated phosphorylation and degradation of Mfn2 can be modulated by certain species of fatty acids (Senyilmaz et al., 2015), opening the possibility for a direct feedback mechanism between lipid metabolism and the regulation of mitochondrial function by altering fusion/fission states. Nevertheless, the role of Mfn2 phosphorylation in metabolic regulation is still poorly understood.

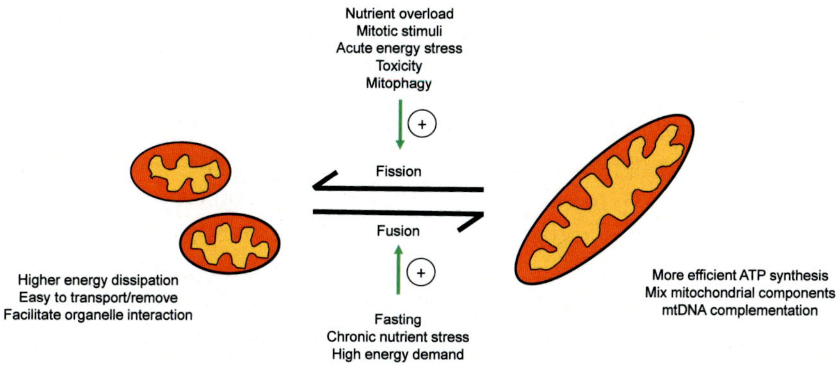

Figure 1 ***Mitochondrial dynamics balance bionergetic and quality control needs.*** Multiple stimuli impact the mitochondrial architecture, prompting it to undergo fusion or fission. Mitochondrial fission will not only facilitate mitochondrial movements across the cell but also enhance energy dissipation at brown adipocytes. In contrast, mitochondrial fusion will favor efficient ATP energy synthesis, as well as the complementation between mitochondrial DNAs and damaged mitochondrial elements.

As a whole, these paragraphs illustrate how mitochondrial architecture is tightly coupled to metabolic needs (Fig. 1). Key signaling paths controlled by nutrient availability can directly impact on the mitochondrial dynamics machinery, albeit these effects can be context or cell specific.

3.3 Mitochondrial Dynamics and Metabolic Control: Lessons From Mouse Models

3.3.1 Combined Mitofusin-Deficient Models

Whole-body KOs for any of the main GTPases involved in mitochondrial fusion or fission processes lead to embryonic lethality (Chen et al., 2003; Davies et al., 2007; Wakabayashi et al., 2009; Ishihara et al., 2009). In particular, the genetic ablation of either *Mfn1* or *Mfn2* leads to a lethal placental defects (Chen et al., 2003). The placental lethality of these models has been overcome by generating conditional KO mice. Initial efforts focused on crossing *Mfn1* or *Mfn2* conditional KO mice with Meox2-Cre mice. Meox2-Cre mice express the Cre recombinase in the embryo by embryonic day 7 but not in cell lineages giving rise to the placenta, hence bypassing the early lethality during the developmental process (Chen et al., 2007). Using this strategy, *Mfn1*-null mice were viable and survived through adulthood with no obvious defects at least until 1 year of age (Chen et al., 2007). In contrast, and while also born at regular Mendelian rations, more than one-third of *Mfn2*-null mice died 1 day after birth (Chen et al., 2007).

The surviving pups displayed severe defects in cerebellar development, movement, and balance, dying before weaning (Chen et al., 2007). Furthermore, pups of *Mfn1*-deficient mice lacking one or both *Mfn2* copies died at birth, suggesting cardiac, circulatory, or pulmonary problems (Chen et al., 2007). When the opposite approach was taken, *Mfn2*-deficient mice lacking one copy of *Mfn1* were indistinguishable from *Mfn2*-null mice (Chen et al., 2007). These results illustrate that Mfn1 and Mfn2 roles are only partially redundant because although Mfn1 was dispensable for mouse development, it critically sensitized against even partial Mfn2 losses (Chen et al., 2007).

David Chan's laboratory developed one of the first tissue-specific KO models for mitofusin deficiency by crossing *Mfn1/Mfn2* floxed mice with MLC-Cre mice to ablate the *Mfn1* and *Mfn2* genes in skeletal muscles (Chen et al., 2010). The muscle-specific double-KO mice were much smaller than control mice and died before 8 weeks of age (Chen et al., 2010). In addition, they displayed abnormally low blood glucose levels under fed and fasting conditions, as well as reduced body temperature (Chen et al., 2010). Even though the origin of these phenotypes is not fully clear, it might be related to an aberrantly high turnover of glucose as a result of a predominant use of glycolysis instead of OXPHOS. In line with this, the mitochondria from these mice showed signs of dysfunction such as swelling, altered cristae morphology, and reduced Cytochrome C oxidase (COX) activity (Chen et al., 2010). The analysis of these mice also revealed that mitofusin deficiency led to mtDNA instability (Chen et al., 2010). Surprisingly, the single muscle-specific deletion of *Mfn1* or *Mfn2* and even that of $Mfn1^{-/-}Mfn2^{+/-}$ and $Mfn1^{+/-}Mfn2^{-/-}$ does not affect mice survival or growth and fail to show the mitochondrial functional deficiencies of the full double KO (Chen et al., 2010).

3.3.2 Single Mitofusin-Deficient Models

The very severe phenotypes derived from the complete loss of mitofusin activity could mask the more subtle influences of these enzymes on metabolic disease. For this purpose, a number of single-KO models for either *Mfn1* or *Mfn2* have been generated, and the *Mfn1* and *Mfn2* floxed mice have been backcrossed to the C57Bl/6 mouse strain, which is more suited for studies on diet-induced metabolic diseases.

The deletion of *Mfn2* specifically in the liver (Mfn2-LKO) does not lead to alterations in body weight or appearance but renders mice with profound abnormalities in glucose homeostasis (Sebastian et al., 2012). Mfn2-LKO mice were characterized by fasting hyperglycemia and glucose intolerance even when fed a regular chow diet (Sebastian et al., 2012). These alterations could be explained by their higher expression of gluconeogenic genes and increased hepatic glucose production (Sebastian et al., 2012). ROS production and ER stress were also prominent in *Mfn2*-deficient livers. Interestingly, alleviating ER stress by treatment with the molecular chaperone tauroursodeoxycholic acid (Ozcan et al., 2004) improved insulin signaling in livers from Mfn2-LKO mice (Sebastian et al., 2012). Whether the increased ER stress in livers from Mfn2-LKO mice is due to a potential direct role of Mfn2 in the mitochondria—ER interaction is not clear. Treatment with *N*-acetylcysteine relieved ER stress and insulin signaling (Sebastian et al., 2012), suggesting that increased ROS production is a key trigger for the phenotypes of the Mfn2-LKO mice.

One could wonder whether the impairment of mitochondrial fusion created by Mfn2 deletion is enough to drive the metabolic phenotype of Mfn2-LKO mice. To answer this question, the Mfn1-LKO model was generated. *Mfn1* deletion in liver cells leads to a largely fragmented hepatic mitochondrial network (Kulkarni et al., 2016). Therefore, if the Mfn2-LKO phenotype was simply due to reduced mitochondrial fusion, it should be phenocopied by the Mfn1-LKO model. However, the Mfn1-LKO phenotype is, in many senses, opposite to that of the Mfn2-LKO phenotype (Kulkarni et al., 2016). *Mfn1* deficiency in the liver led to higher mitochondrial content and whole-body lipid utilization, granting protection against diet-induced insulin resistance. Furthermore, *Mfn1* deletion did not lead to ER stress, despite increased ROS production (Kulkarni et al., 2016). This could be because the increased ROS production upon *Mfn1* or *Mfn2* deletion coexists with different mitochondrial bioenergetic capacities. Thus impaired mitochondrial fusion per se does not lead to mitochondrial disease. In fact, the Mfn1-LKO mouse supports a model in which reduced fusion might be a protective adaptation against lipid overload. However, the possible functions of Mfn2 beyond mitochondrial fusion, such as the coordination of Ca^{2+} fluxes between the ER and the mitochondria, as well as the regulation of Complex I function might drive the detrimental effects on metabolic health.

In the second model, the *Mfn2* floxed mice were crossed with mice expressing the Cre recombinase under the MEF2C promoter. Mfn2 protein levels in the KO group were markedly reduced (80% decrease) in the skeletal muscle, heart, and brain, and a near 50% reduction was detected in adipose tissues, kidney, or liver (Sebastian et al., 2012). Hence, this model is a whole-body *Mfn2* knockdown, with special incidence in muscle tissues.

These mice display insulin resistance and an exacerbated sensitivity to develop diabetes upon high-fat feeding (Sebastian et al., 2012), supporting a causal role of defective Mfn2 function in hepatic or muscle tissue in the development of metabolic diseases.

The impact of Mfn2 in mitochondrial bioenergetics and its role as a metabolic integrator has also been explored in the hypothalamus. The arcuate nucleus is paramount for the control of food intake behavior and satiety (Schneeberger et al., 2014). It contains at least two subsets of neurons with opposite functions in the systemic control of energy balance. One of them is characterized by the expression of the orexigenic neuropeptide agouti-related protein (AgRP), whereas the other is prototypically characterized by expression of the α-melanocyte-stimulating hormone (α-MSH, a product of proopiomelanocortin [POMC] cleavage) (Schneeberger et al., 2014). The α-MSH increases the anorexigenic tone by suppressing appetite and increasing thermogenesis (Poggioli et al., 1986; Olszewski et al., 2001). Conversely, AgRP antagonizes the effects of α-MSH signaling on food intake (Ollmann et al., 1997). Upon lipid overload via high-fat feeding, POMC neurons displayed marked mitochondrial fission, reduced *Mfn2* expression. and increased ER stress (Schneeberger et al., 2013). Consistent with this, conditional deletion of *Mfn2*, but not *Mfn1*, in POMC neurons resulted in spontaneous obesity as a consequence of ER stress–induced leptin resistance (Schneeberger et al., 2013). Surprisingly, AgRP neurons displayed reduced mitochondria number but increased elongation (Dietrich et al., 2013). This is in line with the opposite roles of these two neuronal subsets, even if it is unclear how these opposite outcomes are reached from a molecular perspective.

The latest model of conditional Mfn2 deficiency that has been reported is the adipose tissue–specific *Mfn2* deletion (Mfn2-AKO) (Fig. 2). Electron microscopy not only revealed that mitochondria were rounder and smaller in size in the BAT of Mfn2-AKO mice but also suggested that Mfn2 influenced the relation between the mitochondria and the LD (Boutant et al., 2017). While numerous anecdotal observations from electron microscopic studies have referred for decades to the close physical association between mitochondria and the LD (Bosma et al., 2012; Wang et al., 2011a; Yu et al., 2015), its molecular underpinning has been elusive. Upon cold exposure, mitochondria tether to the LD in BAT (Yu et al., 2015). The work on the Mfn2-AKO mice identified that the interaction between the mitochondria and the LD depends on the presence of Mfn2. When *Mfn2* is deleted, more than 50% of the mitochondria to LD contacts are lost (Boutant et al., 2017).

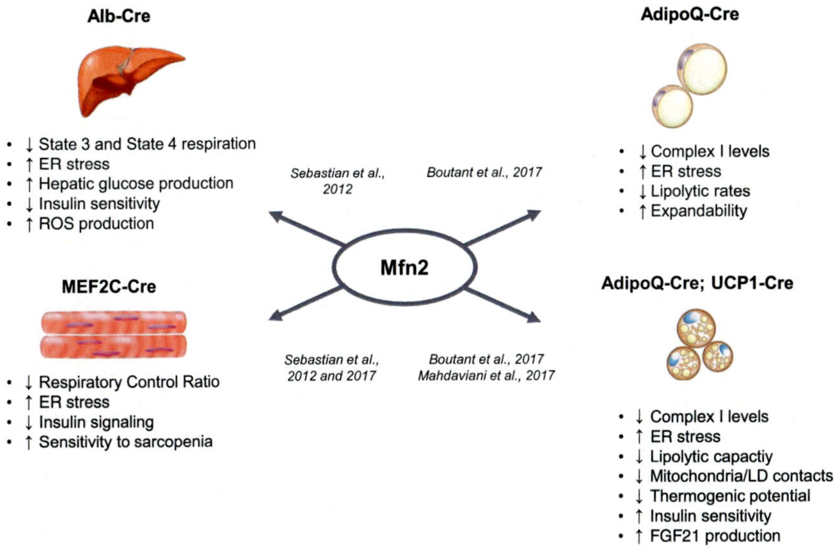

Figure 2 ***The physiological impact of mitofusin 2 (Mfn2) on key tissues for nutrient homeostasis.*** By crossing with different Cre driver lines, a number of tissue-specific knockout models for Mfn2 have been generated. Literature references for each model are included in brackets. Some of the key metabolic phenotypes observed in each conditional knockout are depicted. As a general rule, Mfn2 deficiency compromised mitochondrial functions, albeit at different degrees depending on the tissues. Increased endoplasmic reticulum (ER) stress is also a shared feature in all the Mfn2-deficient tissues examined. In contrast, other aspects, such as insulin sensitivity, respond differently in a tissue-specific fashion. *FGF21*, fibroblast growth factor 21; *LD*, lipid droplet; *MEF2C*, mouse embryonic fibroblast 2C; *ROS*, reactive oxygen species; *UCP1*, uncoupling protein 1.

Experiments in MEF cells demonstrated that both perilipin 1 (PLIN1) and Mfn2 are required for cAMP-induced respiration (Boutant et al., 2017). In line with this, PLIN1 and Mfn2 co-immunoprecipitated. This co-immunoprecipitation was very specific, as Mfn2 did not interact with another protein from the perilipin family, PLIN3. Similarly, PLIN1 did not interact with Mfn1 (Boutant et al., 2017). However, the nature of this interaction and its modulation by adrenergic stimulation is not yet fully solved. In this sense, it is unclear whether Mfn2 participates in the interaction between mitochondria and LDs in other tissues that do not express PLIN1. PLIN5 has been shown to provide physical linkage to mitochondria in heart and muscle cells (Bosma et al., 2012; Wang et al., 2011a), but whether this is based on interactions with Mfn2 is currently unknown. Notably, the interaction between PLIN1 and Mfn2 was regulated by

PLIN1 phosphorylation (Boutant et al., 2017), and PLIN5 function can also be regulated through cAMP-mediated phosphorylation (Pollak et al., 2015). Thus it is tempting to hypothesize that PLIN5 and PLIN1 might facilitate mitochondria—LD interactions through parallel mechanisms. Together, these results align with the hypothesis proposing that the interaction between mitochondria and the LD would allow a more efficient transfer of substrates between both organelles. In line with this concept, the lack of Mfn2 in BAT leads to defective lipolysis and thermogenic function (Boutant et al., 2017; Mahdaviani et al., 2017). The defective thermogenic capacity might not only result in defective lipolysis but also on a dramatically impaired respiratory capacity through Complex I (Boutant et al., 2017; Mahdaviani et al., 2017). Surprisingly, *Mfn2* deletion in the BAT from female mice was associated with increased capacity for lipid oxidation (Mahdaviani et al., 2017). This gender difference could be explained by the fact that in the mouse strain used in these studies, the C57Bl/6 females are resistant to diet- and age-related complications, being less prone to develop inflammation in adipose tissues (Pettersson et al., 2012).

A number of works have proposed positive correlations among BAT content, BMI, and susceptibility to metabolic disease (Peirce et al., 2014). It would therefore be expected that the Mfn2-AKO mice, characterized by a critical impairment in thermogenic function, would quickly develop insulin resistance upon high-fat feeding. In contrast, Mfn2-AKO exhibited a notable protection against diet-induced insulin resistance (Boutant et al., 2017). Hyperinsulinemic-euglycemic clamps revealed that the persistent insulin sensitivity in high-fat fed mice was due to enhanced insulin-stimulated glucose uptake in the BAT (Boutant et al., 2017). This, in turn, was consequent to a glycolytic rewiring to maintain thermogenic activity in order to compensate for their defective lipolytic capacity (Boutant et al., 2017). Hence, thermogenic capacity and insulin sensitivity can be uncoupled in BAT physiology, as confirmed also in a model with BAT-specific deletion of *Mfn2* (Mahdaviani et al., 2017).

3.3.3 Models for Altered Optic Atrophy 1 Function

Given the embryonic lethality of *Opa1*-null mice (Davies et al., 2007; Alavi et al., 2007), several studies have used heterozygous KO mice to analyze the impact of *Opa1* on metabolism. Haploinsufficient mice ($Opa1^{+/-}$) display around a 50% reduction in OPA1 protein levels and show visual impairment. $Opa1^{+/-}$ mice also have altered cardiac mitochondrial structure, albeit a normal respiratory function, making the mice susceptible to cardiac hypertrophy and ventricular dysfunction following pressure overload (Piquereau et al., 2012). Unexpectedly, *Opa1*$^{+/-}$ mice show a comparable training capacity to that control mice (Caffin et al., 2013). After training, *Opa1*$^{+/-}$ mice exhibited altered mitochondrial ultrastructure characterized by a very heterogeneous mitochondrial size, with some large mitochondria harboring altered cristae structure and absent OMM (Caffin et al., 2013). Surprisingly, endurance capacity was increased in *Opa1*$^{+/-}$ trained mice (Caffin et al., 2013). To explain this paradox, the authors proposed that the impairment in mitochondrial biogenesis occurring in *Opa1*$^{+/-}$ mice was compensated by the indirect activation of an adaptive pathway leading to better utilization of lipid energy substrates, ultimately improving endurance.

To this date, only a few conditional *Opa1* KO models for key metabolic tissues have been reported. The muscle-specific deletion of *Opa1* leads to a progressive loss of mitochondrial function and muscle mass (Pereira et al., 2017). However, this activated a number of stress paths, including a dramatic increase in fibroblast growth factor 21 (FGF21), which sustained insulin sensitivity (Pereira et al., 2017). In fact, muscle-specific *Opa1* KO mice are protected against age- and diet-related insulin resistance through a glycolytic rewiring (Pereira et al., 2017). These results echo the observations in the Mfn2-AKO model, where an increase in circulating FGF21 levels was also observed (Boutant et al., 2017), and supports that increased FGF21 gene expression and secretion is a hallmark response to mitochondrial dysfunction (Kim et al., 2013). Although the mechanisms linking mitochondrial dysfunction and *Fgf21* gene expression are still unclear, the activation of the activating transcription factor 4 (ATF4) pathway seems critical (Kim et al., 2013; Quiros et al., 2017). Recent findings, however, indicate that FGF21 might not be the only player influencing whole-body adaptations to muscle mitochondrial dysfunction. This way, mice overexpressing UCP1 in muscle, which leads to a marked decrease in mitochondrial coupling, showed improved whole-body insulin sensitivity irrespective of FGF21 levels (Ost et al., 2016). The quest for additional mechanisms is therefore guaranteed.

The metabolic impact of impaired OPA1 function can also be assessed by altering its two main cleaving activities: OMA1 and YME1L. *Oma1*$^{-/-}$ are viable but exhibit a marked obesity, reduced energy expenditure, and impaired thermogenic response because of deficient OPA1 processing (Quiros et al., 2012). In contrast to *Oma1*, *Yme1l*-deficient mice are not viable (Wai et al., 2015) and conditional deletions will be required to explore the metabolic impacts of YME1L.

Finally, it has been reported that OPA1 overexpression ameliorates muscle performance in mouse models of inherited defects in mitochondrial bioenergetics either at Complex I (*Ndufs4*-null mice) or Complex IV (*Cox15*) levels (Civiletto et al., 2015). This was linked to a correction of cristae ultrastructure and mitochondrial respiration, especially on the *Cox15*-null mice, and ultimately led to prolongation of life span (Civiletto et al., 2015). It would be interesting to understand if OPA1 overexpression also led to changes in mitophagy or mitochondrial quality control mechanisms. Life span, however, was not affected in the absence of deleterious mutations of mitochondrial function (Varanita et al., 2015). It will be interesting to characterize the impact of OPA1 overexpression on energy homeostasis, including glucose and lipid handling. Interestingly, OPA1 transgenesis led to reduced body weight, even though this effect seemed to be mouse strain specific (Varanita et al., 2015).

3.3.4 Models for Defective Fission

Homozygote *Dnm1l* deletion in mice is embryonically lethal because of reduced mitochondrial fission, leading to developmental defects in the brain (Wakabayashi et al., 2009; Ishihara et al., 2009). In contrast, heterozygote *Dnm1l*KO mice have a gross normal appearance in regular housing conditions (Manczak et al., 2012), pointing out once more that haploinsuficiency for genes related to mitochondrial dynamics is compatible with life. The whole body deletion of the MFF also leads to premature lethality, and the mean life span of these mice was reduced to 13 weeks. This suggests that other Drp1 adaptors can only partially compensate for the loss of the MFF. The reduced life span of MFF KO mice was due to dilated cardiomyopathy and cardiac mitochondrial dysfunction (Chen et al., 2015). Interestingly, MFF-deficient mice did not show signs of enhanced glycolysis, despite mitochondrial dysfunction (Chen et al., 2015).The simultaneous ablation of *Mfn1*, but not *Mfn2*, eliminates cardiac-related lethality of *Mff*KO mice, and MFF deletion prevented the perinatal lethality of *Mfn1*-deficient mice (Chen et al., 2015). In contrast, other defects of Mfn1 deficiency, such as small testis size, could not be recovered by the simultaneous KO of the *Mff* (Chen et al., 2015). Altogether, these results suggest that blocking fusion or fission might prevent some pathological states caused by reduction of the opposing process, albeit these compensatory effects might be context/tissue dependent. It must also be reminded that Drp1 or MFF also participate in peroxisomal fission, which could contribute to the overall phenotype.

Conditional KOs for defective fission have also been generated, and the metabolic impact of alterations in Drp1 expression has also been explored in the liver. Under regular feeding conditions the hepatic loss of Drp1 did not lead to any gross abnormality. However, when mice were fed a high-fat diet, Drp1-LKO mice did not gain as much weight as their control littermates (Wang et al., 2015). This was consequent to a dramatic decrease in fat mass and linked to a marked protection against diet-induced glucose intolerance and insulin resistance (Wang et al., 2015). The livers from Drp1-LKO mice displayed lower mitochondrial content, albeit the mitochondria were swollen and longer. Also the ER was swollen, suggesting tissue damage. This was further confirmed upon evaluation of blood biochemistry markers, as serum aspartate aminotransferase and alanine aminotransferase levels were threefold higher in high-fat–fed Drp1-LKO mice (Wang et al., 2015). While the changes in mitochondrial morphology did not lead to major alterations in mitochondrial respiratory complex levels or activity, the alterations in ER led to ER stress and to a marked increase in FGF21 production (Wang et al., 2015). This, in turn, could influence hepatic metabolism and whole-body substrate catabolism, albeit limitations on this hypothesis have been discussed earlier.

A muscle-specific Drp1 gain-of-function model has been reported. Drp1 overexpression exclusively in muscle led to decreased body weight caused by a severe reduction in muscle size linked to impaired postnatal muscle growth (Touvier et al., 2015). While no major differences in the levels of markers for mitochondrial biogenesis were observed, electron microscopic analysis revealed multiple alterations in the mitochondrial network, with virtual absence of intermyofibrillar mitochondria (Touvier et al., 2015). A severe mitochondrial heterogeneity was also observed, where normal mitochondria cohabitated with cristaeless ones (Touvier et al., 2015). Unfortunately, the etiology of the muscle growth defects is not clear, and the impact on metabolic homeostasis was not explored. The impact of Drp1 in muscle metabolism has only been tested in cultured muscle cells, where blocking mitochondrial fission ameliorated insulin-stimulated glucose uptake (Jheng et al., 2012). However, the only in vivo attempt to evaluate this aspect has been done through the use of Mdivi-1 (Jheng et al., 2012), whose specificity is unclear. In this sense, if Mdivi-1 truly acts as a Complex I inhibitor, a metabolic glycolytic rewiring would be certainly expected.

Interestingly, the deficiency of the calcineurin catalytic subunit Ppp3cb protects from obesity induced by high-fat diet (Pfluger et al., 2015). This phenotype stemmed from the muscle, as only muscle-specific (but not adipose- or liver-specific) *Ppp3cb* KO mice can phenocopy the protection against diet-induced obesity (Pfluger et al., 2015). The protection against obesity was intimately linked to the prevention of diet-induced glucose intolerance. At the mitochondrial level, there was a global increase in OXPHOS protein levels and state 3 respiration from Complex I. These effects were closely linked to increased Drp1 S643 phosphorylation and mitochondrial elongation (Pfluger et al., 2015), in line with calcineurin being the main enzyme accounting for S643 dephosphorylation (Cereghetti et al., 2008). Nevertheless, calcineurin deficiency in muscles led to impaired muscle performance upon an exhaustive exercise bout (Pfluger et al., 2015), which suggests that mitochondrial fission is critical to cellular homeostasis under high environmental stress conditions.

4. CONCLUSIONS AND FUTURE PERSPECTIVES

The dynamic change between fused and fission has to be carefully balanced in order to allow bioenergetic and quality control functions. In line with this, most genetically engineered animal models that aimed to lock mitochondria in a particular conformation have consistently led to mitochondrial dysfunction. One exception might be the tissue-specific Mfn1 deletion, as shown in heart and liver (Kulkarni et al., 2016; Papanicolaou et al., 2012). This, however, might be due to either the fragmented nature of the mitochondrial network in these tissues at baseline or the compensatory increases in Mfn2.

Data from multiple cultured cell and animal models support that mitochondrial architecture is influenced by the nutritional environment. New research indicates that also mechanical force suffices to engage fission events (Helle et al., 2017). This could be particularly relevant in contracting tissues such as muscle or heart. It also opens doors to understand how mitochondrial shape might be regulated simply by direct contacts with other mitochondria or cellular organelles. In this sense, it is becoming clear that mitochondrial dynamics—related proteins are critical to ensure the exchange of substrates and cofactors between mitochondria, the LD or the ER.

Mutations in mitochondrial dynamics—related proteins have been related to severe neurological hereditary disorders. Whether changes in mitochondrial architecture are at the root of diet- or age-related diseases is, however, not clear yet. In *Caenorhabditis elegans*, genetic manipulation of mitochondrial morphology is not disruptive to life span (Weir et al., 2017). An important point to consider is that different mitochondrial configurations might be required at different times in life or in response to a particular lifestyle. Illustrating this point, a report indicated that inducing Drp1-mediated mitochondrial fission in midlife increases life span and improves multiple markers of health in aged flies (Rana et al., 2017). Hence, in order to exploit the power of mitochondrial dynamics for therapeutic purposes in age- and diet-related diseases, timing and localization will have to be critically considered.

ACKNOWLEDGMENTS

I would like to thank the Canto laboratory members for inspiring discussions. I would also like to apologize to all the authors whose relevant work was not cited due to space limitations. CC is funded by the NIHS and by the EU Marie Skłodowska-Curie ITN, ChroMe (H2020-MSCA-ITN-2015-ChroMe, project number 675610).

REFERENCES

Adachi, Y., Itoh, K., Yamada, T., Cerveny, K.L., Suzuki, T.L., Macdonald, P., Frohman, M.A., Ramachandran, R., Iijima, M, Sesaki, H., 2016. Coincident phosphatidic acid interaction restrains drp1 in mitochondrial division. Mol. Cell 63, 1034–1043.

Akepati, V.R., Muller, E.C., Otto, A., Strauss, H.M., Portwich, M., Alexander, C., 2008. Characterization of OPA1 isoforms isolated from mouse tissues. J. Neurochem. 106, 372–383.

Alavi, M.V., Bette, S., Schimpf, S., Schuettauf, F., Schraermeyer, U., Wehrl, H.F., Ruttiger, L., Beck, S.C., Tonagel, F., Pichler, B.J., Knipper, M., Peters, T., Laufs, J., Wissinger, B., 2007. A splice site mutation in the murine Opa1 gene features pathology of autosomal dominant optic atrophy. Brain 130, 1029–1042.

Alexander, C., Votruba, M., Pesch, U.E., Thiselton, D.L., Mayer, S., Moore, A., Rodriguez, M., Kellner, U., LEO-Kottler, B., Auburger, G., Bhattacharya, S.S., Wissinger, B., 2000. OPA1, encoding a dynamin-related GTPase, is mutated in autosomal dominant optic atrophy linked to chromosome 3q28. Nat. Genet. 26, 211–215.

Amchenkova, A.A., Bakeeva, L.E., Chentsov, Y.S., Skulachev, V.P., Zorov, D.B., 1988. Coupling membranes as energy-transmitting cables. I. Filamentous mitochondria in fibroblasts and mitochondrial clusters in cardiomyocytes. J. Cell Biol. 107, 481–495.

Anand, R., Wai, T., Baker, M.J., Kladt, N., Schauss, A.C., Rugarli, E., Langer, T., 2014. The i-AAA protease YME1L and OMA1 cleave OPA1 to balance mitochondrial fusion and fission. J. Cell Biol. 204, 919–929.

Bach, D., Pich, S., Soriano, F.X., Vega, N., Baumgartner, B., Oriola, J., Daugaard, J.R., Lloberas, J., Camps, M., Zierath, J.R., Rabasa-Lhoret, R., Wallberg-Henriksson, H., Laville, M., Palacin, M., Vidal, H., Rivera, F., Brand, M., Zorzano, A., 2003. Mitofusin-2 determines mitochondrial network architecture and mitochondrial metabolism. A novel regulatory mechanism altered in obesity. J. Biol. Chem. 278, 17190–17197.

Bach, D., Naon, D., Pich, S., Soriano, F.X., Vega, N., Rieusset, J., Laville, M., Guillet, C., Boirie, Y., Wallberg-Henriksson, H., Manco, M., Calvani, M., Castagneto, M., Palacin, M., Mingrone, G., Zierath, J.R., Vidal, H., Zorzano, A., 2005. Expression of Mfn2, the Charcot-Marie-Tooth neuropathy type 2A gene, in human skeletal muscle: effects of type 2 diabetes, obesity, weight loss, and the regulatory role of tumor necrosis factor alpha and interleukin-6. Diabetes 54, 2685–2693.
Benard, G., Bellance, N., James, D., Parrone, P., Fernandez, H., Letellier, T., Rossignol, R., 2007. Mitochondrial bioenergetics and structural network organization. J. Cell Sci. 120, 838–848.
Benda, C., 1898. Ueber die Spermatogenese der Vertebraten und höherer Evertebraten, II. Theil: die Histiogenese der Spermien (On spermatogenesis of vertebrates and higher invertebrates, Part II: the histogenesis of sperm. Archiv für Anatomie und Physiologie 73, 393–398.
Bordt, E.A., Clerc, P., Roelofs, B.A., Saladino, A.J., Tretter, L., Adam-Vizi, V., Cherok, E., Khalil, A., Yadava, N., Ge, S.X., Francis, T.C., Kennedy, N.W., Picton, L.K., Kumar, T., Uppuluri, S., Miller, A.M., Itoh, K., Karbowski, M., Sesaki, H., Hill, R.B., Polster, B.M., 2017. The putative Drp1 inhibitor Mdivi-1 is a reversible mitochondrial complex I inhibitor that modulates reactive oxygen species. Dev. Cell 40, 583–594 e6.
Bosma, M., Minnaard, R., Sparks, L.M., Schaart, G., Losen, M., De Baets, M.H., Duimel, H., Kersten, S., Bickel, P.E., Schrauwen, P., Hesselink, M.K., 2012. The lipid droplet coat protein perilipin 5 also localizes to muscle mitochondria. Histochem. Cell Biol. 137, 205–216.
Boutant, M., Kulkarni, S.S., Joffraud, M., Ratajczak, J., Valera-Alberni, M., Combe, R., Zorzano, A., Canto, C., 2017. Mfn2 is critical for brown adipose tissue thermogenic function. EMBO J. 36, 1543–1558.
Bui, H.T., Shaw, J.M., 2013. Dynamin assembly strategies and adaptor proteins in mitochondrial fission. Curr. Biol. 23, R891–R899.
Bustillo-Zabalbeitia, I., Montessuit, S., Raemy, E., Basanez, G., Terrones, O., Martinou, J.C., 2014. Specific interaction with cardiolipin triggers functional activation of Dynamin-Related Protein 1. PLoS One 9, e102738.
Caffin, F., Prola, A., Piquereau, J., Novotova, M., David, D.J., Garnier, A., Fortin, D., Alavi, M.V., Veksler, V., Ventura-Clapier, R., Joubert, F., 2013. Altered skeletal muscle mitochondrial biogenesis but improved endurance capacity in trained OPA1-deficient mice. J. Physiol. 591, 6017–6037.
Cantó, C., Gerhart-Hines, Z., Feige, J.N., Lagouge, M., Noriega, L., Milne, J.C., Elliott, P.J., Puigserver, P., Auwerx, J., 2009. AMPK regulates energy expenditure by modulating NAD+ metabolism and SIRT1 activity. Nature 458, 1056–1060.
Canto, C., Jiang, L.Q., Deshmukh, A.S., Mataki, C., Coste, A., Lagouge, M., Zierath, J.R., Auwerx, J., 2010. Interdependence of AMPK and SIRT1 for metabolic adaptation to fasting and exercise in skeletal muscle. Cell Metab. 11, 213–219.
Cartoni, R., Leger, B., Hock, M.B., Praz, M., Crettenand, A., Pich, S., Ziltener, J.L., Luthi, F., Deriaz, O., Zorzano, A., Gobelet, C., Kralli, A., Russell, A.P., 2005. Mitofusins 1/2 and ERRalpha expression are increased in human skeletal muscle after physical exercise. J. Physiol. 567, 349–358.
Cassidy-Stone, A., Chipuk, J.E., Ingerman, E., Song, C., Yoo, C., Kuwana, T., Kurth, M.J., Shaw, J.T., Hinshaw, J.E., Green, D.R., Nunnari, J., 2008. Chemical inhibition of the mitochondrial division dynamin reveals its role in Bax/Bak-dependent mitochondrial outer membrane permeabilization. Dev. Cell 14, 193–204.
Cereghetti, G.M., Stangherlin, A., Martins de Brito, O., Chang, C.R., Blackstone, C., Bernardi, P., Scorrano, L., 2008. Dephosphorylation by calcineurin regulates translocation of Drp1 to mitochondria. Proc. Natl. Acad. Sci. U.S.A 105, 15803–15808.

Chan, D.C., 2012. Fusion and fission: interlinked processes critical for mitochondrial health. Annu. Rev. Genet. 46, 265–287.

Chang, C.R., Blackstone, C., 2007. Cyclic AMP-dependent protein kinase phosphorylation of Drp1 regulates its GTPase activity and mitochondrial morphology. J. Biol. Chem. 282, 21583–21587.

Chen, Y., Dorn 2nd, G.W., 2013. PINK1-phosphorylated mitofusin 2 is a Parkin receptor for culling damaged mitochondria. Science 340, 471–475.

Chen, H., Detmer, S.A., Ewald, A.J., Griffin, E.E., Fraser, S.E., Chan, D.C., 2003. Mitofusins Mfn1 and Mfn2 coordinately regulate mitochondrial fusion and are essential for embryonic development. J. Cell Biol. 160, 189–200.

Chen, H., Chomyn, A., Chan, D.C., 2005. Disruption of fusion results in mitochondrial heterogeneity and dysfunction. J. Biol. Chem. 280, 26185–26192.

Chen, H., Mccaffery, J.M., Chan, D.C., 2007. Mitochondrial fusion protects against neurodegeneration in the cerebellum. Cell 130, 548–562.

Chen, H., Vermulst, M., Wang, Y.E., Chomyn, A., Prolla, T.A., Mccaffery, J.M., Chan, D.C., 2010. Mitochondrial fusion is required for mtDNA stability in skeletal muscle and tolerance of mtDNA mutations. Cell 141, 280–289.

Chen, H., Ren, S., Clish, C., Jain, M., Mootha, V., Mccaffery, J.M., Chan, D.C., 2015. Titration of mitochondrial fusion rescues Mff-deficient cardiomyopathy. J. Cell Biol. 211, 795–805.

Cipolat, S., Martins de Brito, O., Dal Zilio, B., Scorrano, L., 2004. OPA1 requires mitofusin 1 to promote mitochondrial fusion. Proc. Natl. Acad. Sci. U.S.A 101, 15927–15932.

Civiletto, G., Varanita, T., Cerutti, R., Gorletta, T., Barbaro, S., Marchet, S., Lamperti, C., Viscomi, C., Scorrano, L., Zeviani, M., 2015. Opa1 overexpression ameliorates the phenotype of two mitochondrial disease mouse models. Cell Metab. 21, 845–854.

Cogliati, S., Frezza, C., Soriano, M.E., Varanita, T., Quintana-Cabrera, R., Corrado, M., Cipolat, S., Costa, V., Casarin, A., Gomes, L.C., Perales-Clemente, E., Salviati, L., Fernandez-Silva, P., Enriquez, J.A., Scorrano, L., 2013. Mitochondrial cristae shape determines respiratory chain supercomplexes assembly and respiratory efficiency. Cell 155, 160–171.

Cosson, P., Marchetti, A., Ravazzola, M., Orci, L., 2012. Mitofusin-2 independent juxtaposition of endoplasmic reticulum and mitochondria: an ultrastructural study. PLoS One 7, e46293.

Cribbs, J.T., Strack, S., 2007. Reversible phosphorylation of Drp1 by cyclic AMP-dependent protein kinase and calcineurin regulates mitochondrial fission and cell death. EMBO Rep. 8, 939–944.

Davies, V.J., Hollins, A.J., Piechota, M.J., Yip, W., Davies, J.R., White, K.E., Nicols, P.P., Boulton, M.E., Votruba, M., 2007. Opa1 deficiency in a mouse model of autosomal dominant optic atrophy impairs mitochondrial morphology, optic nerve structure and visual function. Hum. Mol. Genet. 16, 1307–1318.

De Brito, O.M., Scorrano, L., 2008. Mitofusin 2 tethers endoplasmic reticulum to mitochondria. Nature 456, 605–610.

Debattisti, V., Pendin, D., Ziviani, E., Daga, A., Scorrano, L., 2014. Reduction of endoplasmic reticulum stress attenuates the defects caused by Drosophila mitofusin depletion. J. Cell Biol. 204, 303–312.

Delettre, C., Lenaers, G., Griffoin, J.M., Gigarel, N., Lorenzo, C., Belenguer, P., Pelloquin, L., Grosgeorge, J., Turc-Carel, C., Perret, E., Astarie-Dequeker, C., Lasquellec, L., Arnaud, B., Ducommun, B., Kaplan, J., Hamel, C.P., 2000. Nuclear gene OPA1, encoding a mitochondrial dynamin-related protein, is mutated in dominant optic atrophy. Nat. Genet. 26, 207–210.

Delettre, C., Griffoin, J.M., Kaplan, J., Dollfus, H., Lorenz, B., Faivre, L., Lenaers, G., Belenguer, P., Hamel, C.P., 2001. Mutation spectrum and splicing variants in the OPA1 gene. Hum. Genet. 109, 584–591.

Dietrich, M.O., Liu, Z.W., Horvath, T.L., 2013. Mitochondrial dynamics controlled by mitofusins regulate Agrp neuronal activity and diet-induced obesity. Cell 155, 188–199.

Ding, H., Jiang, N., Liu, H., Liu, X., Liu, D., Zhao, F., Wen, L., Liu, S., Ji, L.L., Zhang, Y., 2010. Response of mitochondrial fusion and fission protein gene expression to exercise in rat skeletal muscle. Biochim. Biophys. Acta 1800, 250–256.

Ducommun, S., Deak, M., Sumpton, D., Ford, R.J., Nunez Galindo, A., Kussmann, M., Viollet, B., Steinberg, G.R., Foretz, M., Dayon, L., Morrice, N.A., Sakamoto, K., 2015. Motif affinity and mass spectrometry proteomic approach for the discovery of cellular AMPK targets: identification of mitochondrial fission factor as a new AMPK substrate. Cell. Signal. 27, 978–988.

Duvezin-Caubet, S., Jagasia, R., Wagener, J., Hofmann, S., Trifunovic, A., Hansson, A., Chomyn, A., Bauer, M.F., Attardi, G., Larsson, N.G., Neupert, W., Reichert, A.S., 2006. Proteolytic processing of OPA1 links mitochondrial dysfunction to alterations in mitochondrial morphology. J. Biol. Chem. 281, 37972–37979.

Egan, D.F., Shackelford, D.B., Mihaylova, M.M., Gelino, S., Kohnz, R.A., Mair, W., Vasquez, D.S., Joshi, A., Gwinn, D.M., Taylor, R., Asara, J.M., Fitzpatrick, J., Dillin, A., Viollet, B., Kundu, M., Hansen, M., Shaw, R.J., 2011. Phosphorylation of ULK1 (hATG1) by AMP-activated protein kinase connects energy sensing to mitophagy. Science 331, 456–461.

Ehses, S., Raschke, I., Mancuso, G., Bernacchia, A., Geimer, S., Tondera, D., Martinou, J.C., Westermann, B., Rugarli, E.I., Langer, T., 2009. Regulation of OPA1 processing and mitochondrial fusion by m-AAA protease isoenzymes and OMA1. J. Cell Biol. 187, 1023–1036.

Filadi, R., Greotti, E., Turacchio, G., Luini, A., Pozzan, T., Pizzo, P., 2015. Mitofusin 2 ablation increases endoplasmic reticulum-mitochondria coupling. Proc. Natl. Acad. Sci. U.S.A 112, E2174–E2181.

Frezza, C., Cipolat, S., Martins de Brito, O., Micaroni, M., Beznoussenko, G.V., Rudka, T., Bartoli, D., Polishuck, R.S., Danial, N.N., De Strooper, B., Scorrano, L., 2006. OPA1 controls apoptotic cristae remodeling independently from mitochondrial fusion. Cell 126, 177–189.

Gandre-Babbe, S., Van Der Bliek, A.M., 2008. The novel tail-anchored membrane protein Mff controls mitochondrial and peroxisomal fission in mammalian cells. Mol. Biol. Cell 19, 2402–2412.

Garcia-Roves, P.M., Osler, M.E., Holmstrom, M.H., Zierath, J.R., 2008. Gain-of-function R225Q mutation in AMP-activated protein kinase gamma3 subunit increases mitochondrial biogenesis in glycolytic skeletal muscle. J. Biol. Chem. 283, 35724–35734.

Glancy, B., Hartnell, L.M., Malide, D., Yu, Z.X., Combs, C.A., Connelly, P.S., Subramaniam, S., Balaban, R.S., 2015. Mitochondrial reticulum for cellular energy distribution in muscle. Nature 523, 617–620.

Gomes, L.C., Di Benedetto, G., Scorrano, L., 2011. During autophagy mitochondria elongate, are spared from degradation and sustain cell viability. Nat. Cell Biol. 13, 589–598.

Griparic, L., Kanazawa, T., Van der Bliek, A.M., 2007. Regulation of the mitochondrial dynamin-like protein Opa1 by proteolytic cleavage. J. Cell Biol. 178, 757–764.

Guo, Y.H., Chen, K., Gao, W., Li, Q., Chen, L., Wang, G.S., Tang, J., 2007. Overexpression of Mitofusin 2 inhibited oxidized low-density lipoprotein induced vascular smooth muscle cell proliferation and reduced atherosclerotic lesion formation in rabbit. Biochem. Biophys. Res. Commun. 363, 411–417.

Hales, K.G., Fuller, M.T., 1997. Developmentally regulated mitochondrial fusion mediated by a conserved, novel, predicted GTPase. Cell 90, 121–129.

Han, X.J., Lu, Y.F., Li, S.A., Kaitsuka, T., Sato, Y., Tomizawa, K., Nairn, A.C., Takei, K., Matsui, H., Matsushita, M., 2008. CaM kinase I alpha-induced phosphorylation of Drp1 regulates mitochondrial morphology. J. Cell Biol. 182, 573–585.

Head, B., Griparic, L., Amiri, M., Gandre-Babbe, S., Van der Bliek, A.M., 2009. Inducible proteolytic inactivation of OPA1 mediated by the OMA1 protease in mammalian cells. J. Cell Biol. 187, 959–966.

Helle, S.C.J., Feng, Q., Aebersold, M.J., Hirt, L., Gruter, R.R., Vahid, A., Sirianni, A., Mostowy, S., Snedeker, J.G., Saric, A., Idema, T., Zambelli, T., Kornmann, B., 2017. Mechanical force induces mitochondrial fission. Elife 6.

Hirosumi, J., Tuncman, G., Chang, L., Gorgun, C.Z., Uysal, K.T., Maeda, K., Karin, M., Hotamisligil, G.S., 2002. A central role for JNK in obesity and insulin resistance. Nature 420, 333–336.

Huang, P., Yu, T., Yoon, Y., 2007. Mitochondrial clustering induced by overexpression of the mitochondrial fusion protein Mfn2 causes mitochondrial dysfunction and cell death. Eur. J. Cell Biol. 86, 289–302.

Ishihara, N., Eura, Y., Mihara, K., 2004. Mitofusin 1 and 2 play distinct roles in mitochondrial fusion reactions via GTPase activity. J. Cell Sci. 117, 6535–6546.

Ishihara, N., Fujita, Y., Oka, T., Mihara, K., 2006. Regulation of mitochondrial morphology through proteolytic cleavage of OPA1. EMBO J. 25, 2966–2977.

Ishihara, N., Nomura, M., Jofuku, A., Kato, H., Suzuki, S.O., Masuda, K., Otera, H., Nakanishi, Y., Nonaka, I., Goto, Y., Taguchi, N., Morinaga, H., Maeda, M., Takayanagi, R., Yokota, S., Mihara, K., 2009. Mitochondrial fission factor Drp1 is essential for embryonic development and synapse formation in mice. Nat. Cell Biol. 11, 958–966.

James, D.I., Parone, P.A., Mattenberger, Y., Martinou, J.C., 2003. hFis1, a novel component of the mammalian mitochondrial fission machinery. J. Biol. Chem. 278, 36373–36379.

Jheng, H.F., Tsai, P.J., Guo, S.M., Kuo, L.H., Chang, C.S., Su, I.J., Chang, C.R., Tsai, Y.S., 2012. Mitochondrial fission contributes to mitochondrial dysfunction and insulin resistance in skeletal muscle. Mol. Cell Biol. 32, 309–319.

Kaser, M., Kambacheld, M., Kisters-Woike, B., Langer, T., 2003. Oma1, a novel membrane-bound metallopeptidase in mitochondria with activities overlapping with the m-AAA protease. J. Biol. Chem. 278, 46414–46423.

Kemper, C., Habib, S.J., Engl, G., Heckmeyer, P., Dimmer, K.S., Rapaport, D., 2008. Integration of tail-anchored proteins into the mitochondrial outer membrane does not require any known import components. J. Cell Sci. 121, 1990–1998.

Kim, K.H., Jeong, Y.T., Oh, H., Kim, S.H., Cho, J.M., Kim, Y.N., Kim, S.S., Kim Do, H., Hur, K.Y., Kim, H.K., Ko, T., Han, J., Kim, H.L., Kim, J., Back, S.H., Komatsu, M., Chen, H., Chan, D.C., Konishi, M., Itoh, N., Choi, C.S., Lee, M.S., 2013. Autophagy deficiency leads to protection from obesity and insulin resistance by inducing Fgf21 as a mitokine. Nat. Med. 19, 83–92.

Kimberg, D.V., Loeb, J.N., 1972. Effects of cortisone administration on rat liver mitochondria. Support for the concept of mitochondrial fusion. J. Cell Biol. 55, 635–643.

Koshiba, T., Detmer, S.A., Kaiser, J.T., Chen, H., Mccaffery, J.M., Chan, D.C., 2004. Structural basis of mitochondrial tethering by mitofusin complexes. Science 305, 858–862.

Kulkarni, S.S., Joffraud, M., Boutant, M., Ratajczak, J., Gao, A.W., Maclachlan, C., Hernandez-Alvarez, M.I., Raymond, F., Metairon, S., Descombes, P., Houtkooper, R.H., Zorzano, A., Canto, C., 2016. Mfn1 deficiency in the liver protects against diet-induced insulin resistance and enhances the hypoglycemic effect of metformin. Diabetes 65, 3552–3560.

Leboucher, G.P., Tsai, Y.C., Yang, M., Shaw, K.C., Zhou, M., Veenstra, T.D., Glickman, M.H., Weissman, A.M., 2012. Stress-induced phosphorylation and proteasomal degradation of mitofusin 2 facilitates mitochondrial fragmentation and apoptosis. Mol. Cell 47, 547–557.

Lee, J.Y., Kapur, M., Li, M., Choi, M.C., Choi, S., Kim, H.J., Kim, I., Lee, E., Taylor, J.P., Yao, T.P., 2014. MFN1 deacetylation activates adaptive mitochondrial fusion and protects metabolically challenged mitochondria. J. Cell Sci. 127, 4954–4963.

Liesa, M., Shirihai, O.S., 2013. Mitochondrial dynamics in the regulation of nutrient utilization and energy expenditure. Cell Metab. 17, 491–506.
Liesa, M., Palacin, M., Zorzano, A., 2009. Mitochondrial dynamics in mammalian health and disease. Physiol. Rev. 89, 799–845.
Loson, O.C., Song, Z., Chen, H., Chan, D.C., 2013. Fis1, Mff, MiD49, and MiD51 mediate Drp1 recruitment in mitochondrial fission. Mol. Biol. Cell 24, 659–667.
Macvicar, T., Langer, T., 2016. OPA1 processing in cell death and disease - the long and short of it. J. Cell Sci. 129, 2297–2306.
Mahdaviani, K., Benador, I.Y., Su, S., Gharakhanian, R.A., Stiles, L., Trudeau, K.M., Cardamone, M., Enriquez-Zarralanga, V., Ritou, E., Aprahamian, T., Oliveira, M.F., Corkey, B.E., Perissi, V., Liesa, M., Shirihai, O.S., 2017. Mfn2 deletion in brown adipose tissue protects from insulin resistance and impairs thermogenesis. EMBO Rep. 18, 1123–1138.
Malka, F., Guillery, O., Cifuentes-Diaz, C., Guillou, E., Belenguer, P., Lombes, A., Rojo, M., 2005. Separate fusion of outer and inner mitochondrial membranes. EMBO Rep. 6, 853–859.
Manczak, M., Sesaki, H., Kageyama, Y., Reddy, P.H., 2012. Dynamin-related protein 1 heterozygote knockout mice do not have synaptic and mitochondrial deficiencies. Biochim. Biophys. Acta 1822, 862–874.
Mattie, S., Riemer, J., Wideman, J.G., Mcbride, H.M., 2018. A new mitofusin topology places the redox-regulated C terminus in the mitochondrial intermembrane space. J. Cell Biol. 217, 507–515.
Mishra, P., Carelli, V., Manfredi, G., Chan, D.C., 2014. Proteolytic cleavage of Opa1 stimulates mitochondrial inner membrane fusion and couples fusion to oxidative phosphorylation. Cell Metab. 19, 630–641.
Mishra, P., Varuzhanyan, G., Pham, A.H., Chan, D.C., 2015. Mitochondrial dynamics is a distinguishing feature of skeletal muscle fiber types and regulates organellar compartmentalization. Cell Metab. 22, 1033–1044.
Miyakawa, I., Aoi, H., Sando, N., Kuroiwa, T., 1984. Fluorescence microscopic studies of mitochondrial nucleoids during meiosis and sporulation in the yeast, *Saccharomyces cerevisiae*. J. Cell Sci. 66, 21–38.
Molina, A.J., Wikstrom, J.D., Stiles, L., Las, G., Mohamed, H., Elorza, A., Walzer, G., Twig, G., Katz, S., Corkey, B.E., Shirihai, O.S., 2009. Mitochondrial networking protects beta-cells from nutrient-induced apoptosis. Diabetes 58, 2303–2315.
Mourier, A., Motori, E., Brandt, T., Lagouge, M., Atanassov, I., Galinier, A., Rappl, G., Brodesser, S., Hultenby, K., Dieterich, C., Larsson, N.G., 2015. Mitofusin 2 is required to maintain mitochondrial coenzyme Q levels. J. Cell Biol. 208, 429–442.
Munoz, J.P., Ivanova, S., Sanchez-Wandelmer, J., Martinez-Cristobal, P., Noguera, E., Sancho, A., Diaz-Ramos, A., Hernandez-Alvarez, M.I., Sebastian, D., Mauvezin, C., Palacin, M., Zorzano, A., 2013. Mfn2 modulates the UPR and mitochondrial function via repression of PERK. EMBO J. 32, 2348–2361.
Naon, D., Zaninello, M., Giacomello, M., Varanita, T., Grespi, F., Lakshminaranayan, S., Serafini, A., Semenzato, M., Herkenne, S., Hernandez-Alvarez, M.I., Zorzano, A., De Stefani, D., Dorn 2nd, G.W., Scorrano, L., 2016. Critical reappraisal confirms that Mitofusin 2 is an endoplasmic reticulum-mitochondria tether. Proc. Natl. Acad. Sci. U. S. A. 113, 11249–11254.
Olichon, A., Emorine, L.J., Descoins, E., Pelloquin, L., Brichese, L., Gas, N., Guillou, E., Delettre, C., Valette, A., Hamel, C.P., Ducommun, B., Lenaers, G., Belenguer, P., 2002. The human dynamin-related protein OPA1 is anchored to the mitochondrial inner membrane facing the inter-membrane space. FEBS Lett. 523, 171–176.

Olichon, A., Baricault, L., Gas, N., Guillou, E., Valette, A., Belenguer, P., Lenaers, G., 2003. Loss of OPA1 perturbates the mitochondrial inner membrane structure and integrity, leading to cytochrome c release and apoptosis. J. Biol. Chem. 278, 7743–7746.

Olichon, A., Elachouri, G., Baricault, L., Delettre, C., Belenguer, P., Lenaers, G., 2007. OPA1 alternate splicing uncouples an evolutionary conserved function in mitochondrial fusion from a vertebrate restricted function in apoptosis. Cell Death Differ. 14, 682–692.

Ollmann, M.M., Wilson, B.D., Yang, Y.K., Kerns, J.A., Chen, Y., Gantz, I., Barsh, G.S., 1997. Antagonism of central melanocortin receptors in vitro and in vivo by agouti-related protein. Science 278, 135–138.

Olszewski, P.K., Wirth, M.M., Shaw, T.J., Grace, M.K., Billington, C.J., Giraudo, S.Q., Levine, A.S., 2001. Role of alpha-MSH in the regulation of consummatory behavior: immunohistochemical evidence. Am. J. Physiol. Regul. Integr. Comp. Physiol. 281, R673–R680.

Ost, M., Coleman, V., Voigt, A., Van Schothorst, E.M., Keipert, S., Van der Stelt, I., Ringel, S., Graja, A., Ambrosi, T., Kipp, A.P., Jastroch, M., Schulz, T.J., Keijer, J., Klaus, S., 2016. Muscle mitochondrial stress adaptation operates independently of endogenous FGF21 action. Mol. Metab. 5, 79–90.

Otera, H., Wang, C., Cleland, M.M., Setoguchi, K., Yokota, S., Youle, R.J., Mihara, K., 2010. Mff is an essential factor for mitochondrial recruitment of Drp1 during mitochondrial fission in mammalian cells. J. Cell Biol. 191, 1141–1158.

Otera, H., Ishihara, N., Mihara, K., 2013. New insights into the function and regulation of mitochondrial fission. Biochim. Biophys. Acta 1833, 1256–1268.

Ozcan, U., Cao, Q., Yilmaz, E., Lee, A.H., Iwakoshi, N.N., Ozdelen, E., Tuncman, G., Gorgun, C., Glimcher, L.H., Hotamisligil, G.S., 2004. Endoplasmic reticulum stress links obesity, insulin action, and type 2 diabetes. Science 306, 457–461.

Palmer, C.S., Osellame, L.D., Laine, D., Koutsopoulos, O.S., Frazier, A.E., Ryan, M.T., 2011. MiD49 and MiD51, new components of the mitochondrial fission machinery. EMBO Rep. 12, 565–573.

Papanicolaou, K.N., Ngoh, G.A., Dabkowski, E.R., O'connell, K.A., Ribeiro JR., R.F., Stanley, W.C., Walsh, K., 2012. Cardiomyocyte deletion of mitofusin-1 leads to mitochondrial fragmentation and improves tolerance to ROS-induced mitochondrial dysfunction and cell death. Am. J. Physiol. Heart Circ. Physiol. 302, H167–H179.

Peirce, V., Carobbio, S., Vidal-Puig, A., 2014. The different shades of fat. Nature 510, 76–83.

Pereira, R.O., Tadinada, S.M., Zasadny, F.M., Oliveira, K.J., Pires, K.M.P., Olvera, A., Jeffers, J., Souvenir, R., Mcglauflin, R., Seei, A., Funari, T., Sesaki, H., Potthoff, M.J., Adams, C.M., Anderson, E.J., Abel, E.D., 2017. OPA1 deficiency promotes secretion of FGF21 from muscle that prevents obesity and insulin resistance. EMBO J. 36, 2126–2145.

Pettersson, U.S., Walden, T.B., Carlsson, P.O., Jansson, L., Phillipson, M., 2012. Female mice are protected against high-fat diet induced metabolic syndrome and increase the regulatory T cell population in adipose tissue. PLoS One 7, e46057.

Pfluger, P.T., Kabra, D.G., Aichler, M., Schriever, S.C., Pfuhlmann, K., Garcia, V.C., Lehti, M., Weber, J., Kutschke, M., Rozman, J., Elrod, J.W., Hevener, A.L., Feuchtinger, A., Hrabe De Angelis, M., Walch, A., Rollmann, S.M., Aronow, B.J., Muller, T.D., Perez-Tilve, D., Jastroch, M., DE Luca, M., Molkentin, J.D., Tschop, M.H., 2015. Calcineurin links mitochondrial elongation with energy metabolism. Cell Metab. 22, 838–850.

Pich, S., Bach, D., Briones, P., Liesa, M., Camps, M., Testar, X., Palacin, M., Zorzano, A., 2005. The Charcot-Marie-Tooth type 2A gene product, Mfn2, up-regulates fuel oxidation through expression of OXPHOS system. Hum. Mol. Genet. 14, 1405–1415.

Piquereau, J., Caffin, F., Novotova, M., Prola, A., Garnier, A., Mateo, P., Fortin, D., Huynh Le, H., Nicolas, V., Alavi, M.V., Brenner, C., Ventura-Clapier, R., Veksler, V., Joubert, F., 2012. Down-regulation of OPA1 alters mouse mitochondrial morphology, PTP function, and cardiac adaptation to pressure overload. Cardiovasc. Res. 94, 408–417.

Pitts, K.R., Yoon, Y., Krueger, E.W., Mcniven, M.A., 1999. The dynamin-like protein DLP1 is essential for normal distribution and morphology of the endoplasmic reticulum and mitochondria in mammalian cells. Mol. Biol. Cell 10, 4403–4417.

Poggioli, R., Vergoni, A.V., Bertolini, A., 1986. ACTH-(1-24) and alpha-MSH antagonize feeding behavior stimulated by kappa opiate agonists. Peptides 7, 843–848.

Pollak, N.M., Jaeger, D., Kolleritsch, S., Zimmermann, R., Zechner, R., Lass, A., Haemmerle, G., 2015. The interplay of protein kinase A and perilipin 5 regulates cardiac lipolysis. J. Biol. Chem. 290, 1295–1306.

Praefcke, G.J., Mcmahon, H.T., 2004. The dynamin superfamily: universal membrane tubulation and fission molecules? Nat. Rev. Mol. Cell Biol. 5, 133–147.

Pyakurel, A., Savoia, C., Hess, D., Scorrano, L., 2015. Extracellular regulated kinase phosphorylates mitofusin 1 to control mitochondrial morphology and apoptosis. Mol. Cell 58, 244–254.

Quiros, P.M., Ramsay, A.J., Sala, D., Fernandez-Vizarra, E., Rodriguez, F., Peinado, J.R., Fernandez-Garcia, M.S., Vega, J.A., Enriquez, J.A., Zorzano, A., Lopez-Otin, C., 2012. Loss of mitochondrial protease OMA1 alters processing of the GTPase OPA1 and causes obesity and defective thermogenesis in mice. EMBO J. 31, 2117–2133.

Quiros, P.M., Prado, M.A., Zamboni, N., D'amico, D., Williams, R.W., Finley, D., Gygi, S.P., Auwerx, J., 2017. Multi-omics analysis identifies ATF4 as a key regulator of the mitochondrial stress response in mammals. J. Cell Biol. 216, 2027–2045.

Rana, A., Oliveira, M.P., Khamoui, A.V., Aparicio, R., Rera, M., Rossiter, H.B., Walker, D.W., 2017. Promoting Drp1-mediated mitochondrial fission in midlife prolongs healthy lifespan of *Drosophila melanogaster*. Nat. Commun. 8, 448.

Rojo, M., Legros, F., Chateau, D., Lombes, A., 2002. Membrane topology and mitochondrial targeting of mitofusins, ubiquitous mammalian homologs of the transmembrane GTPase Fzo. J. Cell Sci. 115, 1663–1674.

Ruan, Y., Li, H., Zhang, K., Jian, F., Tang, J., Song, Z., 2013. Loss of Yme1L perturbates mitochondrial dynamics. Cell Death Dis. 4, e896.

Ryan, J.J., Marsboom, G., Fang, Y.H., Toth, P.T., Morrow, E., Luo, N., Piao, L., Hong, Z., Ericson, K., Zhang, H.J., Han, M., Haney, C.R., Chen, C.T., Sharp, W.W., Archer, S.L., 2013. PGC1alpha-mediated mitofusin-2 deficiency in female rats and humans with pulmonary arterial hypertension. Am. J. Respir. Crit. Care Med. 187, 865–878.

Santel, A., Fuller, M.T., 2001. Control of mitochondrial morphology by a human mitofusin. J. Cell Sci. 114, 867–874.

Santel, A., Frank, S., Gaume, B., Herrler, M., Youle, R.J., Fuller, M.T., 2003. Mitofusin-1 protein is a generally expressed mediator of mitochondrial fusion in mammalian cells. J. Cell Sci. 116, 2763–2774.

Schneeberger, M., Dietrich, M.O., Sebastian, D., Imbernon, M., Castano, C., Garcia, A., Esteban, Y., Gonzalez-Franquesa, A., Rodriguez, I.C., Bortolozzi, A., Garcia-Roves, P.M., Gomis, R., Nogueiras, R., Horvath, T.L., Zorzano, A., Claret, M., 2013. Mitofusin 2 in POMC neurons connects ER stress with leptin resistance and energy imbalance. Cell 155, 172–187.

Schneeberger, M., Gomis, R., Claret, M., 2014. Hypothalamic and brainstem neuronal circuits controlling homeostatic energy balance. J. Endocrinol. 220, T25–T46.

Schrader, M., Costello, J.L., Godinho, L.F., Azadi, A.S., Islinger, M., 2016. Proliferation and fission of peroxisomes - an update. Biochim. Biophys. Acta 1863, 971–983.

Sebastian, D., Hernandez-Alvarez, M.I., Segales, J., Sorianello, E., Munoz, J.P., Sala, D., Waget, A., Liesa, M., Paz, J.C., Gopalacharyulu, P., Oresic, M., Pich, S., Burcelin, R., Palacin, M., Zorzano, A., 2012. Mitofusin 2 (Mfn2) links mitochondrial and endoplasmic reticulum function with insulin signaling and is essential for normal glucose homeostasis. Proc. Natl. Acad. Sci. U.S.A 109, 5523–5528.

Sebastian, D., Sorianello, E., Segales, J., Irazoki, A., Ruiz-Bonilla, V., Sala, D., Planet, E., Berenguer-Llergo, A., Munoz, J.P., Sanchez-Feutrie, M., Plana, N., Hernandez-Alvarez, M.I., Serrano, A.L., Palacin, M., Zorzano, A., 2016. Mfn2 deficiency links age-related sarcopenia and impaired autophagy to activation of an adaptive mitophagy pathway. EMBO J. 35, 1677–1693.

Segales, J., Paz, J.C., Hernandez-Alvarez, M.I., Sala, D., Munoz, J.P., Noguera, E., Pich, S., Palacin, M., Enriquez, J.A., Zorzano, A., 2013. A form of mitofusin 2 (Mfn2) lacking the transmembrane domains and the COOH-terminal end stimulates metabolism in muscle and liver cells. Am. J. Physiol. Endocrinol. Metab. 305, E1208–E1221.

Senyilmaz, D., Virtue, S., Xu, X., Tan, C.Y., Griffin, J.L., Miller, A.K., Vidal-Puig, A., Teleman, A.A., 2015. Regulation of mitochondrial morphology and function by stearoylation of TFR1. Nature 525, 124–128.

Skulachev, V.P., 2001. Mitochondrial filaments and clusters as intracellular power-transmitting cables. Trends Biochem. Sci. 26, 23–29.

Smirnova, E., Shurland, D.L., Ryazantsev, S.N., Van Der Bliek, A.M., 1998. A human dynamin-related protein controls the distribution of mitochondria. J. Cell Biol. 143, 351–358.

Song, Z., Chen, H., Fiket, M., Alexander, C., Chan, D.C., 2007. OPA1 processing controls mitochondrial fusion and is regulated by mRNA splicing, membrane potential, and Yme1L. J. Cell Biol. 178, 749–755.

Stiburek, L., Cesnekova, J., Kostkova, O., Fornuskova, D., Vinsova, K., Wenchich, L., Houstek, J., Zeman, J., 2012. YME1L controls the accumulation of respiratory chain subunits and is required for apoptotic resistance, cristae morphogenesis, and cell proliferation. Mol. Biol. Cell 23, 1010–1023.

Stojanovski, D., Koutsopoulos, O.S., Okamoto, K., Ryan, M.T., 2004. Levels of human Fis1 at the mitochondrial outer membrane regulate mitochondrial morphology. J. Cell Sci. 117, 1201–1210.

Taguchi, N., Ishihara, N., Jofuku, A., Oka, T., Mihara, K., 2007. Mitotic phosphorylation of dynamin-related GTPase Drp1 participates in mitochondrial fission. J. Biol. Chem. 282, 11521–11529.

Theurey, P., Tubbs, E., Vial, G., Jacquemetton, J., Bendridi, N., Chauvin, M.A., Alam, M.R., Le Romancer, M., Vidal, H., Rieusset, J., 2016. Mitochondria-associated endoplasmic reticulum membranes allow adaptation of mitochondrial metabolism to glucose availability in the liver. J. Mol. Cell Biol. 8, 129–143.

Thomas, D.Y., Wilkie, D., 1968. Recombination of mitochondrial drug-resistance factors in *Saccharomyces cerevisiae*. Biochem. Biophys. Res. Commun. 30, 368–372.

Touvier, T., De Palma, C., Rigamonti, E., Scagliola, A., Incerti, E., Mazelin, L., Thomas, J.L., D'antonio, M., Politi, L., Schaeffer, L., Clementi, E., Brunelli, S., 2015. Muscle-specific Drp1 overexpression impairs skeletal muscle growth via translational attenuation. Cell Death Dis. 6, e1663.

Toyama, E.Q., Herzig, S., Courchet, J., Lewis JR., T.L., Loson, O.C., Hellberg, K., Young, N.P., Chen, H., Polleux, F., Chan, D.C., Shaw, R.J., 2016. Metabolism. AMP-activated protein kinase mediates mitochondrial fission in response to energy stress. Science 351, 275–281.

Twig, G., Graf, S.A., Wikstrom, J.D., Mohamed, H., Haigh, S.E., Elorza, A., Deutsch, M., Zurgil, N., Reynolds, N., Shirihai, O.S., 2006. Tagging and tracking individual networks within a complex mitochondrial web with photoactivatable GFP. Am. J. Physiol. Cell Physiol. 291, C176–C184.

Twig, G., Elorza, A., Molina, A.J., Mohamed, H., Wikstrom, J.D., Walzer, G., Stiles, L., Haigh, S.E., Katz, S., Las, G., Alroy, J., Wu, M., Py, B.F., Yuan, J., Deeney, J.T., Corkey, B.E., Shirihai, O.S., 2008. Fission and selective fusion govern mitochondrial segregation and elimination by autophagy. EMBO J. 27, 433–446.

Varanita, T., Soriano, M.E., Romanello, V., Zaglia, T., Quintana-Cabrera, R., Semenzato, M., Menabo, R., Costa, V., Civiletto, G., Pesce, P., Viscomi, C., Zeviani, M., Di Lisa, F., Mongillo, M., Sandri, M., Scorrano, L., 2015. The OPA1-dependent mitochondrial cristae remodeling pathway controls atrophic, apoptotic, and ischemic tissue damage. Cell Metab. 21, 834–844.

Wai, T., Garcia-Prieto, J., Baker, M.J., Merkwirth, C., Benit, P., Rustin, P., Ruperez, F.J., Barbas, C., Ibanez, B., Langer, T., 2015. Imbalanced OPA1 processing and mitochondrial fragmentation cause heart failure in mice. Science 350, aad0116.

Wakabayashi, T., Green, D.E., 1977. Membrane fusion in mitochondria. I. Ultrastructural basis for fusion. J. Electron. Microsc. 26, 305–320.

Wakabayashi, J., Zhang, Z., Wakabayashi, N., Tamura, Y., Fukaya, M., Kensler, T.W., Iijima, M., Sesaki, H., 2009. The dynamin-related GTPase Drp1 is required for embryonic and brain development in mice. J. Cell Biol. 186, 805–816.

Wang, H., Sreenivasan, U., Hu, H., Saladino, A., Polster, B.M., Lund, L.M., Gong, D.W., Stanley, W.C., Sztalryd, C., 2011a. Perilipin 5, a lipid droplet-associated protein, provides physical and metabolic linkage to mitochondria. J. Lipid Res. 52, 2159–2168.

Wang, Z., Liu, Y., Liu, J., Liu, K., Wen, J., Wen, S., Wu, Z., 2011b. HSG/Mfn2 gene polymorphism and essential hypertension: a case-control association study in Chinese. J. Atheroscler. Thromb. 18, 24–31.

Wang, W., Wang, Y., Long, J., Wang, J., Haudek, S.B., Overbeek, P., Chang, B.H., Schumacker, P.T., Danesh, F.R., 2012. Mitochondrial fission triggered by hyperglycemia is mediated by ROCK1 activation in podocytes and endothelial cells. Cell Metab. 15, 186–200.

Wang, L., Ishihara, T., Ibayashi, Y., Tatsushima, K., Setoyama, D., Hanada, Y., Takeichi, Y., Sakamoto, S., Yokota, S., Mihara, K., Kang, D., Ishihara, N., Takayanagi, R., Nomura, M., 2015. Disruption of mitochondrial fission in the liver protects mice from diet-induced obesity and metabolic deterioration. Diabetologia 58, 2371–2380.

Weir, H.J., Yao, P., Huynh, F.K., Escoubas, C.C., Goncalves, R.L., Burkewitz, K., Laboy, R., Hirschey, M.D., Mair, W.B., 2017. Dietary restriction and AMPK increase lifespan via mitochondrial network and peroxisome remodeling. Cell Metab. 26, 884–896 e5.

Wikstrom, J.D., Mahdaviani, K., Liesa, M., Sereda, S.B., Si, Y., Las, G., Twig, G., Petrovic, N., Zingaretti, C., Graham, A., Cinti, S., Corkey, B.E., Cannon, B., Nedergaard, J., Shirihai, O.S., 2014. Hormone-induced mitochondrial fission is utilized by brown adipocytes as an amplification pathway for energy expenditure. EMBO J. 33, 418–436.

Yoon, Y., Krueger, E.W., Oswald, B.J., Mcniven, M.A., 2003. The mitochondrial protein hFis1 regulates mitochondrial fission in mammalian cells through an interaction with the dynamin-like protein DLP1. Mol. Cell Biol. 23, 5409–5420.

Yu, T., Robotham, J.L., Yoon, Y., 2006. Increased production of reactive oxygen species in hyperglycemic conditions requires dynamic change of mitochondrial morphology. Proc. Natl. Acad. Sci. U.S.A 103, 2653–2658.

Yu, T., Jhun, B.S., Yoon, Y., 2011. High-glucose stimulation increases reactive oxygen species production through the calcium and mitogen-activated protein kinase-mediated activation of mitochondrial fission. Antioxid. Redox Signal. 14, 425–437.

Yu, J., Zhang, S., Cui, L., Wang, W., Na, H., Zhu, X., Li, L., Xu, G., Yang, F., Christian, M., Liu, P., 2015. Lipid droplet remodeling and interaction with mitochondria in mouse brown adipose tissue during cold treatment. Biochim. Biophys. Acta 1853, 918–928.

Zhao, J., Liu, T., Jin, S., Wang, X., Qu, M., Uhlen, P., Tomilin, N., Shupliakov, O., Lendahl, U., Nister, M., 2011. Human MIEF1 recruits Drp1 to mitochondrial outer membranes and promotes mitochondrial fusion rather than fission. EMBO J. 30, 2762–2778.

Zhao, J., Lendahl, U., Nister, M., 2013. Regulation of mitochondrial dynamics: convergences and divergences between yeast and vertebrates. Cell. Mol. Life Sci. 70, 951–976.

Zuchner, S., Mersiyanova, I.V., Muglia, M., Bissar-Tadmouri, N., Rochelle, J., Dadali, E.L., Zappia, M., Nelis, E., Patitucci, A., Senderek, J., Parman, Y., Evgrafov, O., Jonghe, P.D., Takahashi, Y., Tsuji, S., Pericak-Vance, M.A., Quattrone, A., Battaloglu, E., Polyakov, A.V., Timmerman, V., Schroder, J.M., Vance, J.M., 2004. Mutations in the mitochondrial GTPase mitofusin 2 cause Charcot-Marie-Tooth neuropathy type 2A. Nat. Genet. 36, 449–451.

CHAPTER FIVE

Novel Insights Into the Anti-aging Role of Mitophagy

Maria Markaki*,[a], Konstantinos Palikaras*,[a] and Nektarios Tavernarakis*,§,[1]

*Institute of Molecular Biology and Biotechnology, Foundation for Research and Technology-Hellas
§Department of Basic Sciences, Medical School, University of Crete, Heraklion, Greece
[1]Corresponding author: E-mail: tavernarakis@imbb.forth.gr

Contents

Abstract

Aging is a complex biological process affecting almost all living organisms. Although its detrimental effects on animals' physiology have been extensively documented, several aspects of the biology of aging are insufficiently understood. Mitochondria, the central energy producers of the cell, play vital roles in a wide range of cellular processes,

[a] These authors contributed equally

International Review of Cell and Molecular Biology, Volume 340
ISSN 1937-6448
https://doi.org/10.1016/bs.ircmb.2018.05.005

including regulation of bioenergetics, calcium signaling, metabolic responses, and cell death, among others. Thus, proper mitochondrial function is a prerequisite for the maintenance of cellular and organismal homeostasis. Several mitochondrial quality control mechanisms have evolved to allow adaptation to different metabolic conditions, thereby preserving cellular homeostasis and survival. A tight coordination between mitochondrial biogenesis and mitochondrial selective autophagy, known as mitophagy, is a common characteristic of healthy biological systems. The balanced interplay between these two opposing cellular processes dictates stress resistance, healthspan, and lifespan extension. Mitochondrial biogenesis and mitophagy efficiency decline with age, leading to progressive accumulation of damaged and/or unwanted mitochondria, deterioration of cellular function, and ultimately death. Several regulatory factors that contribute to energy homeostasis have been implicated in the development and progression of many pathological conditions, such as neurodegenerative, metabolic, and cardiovascular disorders, among others. Therefore, mitophagy modulation may serve as a novel potential therapeutic approach to tackle age-associated pathologies. Here, we review the molecular signaling pathways that regulate and coordinate mitophagy with mitochondrial biogenesis, highlighting critical factors that hold promise for the development of pharmacological interventions toward enhancing human health and quality of life throughout aging.

1. INTRODUCTION

Aging is associated with marked alterations in cellular and organismal metabolism that are critically linked to whole-organism homeostasis. Research over the last 3 decades has culminated in the identification of conserved signaling pathways and transcription factors that tightly regulate the aging process. Mitochondria, mostly known as the energy-producing organelles in eukaryotic cells, are dynamic structures that play crucial roles in various cellular processes including apoptosis, Ca^{2+} signaling, fatty acid β-oxidation, iron-sulfur cluster synthesis, cell signaling, and interorganellar interactions, particularly with the endoplasmic reticulum (ER) (Saraste, 1999; Parsons and Green, 2010; Seo et al., 2010; Marchi et al., 2012). Accumulating findings from studies in model organisms indicate that changes in mitochondrial network morphology and function influence mitochondrial bioenergetics efficiency in a quite highly conserved manner, contributing to aging and the pathophysiology of age-related diseases such as obesity, diabetes, cancer, neurodegenerative disorders, among others (Liesa and Shirihai, 2013; Sonntag et al., 2017). Although, age-related deterioration of mitochondrial function is a universal phenomenon, its cellular and molecular underpinnings remain largely unknown. Given that the consequences of mitochondrial dysfunction might be particularly detrimental for terminally differentiated cells such as the postmitotic neurons and cardiomyocytes, affected cells develop quality control systems to counteract this damage. Increasing evidence suggests that failure of mitochondrial quality control mechanisms accelerates aging. Macroautophagy (hereafter, referred to as autophagy) is the main process for bulk protein and organelle recycling in cells. It involves the formation of a double-membrane structure, the autophagosome, which engulfs the material destined for degradation in the lysosome (Klionsky and Emr, 2000). Besides the ATG5/ATG7 canonical pathway, autophagy in mammals can occur through an ATG5/ATG7-independent pathway, which also relies on several autophagic proteins, including Unc-51-like kinase 1 (ULK1) and Beclin1. This alternative autophagy requires the small GTPase RAB9 for fusion of isolation membranes with vesicles derived from the *trans*-Golgi and late endosomes. Interestingly, ATG5/ATG7 unconventional autophagy has been shown to mediate mitochondrial elimination during erythrocyte differentiation in vivo (Nishida et al., 2009) and is essential for induced pluripotent stem cell reprogramming (Ma et al., 2015).

Autophagy is considered to be a cellular quality control mechanism whose efficiency declines with age (Rubinsztein et al., 2011). While autophagy is primarily nonselective, selective autophagy pathways that mediate the removal of unwanted cellular cargo or damaged organelles and intracellular pathogens also exist. Selective autophagy relies on the core autophagy machinery and requires specific receptors or adaptors bearing one or more LC3-interacting regions (LIRs) for binding intracellular cargo to proteins of the LC3/GABARAP (microtubule-associated protein 1A/1B-light chain 3/gamma-aminobutyric acid receptor–associated protein) family (Svenning and Johansen, 2013; Stolz et al., 2014). Based on the type of autophagy receptor and the mechanism mediating cargo recognition, selective autophagy can be distinguished into ubiquitin (Ub)-dependent and Ub-independent pathways (Khaminets et al., 2016).

One type of selective autophagy with more or less clearly delineated steps is the selective mitochondrial autophagy (mitophagy). Mitophagy primarily serves a housekeeping function by modulating turnover of mitochondria under steady-state conditions and appropriately adjusting the number of mitochondria according to changing metabolic needs of the cell. In response to stress, mitophagy induction leads to selective elimination of functionally impaired mitochondria, thus contributing to the maintenance of a healthy mitochondrial population. Studies in model organisms have shown that mitophagy is an evolutionarily conserved process subjected to tight regulation (Youle and Narendra, 2011). In mammals, although mitophagy has been studied mainly in cell lines under artificial conditions, conserved molecular mediators of this process have been identified (Table 1) (Liu et al., 2012; Ordureau et al., 2014). Most notably, increasing evidence suggests that impairment of mitophagy is implicated in aging and age-related diseases (Fang et al., 2014; Menzies et al., 2015). Conversely, its coordination with the opposing process of mitochondrial biogenesis is crucial for promoting stress resistance, healthspan and lifespan in the nematode *Caenorhabditis elegans* (Palikaras et al., 2015b). Besides its role in removing damaged or dysfunctional mitochondria, mitophagy is also responsible for mitochondria elimination in diverse developmental contexts such as during red blood cell differentiation (Schweers et al., 2007; Kundu et al., 2008) and brain development in mammalian cells (Deczkowska and Schwartz, 2017; Esteban-Martinez et al., 2017) or following fertilization for degrading sperm-derived paternal mitochondria in the nematode *C. elegans* and in early mouse embryos (Sato and Sato, 2011; Rojansky et al., 2016). In this review, we focus on the role of mitophagy in aging. We first describe the molecular mechanisms that govern the mitophagy process as well as its complex interplay with mitochondrial dynamics. Furthermore, we discuss the age-related decline of mitophagy efficiency and the coordination of mitochondrial biogenesis with mitochondrial degradation to compensate for defects in mitochondria quality control and function. Better understanding of mitochondrial turnover mechanisms is a key requirement for the development of more efficient strategies to battle numerous pathological conditions in humans. Finally, we end with recent advances on the potential of pharmacological and nutritional interventions or the physical exercise to modulate mitophagy for health and/or longevity benefits.

2. THE PATHWAYS OF MITOPHAGY

Different mechanisms of selective mitochondrial autophagy have been identified to date suggesting that different stimuli may induce mitophagy via multiple signaling pathways in distinct cell contexts (Frank et al., 2012; Jin and Youle, 2013; Melser et al., 2013). However, numerous important connections exist between these pathways, and notably, emerging findings indicate that key mitophagy players are functionally conserved (Table 1). In the following sections, we describe the fundamental mechanisms involved in this process.

Table 1 Mitophagy Components are Evolutionarily Conserved in Eukaryotes

Organisms					
Saccharomyces cerevisae	*Caenorhabditis elegans*	*Drosophila melanogaster*	*Mus musculus*	*Homo sapiens*	**Molecular Function**
—	PINK-1	Pink1	PINK1	PINK1	Serine/threonine protein kinase
—	PDR-1	Park	PARK2	PARK2	E3 ubiquitin ligase
ATG32	—	—	—	—	—
ATG8	LGG-1, LGG-2	Atg8a	GABARAP/LC3	GABARAP/LC3	Autophagosomal protein
—	UNC-51	—	ULK1	ULK1	Serine/threonine protein kinase
—	DCT-1	—	BNIP3, NIX/BNIP3L	BNIP3, NIX/BNIP3L	Mitophagy receptor
CG5676	FUNDC-1	—	FUNDC1	FUNDC1	Mitophagy receptor
—	—	—	OPTN	OPTN	Mitophagy receptor
—	—	—	NDP52	NDP52	Mitophagy receptor
—	SQST-1	Ref(2)P	SQST1	p62/SQSTM1	Autophagy adaptor
—	FZO-1	FZO1, DMFN	MFN1/2	MFN1/2	Mitochondrial GTPase
PHB2	PHB-2	PHB2	PHB2	PHB2	Mitophagy receptor
—	FKB-6	CG5482	FKBP8	FKBP8	Mitophagy receptor

2.1 The Phosphatase and Tensin Homolog–Induced Putative Kinase 1/Parkin Pathway

The first pathway shown to mediate mitophagy in animals is the phosphatase and tensin homolog–induced putative kinase 1 (PINK1)/Parkin signaling cascade (Pickrell and Youle, 2015). Several processes, such as mitochondrial dynamics, trafficking, and recruitment of core autophagy components converge into PINK1/Parkin-mediated mitophagy to ensure removal of functionally impaired mitochondria. Its cellular and molecular underpinnings are described below.

The Ser/Thr PINK1 and the E3 ubiquitin ligase Parkin (encoded by *PARK2* in humans) are the key players of this pathway. Parkin is a member of the RING-between-RING E3 ubiquitin family. It consists of an N-terminal ubiquitin-like (Ubl) domain and 4 zinc-coordinating RING-like domains, namely RING0, RING1, a cysteine-rich in-between RING domain (IBR), and RING2 (Pickrell and Youle, 2015). Structural analysis revealed that Parkin normally exists in an autoinhibited state in the cytosol (Chaugule et al., 2011; Trempe et al., 2013; Wauer and Komander, 2013). Following mitochondrial uncoupling, Parkin translocates to damaged mitochondria where it is activated to ubiquitinate mitochondrial substrates for initiating organelle clearance by mitophagy. It is now well established that Parkin recruitment to mitochondria and its subsequent activation depends on the stabilization of PINK1 on the outer membrane of depolarized mitochondria (Matsuda et al., 2010; Narendra et al., 2010). What is the sequence of events that link latent Parkin activation with PINK1 processing and stabilization upon mitochondrial uncoupling or unfolded protein overload?

In healthy mitochondria, PINK1 is imported into the inner mitochondrial membrane, in a membrane potential-dependent manner (Fig. 1). It is processed into a 60-kDa form by the mitochondrial processing peptidase (MPP) (Greene et al., 2012) and then cleaved between Ala103 and Phe104 residues of its amino-terminus, giving rise to an unstable 52-kDa fragment. The cleavage is mediated mainly by the inner membrane rhomboid protease presenilin-associated rhomboid-like protein (PARL) (Jin et al., 2010; Deas et al., 2011) and AFG3L2, a subunit of the m-AAA protease (Greene et al., 2012). The 52-kDa truncated PINK1 is rapidly degraded by the proteasome (Jin et al., 2010; Narendra et al., 2010b).

In depolarized mitochondria, however, the import of proteins, including PINK1, across the inner mitochondrial membrane is abolished.

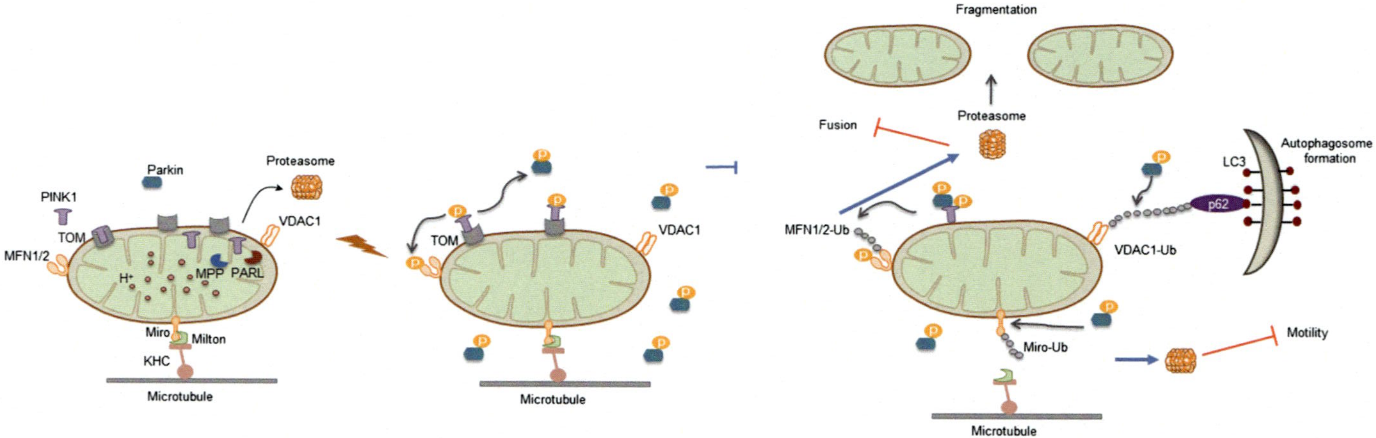

Figure 1 ***The phosphatase and tensin homolog-induced putative kinase 1 (PINK1)/Parkin pathway at a glance.*** In healthy mitochondria, phosphatase and tensin homolog-induced putative kinase 1 (PINK1) is imported across the inner mitochondrial membrane by the translocase of the outer membrane (TOM) complex, processed by mitochondrial processing peptidase (MPP) and presenilins-associated rhomboid-like protein (PARL), and subsequently degraded by the proteasome. Upon mitochondrial membrane depolarization, PINK1 is stabilized on the outer mitochondrial membrane and activated by autophosphorylation, triggering Parkin recruitment and activation. Then, Parkin ubiquitinates several mitochondrial surface proteins, such as voltage-dependent anion-selective channel protein 1 (VDAC1), mitofusins (MFN1/2), and mitochondrial ras homologous (RHO) guanine triphosphate (GTP) ase Miro, coordinating mitochondrial motility, dynamics, and recruitment of autophagic machinery components. *KHC*, kinesin heavy chain; *Ub*, ubiquitin.

As a consequence, unprocessed (63-kDa) PINK1 accumulates on the outer mitochondrial membrane, with its kinase domain facing the cytosol (Fig. 1) (Jin et al., 2010; Matsuda et al., 2010; Narendra et al., 2010b). In turn, PINK1 is assembled in a 700-kDa complex that includes also core components of the translocase of the outer membrane (TOM) complex, but not Parkin (Lazarou et al., 2012). Notably, PINK1 artificially tethered to other compartments that lack TOM complex, such as peroxisomes or lysosomes, is able to induce Parkin recruitment, organelle ubiquitination, and, at least in the case of peroxisomes, autophagic clearance, indicating that the TOM complex is dispensable for PINK1-mediated Parkin recruitment and activation (Lazarou et al., 2012). On the other hand, Parkin artificially recruited to the mitochondria in the absence of PINK1 is unable to ubiquitinate mitochondrial proteins and initiate mitophagic clearance (Matsuda et al., 2010; Narendra et al., 2010b). Hence, PINK1 has a prominent role in activation of Parkin E3 ligase activity and subsequent mitophagy induction.

Besides its ability to detect mitochondrial uncoupling, PINK1 can sense the accumulation of misfolded proteins in the mitochondrial matrix of healthy mitochondria. Its subsequent stabilization on the outer mitochondrial membrane results in Parkin recruitment and mitophagy activation, ultimately leading to a reduction in unfolded protein overload. PINK1/Parkin-mediated mitophagy is enhanced by the knockdown of Lon protease (Jin and Youle, 2013), an adenosine triphosphate (ATP)-stimulated mitochondrial matrix protein, which recognizes and selectively degrades oxidatively damaged proteins (Bota and Davies, 2002). The effect of Lon depletion does not originate from mitochondrial depolarization (Jin and Youle, 2013; Burbulla et al., 2014). Further supporting the protective role of PINK1 and Parkin against intramitochondrial proteotoxicity, a recent study showed that impaired mitochondrial homeostasis caused by the loss of mortalin function is rescued by the PINK1/Parkin-mediated clearance of dysfunctional mitochondria (Burbulla et al., 2014).

The question then arises: how does PINK1 trigger the E3 ligase activity of Parkin? It was shown that PINK1 becomes activated upon membrane depolarization by autophosphorylation at Thr257, Ser228, and Ser402. These phosphorylation events are essential for the recruitment of Parkin to damaged mitochondria (Kondapalli et al., 2012; Okatsu et al., 2012). Then, activated PINK1 directly phosphorylates Parkin on conserved Ser65 of the Ubl domain. Although this phosphorylation event is not sufficient for driving mitochondrial translocation of Parkin, it reportedly releases the autoinhibitory configuration of Parkin, unleashing its enzymatic activity (Sha et al., 2010; Kondapalli et al., 2012; Shiba-Fukushima et al., 2012; Fu et al., 2013; Caulfield et al., 2014). The RING1 domain of Parkin is required for the physical interaction of Parkin with PINK1 (Sha et al., 2010), while the RING2 domain is essential for ubiquitin-thioester transfer through the formation of an ubiquitin-thioester intermediate on Cys431 (Iguchi et al., 2013; Lazarou et al., 2013). Under basal conditions, RING0 occludes the ubiquitin-acceptor site Cys431 in RING2, ensuring autoinhibition of E3 ligase activity (Trempe et al., 2013). Moreover, PINK1 was demonstrated to drive self-association of Parkin through its IBR domain. This self-association is essential for the activation of Parkin ubiquitin ligase activity upstream of its mitochondrial translocation (Lazarou et al., 2013).

Previous studies have shown that ubiquitin itself is phosphorylated by PINK1 on Ser65, similar to the Ubl domain of Parkin (Kane et al., 2014; Kazlauskaite et al., 2014; Koyano et al., 2014; Shiba-Fukushima et al., 2014). Although this phosphorylation was initially suggested to activate the E3 ligase activity of Parkin, it has been later proposed that phosphorylation of polyubiquitin follows ubiquitin conjugation onto Parkin substrates (Ordureau et al., 2014; Stolz et al., 2014). Parkin catalyzes the formation of K6-, K11-, K27-, K48-, and K63-linked ubiquitin chains on its substrates (Geisler et al., 2010; Ordureau et al., 2014). The role of each of these different polyubiquitin structures in substrate degradation and subsequent mitochondrial fate determination remains poorly understood. A recent study identified a binding switch between the phosphoubiquitin (pUb) and the Ubl domain of Parkin as the key in PINK1-mediated Parkin activation on the mitochondria. According to this work, pUb binds to RING1 of Parkin at a site formed by His302 and Arg305, promoting disengagement of the Ubl domain from RING1 and subsequent Parkin phosphorylation, which leads to the activation of its ligase activity (Sauve et al., 2015).

A recent report suggests that PINK1 serves as a pro-fission signal upon uncoupling. By disrupting the protein kinase A (PKA) - A-kinase-anchoring protein 1 (AKAP1) axis, it activates DRP1, promoting fission of damaged organelles for their autophagic clearance, independently of Parkin. Thus, PINK1 is proposed to be a master mitophagy regulator (Pryde et al., 2016).

Following its translocation to damaged mitochondria, Parkin ubiquitinates outer mitochondrial membrane proteins, including mitofusin mitochondrial assembly regulatory factor (MARF) in flies (Poole et al., 2010; Ziviani et al., 2010), mitofusins (MFN1 and MFN2) (Gegg et al., 2010; Tanaka et al., 2010), voltage-dependent anion-selective channel protein 1

(VDAC1) (Geisler et al., 2010), and Miro (Wang et al., 2011; Shlevkov et al., 2016), among others, promoting their degradation. Parkin-mediated polyubiquitination of mitofusins, in particular, leads to their proteasome-dependent degradation (Gegg et al., 2010; Tanaka et al., 2010; Glauser et al., 2011). As a consequence, mitochondrial fusion is blocked, preventing functional organelles to fuse with their damaged counterparts. Another model supports the idea that MFN2 but not MFN1 acts as the receptor for Parkin on the surface of damaged mitochondria. MFN2 —Parkin binding provokes the ubiquitination of mitochondrial proteins, thus targeting defective organelles for autophagic degradation (Chen and Dorn, 2013). Since mitochondrial fission is uninterrupted, it may facilitate mitophagy, presumably by supplying smaller organelles of manageable sizes that can easily be engulfed by autophagosomes. Regarding Miro, a component of the conserved primary motor/adaptor complex that anchors kinesin to the mitochondrial surface, PINK1-mediated phosphorylation triggers its proteosomal degradation in a Parkin-dependent manner and the subsequent release of kinesin from mitochondria. As a consequence, mitochondrial movement is inhibited in both *Drosophila* and mammalian neurons, where damaged organelles may be first sequestered locally within an autophagosome prior to their clearance (Wang et al., 2011). Together, these reports highlight the complex interplay between mitochondrial dynamics and mitophagy, both of which critically contribute to mitochondrial homeostasis.

Interestingly, a new study highlights a previously unappreciated role for reactive oxygen species (ROS) in the completion of PINK/Parkin-mediated mitophagy following Parkin recruitment to damaged mitochondria. Indeed, superoxide rather than hydrogen peroxide is proposed to be the driving force for the execution of mitophagy, although it alone is not able to modulate Parkin dynamics or mitochondrial elimination via autophagy (Xiao et al., 2017).

2.2 Parkin-independent Pathways and More

Although the PINK1/Parkin mitophagy pathway has attracted much attention in recent years, additional pathways capable of mediating the selective clearance of damaged mitochondria have been characterized.

2.2.1 The Phosphatase and Tensin Homolog—Induced Putative Kinase 1—Synphilin-1—Seven in Absentia Homolog 1 Complex

A recent study revealed a novel Parkin-independent mitophagy pathway that involves the PINK1/synphilin-1/seven in absentia homolog 1

(SIAH1) complex. In this specialized case, PINK1 recruits synphilin-1 to the mitochondria both in rat brain tissues and cultured cells, promoting mitochondrial depolarization and stabilization of uncleaved PINK1 at the organelle. In turn, synphilin-1 recruits the E3 ubiquitin ligase SIAH1, which ubiquitinates mitochondrial proteins, resulting in LC3 recruitment for autophagosome formation. Notably, while PINK1 is required for synphilin-1 mitochondrial translocation and mitophagy, the PINK1-synphilin-1 pathway is independent of PINK1 kinase activity. This pathway might represent a new drug target to enhance degradation of damaged mitochondria, thus compensating for mitophagy failure in Parkinson's disease patients carrying mutations in genes encoding PINK1 and Parkin (Szargel et al., 2016).

2.2.2 ARIH1-Mediated Mitophagy

The E3 ubiquitin ligase Ariadne RBR E3 Ub protein ligase 1 (ARIH1) (also known as HHARI), which is extensively expressed in several cancer cells types, especially in breast and lung adenocarcinomas, has recently evolved as a novel mitophagy regulator. In cancer cells where Parkin is frequently downregulated, ARIH1 polyubiquitinates damaged mitochondria, inducing mitophagy. ARIH1-mediated mitophagy depends on its ubiquitin ligase activity and PINK1 stabilization. This mechanism contributes to chemotherapy resistance, challenging the prevailing view that the main mitophagy regulators act as tumor suppressors (Villa et al., 2017).

2.3 Receptor-Mediated Mitophagy Pathways

In recent years, the role of selective autophagy receptors in mammalian mitophagy has been revisited. Based on current evidence, PINK1 phosphorylates ubiquitin, and this phosphorylation event subsequently triggers the recruitment of the autophagy receptors nuclear dot protein 52 (NDP52) and optineurin (OPTN), but not p62, to mitochondria, thereby stimulating mitophagy in a Parkin-independent manner (Fig. 2). Once localized on mitochondria, receptors recruit the autophagy components ULK1, double FYVE-domain—containing protein 1 (DFCP1), and WD repeat domain phosphoinositide—interacting protein 1 (WIPI1) that function upstream of LC3, to initiate autophagosome formation proximal to dysfunctional organelles. When Parkin is present, it acts to amplify the PINK1-generated phosphoubiquitin signal, resulting in robust mitophagy induction (Lazarou et al., 2015). Considerable controversy has ensued concerning the role of p62 in mitophagy (Fig. 2). Certain reports suggested that p62 is crucial for the

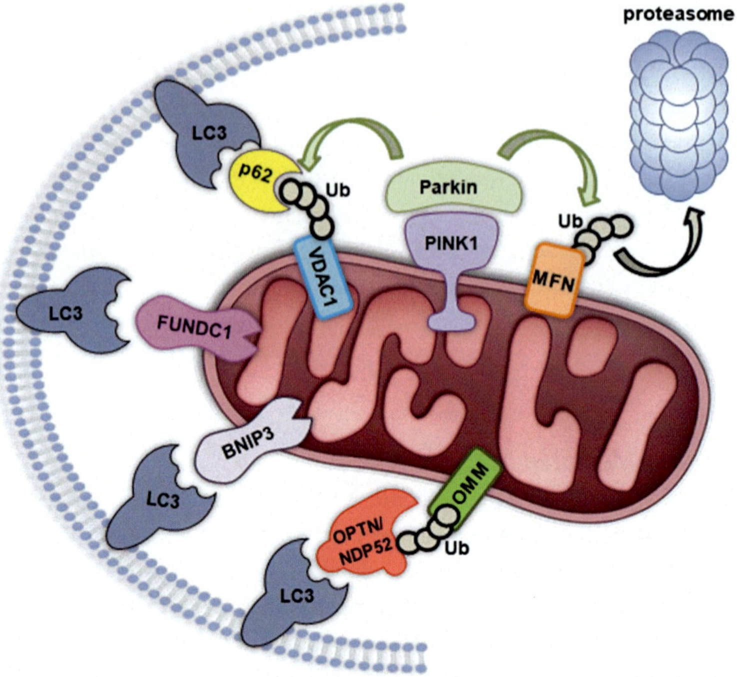

Figure 2 ***Receptor-mediated mitophagy.*** Phosphatase and tensin homolog-induced putative kinase 1 (PINK1) is stabilized on the outer mitochondrial membrane and recruits Parkin in response to mitochondrial damage. In turn, Parkin ubiquitinates several outer mitochondrial membrane proteins leading both to their proteasomal degradation [e.g., mitofusin (MFN)] and association with the adaptor molecule p62. Then, p62 interacts with the autophagosomal protein LC3, mediating mitochondrial removal. In parallel to the PINK1/Parkin signaling pathway, impaired mitochondria could be also removed by receptor-mediated mitophagy. FUN14 domain-containing protein 1 (FUNDC1), BNIP3, optineurin (OPTN), and nuclear dot protein 52 (NDP52) mitophagy receptors interact directly or indirectly with LC3, promoting the sequestration of defective organelles. The existence of multiple mitophagy mechanisms ensures the efficient quality control of mitochondrial population under stress conditions. *OMM*, outer mitochondrial membrane protein; *Ub*, ubiquitin; *VDAC1*, voltage-dependent anion-selective channel protein 1.

elimination of depolarized mitochondria (Ding et al., 2010; Youle and Narendra, 2011; Zhang et al., 2015), whereas other studies have led to the conclusion that p62 is dispensable for mitophagy (Narendra et al., 2010a; Lazarou et al., 2015).

Recently, the serine/threonine kinase TANK-binding kinase 1 (TBK1) has emerged as a critical regulator of the mitophagy network, as it has been shown to phosphorylate the OPTN, NDP52, and p62 autophagy receptors on several autophagy-relevant sites (Wild et al., 2011; Heo et al., 2015; Richter et al., 2016). In mammals, phosphorylation of selective autophagy receptors enhances their binding to both the cargo and the autophagic machinery (Fig. 2) (Heo et al., 2015; Richter et al., 2016). TBK1-mediated phosphorylation of OPTN, in particular, reinforces its binding to the ubiquitin chains of outer mitochondrial membrane proteins, TBK activation, and OPTN retention on damaged mitochondria, thereby promoting their removal (Heo et al., 2015).

As a mode of selective autophagy, mitophagy, in most cases, relies on specific receptors that recognize and direct damaged mitochondria for degradation. In this sense, mitochondrial proteins that engage core autophagy machinery components are considered as mitophagy receptors. Like yeast receptors, which have one or more Atg8 family interacting motifs (AIMs), mammalian receptors have one or more LIRs to associate with LC3/GABARAP family proteins. Functional homologs of mitophagy receptors have been identified in diverse species ranging from yeast to mammals (Table 1). In yeast, two independent screens for mitophagy-defective mutants identified autophagy-related protein 32 (Atg32) as essential for respiratory growth-induced mitophagy in post-log phase cells (Kanki et al., 2009; Okamoto et al., 2009). Atg32 is a mitochondrial-anchor protein with an N-terminal domain facing the cytosol, a transmembrane domain, and a C-terminal domain exposed to mitochondrial intermembrane space. The cytoplasmic domain of Atg32 contains a tetrapeptide sequence, W/YXXI/L/V (AIM), which is conserved among proteins that physically associate with the Atg8 family members. Furthermore, Atg32 has been shown to interact with Atg11, a scaffold protein required for selective autophagy pathways. Together, these findings support the idea that Atg32 forms a complex with Atg8 and Atg11. This interaction precedes and is independent of isolation membrane generation and subsequent autophagosome formation (Kanki et al. 2009, 2010; Kondo-Okamoto et al., 2012). Atg32 is phosphorylated by the conserved serine and/or threonine casein kinase-2 (CK2). This phosphorylation event reinforces Atg32-Atg11 interaction, ultimately inducing mitophagy (Kanki et al., 2013).

Until recently, no Atg32 homolog was known in metazoans (Tolkovsky, 2009). A new study just showed that BCL-2-like protein 13 (BCL2L13) is an Atg32 functional homolog in mammalian cells. Interestingly, BCL2L13 is involved in mitochondrial fragmentation and in mitophagy. It consists of a C-terminal single transmembrane domain, four conserved BCL-2 homology domains (BH1-4), and 2 WXXL/I motifs. The BH domains are crucial for fragmentation, while the WXXI motif mediates the interaction of BCL2L13 with LC3, thus stimulating mitophagy. Notably, BCL2L13 is able to promote mitochondrial fragmentation in DRP1-depleted cells as well as mitophagy in Parkin-deficient cells. However, the molecular mechanism by which BCL2L13 coordinates mitochondrial fission and mitophagy remains elusive (Murakawa et al., 2015).

In mammals, mitochondria are eliminated during terminal differentiation of red blood cells through mitophagy by a mechanism that relies on the outer mitochondrial membrane NIP3-like protein X (NIX; also referred to as BNIP3L). NIX/BNIP3L-deficient mice retain mitochondria in peripheral blood erythrocytes, displaying impaired erythroid maturation and anemia (Schweers et al., 2007; Sandoval et al., 2008). NIX/BNIP3L is a typical BCL-2 homology domain 3 (BH3)-only protein initially suggested to act by engaging the BCL-2 apoptotic pathway. It is now clear that NIX/BNIP3L, which has a WXXL-like motif facing the cytosol, binds to LC3 and GABARAP proteins on autophagosomes, thus mediating, at least partially, mitophagy (Schwarten et al., 2009; Novak et al., 2010). In this case, NIX/BNIP3L functions as a regulated receptor for selective mitochondrial removal (Novak et al., 2010; Rogov et al., 2017). Although NIX/BNIP3L has a crucial role in elimination of mitochondria during reticulocyte maturation, mitophagy can be also induced to some extent without NIX/BNIP3L and also independently of core autophagy proteins such as ATG5 and ATG7 (Nishida et al., 2009; Honda et al., 2014). Similar to NIX/BNIP3L, its relative BNIP3 has been reported to function as a mitophagy receptor. Specifically, it has been shown that phosphorylation of Ser17 and 24 flanking the BNIP3 LIR domain enhances its interaction with specific Atg8 members LC3B and GATE-16, thereby inducing mitophagy versus apoptosis (Zhu et al., 2013). Intriguingly, NIX/BNIP3L was recently identified as a substrate of Parkin in *Drosophila melanogaster*. Moreover, the ubiquitination of NIX/BNIP3L by Parkin on mitochondria recruits the autophagy adaptor neighbor of *BRCA1* gene 1 (NBR1), ultimately leading to NIX/BNIP3L-mediated mitophagy in human cell lines (Gao et al., 2015). Another study showed that DAF-16/FOXO controlled, germline tumor affecting-1 (DCT-1) (Oh et al., 2006; Pinkston-Gosse and Kenyon, 2007) is the *C. elegans* homolog of mammalian BNIP3 and NIX/BNIP3L and serves as a key mediator of mitophagy, promoting survival under stress conditions. Importantly, DCT-1 is ubiquitinated on Lys26, in particular, under mitophagy-inducing conditions. This ubiquitination event requires the activity of PINK-1. Furthermore, DCT-1 ubiquitination depends on PDR-1, the nematode Parkin homolog. Together, these findings indicate that DCT-1, PINK-1, and PDR-1 function in the same genetic pathway to mediate elimination of impaired mitochondria, thus contributing to cell and whole-organism homeostasis (Palikaras et al., 2015b).

The mitochondrial outer membrane protein FUN14 domain-containing protein 1 (FUNDC1) also acts as a mitophagy receptor to recruit autophagosomes for mitochondrial clearance (Fig. 2). FUNDC1 is highly conserved from flies to humans (Table 1). It has three putative transmembrane domains near the C-terminus which face the intermembrane space, and a typical LIR mediating the direct interaction with LC3 under hypoxic conditions (Liu et al., 2012). Further analysis showed that FUNDC1 regulation is quite complex. Specifically, phosphorylation of FUNDC1 LIR motif by Src kinases and CK2 blocks mitophagy under physiological conditions in mammalian cells (Liu et al., 2012). In response to hypoxia or uncoupling, however, the mitochondrial phosphatase phosphoglycerate mutase family member 5 (PGAM5) dephosphorylates FUNDC1, initiating mitophagy (Chen et al., 2014). In a different level of regulation, ULK1-mediated phosphorylation of FUNDC1 at Ser17 renders this receptor capable of responding to a general stimulus for autophagy (Wu et al., 2014). Together, these findings suggest that FUNDC1 is activated upon both general and selective autophagy stimuli through ULK1 and PGAM5 signaling, respectively. As recently shown, the hypoxia-responsive microRNA mir-137 is also implicated in FUNDC1 and NIX/BNIP3L regulation. Downregulation of mir-137 increases FUNDC1 expression, enhancing its interaction with LC3, and ultimately promoting mitophagy during hypoxia (Li et al., 2014). Further delineating the role of FUNDC1 in mitophagy, FUNDC1 is considered as a central node of a newly identified pathway that interfaces mitochondrial dynamics and mitophagy under hypoxic conditions. Specifically in hypoxic cells, FUNDC1 relocates to the ER—mitochondrial contact site (MAM) where it accumulates by binding to the ER membrane protein calnexin. As mitophagy proceeds, FUNDC1/calnexin association attenuates allowing FUNDC1 to interact with DRP1, promoting its recruitment to the MAM and the consequent mitochondrial fission (Wu et al., 2016). A parallel study showed that FUNDC1 interacts with both DRP1 and OPA1, a mitochondrial fusion protein in the intermembrane space to coordinate mitochondrial dynamics and mitophagy. Indeed, FUNDC1 normally interacts with OPA1, but this association is abolished under mitochondrial stress, promoting the interaction of FUNDC1 with DRP1, which then leads to mitochondrial fission. Furthermore, FUNDC1 dephosphorylation under stress conditions promotes the disassembly of the FUNDC1-OPA1 complex while inducing FUNDC1-DRP1 association. Together, these findings suggest that FUNDC1 has a regulatory role in both mitochondrial dynamics and mitophagy (Chen et al., 2016).

Surprisingly, the inner mitochondrial membrane protein prohibitin 2 (PHB2) was lately proposed to act as a mitophagy receptor. Following mitochondrial depolarization or proteasome-dependent outer membrane rupture, PHB2 interacts with LC3 via its LIR domain, inducing mitophagy. This process is responsible for paternal mitochondria elimination after embryonic fertilization in *C. elegans* and Parkin-induced mitophagy in mammalian cells, highlighting its conservation in eukaryotes (Wei et al., 2017).

The anti-apoptotic FK506-binding protein 8 (FKBP8, also known as FKBP38) has been very recently added to the repertoire of known Parkin-independent mitophagy receptors. FKBP8 is an anti-apoptotic protein normally localized to the outer mitochondrial membrane. It contains a canonical N-terminal LIR motif via which it associates with LC3A and related GABARAP members to initiate mitophagy in response to membrane potential dissipation, independently of Parkin. Intriguingly, FKBP8 escapes from the mitochondria to avoid its own degradation, thereby preventing apoptosis as mitophagy progresses (Bhujabal et al., 2017; Lim and Lim, 2017).

3. SIGNALING CASCADES THAT MODULATE AGING THROUGH MITOPHAGY AND MITOCHONDRIAL BIOGENESIS COORDINATION

Emerging observations suggest that alterations in mitophagy leading to accumulation of damaged mitochondria may have an important role in aging and age-related diseases (Madeo et al., 2015). Several molecular mechanisms have shown to mitigate mitophagy, adjusting mitochondrial number in response to energy demands, and intracellular and/or environmental stimuli. Perturbations of these signaling pathways deregulate the tight interplay between mitochondrial biogenesis and mitophagy resulting in impaired energy metabolism and eventually in cellular and tissue degeneration (Palikaras et al., 2015a).

3.1 Mitochondrial Homeostasis Interfaces With Mammalian Target of Rapamycin and AMP-Activated Protein Kinase Signaling

The mammalian target of rapamycin (mTOR) and AMP-activated protein kinase (AMPK) are master regulators of cellular and organismal metabolism, and their activity is associated with the coupling of mitochondrial biogenesis and mitophagy in eukaryotic cells. Previous work has extensively documented that mTOR activity attenuates autophagy (Kim and Guan, 2015). Indeed, several studies have shown an inverse interplay between autophagy initiation and mTOR function (Zoncu et al., 2011). Importantly, it has been reported that mTOR acts as a negative autophagy regulator, since mTOR-dependent ULK1 phosphorylation inhibits autophagosome biogenesis under nutrient deprivation (Ganley et al., 2009; Hosokawa et al., 2009; Jung et al., 2009).

AMPK is a conserved master regulator of cellular and energy metabolism (Herzig and Shaw, 2017). Various stress stimuli, including starvation, hypoxia, and DNA damage, among others, result in energy depletion and AMPK activation (Laderoute et al., 2006). In turn, AMPK promotes an intricate signaling cascade leading to a complex transcriptional network fine-tuning energy homeostasis (Hardie, 2013). Furthermore, AMPK triggers autophagy through the modulation of multiple signaling pathways underlining its pivotal role in autophagy regulation. Firstly, AMPK phosphorylates and inhibits mTOR activity in response to energy depletion, resulting in autophagy induction (Gwinn et al., 2008). Additionally, glucose deprivation leads to AMPK-dependent phosphorylation and activation of ULK1 (Kim et al., 2011). In turn, ULK1 phosphorylates Beclin1 and enhances the activity of VPS34 lipid kinase, promoting selective mitochondrial removal under nutrient limitation and exercise training (Egan et al., 2011; Laker et al., 2017). Although ULK1 mediates autophagy initiation, it serves also as a negative regulator of autophagy. ULK1 phosphorylates AMPK, inhibiting its enzymatic activity and setting up a negative feedback loop, thereby terminating the signaling events and preserving energy homeostasis (Loffler et al., 2011). Thus, AMPK activation could remove defective mitochondria and be beneficial in response to harmful conditions contributing to various age-related pathologies.

Indeed, both genetic and pharmacological AMPK stimulation promotes longevity via systemic remodeling of mitochondrial dynamics and metabolism, highlighting a considerable communication between AMPK activity and mitochondrial homeostasis (Apfeld et al., 2004; Mair et al., 2011; Burkewitz et al., 2015; Fontana and Partridge, 2015; Toyama et al., 2016). Congruently, a recent study demonstrates that AMPK-mediated longevity depends on mitochondrial network morphology, since inhibition of both fission and fusion machinery abolishes lifespan extension driven by AMPK stimulation and caloric restriction (Weir et al., 2017). Interestingly, experimental evidence indicates a direct effect of AMPK on mitochondrial morphology, since it is reported that mitochondrial fission factor (MFF) is phosphorylated and activated by AMPK, which subsequently results in DRP1-mediated mitochondrial fission and mitophagy stimulation (Ducommun et al., 2015; Toyama et al., 2016).

Despite the undoubted association of AMPK and mTOR with mitochondrial metabolism and mitophagy regulation, it is unclear whether they could mediate specific modifications on mitophagy-related proteins, such as mitophagy receptors. To this direction, there is evidence that the BNIP3 mitophagy receptor inactivates mTOR signaling through its direct association with Ras homolog enriched in brain (Rheb), an upstream mTOR activator, upon hypoxia (Li et al., 2007). Interestingly, AMPK phosphorylates and triggers ULK1 activity in response to hypoxic stress. In turn, ULK1 is recruited to dysfunctional mitochondria mediating their removal via mitophagy (Tian et al., 2015). However, whether AMPK alters the phosphorylation status of BNIP3 has not been examined yet. On the contrary, it has been shown that the mitophagy receptor FUNDC1 is phosphorylated by ULK1, enhancing its strong association with the autophagosomal protein LC3 under hypoxia (Wu et al., 2014). However, the exact mechanism of ULK1 mitochondrial transport, whether this translocation occurs under other stress conditions as well, how it triggers mitophagy, and whether it targets other outer mitochondrial membrane proteins are insufficiently understood.

In addition to its role in mitochondrial selective autophagy, AMPK orchestrates also mitochondrial biogenesis in various cell types including muscle cells, adipocytes, hepatic cells, and macrophages, among others (Galic et al., 2011; Hasenour et al., 2014; Mottillo et al., 2016). Increased energy requirements signal for more ATP production leading to mitochondrial biogenesis. Generation of newly synthesized mitochondria requires the upregulation of several mitochondrial proteins and lipids, which in their majority are encoded by nuclear genes. AMPK is shown to be the central regulator of this retrograde signaling cascade (Fig. 3). Many studies have demonstrated that genetic and/or pharmacological AMPK activation promotes mitochondrial biogenesis (Hardie, 2013). Furthermore, exercise training expands mitochondrial content in an AMPK-dependent manner (Narkar et al., 2008). On the contrary, genetic ablation of AMPK subunits results in decreased mitochondrial population and cellular inability to induce mitochondrial biogenesis under energetic stress, highlighting the essential

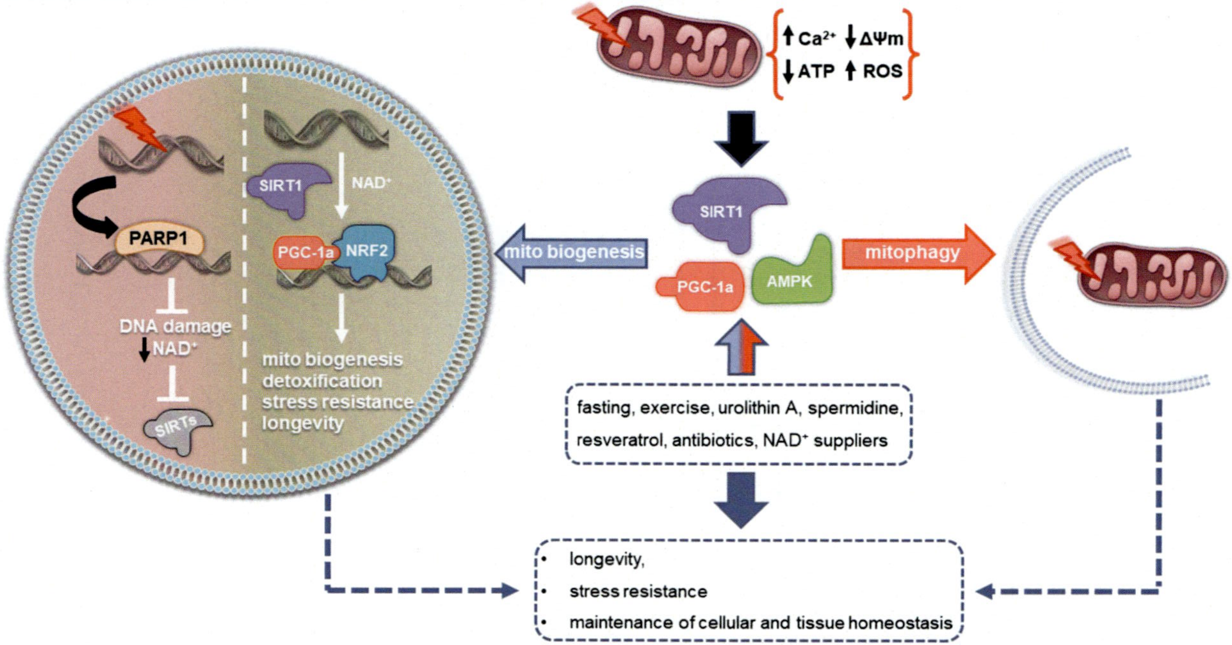

Figure 3 ***Coupling of mitophagy and mitochondrial biogenesis preserves energy metabolism and organismal homeostasis.*** Mitochondrial dysfunction, characterized by elevated cytoplasmic calcium levels, mitochondrial membrane potential collapse, adenosine triphosphate (ATP) depletion, and increased reactive oxygen species (ROS) production, is detrimental for cellular viability. Uncontrolled mitochondrial dysfunction could lead to genotoxic stress and progeria syndromes. Poly (adenosine diphosphate-ribose) polymerase 1 (PARP1) activation protects against DNA damage. However, persistent PARP1 activity results in nicotinamide adenine dinucleotide (NAD^+) depletion and subsequent SIRT1 inhibition. Pharmacological interventions, fasting and exercise promote degradation of impaired organelles through mitophagy. Simultaneously, peroxisome proliferator activated receptor-γ coactivator 1α (PGC-1α) is induced by AMP-activated protein kinase (AMPK) and Sirtuin 1 (SIRT1) activity to mediate mitochondrial biogenesis. PGC-1α serves as a transcriptional co-activator and interacts with multiple transcription factors, such as nuclear factor-erythroid 2-related factor 2 (NRF2), to orchestrate the expression of numerous mitochondrial, detoxification, stress resistance and longevity genes. This intricate interplay between mitophagy and mitochondrial biogenesis upholds energy metabolism, survival and tissue homeostasis during stress.

role of AMPK in mitochondrial metabolism (Jeppesen et al., 2013; Tanner et al., 2013; Lantier et al., 2014).

AMPK mediates mitochondrial biogenesis through the regulation of multiple signaling pathways converging on peroxisome proliferator activated receptor-γ coactivator 1α (PGC-1α) stimulation. PGC-1α serves as a co-activator of numerous transcription factors, including estrogen-related receptors (ERRs), peroxisome proliferator-activated receptors (PPARs), nuclear respiratory factor 1 (NRF1), myocyte enhancer factor 2 (MEF2), and forkhead box O (FOXO), among others. Thus, PGC-1α orchestrates an intricate transcriptional program, which promotes the expression of several mitochondrial-related genes (Fernandez-Marcos and Auwerx, 2011; Scarpulla, 2011; Ploumi et al., 2017). Notably, PGC-1α drives the expression of several genes involved in oxidative metabolism upon AMPK induction. Moreover, AMPK is found to phosphorylate PGC-1α in vitro (Jager et al., 2007). Although, experimental evidence indicates a direct association between AMPK and PGC-1α activation, whether AMPK interacts and phosphorylates PGC-1α directly in vivo is still unclear.

Another layer of mitochondrial biogenesis regulation through an indirect AMPK activity involves Sirtuin 1 (SIRT1). SIRT1 is a nicotinamide adenine dinucleotide (NAD^+)-dependent histone deacetylase, which participates in various cellular processes. AMPK influences SIRT1 enzymatic activity through elevation of intracellular NAD^+ levels (Canto et al., 2009). Recently, it is also demonstrated that AMPK phosphorylates glyceraldehyde 3-phosphate dehydrogenase (GAPDH) leading to SIRT1 activation and autophagy induction upon glucose deprivation (Chang et al., 2015). Additionally, AMPK could also promote directly SIRT1 phosphorylation and stimulation in vitro (Lau et al., 2014). In turn, SIRT1 deacetylates and activates PGC-1α, promoting mitochondrial biogenesis (Canto and Auwerx, 2009).

Collectively, AMPK holds an essential role in mitochondrial homeostasis, and defining its substrates and effects in response to energetic stress would be beneficial with potential therapeutic implications in several pathological states.

3.2 Sirtuins: Critical Regulators of Energy Metabolism

NAD is a crucial metabolite for cellular homeostasis (Fang et al., 2017a). It has been reported that NAD^+/NADH ratio declines with age in multiple organs, such as the brain, liver, muscles, and adipose tissue (Yoshino et al., 2011; Zhu et al., 2015; Zhang et al., 2016). Notably, decreased NAD^+ levels shorten nematodes lifespan, whereas genetic and/or pharmacological augmentation of intracellular NAD^+ levels protects against age-dependent metabolic impairment and promotes lifespan extension (Mouchiroud et al., 2013).

The beneficial effects of NAD^+ on cellular metabolism are mediated, at least in part, by sirtuins (Verdin, 2015; Fang et al., 2017a). As mentioned above, sirtuins are NAD^+-dependent enzymes and their function has been implicated in the regulation of multiple cellular processes, including autophagy, energy metabolism, cell death, and aging. In mammalian cells, the sirtuin family consists of seven proteins, SIRT1−SIRT7. SIRT3, SIRT4, and SIRT5 are mitochondria-targeted proteins, SIRT1, SIRT6, and SIRT7 are mainly nuclear localized, and SIRT2 displays a broad cytoplasmic distribution pattern (Houtkooper et al., 2012). In congruent with the age-associated decrease of NAD^+ levels, sirtuins levels and activity decline also with age. Both mitochondrial SIRT3 and SIRT7 protein levels are reduced in aged tissues (Brown et al., 2013; Mohrin et al., 2015). Upregulation of their expression restores energy metabolism and regenerative capacity, and abolishes age-related deterioration of hematopoietic stem cells (Brown et al., 2013; Mohrin et al., 2015). Additionally, SIRT3 regulates the acetylation status of several mitochondrial proteins mediating adaptive neuronal responses to physiological challenges and protecting against neurodegeneration (Cheng et al., 2016). A recent study has implicated SIRT4 activity in the maintenance of mitochondrial homeostasis via mitophagy stimulation. Interestingly, SIRT4 associates with OPA1, an inner mitochondrial membrane GTPase, influencing mitochondrial morphology in response to mitochondrial damage (Lang et al., 2017).

The most well-studied member of sirtuins is SIRT1. Attenuation of SIRT1 enzymatic activity affects cellular metabolism and aging (Sebastian et al., 2012). Although the lifespan-extending properties of SIRT1 homolog in flies and nematodes have been challenged, its conserved role in longevity is now widely accepted, and has been verified also in mammals (Burnett et al., 2011; Viswanathan and Guarente, 2011; Satoh et al., 2013). In *C. elegans*, NAD^+ supplementation extends lifespan and protects mitochondrial metabolism in an SIR-2.1-dependent manner (Mouchiroud et al., 2013). Furthermore, there is evidence indicating that SIRT1 activity regulates autophagy (Lee et al., 2008). Caloric restriction is shown to mediate longevity through autophagy. Knocking down of SIR-2.1 diminishes autophagy and longevity in response to nutrient deprivation in nematodes (Morselli et al., 2010). Moreover, SIRT1 depletion results in accumulation of dysfunctional mitochondria and increased protein levels of the autophagy substrate p62 in multiple organs (Lee et al., 2008). Additionally, *Sirt1*$^{-/-}$ mice display excessive mitochondrial damage and increased lipid oxidation levels in hepatic cells, indicating a possible role of SIRT1 in mitochondrial selective autophagy (Boily et al., 2008). Indeed, it is demonstrated that defective NAD^+—SIRT1—PGC-1α axis leads to deregulation of PINK1 and subsequently to impaired mitophagy upon genotoxic stress (Fang et al., 2014). SIRT1, initially identified as a histone deacetylase, might serve as a cellular sensor translating metabolic imbalance to transcriptional outputs (Lin et al., 2004). Several transcription factors, including FOXO, NRF1, and PGC-1α, among others, which participate in mitochondrial biogenesis, are modulated by SIRT1 enzymatic activity (Fig. 3) (Brunet et al., 2004; Rodgers et al., 2005; Kim et al., 2010). Notably, SIRT1 physically interacts and deacetylases PGC-1α, enhancing its transcriptional activity during fasting (Rodgers et al., 2005). Several studies underline the pivotal role of SIRT1-dependent modulation of PGC-1α activity as a detrimental regulatory step in cellular adaptation in response to energetic stress (Rodgers et al., 2005; Lagouge et al., 2006; Gerhart-Hines et al., 2007; Rodgers and Puigserver, 2007). Interestingly, impaired SIRT1 function influences also mtDNA metabolism. A recent study demonstrates that SIRT1 depletion reduced mitochondrial transcription factor A (TFAM) activity through its transcriptional regulation (Gomes et al., 2013). Therefore, decreased SIRT1, due to age-dependent decline of NAD^+ intracellular levels, exacerbates deterioration of energy homeostasis.

Altogether, these results demonstrate that influencing the intracellular NAD^+ concentration by using caloric restriction mimetics, fasting, exercise, or extracellular NAD^+ supplementation could stimulate SIRT1—PGC-1α axis, leading to mitophagy induction and subsequent upregulation of mitochondrial biogenesis (Fang et al., 2017a). In turn, the cells could adjust their mitochondrial content and fulfil their energy requirements in response to stress conditions. Although the delineation of SIRT1—PGC-1α interplay represents a milestone in the field of energy metabolism, there are several mechanistic questions to be addressed, including how SIRT1/PGC-1α association is driven in response to various stressors and whether it displays any tissue-specific pattern.

4. MITOPHAGY DEFICIENCY DURING AGING: LESSONS FROM MODEL ORGANISMS

Impaired mitochondrial metabolism, signified by elevated mtDNA mutation rate, decreased electron transport chain function, increased ROS production, defective cytoplasmic calcium buffering, and uncontrolled release of pro-apoptotic factors, is a hallmark of aging, highlighting the pivotal role of mitochondria in organismal fitness (Lopez-Otin et al., 2013). Furthermore, cellular inability to remove dysfunctional organelles, because of the age-dependent autophagy decline, results in accumulation of defective mitochondrial population exacerbating energy metabolism collapse (Palikaras et al., 2017). Therefore, mitochondrial activity is under a constant control to maintain cellular homeostasis and survival.

Recent studies demonstrate the essential role of mitophagy throughout life. For example, it is implicated in paternal mitochondria elimination upon fertilization, in developmental processes such as retina ganglion cell differentiation, erythrocyte, and T lymphocytes maturation, and in the prevention of age-associated pathologies (Pua et al., 2009; Sato and Sato, 2011; Palikaras et al. 2015b, 2017; Esteban-Martinez and Boya, 2017; Fang et al., 2017a). Aberrant accumulation of defective mitochondria has been observed in various tissues and organisms during aging (Wallace, 1999; Preston et al., 2008; Palikaras et al., 2015b). Accelerated mutation rates and deletions of mtDNA have been associated with impairment of energy metabolism resulting in age-related disorders (Vafai and Mootha, 2012; Bratic and Larsson, 2013). Moreover, mitochondrial damage is shown to be accompanied by a senescence response underlining its association with aging phenotypes (Wiley et al., 2016). Thus, elimination of damaged mitochondria seems to be a "safe-guard" process preserving cellular function, tissue homeostasis, and organismal healthspan.

Several conserved longevity pathways modulate aging via mitophagy stimulation. In *C. elegans*, mitophagy induction is required for lifespan extension of several long-lived mutants, including animals with impaired mitochondrial function, reduced activity of insulin/IGF-1 signaling, and caloric restricted mutants, among others (Palikaras et al., 2015b). Furthermore, moderate mitochondrial defects and hypoxia-like responses promote longevity in a mitophagy-dependent manner (Schiavi et al., 2015). In congruent with the cytoprotective role of mitophagy in nematodes, a recent study uncovered that mitochondrial removal is diminished with age in mice (Sun et al., 2015). Cells with high-energy demands, such as hepatic and kidney cells, require constitutively active and fine-tuned mitochondrial function to sustain their homeostasis. Additionally, postmitotic cells, including neurons and cardiomyocytes that need to survive throughout the lifetime of an organism, display increased mitochondrial population and enhanced sensitivity to oxidative damage with age (Palikaras et al., 2017). Notably, old mice display approximately 70% decrease of mitophagy in the hippocampus, a brain area essential for memory and learning (Sun et al., 2015 Mol Cell). Furthermore, mitophagy is reduced in a mouse model of Huntington's disease, highlighting the importance of mitochondrial turnover under pathological conditions (Sun et al., 2015). Satellite cells are muscle-specific stem cells characterized by elevated regenerative capacity and long lifespan. Hence, age-dependent failure of mitochondrial elimination leads to decreased number of muscle satellite stem cells in mice and human due to proteostasis collapse and excessive mitochondrial damage (Garcia-Prat et al., 2016). Age-associated myopathy and sarcopenia are defined by increased mitochondrial dysfunction, impaired energy metabolism, and decreased mitochondrial degradation rate. Chronic inflammation is accompanied by muscle degeneration and diminished anti-inflammatory responses during aging (Woods et al., 2012). Thus, mitophagy levels could be influenced by chronic inflammation leading to sarcopenia. Indeed, a recent study demonstrates that the anti-inflammatory cytokine interleukin 10 (IL-10) regulates immune responses by promoting mitophagy (Ip et al., 2017). IL-10 inhibits mTOR via AMPK activation and mediates the removal of dysfunctional mitochondrial, preventing inflammasome stimulation and the progression of inflammatory diseases (Sag et al., 2008; Ko et al., 2016; Ip et al., 2017).

Mitophagy deficiency results in uncontrolled accumulation of damaged organelles, which are characterized mainly by impaired mitochondrial proteostasis and function (Palikaras et al., 2015b; Schiavi et al., 2015). In turn, defective mitochondria set up an intricate communication with the nucleus sustaining cellular homeostasis and survival. In *C. elegans*, SKN-1, the homolog of the mammalian nuclear factor-erythroid 2-related factor 2 (NRF2), is activated and orchestrates the transcription of several nuclear genes regulating both mitophagy and mitochondrial biogenesis in response to mitochondrial damage (Ghose et al., 2013; Palikaras et al., 2015b). The importance of this intricate interplay between the mitochondria and nucleus is also underlined in progeria diseases, such as Cockayne syndrome, xeroderma pigmentosum group A, and ataxia telangiectasia, characterized by defective DNA repair mechanisms (Fang et al., 2014, 2016). Age-dependent accretion of nuclear DNA damage leads to permanent activation of poly (adenosine diphosphate ribose) polymerase 1 (PARP1), which detects and corrects DNA lesions by recruiting DNA repair complexes (Fig. 3). Although, the role of PARP1 stimulation in DNA maintenance is beneficial, its persistent activation diminishes NAD^+ levels and SIRT1 enzymatic activity, resulting in mitophagy and mitochondrial biogenesis defects

(Fang et al., 2014). Similarly, depletion of PME-1, the nematode homolog of PARP1, improves energy metabolism and promotes longevity by enhancing intracellular NAD^+ concentration and SIR-2.1 activity (Mouchiroud et al., 2013).

5. SYNTHETIC AND NATURAL COMPOUNDS TO MODULATE MITOPHAGY DURING AGING

Since a wide range of human pathologies, including cardiovascular disorders, hepatic failure, autoimmune diseases, and neurodegeneration, among others, are linked with aberrant mitochondrial accumulation and defective energy metabolism, selective priming of impaired organelles could be detrimental for the cellular and tissues physiology. Mitophagy regulation holds a promise for the development of novel therapeutic strategies to tackle several age-associated diseases through the maintenance of mitochondrial homeostasis. Therefore, many synthetic and/or natural compounds have been proposed to regulate mitochondrial elimination promoting cellular viability and enhancing healthspan during aging.

Inducers of sirtuins enzymatic activity, including sirtuin activating compounds (STACs), NAD^+ precursors (e.g., nicotinamide mononucleotide, NMN and nicotinamide riboside, NR) and resveratrol, among others, have been shown to ameliorate age-related pathologies by modulating mitophagy and repairing mitochondrial function (Howitz et al., 2003; Feige et al., 2008; Canto and Auwerx, 2009; Yoshino et al., 2011; Mouchiroud et al., 2013; Bonkowski and Sinclair, 2016; Park et al., 2016). Furthermore, NAD^+ supplementation sustains mitochondrial metabolism, protects against several neurodegenerative phenotypes, and promotes healthspan and lifespan in nematodes and mouse models of progeria syndromes (Mouchiroud et al., 2013; Fang et al. 2014, 2016). Although it has been reported that resveratrol and other sirtuins activators display beneficial effects on cellular metabolism and age-related deterioration, several studies have challenged the lifespan extending properties of these compounds (Pearson et al., 2008; Zarse et al., 2010; Poulsen et al., 2013).

Given the proposed endosymbiotic theory, mitochondria homeostasis is severely affected by antibiotics (Richter et al., 2013). Recent studies report that antibiotics, such as actinonin and doxycycline, interfere with energy metabolism and mediate mitophagy in mammalian cells (Sun et al., 2015; Xing et al., 2017). Thus, controlled use of antibiotics could be beneficial for mitochondrial homeostasis maintenance and prevention of healthspan-limiting diseases.

Furthermore, naturally occurring compounds, including spermidine and urolithin A, preserve mitochondrial function and prolong longevity through mitophagy induction. It is demonstrated that spermidine supplementation promotes lifespan extension in many model organisms, including yeast, nematodes, flies, and mice (Eisenberg et al., 2009). Autophagy has a pivotal role in spermidine-mediated longevity. Notably, dietary supplementation of spermidine induces autophagy, whereas genetic and/or pharmacological inhibition of autophagic pathway abolishes its beneficial effects on animals' physiology during aging (Eisenberg et al., 2009; Morselli et al., 2010). Moreover, spermidine administration ameliorates cardiovascular pathologies and age-associated cardiac alterations, such as hypertrophy, arterial stiffness, impaired diastolic function, maintaining arterial physiology, and cardiac activity in old animals (LaRocca et al., 2013; Eisenberg et al., 2016). Interestingly, the cardioprotective effects of spermidine are mediated through mitophagy stimulation and restoration of cardiomyocytes energy homeostasis (Eisenberg et al., 2016).

An additional first-in-class mitophagy inducing compound is urolithin A. Urolithin A is an ellagitannins-derived metabolite from pomegranate seeds, which initiates mitochondrial removal upon its dietary supplementation (Ryu et al., 2016). The potential therapeutic activity of urolithin A is mainly highlighted by general improvements of mitochondrial metabolism resulting in increased exercise capacity, elevated muscle function, and lifespan extension both in nematodes and rats (Ryu et al., 2016). Interestingly, both SKN-1-dependent lifespan extension and enhanced mitochondrial biogenesis, which are documented in response to long-term urolithin A administration, indicate that urolithin A might coordinate mitochondrial biogenesis and mitophagy to sustain energy homeostasis. In addition to urolithin A, tomatadine, a natural substance found in unripe tomatoes, stimulates mitochondrial elimination leading to increased longevity and improved muscular function both in nematodes and mice (Fang et al., 2017b). Tomatadine treatment mediates mitohormesis resulting in elevated ROS levels and SKN-1/NRF2 activation (Fang et al., 2017b). Thereby, mitophagy and mitochondrial biogenesis in coordination contribute to preserving tissue homeostasis and organismal survival.

6. CONCLUSIONS

Since genetic manipulations could influence the organismal lifespan, awareness has grown that aging is a fine-tuned biological process rather than a passive damage accumulation that occurs in a haphazard manner (Klass, 1983; Friedman and Johnson, 1988; Kenyon et al., 1993). Therefore, there is an extensive effort in the scientific community to delineate the molecular underpinnings and interfere with the rate and quality of organismal aging. The global aged population is projected to substantially rise the next decades. Notably, it is expected that 25% of the world population will be older than 65 years by the year 2100 (Fang et al., 2015). This represents an international concern because of the severe socioeconomic and healthcare challenges associated with aging. Although impressive progress has been made in understanding the physiological and molecular requirements for aging, many questions remain elusive with regard to how this biological process is regulated.

Mitochondria are placed in the center of aging research. Energy homeostasis is impaired during aging, affecting survival and longevity. Furthermore, accrual of defective organelles is underlined as a key feature of multiple age-related pathologies, including neurodegenerative disorders, myopathies, inflammatory diseases, and metabolic syndromes, among others. Thus, maintenance of mitochondrial function, through a fine-tuned mitochondrial quality control system, is a critical factor of cellular and organismal homeostasis. The tight interplay between mitophagy and mitochondrial biogenesis preserves a healthy mitochondrial population, promoting stress resistance and lifespan extension (Fig. 3). Although major milestones in mitophagy research have been achieved, several uncertainties remain to be elucidated, such as the tissue-specificity component and dependency on the disease context.

Taken together, several recent studies suggest that rejuvenation of mitochondrial network or boosting energy metabolism could be an efficient strategy to tackle aging and age-associated disorders. Alongside, numerous screenings are taking place to identify synthetic or natural molecules that can be used to promote healthspan through the coordination of mitophagy and mitochondrial biogenesis (Fig. 3) (Suliman and Piantadosi, 2016; Palikaras et al., 2017). Although the results obtained from recent animal studies by using novel mitophagy modulators are encouraging, the consequences on human physiology and their therapeutic potential remain unclear. Therefore, interventional clinical trials should be organized and promoted to investigate the therapeutic potential of mitophagy- and mitochondrial biogenesis-inducers against age-related pathologies.

ACKNOWLEDGMENTS

The authors apologize to those colleagues whose work could not be referenced owing to space limitations. The author K.P. is supported by an AXA Research Fund long-term fellowship. Work in the authors' laboratory is also funded by grants from the European Commission Framework Programmes and the Greek Ministry of Education.

Conflict of interest
The authors declare no conflict of interest.

REFERENCES

Apfeld, J., O'Connor, G., McDonagh, T., DiStefano, P.S., Curtis, R., 2004. The AMP-activated protein kinase AAK-2 links energy levels and insulin-like signals to lifespan in *C. elegans*. Genes Dev. 18 (24), 3004–3009.

Bhujabal, Z., Birgisdottir, A.B., Sjottem, E., Brenne, H.B., Overvatn, A., Habisov, S., Kirkin, V., Lamark, T., Johansen, T., 2017. FKBP8 recruits LC3A to mediate Parkin-independent mitophagy. EMBO Rep. 18 (6), 947–961.

Boily, G., Seifert, E.L., Bevilacqua, L., He, X.H., Sabourin, G., Estey, C., Moffat, C., Crawford, S., Saliba, S., Jardine, K., Xuan, J., Evans, M., Harper, M.E., McBurney, M.W., 2008. SirT1 regulates energy metabolism and response to caloric restriction in mice. PLoS One 3 (3), e1759.

Bonkowski, M.S., Sinclair, D.A., 2016. Slowing ageing by design: the rise of NAD(+) and sirtuin-activating compounds. Nat. Rev. Mol. Cell Biol. 17 (11), 679–690.

Bota, D.A., Davies, K.J., 2002. Lon protease preferentially degrades oxidized mitochondrial aconitase by an ATP-stimulated mechanism. Nat. Cell Biol. 4 (9), 674–680.

Bratic, A., Larsson, N.G., 2013. The role of mitochondria in aging. J. Clin. Invest. 123 (3), 951–957.

Brown, K., Xie, S., Qiu, X., Mohrin, M., Shin, J., Liu, Y., Zhang, D., Scadden, D.T., Chen, D., 2013. SIRT3 reverses aging-associated degeneration. Cell Rep. 3 (2), 319–327.

Brunet, A., Sweeney, L.B., Sturgill, J.F., Chua, K.F., Greer, P.L., Lin, Y., Tran, H., Ross, S.E., Mostoslavsky, R., Cohen, H.Y., Hu, L.S., Cheng, H.L., Jedrychowski, M.P., Gygi, S.P., Sinclair, D.A., Alt, F.W., Greenberg, M.E., 2004. Stress-dependent regulation of FOXO transcription factors by the SIRT1 deacetylase. Science 303 (5666), 2011–2015.

Burbulla, L.F., Fitzgerald, J.C., Stegen, K., Westermeier, J., Thost, A.K., Kato, H., Mokranjac, D., Sauerwald, J., Martins, L.M., Woitalla, D., Rapaport, D., Riess, O., Proikas-Cezanne, T., Rasse, T.M., Kruger, R., 2014. Mitochondrial proteolytic stress induced by loss of mortalin function is rescued by Parkin and PINK1. Cell Death Dis. 5, e1180.

Burkewitz, K., Morantte, I., Weir, H.J., Yeo, R., Zhang, Y., Huynh, F.K., Ilkayeva, O.R., Hirschey, M.D., Grant, A.R., Mair, W.B., 2015. Neuronal CRTC-1 governs systemic mitochondrial metabolism and lifespan via a catecholamine signal. Cell 160 (5), 842–855.

Burnett, C., Valentini, S., Cabreiro, F., Goss, M., Somogyvari, M., Piper, M.D., Hoddinott, M., Sutphin, G.L., Leko, V., McElwee, J.J., Vazquez-Manrique, R.P., Orfila, A.M., Ackerman, D., Au, C., Vinti, G., Riesen, M., Howard, K., Neri, C., Bedalov, A., Kaeberlein, M., Soti, C., Partridge, L., Gems, D., 2011. Absence of effects of Sir2 overexpression on lifespan in *C. elegans* and Drosophila. Nature 477 (7365), 482–485.

Canto, C., Auwerx, J., 2009. PGC-1alpha, SIRT1 and AMPK, an energy sensing network that controls energy expenditure. Curr. Opin. Lipidol. 20 (2), 98–105.

Canto, C., Gerhart-Hines, Z., Feige, J.N., Lagouge, M., Noriega, L., Milne, J.C., Elliott, P.J., Puigserver, P., Auwerx, J., 2009. AMPK regulates energy expenditure by modulating NAD+ metabolism and SIRT1 activity. Nature 458 (7241), 1056–1060.

Caulfield, T.R., Fiesel, F.C., Moussaud-Lamodiere, E.L., Dourado, D.F., Flores, S.C., Springer, W., 2014. Phosphorylation by PINK1 releases the UBL domain and initializes the conformational opening of the E3 ubiquitin ligase Parkin. PLoS Comput. Biol. 10 (11), e1003935.

Chang, C., Su, H., Zhang, D., Wang, Y., Shen, Q., Liu, B., Huang, R., Zhou, T., Peng, C., Wong, C.C., Shen, H.M., Lippincott-Schwartz, J., Liu, W., 2015. AMPK-dependent phosphorylation of GAPDH triggers Sirt1 activation and is necessary for autophagy upon glucose starvation. Mol. Cell 60 (6), 930–940.

Chaugule, V.K., Burchell, L., Barber, K.R., Sidhu, A., Leslie, S.J., Shaw, G.S., Walden, H., 2011. Autoregulation of Parkin activity through its ubiquitin-like domain. EMBO J. 30 (14), 2853–2867.

Chen, G., Han, Z., Feng, D., Chen, Y., Chen, L., Wu, H., Huang, L., Zhou, C., Cai, X., Fu, C., Duan, L., Wang, X., Liu, L., Liu, X., Shen, Y., Zhu, Y., Chen, Q., 2014. A regulatory signaling loop comprising the PGAM5 phosphatase and CK2 controls receptor-mediated mitophagy. Mol. Cell 54 (3), 362–377.

Chen, M., Chen, Z., Wang, Y., Tan, Z., Zhu, C., Li, Y., Han, Z., Chen, L., Gao, R., Liu, L., Chen, Q., 2016. Mitophagy receptor FUNDC1 regulates mitochondrial dynamics and mitophagy. Autophagy 12 (4), 689–702.

Chen, Y., Dorn 2nd, G.W., 2013. PINK1-phosphorylated mitofusin 2 is a Parkin receptor for culling damaged mitochondria. Science 340 (6131), 471–475.

Cheng, A., Yang, Y., Zhou, Y., Maharana, C., Lu, D., Peng, W., Liu, Y., Wan, R., Marosi, K., Misiak, M., Bohr, V.A., Mattson, M.P., 2016. Mitochondrial SIRT3 mediates adaptive responses of neurons to exercise and metabolic and excitatory challenges. Cell Metab. 23 (1), 128–142.

Deas, E., Plun-Favreau, H., Gandhi, S., Desmond, H., Kjaer, S., Loh, S.H., Renton, A.E., Harvey, R.J., Whitworth, A.J., Martins, L.M., Abramov, A.Y., Wood, N.W., 2011. PINK1 cleavage at position A103 by the mitochondrial protease PARL. Hum. Mol. Genet. 20 (5), 867–879.

Deczkowska, A., Schwartz, M., 2017. NIX-ing mitochondria: from development to pathology. EMBO J. 36 (12), 1650–1652.

Ding, W.X., Ni, H.M., Li, M., Liao, Y., Chen, X., Stolz, D.B., Dorn 2nd, G.W., Yin, X.M., 2010. Nix is critical to two distinct phases of mitophagy, reactive oxygen species-mediated autophagy induction and Parkin-ubiquitin-p62-mediated mitochondrial priming. J. Biol. Chem. 285 (36), 27879–27890.

Ducommun, S., Deak, M., Sumpton, D., Ford, R.J., Nunez Galindo, A., Kussmann, M., Viollet, B., Steinberg, G.R., Foretz, M., Dayon, L., Morrice, N.A., Sakamoto, K., 2015. Motif affinity and mass spectrometry proteomic approach for the discovery of cellular AMPK targets: identification of mitochondrial fission factor as a new AMPK substrate. Cell. Signal. 27 (5), 978–988.

Egan, D.F., Shackelford, D.B., Mihaylova, M.M., Gelino, S., Kohnz, R.A., Mair, W., Vasquez, D.S., Joshi, A., Gwinn, D.M., Taylor, R., Asara, J.M., Fitzpatrick, J., Dillin, A., Viollet, B., Kundu, M., Hansen, M., Shaw, R.J., 2011. Phosphorylation of ULK1 (hATG1) by AMP-activated protein kinase connects energy sensing to mitophagy. Science 331 (6016), 456–461.

Eisenberg, T., Abdellatif, M., Schroeder, S., Primessnig, U., Stekovic, S., Pendl, T., Harger, A., Schipke, J., Zimmermann, A., Schmidt, A., Tong, M., Ruckenstuhl, C., Dammbrueck, C., Gross, A.S., Herbst, V., Magnes, C., Trausinger, G., Narath, S., Meinitzer, A., Hu, Z., Kirsch, A., Eller, K., Carmona-Gutierrez, D., Buttner, S., Pietrocola, F., Knittelfelder, O., Schrepfer, E., Rockenfeller, P., Simonini, C., Rahn, A., Horsch, M., Moreth, K., Beckers, J., Fuchs, H., Gailus-Durner, V., Neff, F., Janik, D., Rathkolb, B., Rozman, J., de Angelis, M.H., Moustafa, T., Haemmerle, G., Mayr, M., Willeit, P., von Frieling-Salewsky, M., Pieske, B.,

Scorrano, L., Pieber, T., Pechlaner, R., Willeit, J., Sigrist, S.J., Linke, W.A., Muhlfeld, C., Sadoshima, J., Dengjel, J., Kiechl, S., Kroemer, G., Sedej, S., Madeo, F., 2016. Cardioprotection and lifespan extension by the natural polyamine spermidine. Nat. Med. 22 (12), 1428–1438.

Eisenberg, T., Knauer, H., Schauer, A., Buttner, S., Ruckenstuhl, C., Carmona-Gutierrez, D., Ring, J., Schroeder, S., Magnes, C., Antonacci, L., Fussi, H., Deszcz, L., Hartl, R., Schraml, E., Criollo, A., Megalou, E., Weiskopf, D., Laun, P., Heeren, G., Breitenbach, M., Grubeck-Loebenstein, B., Herker, E., Fahrenkrog, B., Frohlich, K.U., Sinner, F., Tavernarakis, N., Minois, N., Kroemer, G., Madeo, F., 2009. Induction of autophagy by spermidine promotes longevity. Nat. Cell Biol. 11 (11), 1305–1314.

Esteban-Martinez, L., Boya, P., 2017. BNIP3L/NIX-dependent mitophagy regulates cell differentiation via metabolic reprogramming. Autophagy.

Esteban-Martinez, L., Sierra-Filardi, E., McGreal, R.S., Salazar-Roa, M., Marino, G., Seco, E., Durand, S., Enot, D., Grana, O., Malumbres, M., Cvekl, A., Cuervo, A.M., Kroemer, G., Boya, P., 2017. Programmed mitophagy is essential for the glycolytic switch during cell differentiation. EMBO J. 36 (12), 1688–1706.

Fang, E.F., Kassahun, H., Croteau, D.L., Scheibye-Knudsen, M., Marosi, K., Lu, H., Shamanna, R.A., Kalyanasundaram, S., Bollineni, R.C., Wilson, M.A., Iser, W.B., Wollman, B.N., Morevati, M., Li, J., Kerr, J.S., Lu, Q., Waltz, T.B., Tian, J., Sinclair, D.A., Mattson, M.P., Nilsen, H., Bohr, V.A., 2016. NAD(+) Replenishment improves lifespan and healthspan in Ataxia Telangiectasia models via mitophagy and DNA repair. Cell Metab. 24 (4), 566–581.

Fang, E.F., Lautrup, S., Hou, Y., Demarest, T.G., Croteau, D.L., Mattson, M.P., Bohr, V.A., 2017a. NAD(+) in aging: molecular mechanisms and translational implications. Trends Mol. Med. 23 (10), 899–916.

Fang, E.F., Scheibye-Knudsen, M., Brace, L.E., Kassahun, H., SenGupta, T., Nilsen, H., Mitchell, J.R., Croteau, D.L., Bohr, V.A., 2014. Defective mitophagy in XPA via PARP-1 hyperactivation and NAD(+)/SIRT1 reduction. Cell 157 (4), 882–896.

Fang, E.F., Scheibye-Knudsen, M., Jahn, H.J., Li, J., Ling, L., Guo, H., Zhu, X., Preedy, V., Lu, H., Bohr, V.A., Chan, W.Y., Liu, Y., Ng, T.B., 2015. A research agenda for aging in China in the 21st century. Ageing Res. Rev. 24 (Pt B), 197–205.

Fang, E.F., Waltz, T.B., Kassahun, H., Lu, Q., Kerr, J.S., Morevati, M., Fivenson, E.M., Wollman, B.N., Marosi, K., Wilson, M.A., Iser, W.B., Eckley, D.M., Zhang, Y., Lehrmann, E., Goldberg, I.G., Scheibye-Knudsen, M., Mattson, M.P., Nilsen, H., Bohr, V.A., Becker, K.G., 2017b. Tomatidine enhances lifespan and healthspan in *C. elegans* through mitophagy induction via the SKN-1/Nrf2 pathway. Sci. Rep. 7, 46208.

Feige, J.N., Lagouge, M., Canto, C., Strehle, A., Houten, S.M., Milne, J.C., Lambert, P.D., Mataki, C., Elliott, P.J., Auwerx, J., 2008. Specific SIRT1 activation mimics low energy levels and protects against diet-induced metabolic disorders by enhancing fat oxidation. Cell Metab. 8 (5), 347–358.

Fernandez-Marcos, P.J., Auwerx, J., 2011. Regulation of PGC-1alpha, a nodal regulator of mitochondrial biogenesis. Am. J. Clin. Nutr. 93 (4), 884S–890S.

Fontana, L., Partridge, L., 2015. Promoting health and longevity through diet: from model organisms to humans. Cell 161 (1), 106–118.

Frank, M., Duvezin-Caubet, S., Koob, S., Occhipinti, A., Jagasia, R., Petcherski, A., Ruonala, M.O., Priault, M., Salin, B., Reichert, A.S., 2012. Mitophagy is triggered by mild oxidative stress in a mitochondrial fission dependent manner. Biochim. Biophys. Acta 1823 (12), 2297–2310.

Friedman, D.B., Johnson, T.E., 1988. A mutation in the age-1 gene in *Caenorhabditis elegans* lengthens life and reduces hermaphrodite fertility. Genetics 118 (1), 75–86.

Fu, M., St-Pierre, P., Shankar, J., Wang, P.T., Joshi, B., Nabi, I.R., 2013. Regulation of mitophagy by the Gp78 E3 ubiquitin ligase. Mol. Biol. Cell 24 (8), 1153–1162.

Galic, S., Fullerton, M.D., Schertzer, J.D., Sikkema, S., Marcinko, K., Walkley, C.R., Izon, D., Honeyman, J., Chen, Z.P., van Denderen, B.J., Kemp, B.E., Steinberg, G.R., 2011. Hematopoietic AMPK beta1 reduces mouse adipose tissue macrophage inflammation and insulin resistance in obesity. J. Clin. Invest. 121 (12), 4903–4915.

Ganley, I.G., Lam du, H., Wang, J., Ding, X., Chen, S., Jiang, X., 2009. ULK1.ATG13.-FIP200 complex mediates mTOR signaling and is essential for autophagy. J. Biol. Chem. 284 (18), 12297–12305.

Gao, F., Chen, D., Si, J., Hu, Q., Qin, Z., Fang, M., Wang, G., 2015. The mitochondrial protein BNIP3L is the substrate of PARK2 and mediates mitophagy in PINK1/PARK2 pathway. Hum. Mol. Genet. 24 (9), 2528–2538.

Garcia-Prat, L., Martinez-Vicente, M., Perdiguero, E., Ortet, L., Rodriguez-Ubreva, J., Rebollo, E., Ruiz-Bonilla, V., Gutarra, S., Ballestar, E., Serrano, A.L., Sandri, M., Munoz-Canoves, P., 2016. Autophagy maintains stemness by preventing senescence. Nature 529 (7584), 37–42.

Gegg, M.E., Cooper, J.M., Chau, K.Y., Rojo, M., Schapira, A.H., Taanman, J.W., 2010. Mitofusin 1 and mitofusin 2 are ubiquitinated in a PINK1/parkin-dependent manner upon induction of mitophagy. Hum. Mol. Genet. 19 (24), 4861–4870.

Geisler, S., Holmstrom, K.M., Skujat, D., Fiesel, F.C., Rothfuss, O.C., Kahle, P.J., Springer, W., 2010. PINK1/Parkin-mediated mitophagy is dependent on VDAC1 and p62/SQSTM1. Nat. Cell Biol. 12 (2), 119–131.

Gerhart-Hines, Z., Rodgers, J.T., Bare, O., Lerin, C., Kim, S.H., Mostoslavsky, R., Alt, F.W., Wu, Z., Puigserver, P., 2007. Metabolic control of muscle mitochondrial function and fatty acid oxidation through SIRT1/PGC-1alpha. EMBO J. 26 (7), 1913–1923.

Ghose, P., Park, E.C., Tabakin, A., Salazar-Vasquez, N., Rongo, C., 2013. Anoxia-reoxygenation regulates mitochondrial dynamics through the hypoxia response pathway, SKN-1/Nrf, and stomatin-like protein STL-1/SLP-2. PLoS Genet. 9 (12), e1004063.

Glauser, L., Sonnay, S., Stafa, K., Moore, D.J., 2011. Parkin promotes the ubiquitination and degradation of the mitochondrial fusion factor mitofusin 1. J. Neurochem. 118 (4), 636–645.

Gomes, A.P., Price, N.L., Ling, A.J., Moslehi, J.J., Montgomery, M.K., Rajman, L., White, J.P., Teodoro, J.S., Wrann, C.D., Hubbard, B.P., Mercken, E.M., Palmeira, C.M., de Cabo, R., Rolo, A.P., Turner, N., Bell, E.L., Sinclair, D.A., 2013. Declining NAD(+) induces a pseudohypoxic state disrupting nuclear-mitochondrial communication during aging. Cell 155 (7), 1624–1638.

Greene, A.W., Grenier, K., Aguileta, M.A., Muise, S., Farazifard, R., Haque, M.E., McBride, H.M., Park, D.S., Fon, E.A., 2012. Mitochondrial processing peptidase regulates PINK1 processing, import and Parkin recruitment. EMBO Rep. 13 (4), 378–385.

Gwinn, D.M., Shackelford, D.B., Egan, D.F., Mihaylova, M.M., Mery, A., Vasquez, D.S., Turk, B.E., Shaw, R.J., 2008. AMPK phosphorylation of raptor mediates a metabolic checkpoint. Mol. Cell 30 (2), 214–226.

Hardie, D.G., 2013. AMPK: a target for drugs and natural products with effects on both diabetes and cancer. Diabetes 62 (7), 2164–2172.

Hasenour, C.M., Ridley, D.E., Hughey, C.C., James, F.D., Donahue, E.P., Shearer, J., Viollet, B., Foretz, M., Wasserman, D.H., 2014. 5-Aminoimidazole-4-carboxamide-1-beta-D-ribofuranoside (AICAR) effect on glucose production, but not energy metabolism, is independent of hepatic AMPK in vivo. J. Biol. Chem. 289 (9), 5950–5959.

Heo, J.M., Ordureau, A., Paulo, J.A., Rinehart, J., Harper, J.W., 2015. The PINK1-parkin mitochondrial ubiquitylation pathway drives a program of OPTN/NDP52 recruitment and TBK1 activation to promote mitophagy. Mol. Cell 60 (1), 7–20.

Herzig, S., Shaw, R.J., 2017. AMPK: guardian of metabolism and mitochondrial homeostasis. Nat. Rev. Mol. Cell Biol.

Honda, S., Arakawa, S., Nishida, Y., Yamaguchi, H., Ishii, E., Shimizu, S., 2014. Ulk1-mediated Atg5-independent macroautophagy mediates elimination of mitochondria from embryonic reticulocytes. Nat. Commun. 5, 4004.

Hosokawa, N., Hara, T., Kaizuka, T., Kishi, C., Takamura, A., Miura, Y., Iemura, S., Natsume, T., Takehana, K., Yamada, N., Guan, J.L., Oshiro, N., Mizushima, N., 2009. Nutrient-dependent mTORC1 association with the ULK1-Atg13-FIP200 complex required for autophagy. Mol. Biol. Cell 20 (7), 1981–1991.

Houtkooper, R.H., Pirinen, E., Auwerx, J., 2012. Sirtuins as regulators of metabolism and healthspan. Nat. Rev. Mol. Cell Biol. 13 (4), 225–238.

Howitz, K.T., Bitterman, K.J., Cohen, H.Y., Lamming, D.W., Lavu, S., Wood, J.G., Zipkin, R.E., Chung, P., Kisielewski, A., Zhang, L.L., Scherer, B., Sinclair, D.A., 2003. Small molecule activators of sirtuins extend *Saccharomyces cerevisiae* lifespan. Nature 425 (6954), 191–196.

Iguchi, M., Kujuro, Y., Okatsu, K., Koyano, F., Kosako, H., Kimura, M., Suzuki, N., Uchiyama, S., Tanaka, K., Matsuda, N., 2013. Parkin-catalyzed ubiquitin-ester transfer is triggered by PINK1-dependent phosphorylation. J. Biol. Chem. 288 (30), 22019–22032.

Ip, W.K.E., Hoshi, N., Shouval, D.S., Snapper, S., Medzhitov, R., 2017. Anti-inflammatory effect of IL-10 mediated by metabolic reprogramming of macrophages. Science 356 (6337), 513–519.

Jager, S., Handschin, C., St-Pierre, J., Spiegelman, B.M., 2007. AMP-activated protein kinase (AMPK) action in skeletal muscle via direct phosphorylation of PGC-1alpha. Proc. Natl. Acad. Sci. U.S.A 104 (29), 12017–12022.

Jeppesen, J., Maarbjerg, S.J., Jordy, A.B., Fritzen, A.M., Pehmoller, C., Sylow, L., Serup, A.K., Jessen, N., Thorsen, K., Prats, C., Qvortrup, K., Dyck, J.R., Hunter, R.W., Sakamoto, K., Thomson, D.M., Schjerling, P., Wojtaszewski, J.F., Richter, E.A., Kiens, B., 2013. LKB1 regulates lipid oxidation during exercise independently of AMPK. Diabetes 62 (5), 1490–1499.

Jin, S.M., Lazarou, M., Wang, C., Kane, L.A., Narendra, D.P., Youle, R.J., 2010. Mitochondrial membrane potential regulates PINK1 import and proteolytic destabilization by PARL. J. Cell Biol. 191 (5), 933–942.

Jin, S.M., Youle, R.J., 2013. The accumulation of misfolded proteins in the mitochondrial matrix is sensed by PINK1 to induce PARK2/Parkin-mediated mitophagy of polarized mitochondria. Autophagy 9 (11), 1750–1757.

Jung, C.H., Jun, C.B., Ro, S.H., Kim, Y.M., Otto, N.M., Cao, J., Kundu, M., Kim, D.H., 2009. ULK-Atg13-FIP200 complexes mediate mTOR signaling to the autophagy machinery. Mol. Biol. Cell 20 (7), 1992–2003.

Kane, L.A., Lazarou, M., Fogel, A.I., Li, Y., Yamano, K., Sarraf, S.A., Banerjee, S., Youle, R.J., 2014. PINK1 phosphorylates ubiquitin to activate Parkin E3 ubiquitin ligase activity. J. Cell Biol. 205 (2), 143–153.

Kanki, T., Kurihara, Y., Jin, X., Goda, T., Ono, Y., Aihara, M., Hirota, Y., Saigusa, T., Aoki, Y., Uchiumi, T., Kang, D., 2013. Casein kinase 2 is essential for mitophagy. EMBO Rep. 14 (9), 788–794.

Kanki, T., Wang, K., Cao, Y., Baba, M., Klionsky, D.J., 2009. Atg32 is a mitochondrial protein that confers selectivity during mitophagy. Dev. Cell 17 (1), 98–109.

Kanki, T., Wang, K., Klionsky, D.J., 2010. A genomic screen for yeast mutants defective in mitophagy. Autophagy 6 (2), 278–280.

Kazlauskaite, A., Kondapalli, C., Gourlay, R., Campbell, D.G., Ritorto, M.S., Hofmann, K., Alessi, D.R., Knebel, A., Trost, M., Muqit, M.M., 2014. Parkin is activated by PINK1-dependent phosphorylation of ubiquitin at Ser65. Biochem. J. 460 (1), 127–139.

Kenyon, C., Chang, J., Gensch, E., Rudner, A., Tabtiang, R., 1993. A *C. elegans* mutant that lives twice as long as wild type. Nature 366 (6454), 461–464.

Khaminets, A., Behl, C., Dikic, I., 2016. Ubiquitin-dependent and independent signals in selective autophagy. Trends Cell Biol. 26 (1), 6–16.

Kim, H.S., Xiao, C., Wang, R.H., Lahusen, T., Xu, X., Vassilopoulos, A., Vazquez-Ortiz, G., Jeong, W.I., Park, O., Ki, S.H., Gao, B., Deng, C.X., 2010. Hepatic-specific disruption of SIRT6 in mice results in fatty liver formation due to enhanced glycolysis and triglyceride synthesis. Cell Metab. 12 (3), 224–236.

Kim, J., Kundu, M., Viollet, B., Guan, K.L., 2011. AMPK and mTOR regulate autophagy through direct phosphorylation of Ulk1. Nat. Cell Biol. 13 (2), 132–141.

Kim, Y.C., Guan, K.L., 2015. mTOR: a pharmacologic target for autophagy regulation. J. Clin. Invest. 125 (1), 25–32.

Klass, M.R., 1983. A method for the isolation of longevity mutants in the nematode *Caenorhabditis elegans* and initial results. Mech. Ageing Dev. 22 (3–4), 279–286.

Klionsky, D.J., Emr, S.D., 2000. Autophagy as a regulated pathway of cellular degradation. Science 290 (5497), 1717–1721.

Ko, F., Abadir, P., Marx, R., Westbrook, R., Cooke, C., Yang, H., Walston, J., 2016. Impaired mitochondrial degradation by autophagy in the skeletal muscle of the aged female interleukin 10 null mouse. Exp. Gerontol. 73, 23–27.

Kondapalli, C., Kazlauskaite, A., Zhang, N., Woodroof, H.I., Campbell, D.G., Gourlay, R., Burchell, L., Walden, H., Macartney, T.J., Deak, M., Knebel, A., Alessi, D.R., Muqit, M.M., 2012. PINK1 is activated by mitochondrial membrane potential depolarization and stimulates Parkin E3 ligase activity by phosphorylating Serine 65. Open Biol. 2 (5), 120080.

Kondo-Okamoto, N., Noda, N.N., Suzuki, S.W., Nakatogawa, H., Takahashi, I., Matsunami, M., Hashimoto, A., Inagaki, F., Ohsumi, Y., Okamoto, K., 2012. Autophagy-related protein 32 acts as autophagic degron and directly initiates mitophagy. J. Biol. Chem. 287 (13), 10631–10638.

Koyano, F., Okatsu, K., Kosako, H., Tamura, Y., Go, E., Kimura, M., Kimura, Y., Tsuchiya, H., Yoshihara, H., Hirokawa, T., Endo, T., Fon, E.A., Trempe, J.F., Saeki, Y., Tanaka, K., Matsuda, N., 2014. Ubiquitin is phosphorylated by PINK1 to activate parkin. Nature 510 (7503), 162–166.

Kundu, M., Lindsten, T., Yang, C.Y., Wu, J., Zhao, F., Zhang, J., Selak, M.A., Ney, P.A., Thompson, C.B., 2008. Ulk1 plays a critical role in the autophagic clearance of mitochondria and ribosomes during reticulocyte maturation. Blood 112 (4), 1493–1502.

Laderoute, K.R., Amin, K., Calaoagan, J.M., Knapp, M., Le, T., Orduna, J., Foretz, M., Viollet, B., 2006. 5'-AMP-activated protein kinase (AMPK) is induced by low-oxygen and glucose deprivation conditions found in solid-tumor microenvironments. Mol. Cell Biol. 26 (14), 5336–5347.

Lagouge, M., Argmann, C., Gerhart-Hines, Z., Meziane, H., Lerin, C., Daussin, F., Messadeq, N., Milne, J., Lambert, P., Elliott, P., Geny, B., Laakso, M., Puigserver, P., Auwerx, J., 2006. Resveratrol improves mitochondrial function and protects against metabolic disease by activating SIRT1 and PGC-1alpha. Cell 127 (6), 1109–1122.

Laker, R.C., Drake, J.C., Wilson, R.J., Lira, V.A., Lewellen, B.M., Ryall, K.A., Fisher, C.C., Zhang, M., Saucerman, J.J., Goodyear, L.J., Kundu, M., Yan, Z., 2017. Ampk phosphorylation of Ulk1 is required for targeting of mitochondria to lysosomes in exercise-induced mitophagy. Nat. Commun. 8 (1), 548.

Lang, A., Anand, R., Altinoluk-Hambuchen, S., Ezzahoini, H., Stefanski, A., Iram, A., Bergmann, L., Urbach, J., Bohler, P., Hansel, J., Franke, M., Stuhler, K., Krutmann, J., Scheller, J., Stork, B., Reichert, A.S., Piekorz, R.P., 2017. SIRT4 interacts with OPA1 and regulates mitochondrial quality control and mitophagy. Aging (Albany NY) 9 (10), 2163–2189.

Lantier, L., Fentz, J., Mounier, R., Leclerc, J., Treebak, J.T., Pehmoller, C., Sanz, N., Sakakibara, I., Saint-Amand, E., Rimbaud, S., Maire, P., Marette, A., Ventura-Clapier, R., Ferry, A., Wojtaszewski, J.F., Foretz, M., Viollet, B., 2014. AMPK controls exercise endurance, mitochondrial oxidative capacity, and skeletal muscle integrity. FASEB J. 28 (7), 3211–3224.

LaRocca, T.J., Gioscia-Ryan, R.A., Hearon Jr., C.M., Seals, D.R., 2013. The autophagy enhancer spermidine reverses arterial aging. Mech. Ageing Dev. 134 (7–8), 314–320.

Lau, A.W., Liu, P., Inuzuka, H., Gao, D., 2014. SIRT1 phosphorylation by AMP-activated protein kinase regulates p53 acetylation. Am. J. Cancer Res. 4 (3), 245–255.

Lazarou, M., Jin, S.M., Kane, L.A., Youle, R.J., 2012. Role of PINK1 binding to the TOM complex and alternate intracellular membranes in recruitment and activation of the E3 ligase Parkin. Dev. Cell 22 (2), 320–333.

Lazarou, M., Narendra, D.P., Jin, S.M., Tekle, E., Banerjee, S., Youle, R.J., 2013. PINK1 drives Parkin self-association and HECT-like E3 activity upstream of mitochondrial binding. J. Cell Biol. 200 (2), 163–172.

Lazarou, M., Sliter, D.A., Kane, L.A., Sarraf, S.A., Wang, C., Burman, J.L., Sideris, D.P., Fogel, A.I., Youle, R.J., 2015. The ubiquitin kinase PINK1 recruits autophagy receptors to induce mitophagy. Nature 524 (7565), 309–314.

Lee, I.H., Cao, L., Mostoslavsky, R., Lombard, D.B., Liu, J., Bruns, N.E., Tsokos, M., Alt, F.W., Finkel, T., 2008. A role for the NAD-dependent deacetylase Sirt1 in the regulation of autophagy. Proc. Natl. Acad. Sci. U.S.A 105 (9), 3374–3379.

Li, W., Zhang, X., Zhuang, H., Chen, H.G., Chen, Y., Tian, W., Wu, W., Li, Y., Wang, S., Zhang, L., Li, L., Zhao, B., Sui, S., Hu, Z., Feng, D., 2014. MicroRNA-137 is a novel hypoxia-responsive microRNA that inhibits mitophagy via regulation of two mitophagy receptors FUNDC1 and NIX. J. Biol. Chem. 289 (15), 10691–10701.

Li, Y., Wang, Y., Kim, E., Beemiller, P., Wang, C.Y., Swanson, J., You, M., Guan, K.L., 2007. Bnip3 mediates the hypoxia-induced inhibition on mammalian target of rapamycin by interacting with Rheb. J. Biol. Chem. 282 (49), 35803–35813.

Liesa, M., Shirihai, O.S., 2013. Mitochondrial dynamics in the regulation of nutrient utilization and energy expenditure. Cell Metab. 17 (4), 491–506.

Lim, G.G., Lim, K.L., 2017. Parkin-independent mitophagy-FKBP8 takes the stage. EMBO Rep. 18 (6), 864–865.

Lin, S.J., Ford, E., Haigis, M., Liszt, G., Guarente, L., 2004. Calorie restriction extends yeast life span by lowering the level of NADH. Genes Dev. 18 (1), 12–16.

Liu, L., Feng, D., Chen, G., Chen, M., Zheng, Q., Song, P., Ma, Q., Zhu, C., Wang, R., Qi, W., Huang, L., Xue, P., Li, B., Wang, X., Jin, H., Wang, J., Yang, F., Liu, P., Zhu, Y., Sui, S., Chen, Q., 2012. Mitochondrial outer-membrane protein FUNDC1 mediates hypoxia-induced mitophagy in mammalian cells. Nat. Cell Biol. 14 (2), 177–185.

Loffler, A.S., Alers, S., Dieterle, A.M., Keppeler, H., Franz-Wachtel, M., Kundu, M., Campbell, D.G., Wesselborg, S., Alessi, D.R., Stork, B., 2011. Ulk1-mediated phosphorylation of AMPK constitutes a negative regulatory feedback loop. Autophagy 7 (7), 696–706.

Lopez-Otin, C., Blasco, M.A., Partridge, L., Serrano, M., Kroemer, G., 2013. The hallmarks of aging. Cell 153 (6), 1194–1217.

Ma, T., Li, J., Xu, Y., Yu, C., Xu, T., Wang, H., Liu, K., Cao, N., Nie, B.M., Zhu, S.Y., Xu, S., Li, K., Wei, W.G., Wu, Y., Guan, K.L., Ding, S., 2015. Atg5-independent autophagy regulates mitochondrial clearance and is essential for iPSC reprogramming. Nat. Cell Biol. 17 (11), 1379–1387.

Madeo, F., Zimmermann, A., Maiuri, M.C., Kroemer, G., 2015. Essential role for autophagy in life span extension. J. Clin. Invest. 125 (1), 85–93.

Mair, W., Morantte, I., Rodrigues, A.P., Manning, G., Montminy, M., Shaw, R.J., Dillin, A., 2011. Lifespan extension induced by AMPK and calcineurin is mediated by CRTC-1 and CREB. Nature 470 (7334), 404–408.

Marchi, S., Giorgi, C., Suski, J.M., Agnoletto, C., Bononi, A., Bonora, M., De Marchi, E., Missiroli, S., Patergnani, S., Poletti, F., Rimessi, A., Duszynski, J., Wieckowski, M.R., Pinton, P., 2012. Mitochondria-ros crosstalk in the control of cell death and aging. J. Signal. Transduct. 2012, 329635.

Matsuda, N., Sato, S., Shiba, K., Okatsu, K., Saisho, K., Gautier, C.A., Sou, Y.S., Saiki, S., Kawajiri, S., Sato, F., Kimura, M., Komatsu, M., Hattori, N., Tanaka, K., 2010. PINK1 stabilized by mitochondrial depolarization recruits Parkin to damaged mitochondria and activates latent Parkin for mitophagy. J. Cell Biol. 189 (2), 211–221.

Melser, S., Chatelain, E.H., Lavie, J., Mahfouf, W., Jose, C., Obre, E., Goorden, S., Priault, M., Elgersma, Y., Rezvani, H.R., Rossignol, R., Benard, G., 2013. Rheb regulates mitophagy induced by mitochondrial energetic status. Cell Metab. 17 (5), 719–730.

Menzies, F.M., Fleming, A., Rubinsztein, D.C., 2015. Compromised autophagy and neurodegenerative diseases. Nat. Rev. Neurosci. 16 (6), 345–357.

Mohrin, M., Shin, J., Liu, Y., Brown, K., Luo, H., Xi, Y., Haynes, C.M., Chen, D., 2015. Stem cell aging. A mitochondrial UPR-mediated metabolic checkpoint regulates hematopoietic stem cell aging. Science 347 (6228), 1374–1377.

Morselli, E., Maiuri, M.C., Markaki, M., Megalou, E., Pasparaki, A., Palikaras, K., Criollo, A., Galluzzi, L., Malik, S.A., Vitale, I., Michaud, M., Madeo, F., Tavernarakis, N., Kroemer, G., 2010. Caloric restriction and resveratrol promote longevity through the Sirtuin-1-dependent induction of autophagy. Cell Death Dis. 1, e10.

Mottillo, E.P., Desjardins, E.M., Crane, J.D., Smith, B.K., Green, A.E., Ducommun, S., Henriksen, T.I., Rebalka, I.A., Razi, A., Sakamoto, K., Scheele, C., Kemp, B.E., Hawke, T.J., Ortega, J., Granneman, J.G., Steinberg, G.R., 2016. Lack of adipocyte AMPK exacerbates insulin resistance and hepatic steatosis through brown and beige adipose tissue function. Cell Metab. 24 (1), 118–129.

Mouchiroud, L., Houtkooper, R.H., Moullan, N., Katsyuba, E., Ryu, D., Canto, C., Mottis, A., Jo, Y.S., Viswanathan, M., Schoonjans, K., Guarente, L., Auwerx, J., 2013. The NAD(+)/sirtuin pathway modulates longevity through activation of mitochondrial UPR and FOXO signaling. Cell 154 (2), 430–441.

Murakawa, T., Yamaguchi, O., Hashimoto, A., Hikoso, S., Takeda, T., Oka, T., Yasui, H., Ueda, H., Akazawa, Y., Nakayama, H., Taneike, M., Misaka, T., Omiya, S., Shah, A.M., Yamamoto, A., Nishida, K., Ohsumi, Y., Okamoto, K., Sakata, Y., Otsu, K., 2015. Bcl-2-like protein 13 is a mammalian Atg32 homologue that mediates mitophagy and mitochondrial fragmentation. Nat. Commun. 6, 7527.

Narendra, D., Kane, L.A., Hauser, D.N., Fearnley, I.M., Youle, R.J., 2010a. p62/SQSTM1 is required for Parkin-induced mitochondrial clustering but not mitophagy; VDAC1 is dispensable for both. Autophagy 6 (8), 1090–1106.

Narendra, D.P., Jin, S.M., Tanaka, A., Suen, D.F., Gautier, C.A., Shen, J., Cookson, M.R., Youle, R.J., 2010b. PINK1 is selectively stabilized on impaired mitochondria to activate Parkin. PLoS Biol. 8 (1), e1000298.

Narkar, V.A., Downes, M., Yu, R.T., Embler, E., Wang, Y.X., Banayo, E., Mihaylova, M.M., Nelson, M.C., Zou, Y., Juguilon, H., Kang, H., Shaw, R.J., Evans, R.M., 2008. AMPK and PPARdelta agonists are exercise mimetics. Cell 134 (3), 405–415.

Nishida, Y., Arakawa, S., Fujitani, K., Yamaguchi, H., Mizuta, T., Kanaseki, T., Komatsu, M., Otsu, K., Tsujimoto, Y., Shimizu, S., 2009. Discovery of Atg5/Atg7-independent alternative macroautophagy. Nature 461 (7264), 654–658.

Novak, I., Kirkin, V., McEwan, D.G., Zhang, J., Wild, P., Rozenknop, A., Rogov, V., Lohr, F., Popovic, D., Occhipinti, A., Reichert, A.S., Terzic, J., Dotsch, V., Ney, P.A., Dikic, I., 2010. Nix is a selective autophagy receptor for mitochondrial clearance. EMBO Rep. 11 (1), 45–51.

Oh, S.W., Mukhopadhyay, A., Dixit, B.L., Raha, T., Green, M.R., Tissenbaum, H.A., 2006. Identification of direct DAF-16 targets controlling longevity, metabolism and diapause by chromatin immunoprecipitation. Nat. Genet. 38 (2), 251–257.

Okamoto, K., Kondo-Okamoto, N., Ohsumi, Y., 2009. Mitochondria-anchored receptor Atg32 mediates degradation of mitochondria via selective autophagy. Dev. Cell 17 (1), 87–97.

Okatsu, K., Oka, T., Iguchi, M., Imamura, K., Kosako, H., Tani, N., Kimura, M., Go, E., Koyano, F., Funayama, M., Shiba-Fukushima, K., Sato, S., Shimizu, H., Fukunaga, Y., Taniguchi, H., Komatsu, M., Hattori, N., Mihara, K., Tanaka, K., Matsuda, N., 2012. PINK1 autophosphorylation upon membrane potential dissipation is essential for Parkin recruitment to damaged mitochondria. Nat. Commun. 3, 1016.

Ordureau, A., Sarraf, S.A., Duda, D.M., Heo, J.M., Jedrychowski, M.P., Sviderskiy, V.O., Olszewski, J.L., Koerber, J.T., Xie, T., Beausoleil, S.A., Wells, J.A., Gygi, S.P., Schulman, B.A., Harper, J.W., 2014. Quantitative proteomics reveal a feedforward mechanism for mitochondrial PARKIN translocation and ubiquitin chain synthesis. Mol. Cell 56 (3), 360–375.

Palikaras, K., Daskalaki, I., Markaki, M., Tavernarakis, N., 2017. Mitophagy and age-related pathologies: development of new therapeutics by targeting mitochondrial turnover. Pharmacol. Ther. 178, 157–174.

Palikaras, K., Lionaki, E., Tavernarakis, N., 2015a. Balancing mitochondrial biogenesis and mitophagy to maintain energy metabolism homeostasis. Cell Death Differ. 22 (9), 1399–1401.

Palikaras, K., Lionaki, E., Tavernarakis, N., 2015b. Coordination of mitophagy and mitochondrial biogenesis during ageing in *C. elegans*. Nature 521 (7553), 525–528.

Park, D., Jeong, H., Lee, M.N., Koh, A., Kwon, O., Yang, Y.R., Noh, J., Suh, P.G., Park, H., Ryu, S.H., 2016. Resveratrol induces autophagy by directly inhibiting mTOR through ATP competition. Sci. Rep. 6, 21772.

Parsons, M.J., Green, D.R., 2010. Mitochondria in cell death. Essays Biochem. 47, 99–114.

Pearson, K.J., Baur, J.A., Lewis, K.N., Peshkin, L., Price, N.L., Labinskyy, N., Swindell, W.R., Kamara, D., Minor, R.K., Perez, E., Jamieson, H.A., Zhang, Y., Dunn, S.R., Sharma, K., Pleshko, N., Woollett, L.A., Csiszar, A., Ikeno, Y., Le Couteur, D., Elliott, P.J., Becker, K.G., Navas, P., Ingram, D.K., Wolf, N.S., Ungvari, Z., Sinclair, D.A., de Cabo, R., 2008. Resveratrol delays age-related deterioration and mimics transcriptional aspects of dietary restriction without extending life span. Cell Metab. 8 (2), 157–168.

Pickrell, A.M., Youle, R.J., 2015. The roles of PINK1, parkin, and mitochondrial fidelity in Parkinson's disease. Neuron 85 (2), 257–273.

Pinkston-Gosse, J., Kenyon, C., 2007. DAF-16/FOXO targets genes that regulate tumor growth in *Caenorhabditis elegans*. Nat. Genet. 39 (11), 1403–1409.

Ploumi, C., Daskalaki, I., Tavernarakis, N., 2017. Mitochondrial biogenesis and clearance: a balancing act. FEBS J. 284 (2), 183–195.

Poole, A.C., Thomas, R.E., Yu, S., Vincow, E.S., Pallanck, L., 2010. The mitochondrial fusion-promoting factor mitofusin is a substrate of the PINK1/parkin pathway. PLoS One 5 (4), e10054.

Poulsen, M.M., Vestergaard, P.F., Clasen, B.F., Radko, Y., Christensen, L.P., Stodkilde-Jorgensen, H., Moller, N., Jessen, N., Pedersen, S.B., Jorgensen, J.O., 2013. High-dose resveratrol supplementation in obese men: an investigator-initiated, randomized, placebo-controlled clinical trial of substrate metabolism, insulin sensitivity, and body composition. Diabetes 62 (4), 1186–1195.

Preston, C.C., Oberlin, A.S., Holmuhamedov, E.L., Gupta, A., Sagar, S., Syed, R.H., Siddiqui, S.A., Raghavakaimal, S., Terzic, A., Jahangir, A., 2008. Aging-induced alterations in gene transcripts and functional activity of mitochondrial oxidative phosphorylation complexes in the heart. Mech. Ageing Dev. 129 (6), 304–312.

Pryde, K.R., Smith, H.L., Chau, K.Y., Schapira, A.H., 2016. PINK1 disables the anti-fission machinery to segregate damaged mitochondria for mitophagy. J. Cell Biol. 213 (2), 163–171.

Pua, H.H., Guo, J., Komatsu, M., He, Y.W., 2009. Autophagy is essential for mitochondrial clearance in mature T lymphocytes. J. Immunol. 182 (7), 4046–4055.

Richter, B., Sliter, D.A., Herhaus, L., Stolz, A., Wang, C., Beli, P., Zaffagnini, G., Wild, P., Martens, S., Wagner, S.A., Youle, R.J., Dikic, I., 2016. Phosphorylation of OPTN by TBK1 enhances its binding to Ub chains and promotes selective autophagy of damaged mitochondria. Proc. Natl. Acad. Sci. U.S.A 113 (15), 4039–4044.

Richter, U., Lahtinen, T., Marttinen, P., Myohanen, M., Greco, D., Cannino, G., Jacobs, H.T., Lietzen, N., Nyman, T.A., Battersby, B.J., 2013. A mitochondrial ribosomal and RNA decay pathway blocks cell proliferation. Curr. Biol. 23 (6), 535–541.

Rodgers, J.T., Lerin, C., Haas, W., Gygi, S.P., Spiegelman, B.M., Puigserver, P., 2005. Nutrient control of glucose homeostasis through a complex of PGC-1alpha and SIRT1. Nature 434 (7029), 113–118.

Rodgers, J.T., Puigserver, P., 2007. Fasting-dependent glucose and lipid metabolic response through hepatic sirtuin 1. Proc. Natl. Acad. Sci. U.S.A 104 (31), 12861–12866.

Rogov, V.V., Suzuki, H., Marinkovic, M., Lang, V., Kato, R., Kawasaki, M., Buljubasic, M., Sprung, M., Rogova, N., Wakatsuki, S., Hamacher-Brady, A., Dotsch, V., Dikic, I., Brady, N.R., Novak, I., 2017. Phosphorylation of the mitochondrial autophagy receptor Nix enhances its interaction with LC3 proteins. Sci. Rep. 7 (1), 1131.

Rojansky, R., Cha, M.Y., Chan, D.C., 2016. Elimination of paternal mitochondria in mouse embryos occurs through autophagic degradation dependent on PARKIN and MUL1. Elife 5.

Rubinsztein, D.C., Marino, G., Kroemer, G., 2011. Autophagy and aging. Cell 146 (5), 682–695.

Ryu, D., Mouchiroud, L., Andreux, P.A., Katsyuba, E., Moullan, N., Nicolet-Dit-Felix, A.A., Williams, E.G., Jha, P., Lo Sasso, G., Huzard, D., Aebischer, P., Sandi, C., Rinsch, C., Auwerx, J., 2016. Urolithin A induces mitophagy and prolongs lifespan in *C. elegans* and increases muscle function in rodents. Nat. Med. 22 (8), 879–888.

Sag, D., Carling, D., Stout, R.D., Suttles, J., 2008. Adenosine 5'-monophosphate-activated protein kinase promotes macrophage polarization to an anti-inflammatory functional phenotype. J. Immunol. 181 (12), 8633–8641.

Sandoval, H., Thiagarajan, P., Dasgupta, S.K., Schumacher, A., Prchal, J.T., Chen, M., Wang, J., 2008. Essential role for Nix in autophagic maturation of erythroid cells. Nature 454 (7201), 232–235.

Saraste, M., 1999. Oxidative phosphorylation at the fin de siecle. Science 283 (5407), 1488–1493.

Sato, M., Sato, K., 2011. Degradation of paternal mitochondria by fertilization-triggered autophagy in *C. elegans* embryos. Science 334 (6059), 1141–1144.

Satoh, A., Brace, C.S., Rensing, N., Cliften, P., Wozniak, D.F., Herzog, E.D., Yamada, K.A., Imai, S., 2013. Sirt1 extends life span and delays aging in mice through the regulation of Nk2 homeobox 1 in the DMH and LH. Cell Metab. 18 (3), 416–430.

Sauve, V., Lilov, A., Seirafi, M., Vranas, M., Rasool, S., Kozlov, G., Sprules, T., Wang, J., Trempe, J.F., Gehring, K., 2015. A Ubl/ubiquitin switch in the activation of Parkin. EMBO J. 34 (20), 2492–2505.

Scarpulla, R.C., 2011. Metabolic control of mitochondrial biogenesis through the PGC-1 family regulatory network. Biochim. Biophys. Acta 1813 (7), 1269–1278.

Schiavi, A., Maglioni, S., Palikaras, K., Shaik, A., Strappazzon, F., Brinkmann, V., Torgovnick, A., Castelein, N., De Henau, S., Braeckman, B.P., Cecconi, F., Tavernarakis, N., Ventura, N., 2015. Iron-starvation-induced mitophagy mediates lifespan extension upon mitochondrial stress in *C. elegans*. Curr. Biol. 25 (14), 1810–1822.

Schwarten, M., Mohrluder, J., Ma, P., Stoldt, M., Thielmann, Y., Stangler, T., Hersch, N., Hoffmann, B., Merkel, R., Willbold, D., 2009. Nix directly binds to GABARAP: a possible crosstalk between apoptosis and autophagy. Autophagy 5 (5), 690–698.

Schweers, R.L., Zhang, J., Randall, M.S., Loyd, M.R., Li, W., Dorsey, F.C., Kundu, M., Opferman, J.T., Cleveland, J.L., Miller, J.L., Ney, P.A., 2007. NIX is required for programmed mitochondrial clearance during reticulocyte maturation. Proc. Natl. Acad. Sci. U.S.A 104 (49), 19500–19505.

Sebastian, C., Satterstrom, F.K., Haigis, M.C., Mostoslavsky, R., 2012. From sirtuin biology to human diseases: an update. J. Biol. Chem. 287 (51), 42444–42452.

Seo, A.Y., Joseph, A.M., Dutta, D., Hwang, J.C., Aris, J.P., Leeuwenburgh, C., 2010. New insights into the role of mitochondria in aging: mitochondrial dynamics and more. J. Cell Sci. 123 (Pt 15), 2533–2542.

Sha, D., Chin, L.S., Li, L., 2010. Phosphorylation of parkin by Parkinson disease-linked kinase PINK1 activates parkin E3 ligase function and NF-kappaB signaling. Hum. Mol. Genet. 19 (2), 352–363.

Shiba-Fukushima, K., Arano, T., Matsumoto, G., Inoshita, T., Yoshida, S., Ishihama, Y., Ryu, K.Y., Nukina, N., Hattori, N., Imai, Y., 2014. Phosphorylation of mitochondrial polyubiquitin by PINK1 promotes Parkin mitochondrial tethering. PLoS Genet. 10 (12), e1004861.

Shiba-Fukushima, K., Imai, Y., Yoshida, S., Ishihama, Y., Kanao, T., Sato, S., Hattori, N., 2012. PINK1-mediated phosphorylation of the Parkin ubiquitin-like domain primes mitochondrial translocation of Parkin and regulates mitophagy. Sci. Rep. 2, 1002.

Shlevkov, E., Kramer, T., Schapansky, J., LaVoie, M.J., Schwarz, T.L., 2016. Miro phosphorylation sites regulate Parkin recruitment and mitochondrial motility. Proc. Natl. Acad. Sci. U.S.A. 113 (41), E6097–E6106.

Sonntag, K.C., Ryu, W.I., Amirault, K.M., Healy, R.A., Siegel, A.J., McPhie, D.L., Forester, B., Cohen, B.M., 2017. Late-onset Alzheimer's disease is associated with inherent changes in bioenergetics profiles. Sci. Rep. 7 (1), 14038.

Stolz, A., Ernst, A., Dikic, I., 2014. Cargo recognition and trafficking in selective autophagy. Nat. Cell Biol. 16 (6), 495–501.

Suliman, H.B., Piantadosi, C.A., 2016. Mitochondrial quality control as a therapeutic target. Pharmacol. Rev. 68 (1), 20–48.

Sun, N., Yun, J., Liu, J., Malide, D., Liu, C., Rovira II, Holmstrom, K.M., Fergusson, M.M., Yoo, Y.H., Combs, C.A., Finkel, T., 2015. Measuring in vivo mitophagy. Mol. Cell 60 (4), 685–696.

Svenning, S., Johansen, T., 2013. Selective autophagy. Essays Biochem. 55, 79–92.

Szargel, R., Shani, V., Abd Elghani, F., Mekies, L.N., Liani, E., Rott, R., Engelender, S., 2016. The PINK1, synphilin-1 and SIAH-1 complex constitutes a novel mitophagy pathway. Hum. Mol. Genet. 25 (16), 3476–3490.

Tanaka, A., Cleland, M.M., Xu, S., Narendra, D.P., Suen, D.F., Karbowski, M., Youle, R.J., 2010. Proteasome and p97 mediate mitophagy and degradation of mitofusins induced by Parkin. J. Cell Biol. 191 (7), 1367–1380.

Tanner, C.B., Madsen, S.R., Hallowell, D.M., Goring, D.M., Moore, T.M., Hardman, S.E., Heninger, M.R., Atwood, D.R., Thomson, D.M., 2013. Mitochondrial and performance adaptations to exercise training in mice lacking skeletal muscle LKB1. Am. J. Physiol. Endocrinol. Metab. 305 (8), E1018–E1029.

Tian, W., Li, W., Chen, Y., Yan, Z., Huang, X., Zhuang, H., Zhong, W., Chen, Y., Wu, W., Lin, C., Chen, H., Hou, X., Zhang, L., Sui, S., Zhao, B., Hu, Z., Li, L., Feng, D., 2015. Phosphorylation of ULK1 by AMPK regulates translocation of ULK1 to mitochondria and mitophagy. FEBS Lett. 589 (15), 1847–1854.

Tolkovsky, A.M., 2009. New gene on the block: Atg32—a specific receptor for selective mitophagy in *S. cerevisiae*. Autophagy 5 (8), 1077–1078.

Toyama, E.Q., Herzig, S., Courchet, J., Lewis Jr., T.L., Loson, O.C., Hellberg, K., Young, N.P., Chen, H., Polleux, F., Chan, D.C., Shaw, R.J., 2016. Metabolism. AMP-activated protein kinase mediates mitochondrial fission in response to energy stress. Science 351 (6270), 275–281.

Trempe, J.F., Sauve, V., Grenier, K., Seirafi, M., Tang, M.Y., Menade, M., Al-Abdul-Wahid, S., Krett, J., Wong, K., Kozlov, G., Nagar, B., Fon, E.A., Gehring, K., 2013. Structure of parkin reveals mechanisms for ubiquitin ligase activation. Science 340 (6139), 1451–1455.

Vafai, S.B., Mootha, V.K., 2012. Mitochondrial disorders as windows into an ancient organelle. Nature 491 (7424), 374–383.

Verdin, E., 2015. NAD(+) in aging, metabolism, and neurodegeneration. Science 350 (6265), 1208–1213.

Villa, E., Proics, E., Rubio-Patino, C., Obba, S., Zunino, B., Bossowski, J.P., Rozier, R.M., Chiche, J., Mondragon, L., Riley, J.S., Marchetti, S., Verhoeyen, E., Tait, S.W.G., Ricci, J.E., 2017. Parkin-independent mitophagy controls chemotherapeutic response in cancer cells. Cell Rep. 20 (12), 2846–2859.

Viswanathan, M., Guarente, L., 2011. Regulation of *Caenorhabditis elegans* lifespan by sir-2.1 transgenes. Nature 477 (7365), E1–E2.

Wallace, D.C., 1999. Mitochondrial diseases in man and mouse. Science 283 (5407), 1482–1488.

Wang, X., Winter, D., Ashrafi, G., Schlehe, J., Wong, Y.L., Selkoe, D., Rice, S., Steen, J., LaVoie, M.J., Schwarz, T.L., 2011. PINK1 and Parkin target Miro for phosphorylation and degradation to arrest mitochondrial motility. Cell 147 (4), 893–906.

Wauer, T., Komander, D., 2013. Structure of the human Parkin ligase domain in an autoinhibited state. EMBO J. 32 (15), 2099–2112.

Wei, Y., Chiang, W.C., Sumpter Jr., R., Mishra, P., Levine, B., 2017. Prohibitin 2 is an inner mitochondrial membrane mitophagy receptor. Cell 168 (1–2), 224–238 e10.

Weir, H.J., Yao, P., Huynh, F.K., Escoubas, C.C., Goncalves, R.L., Burkewitz, K., Laboy, R., Hirschey, M.D., Mair, W.B., 2017. Dietary restriction and AMPK increase lifespan via mitochondrial network and peroxisome remodeling. Cell Metab.

Wild, P., Farhan, H., McEwan, D.G., Wagner, S., Rogov, V.V., Brady, N.R., Richter, B., Korac, J., Waidmann, O., Choudhary, C., Dotsch, V., Bumann, D., Dikic, I., 2011. Phosphorylation of the autophagy receptor optineurin restricts Salmonella growth. Science 333 (6039), 228–233.

Wiley, C.D., Velarde, M.C., Lecot, P., Liu, S., Sarnoski, E.A., Freund, A., Shirakawa, K., Lim, H.W., Davis, S.S., Ramanathan, A., Gerencser, A.A., Verdin, E., Campisi, J., 2016. Mitochondrial dysfunction induces senescence with a distinct secretory phenotype. Cell Metab. 23 (2), 303–314.

Woods, J.A., Wilund, K.R., Martin, S.A., Kistler, B.M., 2012. Exercise, inflammation and aging. Aging Dis. 3 (1), 130–140.

Wu, W., Lin, C., Wu, K., Jiang, L., Wang, X., Li, W., Zhuang, H., Zhang, X., Chen, H., Li, S., Yang, Y., Lu, Y., Wang, J., Zhu, R., Zhang, L., Sui, S., Tan, N., Zhao, B., Zhang, J., Li, L., Feng, D., 2016. FUNDC1 regulates mitochondrial dynamics at the ER-mitochondrial contact site under hypoxic conditions. EMBO J. 35 (13), 1368–1384.

Wu, W., Tian, W., Hu, Z., Chen, G., Huang, L., Li, W., Zhang, X., Xue, P., Zhou, C., Liu, L., Zhu, Y., Li, L., Zhang, L., Sui, S., Zhao, B., Feng, D., 2014. ULK1 translocates to mitochondria and phosphorylates FUNDC1 to regulate mitophagy. EMBO Rep. 15 (5), 566–575.

Xiao, B., Deng, X., Lim, G.G.Y., Xie, S., Zhou, Z.D., Lim, K.L., Tan, E.K., 2017. Superoxide drives progression of Parkin/PINK1-dependent mitophagy following translocation of Parkin to mitochondria. Cell Death Dis. 8 (10), e3097.

Xing, Y., Liqi, Z., Jian, L., Qinghua, Y., Qian, Y., 2017. Doxycycline induces mitophagy and suppresses production of interferon-beta in IPEC-J2 cells. Front Cell Infect. Microbiol. 7, 21.

Yoshino, J., Mills, K.F., Yoon, M.J., Imai, S., 2011. Nicotinamide mononucleotide, a key NAD(+) intermediate, treats the pathophysiology of diet- and age-induced diabetes in mice. Cell Metab. 14 (4), 528–536.

Youle, R.J., Narendra, D.P., 2011. Mechanisms of mitophagy. Nat. Rev. Mol. Cell Biol. 12 (1), 9–14.

Zarse, K., Schmeisser, S., Birringer, M., Falk, E., Schmoll, D., Ristow, M., 2010. Differential effects of resveratrol and SRT1720 on lifespan of adult *Caenorhabditis elegans*. Horm. Metab. Res. 42 (12), 837–839.

Zhang, H., Ryu, D., Wu, Y., Gariani, K., Wang, X., Luan, P., D'Amico, D., Ropelle, E.R., Lutolf, M.P., Aebersold, R., Schoonjans, K., Menzies, K.J., Auwerx, J., 2016. NAD(+) repletion improves mitochondrial and stem cell function and enhances life span in mice. Science 352 (6292), 1436–1443.

Zhang, Q., Kuang, H., Chen, C., Yan, J., Do-Umehara, H.C., Liu, X.Y., Dada, L., Ridge, K.M., Chandel, N.S., Liu, J., 2015. The kinase Jnk2 promotes stress-induced mitophagy by targeting the small mitochondrial form of the tumor suppressor ARF for degradation. Nat. Immunol. 16 (5), 458–466.

Zhu, X.H., Lu, M., Lee, B.Y., Ugurbil, K., Chen, W., 2015. In vivo NAD assay reveals the intracellular NAD contents and redox state in healthy human brain and their age dependences. Proc. Natl. Acad. Sci. U. S. A. 112 (9), 2876–2881.

Zhu, Y., Massen, S., Terenzio, M., Lang, V., Chen-Lindner, S., Eils, R., Novak, I., Dikic, I., Hamacher-Brady, A., Brady, N.R., 2013. Modulation of serines 17 and 24 in the LC3-interacting region of Bnip3 determines pro-survival mitophagy versus apoptosis. J. Biol. Chem. 288 (2), 1099–1113.

Ziviani, E., Tao, R.N., Whitworth, A.J., 2010. Drosophila parkin requires PINK1 for mitochondrial translocation and ubiquitinates mitofusin. Proc. Natl. Acad. Sci. U.S.A 107 (11), 5018–5023.

Zoncu, R., Efeyan, A., Sabatini, D.M., 2011. mTOR: from growth signal integration to cancer, diabetes and ageing. Nat. Rev. Mol. Cell Biol. 12 (1), 21–35.

CHAPTER SIX

Mitochondria and Reactive Oxygen Species in Aging and Age-Related Diseases

Carlotta Giorgi[*,a], Saverio Marchi[*,a], Ines C.M. Simoes[§,a], Ziyu Ren[¶,a], Giampaolo Morciano[*,||,###,a], Mariasole Perrone[*], Paulina Patalas-Krawczyk[§], Sabine Borchard[#], Paulina Jędrak[**], Karolina Pierzynowska[**], Jędrzej Szymański[§], David Q. Wang[§§], Piero Portincasa[¶¶], Grzegorz Węgrzyn[**], Hans Zischka[#,||||], Pawel Dobrzyn[§], Massimo Bonora[##], Jerzy Duszynski[§], Alessandro Rimessi[*], Agnieszka Karkucinska-Wieckowska[***], Agnieszka Dobrzyn[§§§], Gyorgy Szabadkai[¶,§§§,¶¶¶], Barbara Zavan[||,¶¶¶], Paulo J. Oliveira[||||||], Vilma A. Sardao[||||||], Paolo Pinton[*,||,b] and Mariusz R. Wieckowski[§,1,b]

[*]Department of Morphology Surgery and Experimental Medicine, Section of Pathology Oncology and Experimental Biology, Interdisciplinary Center for the Study of Inflammation (ICSI), Laboratory for Technologies of Advanced Therapies (LTTA), University of Ferrara, Ferrara, Italy
[§]Department of Biochemistry, Nencki Institute of Experimental Biology, Polish Academy of Sciences, Warsaw, Poland
[¶]Department of Cell and Developmental Biology, Consortium for Mitochondrial Research, University College London, London, United Kingdom
[||]Cecilia Hospital, GVM Care & Research, 48033 Cotignola, Ravenna, Italy
[#]Institute of Molecular Toxicology and Pharmacology, Helmholtz Center Munich, German Research Center for Environmental Health, Neuherberg, Germany
[**]Department of Molecular Biology, University of Gdańsk, Gdańsk, Poland
[§§]Department of Medicine, Division of Gastroenterology and Liver Diseases, Marion Bessin Liver Research Center, Albert Einstein College of Medicine, Bronx, NY, United States
[¶¶]Clinica Medica "A. Murri", Dept. of Biomedical Sciences & Human Oncology, University of Bari "Aldo Moro" Medical School, Bari, Italy
[||||]Institute of Toxicology and Environmental Hygiene, Technical University Munich, Munich, Germany
[##]Departments of Cell Biology and Gottesman Institute for Stem Cell & Regenerative Medicine Research, Albert Einstein College of Medicine, Bronx, NY, United States
[***]Department of Pathology, The Children's Memorial Health Institute, Warsaw, Poland
[§§§]The Francis Crick Institute, London, United Kingdom
[¶¶¶]Department of Biomedical Sciences, University of Padua, Padua, Italy
[||||||]CNC - Center for Neuroscience and Cell Biology, UC-Biotech, Biocant Park, University of Coimbra, Cantanhede, Portugal
[###]Maria Pia Hospital, GVM Care & Research, Torino, Italy
[1]Corresponding author: E-mail: m.wieckowski@nencki.gov.pl

[a] These authors contributed equally to this work.
[b] These authors share senior authorship.

International Review of Cell and Molecular Biology, Volume 340
ISSN 1937-6448
https://doi.org/10.1016/bs.ircmb.2018.05.006

Contents

Abstract

Aging has been linked to several degenerative processes that, through the accumulation of molecular and cellular damage, can progressively lead to cell dysfunction and organ failure. Human aging is linked with a higher risk for individuals to develop cancer, neurodegenerative, cardiovascular, and metabolic disorders. The understanding of the molecular basis of aging and associated diseases has been one major challenge of scientific research over the last decades. Mitochondria, the center of oxidative metabolism and principal site of reactive oxygen species (ROS) production, are crucial both in health and in pathogenesis of many diseases. Redox signaling is important for the modulation of cell functions and several studies indicate a dual role for ROS in cell physiology. In fact, high concentrations of ROS are pathogenic and can cause severe damage to cell and organelle membranes, DNA, and proteins. On the other hand, moderate amounts of ROS are essential for the maintenance of several biological processes, including gene expression. In this review, we provide an update regarding the key roles of ROS—mitochondria cross talk in different fundamental physiological or pathological situations accompanying aging and highlighting that mitochondrial ROS may be a decisive target in clinical practice.

1. REACTIVE OXYGEN SPECIES AND AGING

Proposed in 1954, by Denham Harman, the free radical theory of aging (FRTA) was the first attempt to link aging and oxidative stress (Harman, 1956). Later on, in 1972, this theory was revised, and the same author developed the mitochondrial free radical theory of aging (MFRTA) (Harman, 1972; Schriner et al., 2005), which states that mitochondrial dysfunction and consequent increased reactive oxygen species (ROS) production result in a vicious cycle contributing to cellular damage and consequent cell death. Although his theory was initially received by his peers with indifference and sometimes rebuttal, we know nowadays that ROS are important during the aging process. In fact, these highly reactive oxygen-derived molecules produced during aerobic metabolism can interact with cellular components, causing cumulative oxidative damage along time that may thus plausibly reduce life span (Harman, 1956). Oxidative damage to DNA genomes, proteins, and lipids has been associated with elevated ROS production, mitochondrial function impairment, and ultimately cell senescence or death (Bokov et al., 2004; Sohal, Weindruch, 1996). Of particular importance, the close proximity between ROS production sites and mitochondrial DNA (mtDNA) can favor the accumulation of oxidative stress—associated DNA damages. Elevated ROS production has been correlated with mitochondrial oxidative damage, along with a reduction of mitochondrial copy number (Cocheme et al., 2011; Herbener, 1976; Lambert et al., 2007; Yen et al., 1989). These alterations are associated with an increased mutation rate of mtDNA in the brain, liver, and muscle fibers of aged individuals (Cahill et al., 2005; Corral-Debrinski et al., 1992; Fayet et al., 2002; Rahaet al., 2000; Yen et al., 1991). Interestingly, the establishment of the mutator mouse model allowed the demonstration of a direct correlation between an increased number of mtDNA mutations and a decreased mitochondrial respiratory chain activity (Trifunovic et al., 2004). For instance, these alterations were accompanied by the development of typical symptoms of aging in humans, namely hair loss, weight and fat reduction, decreased bone density, and cardiomyopathy (Trifunovic et al., 2004).

Aging has been also associated with a decline of antioxidant defense efficiency, which together with increased ROS production significantly contributes to a manifestation of an oxidative stress state. This in turn can initially disturb enzyme activity through reversible oxidation of thiol groups, but which ultimately can lead to a more profound alteration in biomolecule structure and integrity (Freitas et al., 2016). Consistent with this, overexpression of antioxidant enzymes decreases ROS production and protects DNA from harmful ROS effects, which is associated with a prolonged life span in *Drosophila melanogaster* (Orr, Sohal, 1994; Schriner et al., 2005). Moreover, it has been found that long-lived mice strains possess higher level of antioxidant enzymes and have reduced oxidative damage of proteins and lipids (Pamplona et al., 2002; Rebrin, Sohal, 2004). Interestingly, the reduced oxidative damage in long-lived species could be explained by an adaptive mechanism of cysteine depletion in mitochondria (Moosmann, Behl, 2008).

Despite the numerous studies supporting Harman's ROS theory of aging, other discoveries are questioning a direct correlation between oxidative stress damages and the life span. Using *Caenorhabditis elegans* as a model, mitochondrial mutations had no effect on overall ROS despite an increase of mitochondrial superoxide level (Yanget al., 2010). Surprisingly, the above-mentioned study reported a positive correlation between mitochondrial oxidative stress and the extension of life span (antioxidants supplementation shortened life span of mutants). Similarly, a number of recent works using mice models have also questioned the validity of ROS as the cause of an aged phenotype. Lapointe and Hekimi showed that a reduced level of mitochondrial enzyme MCLK1 causes mitochondrial dysfunction manifested as a reduction of electrons transport through mitochondrial respiratory chain and decrease of tricarboxylic acid (TCA) cycle activity. All these events are accompanied by increased mitochondrial oxidative stress but decreased oxidative damage to cytosolic proteins and reduced level of isoprostanes in plasma (systemic biomarker of aging and oxidative stress) (Lapointe, Hekimi, 2008). Additionally, the silencing of antioxidant enzymes, such as mitochondrial SOD2 (manganese superoxide dismutase; also called MnSOD) and GPx-1 (glutathione peroxidase-1), did not affect longevity in spite of increased oxidative stress (Perez et al., 2009a; Zhang et al., 2009). It thus seems that there is not sufficient evidence to undermine credibility of the FRTA; however, the contradictory studies have been rather supporting for a new recent theory named mitohormesis. According to this theory, moderate levels of mitochondrial ROS could activate compensatory mechanisms that protect cellular organelles from the deleterious effects of ROS and ultimately, delaying the appearance of an aging phenotype (Ristow, Zarse, 2010). For instance, moderately increased levels of ROS have been linked to an extension of longevity in *D. melanogaster* and in young mice (Copeland et al., 2009) (Basisty et al., 2016; Csiszar et al., 2008). The discovery that the reduction of elevated mitochondrial ROS levels protects against age-related decline in old mice (Basisty et al., 2016) implies that a decrease of ROS levels could be a determinant factor to delay progression of diseases parallel to the extension of life span in mammals in more advanced ages (Schriner et al., 2005). For example, administration of an antioxidant N-acetylcysteine (NAC) has been shown to prevent the loss of activity (observed during aging) of complexes I and IV (Miquel et al., 1995). Moreover, supplementation with antioxidant compounds selegiline and vitamin E (Vit-E) alone or in combined therapy showed to delay Alzheimer disease (AD) progression in human subjects (Sano et al., 1997). In conflict with the mentioned studies, some evidence reported that antioxidant therapies may not be universally beneficial in the prevention of age-related diseases. While Vit-E did not show to protect or delay Parkinson's disease (PD) progression (Parkinson Study, 1993), this antioxidant compound was even deleterious in AD patients (Lloret et al., 2009). The finding that not all patients respond similarly to the antioxidant therapy is consistent with the requisite of moderate level of ROS to induce stress resistance adaptation. As opposed to the controversial effects of dietary antioxidant compounds, caloric restriction (CR) is a promising therapeutic strategy able to retard or prevent aging in several species ranging from worms to humans (Hekimi, Guarente, 2003; Sohal, Weindruch, 1996). The mechanism underlying these effects is not completely understood. Although, evidence supported the role of ROS as inducers of mitochondrial oxidative stress adaptations, including a marked increase in mitochondrial function through peroxisome proliferator—activated receptor gamma coactivator 1-alpha (PGC-1α) and mitochondrial NAD—dependent deacetylase sirtuin-1 (SIRT1) activation (Nisoli et al., 2005). Likewise, endurance training may cause increased levels of ROS, which induce cellular signaling pathways associated to the function and turnover of mitochondria, hence contributing to the extension of life span (Lanza et al., 2008; Ristow, Zarse, 2010).

1.1 Intracellular Sources of Reactive Oxygen Species

Uncontrolled ROS production may lead to the oxidation of fundamental cellular components, such as proteins, phospholipids, and nucleic acids. Ultimately, ROS (hydroxyl radical, •OH; superoxide anion, $O_2^{\bullet -}$; hydrogen peroxide, H_2O_2; alkoxyl and peroxyl radicals, as well as singlet oxygen) can not only modify enzyme activity but also result in profound alterations in biomolecular structure. Cellular components can be also modified by products of free radical reaction intermediates such as peroxynitrite (formed by the reaction of nitric oxide with superoxide anion) or lipid hydroperoxides (prominent nonradical intermediates of lipid peroxidation produced by the reaction of a hydroxyl radical with unsaturated fatty acids). Although ROS are generally seen as harmful agents that need to be removed by detoxification mechanisms, the truth is that some of these species, most notably H_2O_2, play a physiological role in cell homeostasis, functioning as signaling molecules. This is still a controversial concept because the specificity of ROS action is unclear due to the high reactivity of some of the species toward many macromolecules, the covalent nature of modifications they bring, and the limited (in some cases) spatial effects of some of the species. Nevertheless, it is accepted that some ROS regulate their own demise through upregulation of ROS detoxification enzymes (D'Autreaux, Toledano, 2007). Under physiological conditions, ROS can act as mediators and regulators of cell metabolism, by interfering with the transmission of signals to and throughout the cell. Specific ROS such as H_2O_2 and $O_2^{\bullet -}$ are important second messengers in growth, differentiation, and cell death, activating proteins involved in cell division (mitogenic-activated protein) and participating in the immune response of the organism. By affecting the synthesis, release, or inactivation of the endothelium-derived relaxing factor (EDRF), ROS may cause the relaxation or contraction of the vascular wall. In addition, ROS can increase the permeability of the capillary walls and stimulate transport of glucose into cells and of serotonin into platelets (Droge, 2002). Furthermore, H_2O_2 regulates the expression of many genes, including AP-1, CREB, HSF1, NRF2, HIF-1, TP53, NF-κB, NOTCH, SP1, or SCREB-1 (Marinho et al., 2014; Sies, 2017). Finally, it has been demonstrated that peroxides may regulate the synthesis of prostanoids (Korbecki et al., 2013).

Mitochondria are considered one of the important sources of ROS and these, when produced extensively during pathological conditions, can evoke intracellular oxidative stress, leading to the aforementioned damage. ROS overproduction in cells may cause disruption of tissue and organ function, leading to different pathologies or even premature death of the organism. Not surprisingly, mitochondria are both producers and targets of ROS. So far, several distinct sites of ROS production in mammalian mitochondria have been identified. The two sites that have been most extensively studied are complexes I and III of the mitochondrial respiratory chain, with the focus on the mechanistic role of the ubiquinone cycle in promoting univalent oxygen reduction (Brand, 2010; St-Pierre et al., 2002). Traditionally, complex II was not considered a source of ROS per se, instead it was described to contribute to their formation via its substrate, succinate. In many tissues, succinate plays a role in reverse electron transfer, the process in which electrons are transferred from succinate to ubiquinone via complex II and then back to complex I (Liu et al., 2002; Yankovskaya et al., 2003). Despite this, it has been suggested that complex II alterations with tissue aging would be responsible for $O_2^{\bullet -}$ production (Ishii et al., 2011). The hypothesis concerning the involvement of complex II in ROS production is discussed later in the context of diabetes (Nishikawa et al., 2000) and skin aging (Anderson et al., 2014). Moreover, it was suggested that mutation in complex II might also result in $O_2^{\bullet -}$ overproduction (Ishii et al., 2005). Additionally, Paddenberg et al. investigated the role of mitochondrial complex II in ROS production, showing that complex II plays an essential role during hypoxia. At reduced oxygen tension, catalytic activity of complex II switches from succinate dehydrogenase to fumarate reductase, with this alteration being associated with increased ROS production (Paddenberg et al., 2003a, 2003b; Yankovskaya et al., 2003). Reports indicate that the magnitude of the transmembrane electric potential regulates ROS generation by the respiratory chain (Korshunov et al., 1998), which has been shown to depend on the AMP-activated protein kinase (AMPK) activity (Weisova et al., 2012), whereas others presented evidence against this relationship between mitochondrial polarization and ROS production (Shabalina, Nedergaard, 2011).

It is important to note that mitochondria are not the only ROS-producing organelles in the cell. Microsomal enzymes, including the cytochrome P450 system (Bhattacharyya et al., 2014), peroxisomal enzymes, xanthine oxidase, polyamine oxidase, sarcosine oxidase, and different types of acyl-CoA oxidases (Bonekamp et al., 2009), as well as some plasma membrane enzymes (NADPH oxidase and lipoxygenase) (Bedard, Krause, 2007; Shintoku et al., 2017), have been identified as nonmitochondrial ROS generators. Despite the fact that Brown and Borutaite presented a number of examples supporting the hypothesis that mitochondria are not the primary source of ROS (Brown; Borutaite, 2011), oxidative phosphorylation accounts for 90%—95% of cellular oxygen consumption. Although it is difficult to make an exact assessment because of frequent artifacts with the use of fluorescence-based redox-sensitive dyes, it is now considered that the initial idea that 1%—4% of oxygen consumption is converted into $O_2^{\bullet -}$ is wrong, as most of the original works were performed with mitochondrial inhibitors (Chance et al., 1979). More recent work brought down the value to 0.15%, with $O_2^{\bullet -}$ being generated at distinct topologies at the respiratory chain, notably at complex I and III (Quinlan et al., 2013; St-Pierre et al., 2002). Although it may seem a very small amount, 0.15% of total oxygen consumed represents a significant amount of $O_2^{\bullet -}$ produced and therefore should not be downplayed, when considering mitochondria as an ROS producer under physiological and pathological situations (Fridovich, 2004). Other documented sources of ROS in mitochondria include monoamine oxidase and dihydroorotate dehydrogenase (Cadenas, Davies, 2000; Lenaz, 2001). The former enzyme was previously demonstrated to be involved in oxidative damage in myocytes from patients with collagen V myopathies (Sorato et al., 2014). In addition, the flavoproteins acyl-CoA dehydrogenase and glycerol phosphate dehydrogenase can produce ROS in tissues during the oxidation of lipid-derived substrates (Lambertucci et al., 2008; St-Pierre et al., 2002). Both pyruvate and α-ketoglutarate dehydrogenase contain flavoenzyme dihydrolipoyl dehydrogenase subunits and are additional ROS sources (Starkov et al., 2004; Tahara et al., 2007). Mitochondria, as both generators and targets of ROS, accumulate some of the damage that can initiate a vicious circle of further ROS formation. The age-dependent handicapping of mitochondrial energetics is related to the accumulation of defective mtDNA and defective respiratory chain complexes that are prone to electron leakage (Linnane et al., 1989; Wei, 1992).

1.2 Mitochondria as a Source and Target for Reactive Oxygen Species in Aging: An Interventional Review

As mentioned above, approximately five decades ago, coincident with the postulation of the "FRTA" increased formation of ROS was proposed to be the major factor responsible for the aging process and decreased life span (Harman, 1956). The continuous generation of ROS by mitochondria throughout cell life produces an age-related chronic oxidative stress, especially on mtDNA, resulting in oxidative modification of bases or deletions

(Santos et al., 2013). As a consequence, mitochondria have been identified as key players in the aging process (Miquel et al., 1980). However, new findings in the last years suggested that ROS generation cannot be the initial trigger of the aging process, providing an alternative point of view to the Harman's hypothesis. One of the stronger evidence in this lack of mechanistical linkage is the lack of effect on life span of under- or overexpressing a large number and wide variety of genes coding for antioxidant enzymes (Perez et al., 2009a). In addition, a recent study showed that oxidative damage of cardiomyocytes did not positively correlate with age in human beings, although the samples were obtained from a restricted age span (<2 years old) (Huang et al., 2017). To have a thorough knowledge on this topic, we refer the readers to very relevant review articles (Gems, Partridge, 2008; Hekimi et al., 2011; Ristow, Schmeisser, 2011). In this section, we want to discuss the "canonical" association between ROS production and aging, with particular relevance on CR, which represents the most convincing intervention to delay aging and attenuate age-related disease in multiple species.

Mitochondrial function during aging has been described to decrease, especially at advanced ages. Different studies showed a decline at multiple levels, ranging from decreased activity of the respiratory chain and ATP synthase, Krebs cycle fluxes, oxidative alterations of cardiolipin, disrupted regulation, and defective mtDNA regulation and activity, among other described effects (de Almeida et al., 1989; Emelyanova et al., 2017; Petrosillo et al., 2009; Rottenberg, Hoek, 2017). A recent model describes a biphasic model in which an initial increase in mitochondrial function in middle age is followed by a fast decline at older ages (Baker, Peleg, 2017). As already mentioned, a significant number of studies in different model organisms suggest that inhibition of oxidative stress contributes to an increase in life span. Administration of Vit-E was previously shown to extend the life span of many animals, including the nematode *C. elegans* (Harrington, Harley, 1988); male mice receiving Vit-E from 28 weeks of age showed a 40% increased median life span, with a beneficial effect on aging-related decline in neurological performance and mitochondrial function. The activities of mitochondrial nitric oxide synthase and SOD2 decrease with age, whereat these effects are ameliorated by Vit-E treatment (Navarro et al., 2005). A class of Vit-E analogues, called tocotrienols, shows excellent antioxidant activity in vitro and have been suggested to suppress ROS production more efficiently than tocopherols (Schaffer et al., 2005); tocotrienols extend life span by reducing ROS damage (Collins et al., 2006). Despite this, antioxidants have had limited success in preventing the progression of diseases involving mitochondrial oxidative damage probably because they distribute around the body, with only a small fraction being accumulated by mitochondria (Serviddio et al., 2011). To overcome this problem, lipophilic cations have been conjugated with several antioxidants to allow their specific accumulation inside mitochondria. MitoVitE is one of the first mitochondria-targeted antioxidants, rapidly taken up by mitochondria (Smith et al., 2003). In cerebellar granule cells, MitoVitE mitigated EtOH-induced accumulation of intracellular oxidants and counteracts suppression of glutathione peroxidase/glutathione reductase functions and overall cellular glutathione levels (Siler-Marsiglio et al., 2005). MitoQ was reported to significantly increase the life span of SOD-deficient flies and to improve their tolerance to paraquat stress, but it could neither increase the life span nor rescue the paraquat sensitivity of wild-type *D. melanogaster* (Magwere et al., 2006). Moreover, MitoQ was shown to be effective as an antioxidant when complex I-derived superoxide generation is already elevated due to disrupted electron flow, whereas it has a prooxidant role in intact cells with normal complex I activity. Consequently, MitoQ may be useful in the treatment of diseases originating from impairment of respiratory chain complex I due to oxidatively damaged mtDNA, when its targeted delivery to pathogenic tissues is ensured (Plecita-Hlavata et al., 2009). SkQ molecules, in which a plastoquinone molecule is bound to a positively charged carrier, have also been attempted in the context of delaying aging/senescence effects in cells and tissues, with positive effects being obtained in the eye, heart, and kidney (Skulachev et al., 2009).

For several years, coenzyme Q 10 (CoQ10), whose levels are affected during aging and neurodegenerative diseases, has been considered a key factor in the progression of aging-associated complications (Lopez-Lluch et al., 2010). In rats fed a diet enriched in polyunsaturated fatty acids (PUFAs), supplementation with CoQ10 produces significant increases of mean and maximum life span, attenuating oxidative alterations related to this specific kind of diet (Quiles et al., 2004). Furthermore, enrichment of cells with CoQ10 resulted in an ordering and condensing effect on cell membranes, leading to a decrease in ROS generation and to a protective benefit on DNA integrity (Tomasetti et al., 2001). On the other hand, different studies in *C. elegans* demonstrated that lowering CoQ10 content, by inactivating genes involved in ubiquinone biosynthesis (Asencio et al., 2003) or by dietary deprivation (Larsen, Clarke, 2002), induces a significant life span increase. This discrepancy may be explained by considering that life span extension occurs with moderately low levels of global CoQ10 content (up to 50%), whereas severe CoQ10 depletion leads to developmental and reproductive inefficiency, with no extension in longevity. This interpretation is supported by the observation that the hallmark of CoQ10 deficiency syndrome is, obviously, a decreased CoQ10 concentration (about 30% or lower of the total coenzyme content, compared to healthy individuals) in human muscle and/or fibroblasts (Montero et al., 2007). Moreover, CoQ10 concentration progressively declines after the age of 40, and, in rodents, this drop occurs even under dietary CoQ10 supplementation, suggesting a higher CoQ10 consumption. Thus, although lifelong CoQ10 supplementation did not prolong or shorten the life span of either wild-type rats or mice (Lonnrot et al., 1998), it may help to prevent life span shortening due to cumulative oxidative insults.

Use of antioxidants or targeting antioxidants to mitochondria by conjugation to lipophilic cations is not the only strategy to reduce oxidative damage and prolong life span (Fig. 1). Dietary factors, including restriction of caloric intake, restriction of protein or methionine intake, or the ingestion of specific nutrients, have been shown to alter mitochondrial redox metabolism, cellular oxidative stress, and animal life span (Page et al., 2010). An inverse correlation between longevity and mitochondrial

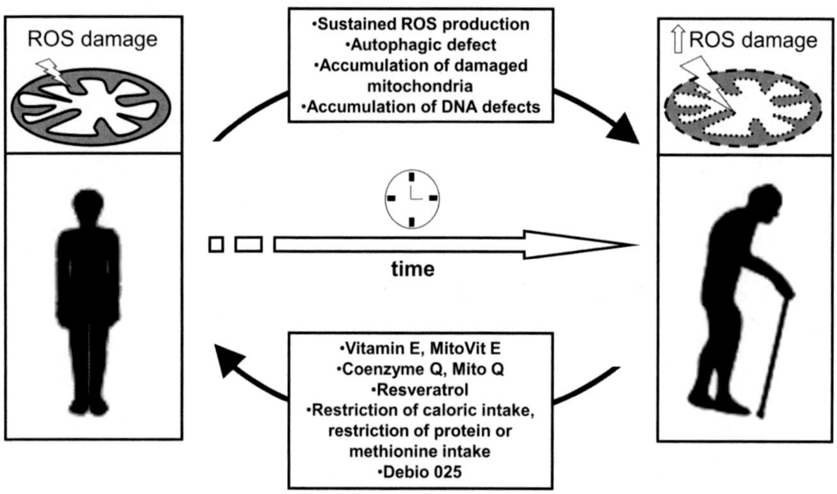

Figure 1 ***Potential role of mitochondrial ROS increase in aging.*** During aging, mitochondrial ROS production steadily increases, leading to mitochondrial damage and decreased life span. Here we report the major events contributing to aging (upper panel), or most important chemicals and experimental interventions, which may promote longevity (bottom panel).

ROS generation is demonstrated by three different studies in which H_2O_2 production by mitochondria isolated from the liver, skeletal muscle, and brain was reduced in calorie-restricted rats (Bevilacqua et al., 2004; Hagopian et al., 2005; Sanz et al., 2005). In particular, CR increased the efficiency of brain mitochondria in electron transfer in complex I, avoiding electron leak in that complex. Superoxide anion generated by complex I is specifically directed to the mitochondrial matrix, with consequent mtDNA damage and resulting bioenergetic deficits (Stefanatos, Sanz, 2011). The relationship between longevity and complex I is described in a report by Ayala et al., in which liver mitochondria of calorie-restricted rats demonstrated reduced levels of complex I (Ayala et al., 2007). Moreover, analysis of rat liver samples revealed a significant change in abundance in specific subunits of respiratory chain complexes I and IV with lifelong CR, allegedly to minimize the electron leak and subsequent ROS formation (Dani et al., 2010). Conversely, dietary restriction did not affect the activity of the oxidative-phosphorylation system or the mitochondrial H_2O_2 production in a similar rat strain (Valle et al., 2007). Thus, more work is required to confirm whether modulation of complex I levels represents one mechanism by which mitochondrial ROS production is reduced in parallel with extended longevity.

CR also reduces oxidative stress through a mechanism involving the mitochondrial deacetylase sirtuin-3, SIRT3. Expression of SIRT3 is increased during CR, and SIRT3 reduces cellular ROS levels by regulating SOD2 through deacetylation of two critical lysine residues, promoting its antioxidant activity (Qiu et al., 2010). However, there is some uncertainty regarding which acetylated lysine residue regulates SOD2 activity, and different groups proposed different site-specific regulation of SOD2 by SIRT3 (for a complete review on sirtuins and redox stressors, see (Webster et al., 2012)). Although the precise site(s) of regulation remains unclear, ROS levels are tightly controlled by SIRT3. Beyond SOD2, another important redox target of SIRT3 activity is represented by $NADP^+$-dependent isocitrate dehydrogenase 2 (IDH2), found in mitochondria that catalyzes the oxidative decarboxylation of isocitrate to 2-oxoglutarate. In response to CR, SIRT3 activates IDH2, thereby increasing NADPH levels in mitochondria. This in turn leads to an increased ratio of reduced-to-oxidized glutathione and decreased levels of ROS (Someya et al., 2010). The sites of SIRT3 deacetylation (K211 and K212) were found by Schlicker et al., who showed that in the presence of NAD^+ purified SIRT3, but not SIRT5, deacetylated IDH2 and increased its activity (Schlicker et al., 2008).

Importantly, IDH2 deacetylation/activation mediated by SIRT3 has been linked to age-related hearing loss, whereat a calorie restricted diet reduces the age-related loss of neurons and hair cells, whereas this effect is abrogated in SIRT3-deficient mice (Someya et al., 2010). Association between CR, SIRT3, and IDH2 sustains the concept that oxidative stress is a major component of aging and that nutrient status can regulate the cellular response to degenerative pathologies.

A report by Morselli et al., demonstrates how another component of the Sirtuin family, SIRT1, is required for the life span-prolonging effects of CR and resveratrol, through a mechanism that involves autophagy (Morselli et al., 2010). Dietary delivery of resveratrol increases mitochondrial abundance and aerobic capacity in cultured endothelial cells and mice (Lagouge et al., 2006). Interestingly, resveratrol is able to interact with different components of the respiratory chain: by competition with coenzyme Q, resveratrol is able to decrease complex III activity (Zini et al., 1999), and a binding site on complex V/ATP synthase has been observed (Gledhill, Walker, 2005). In two different cellular settings, cardiomyocytes and dopaminergic neurons, resveratrol protected against oxidative stress and was able to maintain mitochondrial membrane potential (MMP), with both effects directly related to resveratrol-dependent increase in SOD2 activity (Danz et al., 2009; Okawara et al., 2007). In fact, resveratrol supplementation in the context of a high-fat diet proved to be effective at elevating antioxidant capacity in the brain, resulting in an increase in both SOD2 protein levels and activity (Robb et al., 2008). Interestingly, recent data showed regulation of SIRT3 activity by SIRT1-mediated deacetylation, with aging demonstrated to be related with SIRT3 acetylation (Kwon et al., 2017). Considering that SIRT3 regulates several mitochondrial metabolic pathways (Hirschey et al., 2010; Pereira et al., 2012), this discovery sheds light on a cytosolic-mitochondrial sirtuin-based cross talk with important roles in mitochondrial alterations during aging.

Overexpression of SOD2 confers enhanced oxidative capacity and greater resistance against inducers of mitochondrial permeability transition (Silva et al., 2005). Moreover, flies with severe reductions in SOD2 expression exhibited accelerated senescence of olfactory behavior as well as precocious neurodegeneration and neuronal DNA strand breakage (Paul et al., 2007). Antioxidant supplementation, such as with Vit-E and Vitamin-C (Vit-C), reduces oxidative stress, improves SOD2 activity, with consequent positive muscle work in chronically loaded muscles of aged rats (Ryan et al., 2010). Furthermore, due to increased SOD2 expression, melatonin-treated animals showed an increase in active mitochondria population and the ability to restore the mitochondrial potential of age-damaged neurons (Garcia-Macia et al., 2011). Interestingly, mice receiving intravenous SOD2-plasmid liposome prior to total-body irradiation show increased survival from the acute hematopoietic syndrome, and males demonstrate improved long-term survival (Epperly et al., 2011). Given these observations, it appears surprising that different publications reported the failure of SOD2 overexpression in prolonging life span (Perez et al., 2009b; Zhang et al., 2009), a discrepancy that future research on aging must address. Still, the controversy supports the notion that mitochondrial-produced ROS may have a duality of effects, which are at present difficult to fully understand.

SOD2 expression is also increased in p66Shc knockout mice, which exhibit prolonged life span (Haga et al., 2008). Other genetic mouse models of longevity have been reported, such as Ames and Snell dwarf mice and $Igf1r^{+/-}$ female mice (for a review, see Liang et al., 2003), and the increased life span of these models has been correlated to increased resistance to oxidative stress. To test the causative role of mtDNA mutations in aging, the mtDNA mutator mouse, which accumulates high levels of point mutations due to a proofreading deficiency of the mtDNA polymerase γ (POLG) has been developed. In this model, mtDNA mutations resulted in a variety of aging phenotypes, i.e., weight loss, alopecia, osteoporosis, anemia, reduced fertility, heart disease, progressive hearing loss and decreased spontaneous activity, but without inducing ROS production or increasing oxidative stress (Edgar, Trifunovic, 2009). This is not the case in p66Shc KO mice, in which oxidative stress plays a crucial role. p66Shc is localized to mitochondria in about 20% of fibroblasts of higher organisms, and oxidative stress promotes a translocation of part of the cytosolic pool of p66Shc to mitochondria (Orsini et al., 2004). Within mitochondria, inner mitochondrial membrane p66shc acts as a redox enzyme, with a consequent increment of ROS production and aging. The molecular route that leads to p66Shc activation and mitochondrial import was identified by our group in 2006 (Pinton et al., 2007). p66Shc must be phosphorylated at serine 36 to be active (Migliaccio et al., 1999), and this phosphorylation is mediated by PKCβ, a kinase of the PKC family, activated after an oxidative challenge. Once phosphorylated, p66Shc can be recognized by Pin1, a peptidyl-prolyl isomerase that induces cis-trans isomerization of phosphorylated Ser-Pro bonds, causing mitochondrial translocation of p66Shc (Pinton et al., 2007). At this point, p66Shc can exert its oxidoreductase

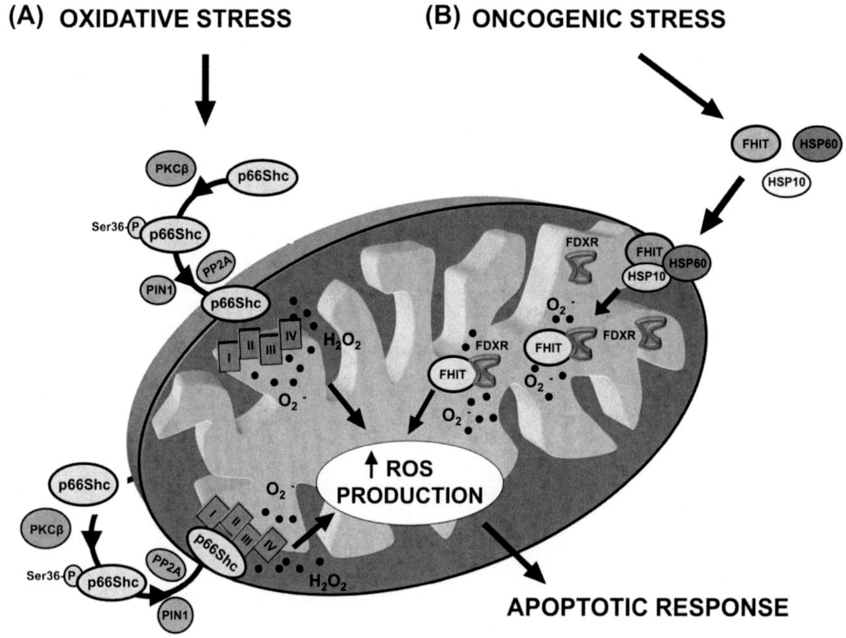

Figure 2 ***Schematic representations of two pivotal apoptotic molecular routes, involved during oxidative stress and oncogenic stress, respectively.*** (A) Oxidative stress-induced activation of PKCβ leads to phosphorylation of p66shc at Serine 36 residue, allowing translocation of protein to mitochondria by PIN1-dependent mechanism. The mitochondrial pool of p66shc oxidizes cytochrome c and catalyzes the reduction of O_2 to H_2O_2, inducing ROS production and successively apoptotic induction. (B) Oncogenic stress promotes the binding between FHIT and mitochondrial-import complex Hsp60/Hsp10. In mitochondria FHIT interacts with FDXR promoting ROS production and cytochrome c release, respectively, leading to apoptotic response.

activity, generating H_2O_2 and inducing the opening of the permeability transition pore (PTP) (Fig. 2). In turn, this event perturbs mitochondrial structure and function (as revealed by the reduced Ca^{2+} responsiveness and the alteration of mitochondrial three-dimensional structure (Pinton, Rizzuto, 2008)). A novel cyclophilin-binding agent, Debio 025, was demonstrated to inhibit cyclophilin D (a component of PTP) without having immunosuppressive effects (Ptak et al., 2008). It has been recently reported that Debio 025 was able to normalize mitochondrial function, muscle apoptosis, and ultrastructural defects in Col6a1$^{-/-}$ myopathic mice (Tiepolo et al., 2009), and it may represent a novel therapeutic opportunity to extend life span, minimizing oxidative stress—induced damages typical of aging.

1.2.1 ROS, Mitochondrial DNA, and Aging

Human mtDNA is a circular, double-stranded molecule consisting of 16,569 base pairs (for a review see (Lauri et al., 2014)). 13 mitochondrial proteins (subunits of the electron transport chain, ETC), 22 tRNA molecules, and 2 rRNA species are encoded in this DNA molecule. Depending on the cell type, there are a few to several hundreds of mitochondria per cell, and from a few to several (on average) mtDNA copies in each mitochondrion. This gives the number of about 10^3–10^4 mtDNA molecules per cell, on average (Lauri et al., 2014). The mtDNA is replicated by DNA polymerase γ that is encoded by the chromosomal *POLG* gene. All mtDNA molecules in one mitochondrion and in the cell can have exactly the same nucleotide sequence, which is called homoplasmy. However, occurrence of different variants of mtDNA (i.e., both wild-type and those containing point mutations or deletions) is referred to as heteroplasmy.

In evolution, mitochondria appeared as a result of a symbiosis between α-proteobacterium and a eubacterium, which was a milestone in the formation of eukaryotic organisms (Otten, Smeets, 2015 and references therein). Creation of mitochondria, which still, after billions of years of evolution, contain their own DNA, allowed producing a large amount of ATP in the cell. However, the price for this advantage is the production of large amounts of ROS, which can be deleterious for mitochondria and the whole cells. Recent analysis (Otten, Smeets, 2015) provided interesting conclusions about differences in evolutionary strategies of mtDNA between plants, fungi, and animals. Plants have much larger mtDNAs than animals, with fungal mitochondrial genomes being in between. In plant cells, mtDNA occurs in a low copy number while recombination is quite efficient. Moreover, anti-ROS mechanisms are efficient. Contrary to plants, animals have small mitochondrial genomes, recombination is very rare if any, and mtDNA is prone to mutagenesis. However, relatively high copy number of mtDNA per cell allows to compensate for effects of deleterious mutations in a few mtDNA copies, while providing opportunity to adapt animals, the active creatures, efficiently to various and changing environmental conditions. Nevertheless, the cost of such a strategy is accumulation of mutations in mtDNA during the life. As a consequence, mitochondrial diseases can appear in further generations. Since such diseases are generally rare, this is an acceptable event for a population, which benefits from increased adaptive possibilities. However, several mutations in mtDNA contribute also to aging of cells and organisms.

Again, this is not dangerous for a population, as aging occurs after the time of a reproductive peak, but has a serious consequence for elderly individuals.

The MFRTA hypothesis has been proposed over 60 years ago and has been reviewed many times (see, for example (Lee, Wei, 1997; Wei, 1998; Wei et al., 2001)). The main assumption of this theory is that mtDNA is not protected by histones, thus, it is more prone to lesions caused by ROS. Because mitochondria are one major source of ROS, actions of these agents lead to accumulation of mutations in mtDNA. This, in turn, may cause dysfunctions of ETC, which result in even more efficient production of ROS. Such a positive feedback, called also the vicious cycle, has been considered the major cause of mitochondrial destruction, then cell functions' inefficiency, and finally apoptosis of cells and aging of the organism (summarized by (Guest, Russell, 1992; Lee, Wei, 1997; Wei, 1998)). The importance of the role of ROS production in aging may be also supported by the observed age-related decreases in activities of several antioxidant enzymes, which could reflect overall accumulation of oxidized proteins with age (Stadtman, 1992, 2006). Nevertheless, even at the mature state of this theory, there were some problems that were difficult to explain. For example, considering that phenotypic manifestation of the mtDNA mutation occurs only when a threshold level is exceeded, usually 60%–80% heteroplasmy, depending on the type of mutation (Rossignol et al., 2003), it was difficult to explain how aging can be provoked by mutations in a low fraction of damaged or mutated mtDNA molecules of about 1%–5%, which was determined experimentally (Lee, Wei, 1997).

What kind of mutations occurs predominantly in mtDNA? There are both base pair substitutions and deletions (reviewed by (DeBalsi et al., 2017; Lauri et al., 2014)). Molecular mechanisms of appearance of such mutations in mtDNA have been excellently summarized and deeply discussed recently (Szczepanowska, Trifunovic, 2017), thus, they will be mentioned only shortly here. Nevertheless, one should also note that mtDNA depletion (a decrease in the copy number of mtDNA, not necessarily associated with mutations in this molecule; see, for example (Weglewska et al., 2005)) may also occur is cells of aging organisms. It was considered that formation of 8-oxo-7,8-dihydro-2′-deoxyguanosine (8-oxodG), caused by ROS production by Complex I, is the major mutator causing base substitutions in mtDNA. ROS-mediated conversion of guanosine to 8-oxodG appears to be a particularly efficient process, and since 8-oxodG pairs with adenine instead of cytosine, the next replication round results in creation of the

G to T transversion. Moreover, ROS cause breaks in DNA strands, which is a prerequisite to formation of deletions. In fact, early studies on mtDNA in a mouse model indicated that a particular deletion of 4236 base pairs in mtDNA and 8-oxodG were abundant in mitochondria from old mice while absent in young animals (Muscari et al., 1996). Accumulation of mtDNA deletion of 4977 bp was reported to increase with age (between 32 and 82 years) in neurons from the human brain (Soong et al., 1992). It was supposed that mtDNA should be prone to ROS-mediated mutagenesis more than nuclear DNA because it is not protected by histones, while it is located close to the ROS production site, i.e., complex I in mitochondrion. However, more recent studies, in which the duplex sequencing method was used, allowing to detect one mutation per 10^7 DNA molecules, provided evidence against these early assumptions (Kennedy et al., 2013). When mtDNA was isolated from human brains of old (over 80 years old) and young individuals, the frequency of substitutions in DNA was about fivefold higher in elderly people. However, the frequency of the G to T transition, which is often used as a marker of ROS-mediated mutations, did not increase with age. On the other hand, in both young and old individuals, transition mutations predominated (Kennedy et al., 2013). Moreover, similar analyses performed in mice indicated that the frequency of mutations in mtDNA appears to be an order of magnitude lower than that reported in earlier studies with the use of significantly less sensitive methods (Vermulst et al., 2007). Such results suggest that errors made by DNA polymerase γ, the only DNA polymerase present in mitochondria (though encoded by nuclear DNA) and responsible for replication of mtDNA, may be responsible for most of the mutations in mtDNA appearing during the whole life.

Other results that are against the early MFRTA theory came from advanced microscopic studies. Contrary to previous assumptions, it appeared that mtDNA is not "naked," but instead, it is covered by the TFAM (transcription factor A) protein which together with DNA forms a nucleoid-like structure (Kukat et al., 2011). Moreover, it was reported that mtDNA may be separated from sites of ROS production in mitochondria due to microcompartmentalization of the matrix (Appelhans et al., 2012). In fact, by using sensitive methods for detection of mutations, it was found that frequencies of stable genetic changes in mtDNA are similar to those detected in nuclear DNA (Anson et al., 2000; Lim et al., 2005), rather than higher as presumed on the basis of earlier studies in which small amount of mtDNA was a limiting factor for detailed analyses.

The milestone in studies on mutations in mtDNA was construction of mice, which produce DNA polymerase γ with deficiency in the proof-reading function. Such mice constructed independently by two teams (Kujoth et al., 2005; Trifunovic et al., 2004) accumulate highly elevated number of mutations relative to wild-type animals, which allow detailed studies on mitochondrial mutagenesis in vivo. In these animals, no correlation could be found between number of accumulated mutations in mtDNA and oxidative stress markers (Kujoth et al., 2005). Rather, an increasing number of mutations in mtDNA correlate with aging without influence of the oxidative stress (Trifunovic et al., 2004). Subsequent studies indicated that accumulation of mutations in mtDNA by mice with mutator DNA polymerase γ occurred linearly with age, while production of ROS was at the level comparable to that in wild-type animals (Trifunovic et al., 2005). Therefore, one might suppose that errors made by DNA polymerase γ, rather than ROS-mediated DNA lesions, are the major cause of mtDNA mutations. It was proposed that during aging, their accumulation arise due to clonal expansions of particular mutated mtDNA molecules, rather than due to ongoing ROS-mediated mutation events (Wiesner et al., 2006).

This theory, alternative to MFRTA, received some additional experimental support. Studies on cultured fibroblasts indicated that the T414G mutation in mtDNA has little effect on ROS production and cell aging, which is against the MFRTA theory (Birket et al., 2009). On the other hand, another study on cultured fibroblasts demonstrated that a higher level of a deletion in mtDNA was accompanied with elevated concentrations of ROS (Quan et al., 2015). In fact, there is still an extensive debate on which theory is valid. Still some experimental results might suggest that the old MFRTA is valid, i.e., that ROS cause significant numbers of mtDNA lesions, which are accumulated in time due to defects in ETC and result in a constantly increasing ROS production, whereas other studies indicated that errors made by DNA polymerase γ during replication of mtDNA are the main source of mutations in the mitochondrial genome. The debate is ongoing, and definitely it is not finished. The reader is referred to recent review articles on this topic for detailed discussions on the mechanisms which lead to accumulation of mutations in mtDNA during aging (Bautista-Niño et al., 2016; Chih-Hao et al., 2013; DeBalsi et al., 2017; Edgar, Trifunovic, 2009; Kim et al., 2015; Kirkwood, Kowald, 2012; Lagouge, Larsson, 2013; Lauri et al., 2014; Lee, Wei, 2012; Pamplona, 2011; Szczepanowska, Trifunovic, 2017; Zapico, Ubelaker, 2013). In addition to the two theories mentioned in the preceding paragraph, a third one has been proposed recently. Studies based on the deep sequence analysis indicated that in the absence of overproduction of ROS and with no accumulation of somatic mutations in mtDNA, the respiration defects associated with aging of human fibroblasts are still evident (Hashizume et al., 2015). Moreover, the previously reduced respiratory function could be restored if aging fibroblasts were reprogrammed by formation of iPSCs (induced pluripotent stem cells). In fact, reprogramming of cells could also recover some other phenotypes related to aging (Lapasset et al., 2011). Therefore, a hypothesis has been published that mitochondrial defects associated with aging may be due to epigenetic changes in nuclear genes, rather than due to accumulation of mutations in mtDNA (Hayashi et al., 2016).

In conclusion, three different theories have been proposed to explain the cause and the role of mutations in mtDNA in aging. It is documented that number of mutations in mtDNA increases significantly with age. However, it is still unclear whether the major cause of senescence might be (1) ROS-mediated lesions in mtDNA and the positive feedback leading to enhanced production of ROS and more and more elevated mutagenesis, (2) errors made by DNA polymerase γ and subsequent clonal expansion of mutated mtDNA molecules, or (3) epigenetic changes in nuclear genes affecting mitochondrial metabolism, with little contribution of mtDNA mutations. It is still possible that each of these mechanisms may contribute to the total picture of the senescence in relation to mtDNA and ROS.

1.3 Intracellular Defense Mechanisms Against ROS: Regulation on Aging and Disease

As protection against the deleterious effects of excessive ROS production, cells developed several enzymatic and nonenzymatic antioxidant defenses. The enzymatic system consists in a number of antioxidant enzymes such as superoxide dismutases, glutathione peroxidase, and catalase, localized in distinct cellular compartments. Superoxide anion is a by-product of oxidative phosphorylation, generated by the leak of electrons from the respiratory chain complexes. $O_2^{\bullet -}$, which is generated in the respiratory chain, especially in a pathological state (Ide et al., 1999; Raha et al., 2000), can be converted to H_2O_2 by SOD2 in the mitochondrial matrix or SOD1 (zinc—copper superoxide dismutase; also called Zn-CuSOD) in the intermembrane space of mitochondria (IMS) and cytosol (Sturtz et al., 2001). Subsequently, H_2O_2 can be converted to water by GPx (a selenocysteine-containing enzyme specific to organic peroxides) and peroxiredoxins (Prx 3 and 5; also controlling the level of peroxynitrite) directly in mitochondria (Cao et al., 2007).

Finally, H_2O_2 can diffuse through the mitochondrial membranes to the cytosol, where peroxisomal catalase or cytosolic GPx convert it to water. Besides the antioxidant enzymes, cells and specifically mitochondria also possess other nonenzymatic antioxidant system that comprises small molecules such as ascorbate (Vit-C), glutathione, tocopherol (Vit-E), retinoic acid, uric acid, pyridine nucleotides and thioredoxin, which also provide efficient cellular protection against excessive oxidative stress. These molecules can act directly as free radical scavengers or by modulating the activity of enzymatic systems (Lu et al., 2010). Examples of free radical scavengers include ascorbate and tocopherol (Lu et al., 2010). Ascorbate, present in aqueous phase, becomes a very stable and nonreactive radical and can be subsequently regenerated by pyridine nucleotide-dependent reductases. Tocopherol, present in the cellular lipid phase, is able to neutralize lipid peroxyl radicals, becoming a less reactive phenyl radical (Lu et al., 2010). Tocopherol radicals can then be regenerated by ascorbate. Uric acid is a strong scavenger of peroxynitrite in extracellular fluid but requires the presence of ascorbic acid and thiols for a complete scavenging (Nimse, Pal, 2015). Reduced glutathione, pyridine nucleotides, and thioredoxin work together with antioxidant enzymatic systems, donating reducing equivalents to neutralize ROS (Lu et al., 2010; Nimse, Pal, 2015). Melatonin, the sleep hormone produced by the pineal gland, was also demonstrated to have intrinsic antioxidant protective effects on mitochondria, especially by preventing cardiolipin oxidation and limiting the loss of activity of the mitochondrial respiratory chain (Paradies et al., 2010, 2017; Petrosillo et al., 2008).

The relationship between the MMP and mitochondrial ROS production has been described by Korshunov et al (Korshunov et al., 1997), as a potential mechanism to regulate oxidative stress. Increased ROS production occurring in the presence of mitochondrial hyperpolarisation relates to the higher NADH/NAD ratio indicating a more reduced NADH pool (Adam-Vizi, Chinopoulos, 2006; Aon et al., 2010; Kushnareva et al., 2002). A protection mechanism against excessive ROS production is proposed to be a mild uncoupling of MMP. This process is catalyzed both by a group of uncoupling proteins (UCP2-5) and free fatty acids (FFAs). In the case of hyperpolarization, mitochondrial carriers (e.g., adenine nucleotide translocase, dicarboxylate, and glutamate/aspartate carriers), UCPs, and FFA have been shown to be able to partly "discharge" the high proton gradient to a physiological level (Wieckowski, Wojtczak, 1997). It has been demonstrated that (mild) uncoupling of the MMP (by 10%) enables the reduction of ROS generation by approximately 90% (Korshunov et al., 1997). Moreover, superoxide activates UCPs (UCP1, UCP2, and UCP3), causing increased proton leakage, mitochondrial depolarization, and decreased ROS production (Echtay et al., 2002; Mailloux, Harper, 2011). In the rat heart and skeletal muscle, UCPs have been proposed to remove the superoxide anion radical from the mitochondrial matrix (Wojtczak et al., 2011). There is an on-going debate regarding the physical and biological functions of UCP, a number of associations of UCPs against ROS production have up-to-date been proposed, elucidating this group of proteins as potentially significant in a number of pathophysiological situations (Bugger et al., 2011; Diao et al., 2008; Prakash et al., 2015). The role of mitochondrial uncoupling in preserving muscle fibers from the aging process was demonstrated by Amara et al. In this work, the authors demonstrated that mild uncoupling serves to protect mitochondrial function and contribute to the longevity of the most active muscle fibers, i.e., those with higher oxidative capacity (Amara et al., 2007). Interestingly, opposite effects were found in fibroblasts and yeast cells, where mild-uncoupling led to fibroblast senescence and decrease life span in yeast (Stockl et al., 2007).

It has been repeatedly demonstrated that a defective antioxidant defense system may lead to serious ROS-related human pathologies, often associated with the aging process, such as AD, PD, and diabetes. Oxidative stress caused by disruption of the antioxidant defense and excessive ROS production is closely linked to the pathogenesis of neurodegeneration (Uttara et al., 2009). For example, the levels of SOD1 in AD patients were significantly decreased. Interestingly, the levels of mitochondrial SOD2 and extracellular SOD3 were not changed (Murakami et al., 2011). These data can be supported by the observations that the antioxidant defense of cells from familial AD patients was weaker than in cells from healthy individuals (Cecchi et al., 2002). In another context, Hwang et al. presented evidence that catalase plays an important role in kidney protection during hyperglycemia. Deficiency of catalase, in catalase knockout (KO) mice accelerated diabetic nephropathy observed in streptozotocin-induced diabetes (Hwang et al., 2012). Liu et al. (2017) also demonstrated an increased ROS levels in oocytes from diabetic mice resulting from altered acetylation status of SOD2 in lysine 68. Also, supporting the antioxidant role of UCP and their importance for the development of diabetes, Robson-Doucette et al. (2011) demonstrated that UCP2 regulates ROS production and affects insulin and glucagon secretion by pancreatic cells. Broche et al. (2018) demonstrated that UCP2 regulates pancreas development during embryogenesis through ROS-AKT mediated signaling pathway, evidencing the importance of the antioxidant role of the UCP in the development of diabetes.

Increased ROS formation is also a common intracellular stress that effectively leads to induction of autophagy, which acts as a protective mechanism, at least up to a certain degree (Martinet et al., 2009). The principal link between ROS and autophagy is the cysteine protease Atg4 (autophagy-related gene 4), which is a direct target of H_2O_2. Oxidation of cysteine promotes lipidation of LC3 (MAP1 light chain 3)/Atg8, essential for autophagosome maturation (Scherz-Shouval et al., 2007). On the other hand, it has been demonstrated that autophagy can be a crucial mechanism for preventing the accumulation of ROS by removing damaged mitochondria (Kroemer et al., 2010). This mitochondrial quality control is named mitophagy and serves to eliminate the fraction of damaged mitochondria normally suffering from mitochondrial membrane permeabilization. The recognition of this subset of mitochondria is provided by the involvement of the mitochondrial kinase PINK1 (PTEN-induced putative kinase protein 1). When mitochondria lose membrane potential, PINK1 rapidly accumulates on the mitochondrial surface, leads to the recruitment of the cytosolic protein Parkin, which mediates the ubiquitination of mitochondrial protein with consequent engulfment of damaged mitochondrial by membranes that then fuse with lysosomes (Narendra et al., 2010), a process called mitophagy. In fact, mitophagy has been considered a process by which mitochondrial quality is maintained during the course of aging, avoiding excessive oxidative damage to these organelles (Shi et al., 2017).

Although it exceeds boundaries of this review, it is worth to mention that a number of regulatory programs exist capable of modulating the intrinsic antioxidant defenses. Calorie restriction is known to be one of the strongest life span extending interventions and to have a positive impact in different pathologies (Colman et al., 2009). However, the mechanisms underlying CR remain largely elusive and have become the point of interest of many research groups. Among the proposed mechanisms, the activation of members of the Sirt2 family proteins has gained much interest (Rogina, Helfand, 2004). There are a number of experimental works that associate the activation of Sirt2 orthologues with upregulation of cellular antioxidant defense systems (Qiu et al., 2010; Rahman et al., 2009). Interestingly, the positive effects of CR were mimicked in mice through the administration of low dosages of the protonophore 2,4-dinitrophenol, suggesting that mild uncoupling (see above) can indeed be a protective strategy by upregulating the antioxidant network (Caldeira da Silva et al., 2008). In addition, it has been proposed that CR is able to activate the Nrf2/antioxidant response element (ARE) pathway, inducing ROS detoxification systems, exert antiinflammatory effects, and, thereby, suppress initiation/progression of vascular disease (Ungvari et al., 2008). Next to CR, another important antioxidative intervention is physical activity. It is believed that the two mechanisms together are capable of restoring age-dependent reductions of critical endogenous protective mechanisms such as ischemic preconditioning (Abete et al., 2011). This adaptive mechanism in response to brief episodes of myocardial ischemia enables the reduction of cellular damage due to a prolonged ischemic insult (Murry et al., 1986). Increased ROS production observed on physical exertion may induce compensational increase of the antioxidant defense efficiency (Ji, 1993). Growing evidence indicates that exercise training can result in an elevation in the activity of antioxidant enzymes. Similarly, long-term physical activity is related to the increase of catalase, SOD, and glutathione peroxidase activities in muscles of trained animals (Ascensao et al., 2013; Goncalves et al., 2013; Ji, 1993; Laughlin et al., 1990; Powers et al., 1999).

Dietary antioxidant supplementation is another strategy to boost antioxidant defenses in different cell types, with possible positive impacts during aging. Phenolic and thiol compounds, flavonoids, and carotenoids are examples of antioxidant compounds that can be obtained from fruit, vegetables, spices, grain, and herbs (Nimse, Pal, 2015). A well-known dietary antioxidant supplement is resveratrol. Resveratrol is a stilbenoid phenolic compound, with its antioxidant capacity giving it several possible therapeutic applications (Cho et al., 2017b; Ko et al., 2017) (Li et al., 2017; Sawda et al., 2017; Truong et al., 2017), including age-related diseases (Lange, Li, 2017; Li et al., 2017; Navarro-Cruz et al., 2017) and cancer (Deus et al., 2017). In fact, resveratrol appears to improve mitochondrial function and biogenesis in skeletal muscle of aged animals (Muhammad, Allam, 2017). Also, due to its beneficial effects on mitochondria, resveratrol was proposed as a promising nutraceutical supplement in the treatment of mitochondrial disorders (De Paepe, Van Coster, 2017). Also, due to its antioxidant properties, flavonoids have demonstrated neuroprotective actions against neurodegeneration (Frandsen, Narayanasamy, 2018). Equally, high dietary intake of carotenoids appears to reduce the risk of stroke and stroke mortality (Bahonar et al., 2017). The effects of some approaches have been summarized in a recent review by Suski et al. (2011). However, although having good antioxidant properties, some compounds are not able to reach mitochondria, where the majority of ROS are formed. Thereby, to observe any effect, a larger amount of those compounds is required, which may lead to undesirable side effects. Thus, new strategies are required to improve the delivery of those compounds.

Another interesting antioxidant strategy can be assigned to heme oxygenase-1 (HO-1) (Otterbein, Choi, 2000). The expression of this enzyme is increased as a response to oxidative and heat stress and its role is to degrade free heme, originating from the denaturation and proteolysis of hemoproteins, to release biliverdin, carbon monoxide, and iron (Baranano et al., 2002; Liu et al., 2006; Morse, Choi, 2005). The iron released by HO-1 increases the synthesis of ferritin, what minimizes the probability of Fenton reaction initiation. Furthermore, biliverdin can be converted by the enzyme biliverdin reductase into bilirubin, a potent antioxidant, capable of protecting cells from 10,000-fold higher concentrations of H_2O_2 (Morse, Choi, 2005). Activation of HO-1 was described already as a possible mechanism by which different natural compounds delay skin aging (Park et al., 2016).

In summary, although cells possess several mechanisms of protection against oxidative stress, impairment in mitochondrial function and antioxidant defenses systems during the normal aging phenotype promote an imbalance between ROS formation and cleansing, increasing cellular oxidative stress. It remains to be determined which are the best strategies to decrease cell and mitochondrial oxidative stress that accompanies the aging process without disturbing the physiological role of ROS. Such strategies may pass through the modulation of intrinsic antioxidant mechanisms.

1.4 Mitochondrial Morphology, Calcium Homeostasis, and Dynamics in Aging

1.4.1 Mitochondrial Dynamics

Unlike ROS, mitochondrial dynamics has only recently been studied as a possible player in aging, and specific links between these three processes have been scarcely documented. Thus, here we will give only a brief overview of the studies giving insight into the relationship between mitochondrial dynamics and aging and attempt to highlight the possible trends, underlying mechanisms and links with ROS production. Mitochondrial morphology in living cells is heterogeneous and can range from small spheres to interconnected tubules (Rizzuto et al., 1998) (Fig. 3). Growing evidence indicates that mitochondrial morphology is critical for the physiology of the cell, and changes in mitochondrial shape have been related to many different processes such as development, neurodegeneration,

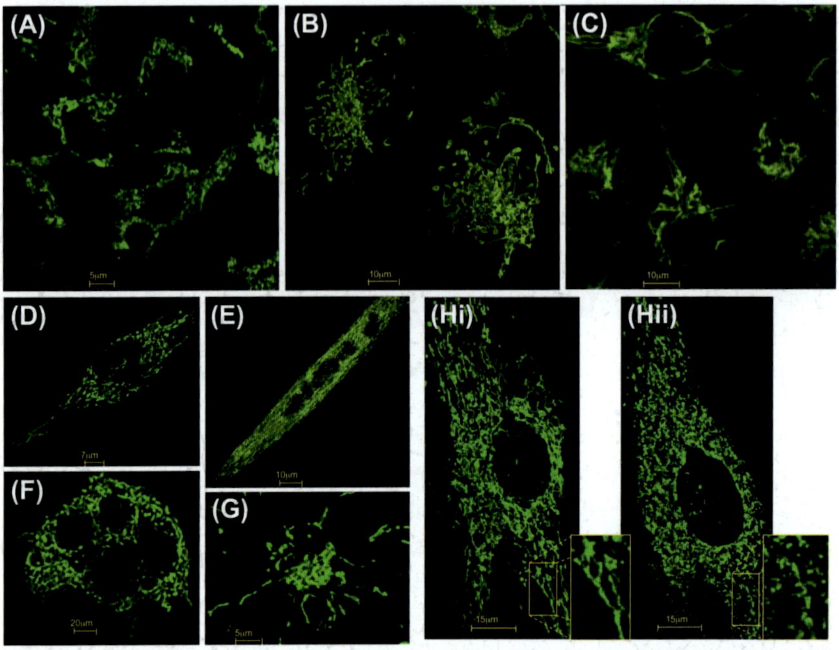

Figure 3 ***Mitochondrial network displays huge heterogeneity and shape rearrangements.*** Examples of mitochondrial network are shown in both immortalized [(A) HEK, (B) Cos7, (C) IB3] and primary cultured cell lines (D) human fibroblast, (E) rat myotube, (F) rat adipocyte, (G) rat oligodendrocyte progenitor). Rearrangement of mitochondrial network is also a typical adaptation to stress stimulus such as oxidative stress as shown for mouse embryonic fibroblast before (Hi) and after (Hii) exposure to H_2O_2.

calcium (Ca^{2+}) signaling, ROS production, cell division, and apoptotic cell death (Cereghetti, Scorrano, 2006). Mitochondrial dynamics in a broad sense involves the processes of fission, fusion, mitochondrial movement or transport, and interactions with other organelles (Benard, Rossignol, 2008). Mitochondrial fission and fusion regulate mitochondrial morphology, branching, and network formation and also determines individual mitochondrion size. Mitochondrial transport determines mitochondrial localization and overall distribution within the cell, this is especially important in highly polarized and large cells, such as neurons or oocytes (Barnhart, 2016; Frederick, Shaw, 2007; Mishra, Chan, 2014; Pernas, Scorrano, 2016). In the last two decades, we have learned intricate molecular details about these processes and their regulation, which now appears to be concertedly controlled by cellular activity. Mitochondria can associate with different cytoskeletal filaments to facilitate intracellular movement.

In mammalian cells microtubule filaments and dynein/kinesin motors are often used for movement purposes, but actin and actin nucleation factors recently emerged as essential players in determining mitochondrial positioning (Barnhart, 2016; Frederick, Shaw, 2007; Kanfer, Kornmann, 2016; Melkov, Abdu, 2018; Pathak et al., 2010). Moreover, mitochondrial interactions with the cytoskeleton are also important to link localization and movement with mitochondrial shape, by interactions with an elaborated machinery on the outer and inner mitochondrial membrane (OMM and IMM) accomplishing the fusion and fission of mitochondria (Chakrabarti et al., 2017; Hatch et al., 2016; Korobova et al., 2014; Korobova et al., 2013; Manor et al., 2015; Prudent, McBride). The cyclic rearrangement of the interconnected dynamic mitochondrial network, by individual mitochondria constantly undergoing fission and then fuse with each other has been studied in much molecular detail, but we still have no consensus on the functional consequences, the purpose and the exact regulation of the entire process (Cho et al., 2017a; Kanfer, Kornmann, 2016; Lee et al., 2016a; Misgeld, Schwarz 2017; Pernas, Scorano, 2016; Yamada et al., 2016). Fusion is an event where the outer and inner membrane of a mitochondrion fuses with the outer and inner membrane of another mitochondrion, respectively, allowing the matrix content of the two mitochondria to mix freely and form a single mitochondrion, whereas fission is the reverse of this event, and the two processes are regulated independently. Intriguingly, fission occurs preferably at sites where mitochondria interact with other organelles, in particular the endoplasmic reticulum (ER) (Friedman et al., 2011). These interactions involve a particular machinery present in subdomains of the ER (mitochondria-associated membranes—MAMs), further interacting with the actin cytoskeleton and the fission machinery of the OMM (Cho et al., 2017a; Hatch et al., 2016; Korobova et al., 2013; Kraus, Ryan, 2017; Li et al., 2015b; Manor et al., 2015; Moore et al., 2016). OMM fission is then promoted by dynamin-related protein 1 (Drp1), a dynamin-related GTPase, and Drp1 is recruited to the mitochondria by fission protein 1 (Fis1), mitochondrial fission factor (Mff), mitochondrial dynamics protein of 49 kDa (Mid49), and 51 kDa (Mid51) (Losón et al., 2013). While the molecular details of IMM fission is less known, both OMM and IMM fusion, facilitated by three dynamin-related GTPases, by mitofusin 1 and 2 (Mfn1 and Mfn2) located on the outer membrane, and by optic atrophy 1 (Opa1) located on the inner membrane (Chen et al., 2003; Cipolat et al., 2004), is reasonably well characterized. With interactions with all components of mitochondrial dynamics, the ER–mitochondrial interaction sites are thus central hubs for a concerted regulation of both intracellular networks, probably executed by cellular and mitochondrial Ca^{2+} signals (Chakrabarti et al., 2017), and other processes regulating cell activity and shape (Shao et al., 2015; Wales et al., 2016). Here, an important link between local ROS production in this subdomain and mitochondrial dynamics has also been recently demonstrated (Debattisti et al., 2017; Norton et al., 2014). Interestingly, MAMs appointed to the modulation of calcium (Bononi et al., 2017; Kuchay et al., 2017; Marchi et al., 2018) and ROS (Verfaillie et al., 2012) signaling in health and disease appear to be altered in aged mice hearts in which Ca^{2+} transients, NAD(P)H regeneration, glutathione levels, and ER—mitochondria contact sites are significantly reduced compared with the young ones with a concomitant increase in ROS generation and mitochondrial protein oxidation (Fernandez-Sanz et al., 2014). Interestingly, recently it was demonstrated that knockdown of MCU and inositol 1,4,5-trisphosphate receptor type 2 (ITPR2), both involved in the accumulation of calcium in mitochondria, resulted in senescence escape, indicating the role of mitochondrial calcium accumulation in senescence induction (Wiel et al., 2014). Similarly, lower number of contacts between mitochondria and the ER in senescent human fibroblasts (Fig. 4) can also be responsible for the compromised mitochondrial calcium uptake in senescent cells. However, additional studies are needed to validate the alterations in the number of contacts between mitochondria and ER during aging or senescence to identify which factors have the highest influence of the regulation of Ca^{2+} fluxes through mitochondria—ER contacts sites in aging cells. The structure and function of MAM in the aspect of aging and senescence has been recently reviewed by Janikiewicz et al. (2015).

1.4.2 Mitochondrial Dynamics and Life Span in Model Organisms

Mitochondrial dynamics has been linked to existing pathways that regulate life span in *C. elegans*. It was shown that mitochondrial trafficking in distal neuronal processes declines progressively with age, and long-lived daf-2 mutants with reduced insulin signaling (IIS) show resistance to this decline (Morsci et al., 2016). Neuron-specific activation of CREB-regulated transcriptional coactivator 1 (CRTC-1), which promotes mitochondrial network fragmentation, is able to suppresses both AMPK and calcineurin-mediated life span extension in *C. elegans* (Burkewitz et al., 2015). Similarly, inactivation of *C. elegans* Drp1 significantly enhanced the ability of IIS to extend life span (Yang et al., 2011), suggesting a correlation between increased fusion/fission ratio and life span extension. Certain mutant bacteria

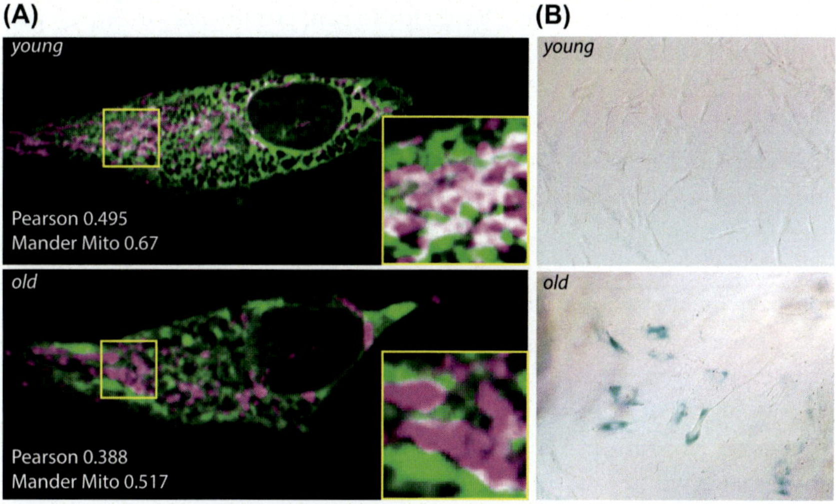

Figure 4 ***The number of contact sites between mitochondria and the ER in young and senescent human fibroblasts.*** (A) Image of the endoplasmic reticulum (ER) (green) and mitochondria (red) in young and senescent human fibroblasts. Maximum intensity projections of confocal micrographs from young and old fibroblasts expressing mitochondria-targeted Cherry (magenta) and ER-targeted GFP (green) contacts sites are represented by colocalization areas (white). Colocalization extents were quantified using Pearson's and Mander's coefficients. (B) Activity of senescence-associated β-galactosidase in young and senescent human fibroblasts. Cells referred as "young" fibroblasts were at 4th passage, and the "old" ones (senescent) at 16th passage.

of the *C. elegans* microbiome are able to extend life span via increased secretion of the polysaccharide colanic acid, and this extension is dependent on Drp1 and mitochondrial fission (Han et al.). Thus, existing pathways of life span extension such as the AMPK and IIS pathways are closely linked to mitochondrial dynamics but the causative role of any of these processes has not yet been demonstrated and is being debated. Indeed, it has been suggested that mitochondrial mass and fragmentation do not affect life span and are merely changes associated with aging (Regmi et al., 2014). However, it is still possible that mitochondrial dynamics plays important roles in the regulation of aging when working in coherence with or as part of other aging pathways such as IIS or metabolic regulation of life span via the AMPK pathway.

Fusion and fission balance has also been studied in fungal and *Drosophila* models. In *Podospora anserina* and *Saccharomyces cerevisiae*, deletion of dynamin-related protein 1 (Dnm1p), mediating fission, retards aging and extends life span (Scheckhuber et al., 2006). A double deletion mutant of

Saccharomyces cerevisiae where Dnm1 (the yeast orthologue of Drp1) and Mgm1 (the yeast orthologue of Opa1), are both deleted, contain wild-type like filamentous mitochondria, but a decrease in mitophagy and replicative life span (Bernhardt et al., 2015). While these models would support a generalization that fusion is associated with increased life span, the *Drosophila* model shows opposite trend. Upregulating Drp1 expression in midlife extends *Drosophila* life span; this is likely linked to autophagy as autophagy is required for the life span extending effect to occur (Rana et al., 2017). Similarly, overexpression of parkin in *Drosophila* extended life span and reduced the level of *Drosophila* orthologue of mitofusin, mitochondrial assembly regulatory factor (Marf), which typically increases with age in flies (Rana et al., 2013).

Overall, from these studies it appears that there is no clear association between either fusion or fission and life span, and promoting either process will have an effect that depends on the signaling and metabolic context. Importantly, a few studies pointed to the importance of the interplay between autophagy and mitochondrial-specific mitophagy in determining cellular homeostasis, affecting life span of whole organisms. This suggests that it is rather the homeostatic function of mitochondrial quality control events, mediated by mitochondrial fusion and fission, bearing the importance for determining or influencing the aging process. Indeed, in mammalian models it is now well documented that any disturbance of the homeostatic circuit maintaining the functional mitochondrial network has a profound effect on the healthy life span of the animals. Mice knockouts for essential components of mitochondrial dynamics (Pernas, Scorrano, 2016) develop pleiotropic symptoms in many organ systems reflecting loss of cellular function or the ability to cope with cellular stress, leading to indicators of early aging or leading to at least a reduced healthy life span. For example, Fis1 KO mice develop multiple early aging signs including lordokyphosis, lack of vigor, inability to accumulate fat, reduced ability to tolerate stress, perturbed Ca^{2+} dynamics, and decreased life span (Uzhachenko et al., 2017). Transmembrane protein 135 (TMEM135) is a protein likely involved in mitochondrial fission, and mice with mutated TMEM135 display abnormal mitochondrial dynamics and accelerated aging in the retina as well as pathologies observed in age-dependent retinal diseases (Lee et al., 2016b). Furthermore, it has been long known that mTOR plays an important role in aging in a range of different organisms (Johnson et al., 2013). mTORC1 has been linked to regulation of mitochondrial dynamics by stimulating translation of mitochondrial fission process 1 (Mtfp1), promoting mitochondrial fission. Potent active-site mTOR inhibitors promote mitochondrial fusion over fission events (Morita et al., 2017).

We have to emphasize that the cellular consequences of perturbed mitochondrial dynamics are often associated with loss of redox homeostasis (Abeti et al., 2011) and increased ROS production. In this context, ROS is often synonymous with cellular damage (Röth et al., 2014; Willems et al., 2015), but mitochondrial or ER stress—related ROS production might play a role in negative feedback regulation of organelles homeostasis, e.g., by altering gene expression via mitohormetic and mitochondrial unfolded protein responses (Shpilka, Haynes, 2017; Yun, Finkel, 2014).

1.5 Mitochondrial Dynamics and Age-Related Diseases in Humans

Mammalian models, such as mice with genetically altered mitochondrial dynamics, develop symptoms that resemble human age-related disease, affecting critical organs (nervous system, liver, and endocrine and cardiovascular systems) or increasing incidence of cancer (Altieri, 2017; Mishra, Chan, 2014; Pernas, Scorrano, 2016; Senft, Ronai, 2016; Youle, van der Bliek, 2012). Most deaths in old age are still due to diseases and it is not possible to extend life span without reducing the effects of age-related diseases. Here, we summarize a series of studies that investigated the relationship between mitochondrial dynamics and age-related diseases.

Cardiovascular disease is a range of diseases whose incidence of occurrence increases with age and is one of the leading causes of death in the developed world (Rapsomaniki et al., 2014). In humans and mice, heart failure is linked to decreased mitochondrial fusion, fragmentation of the mitochondrial network, and lower levels of Opa1 expression (Chen et al., 2009). In *Drosophila*, Marf and Opa1 are essential for proper cardiomyocyte function and fusion defects are associated with cardiomyopathy (Dorn, Scorrano, 2010). Opa1 mutation heterozygotes have late-onset cardiomyopathy in mice (Chen et al., 2012a). Furthermore, unbalanced Opa1 processing and a decrease in mitochondrial fusion are linked to fragmentation results in heart failure in mice (Wai et al., 2015). It has been shown that endothelial cells (HUVECs) maintain a tubular mitochondrial network, but senescent cells have more elongated, interconnected mitochondria, and the change in mitochondrial morphology is caused by downregulation of Fis1 and Drp1 (Mai et al., 2010). However, another group found that in HUVECs, mitochondria of old cells showed a significant and equal decrease of both fusion and fission activity (Jendrach et al., 2005). These results are relatively consistent and suggest that mitochondrial fusion, especially Opa1, is vital for proper heart function and preventing heart failure. This seems to be conserved throughout flies, mice, and humans. Expanding on this, it would be interesting to see if overexpressing of Opa1 or increasing mitochondrial fusion can improve declining heart function. In addition, differentiation of arterial smooth muscle cells and of cardiomyocytes is dependent of reduction of fusion, thus prone to disorders related to altered mitochondrial fusion/fission ratios (Chalmers et al., 2016; Kasahara et al., 2013). Altered ROS production was also found associated with disturbances of mitochondrial dynamics in this disease (Chen et al., 2012a; Mai et al., 2010).

Neurodegenerative diseases such as AD and PD are probably the most prevalent age-related neurodegenerative diseases whose risks increase dramatically with age (Fjell et al., 2014; Rodriguez et al., 2015). AD and PD have been linked to mitochondrial dynamics and mitophagy, as parts of the mitochondrial quality control machinery, and supposed to play important roles in these diseases (Chenet al., 2009). The vast literature on the overall importance of mitochondrial quality control in neurodegenerative disease has been recently reviewed, thus here we will focus on the less covered specific functions of mitochondrial fusion and fission. A large number of studies found that increased mitochondrial fission and fragmentation are linked to AD. Tau mice have higher levels of fission proteins Drp1, Fis1, and lower levels of fusion proteins Mfn1, Mfn2, Opa1 compared with WT mice (Kandimalla et al., 2018). In AD, amyloid-β (Aβ) is linked to Drp1-induced excessive mitochondria network fragmentation in AD progression (Reddy et al., 2017). Drp1 interacts with Aβ and phosphorylated tau, leading to mitochondrial fragmentation, abnormal mitochondrial dynamics, and synaptic damage (Manczak et al., 2011; Manczak, Reddy, 2012). Aβ precursor protein transgenic (APP) mice have significantly decreased anterograde mitochondrial movement, increased mitochondrial fission, and decreased fusion in neurons (Calkins et al., 2011). Furthermore, reducing levels of Drp1 decreases the amount of soluble Aβ production in AD progression and protects against Aβ-induced mitochondrial and synaptic toxicities in AD progression and pathogenesis (Manczak et al., 2016). However, sporadic AD is associated with a significantly lower level of Drp1 in fibroblasts (Wang et al., 2008a). Decreased Drp1 levels and mitochondrial localization, as well as reduced stomatin-like protein 2 (STOML2) and Mfn2 fusion protein levels, are observed in the fibroblasts of sporadic AD patients (Martín-Maestro et al., 2017). In cells overexpressing APP, a fragmented mitochondria structure and abnormal distribution is observed. Moreover, levels of Drp1 and Opa1 were significantly decreased, whereas levels of Fis1 were significantly increased, and increasing Drp1 and Opa1 were both able to rescue some mitochondrial defects (Wang et al., 2008b). Thus, current evidence fails to reach consensus, but since alterations in ER–mitochondrial contacts, which appear as central organizer of fusion and fission events (see above), are implicated in AD (Area-Gomez, Schon, 2017; Filadi et al., 2017), one can speculate that deregulation of these membrane fusion and fission events is linked to disturbances of the ER–mitochondrial contact sites with variable outcome. Then again, mitochondrial ROS has been described as culprit for neuronal damage (Angelova, Abramov, 2018), but it also appears the local ROS in the contact sites is also a cellular signal at this interface and regulatory hub (Debattisti et al., 2017). Similarly, altered fusion/fission ratios accompany PD. In respiratory chain–deficient dopaminergic neurons fragmentation of the mitochondrial network and impaired anterograde axonal transport of mitochondria have been observed (Sterky et al., 2011). Alpha-synuclein (αS) has an inhibitory function on membrane fusion. On increased expression in cultured cells and in *C. elegans*, αS shift mitochondrial shape toward reduced fusion, leading to a fragmented mitochondrial structure (Kamp et al., 2010). Elegant studies demonstrated the ROS dependence of neuronal dysfunction in vivo in dopaminergic (DA) neurons (Guzman et al., 2010).

Another system affected by degenerative age-related symptoms is muscle. Muscle wasting is a hallmark of aging and the primary reason of declining physical abilities with age (Kalyani et al., 2014). Mitochondrial network in skeletal muscle has a complex and seemingly rigid organization, but it has been shown that mitochondrial fission is important for muscle atrophy to occur (Romanello et al., 2010). However, in aged muscle in mice, intermyofibrillar mitochondria in skeletal muscle were longer and more branched, suggesting increased fusion and/or decreased fission, and mitochondrial fusion index (Mfn2-to-Drp1 ratio) was significantly increased in aged muscles (Leduc-Gaudet et al., 2015). In human participants, the levels of the fusion protein Opa1 were lower in muscle from elderly subjects; however, no changes were detected in Mfn2, Drp1, or Fis1 among the groups (Joseph et al., 2012). Thus, studies on muscle atrophy do not have consistent results (Joseph et al., 2012; Leduc-Gaudet et al., 2015), and proper understanding of the organizational principles of the skeletal muscle mitochondrial network will be required to reveal the connection between muscle growth and function in aging.

Metabolic disorders are inherently linked to mitochondrial dysfunction, which can follow alterations in mitochondrial dynamics. This can be extended to insulin resistance and type 2 diabetes mellitus, which are strongly linked to aging and incidence of occurrence increases with age (Facchini et al., 2001; Meigs et al., 2003). While both pathologies have been linked to mitochondrial dysfunction (Mootha et al., 2003), the participation of mitochondrial dynamics is less known in the pathomechanism. Increased mitochondrial network fragmentation and Fis1 expression is observed in venous endothelial cells of type 2 diabetes mellitus patients compared to control (Shenouda et al., 2011). A shift toward mitochondrial fission with reduction of fusion protein, mainly Mfn2, has been associated with reduced insulin sensitivity and inflammation in obesity and insulin resistance development (Putti et al., 2015), which can be associated with altered ROS production.

Finally, although not primarily caused by degenerative age-related processes such as the abovediscussed diseases, in cancer, age is a major risk factor. The role of mitochondria in tumorigenesis is extensively researched (Gasparre et al., 2017; Valcarcel-Jimenez et al. 2017), and a few studies established links between mitochondrial dynamics and cancer cell biology. Some suggest a link between mitochondrial dynamics and development and metastasis (Altieri, 2017; Caino et al., 2016; Senft, Ronai, 2016), but not cancer causation. As such, the link between cancer and mitochondrial dynamics is less relevant toward aging for the purpose of this review but nonetheless could yield interesting insight into cellular transformation.

2. MITOCHONDRIAL DYSFUNCTION AND INCREASED ROS-RELATED/ACCOMPANIED PATHOLOGIES IN THE CONTEXT OF AGING

In this section, we provide some examples of pathological situations illustrating important role of ROS, oxidative stress, and mitochondrial dysfunction in the pathogenesis of described below abnormalities in the context of aging.

2.1 Liver, Mitochondria, and Aging

The process of aging develops in parallel with a gradual deterioration of cell functions, including the hepatocytes, in the body. The liver is a vital organ with a full battery of crucial functions that include the regulation of cholesterol, bile acid, triglyceride, protein, glucose, and energy metabolism, as well as detoxification, and production of bile. This latter function is essential for digestion and absorption of intestinal cholesterol, triglycerides, and fat-soluble vitamins and is involved in hepatic secretion of three lipid components, i.e., bile acids, cholesterol, and phospholipids, which helps the body to excrete excessive cholesterol into the feces (Di Ciaula et al., 2017; Wang et al., 2017b). Although the liver has a major capacity to naturally regenerate, aging induces progressive "physiological" changes of the liver in the structural and functional aspects (Table 1). Whether the liver function becomes seriously compromised in elderly subjects remains largely contradictory because few studies have examined the aging process of the liver from a structural and functional point of view (Schmucker, 1998). The aging liver, however, indeed shows an impaired ability to counteract the hepatic insults, which is a situation often experienced by the elderly.

It has been repeatedly demonstrated that mitochondria play a critical role in driving the age-dependent oxidative lesions with ROS increase (Ames et al., 1995; Pamplona et al., 1998; Sastre et al., 1996). Liver mitochondria contain *manganese*-dependent superoxide dismutase (SOD) with antioxidant function. With aging, however, endogenous, mitochondria-derived free radicals might overwhelm the endogenous defensive response. Bejma et al. (2000) investigated the effects of aging and an acute bout of exercise on intracellular oxidant generation, lipid peroxidation, protein oxidation, and glutathione (GSH) status in the heart and liver of young adult (8 month) and old (24 month) male rats. In the whole liver homogenates and in the mitochondria (and also the heart), the rate of dichlorofluorescin oxidation, an indication of intracellular oxidant production, was higher in the homogenates of aged rats. Lipid peroxidation was also increased in the aged liver and exercised aged heart. Both electron transport chain and NADPH oxidase were two major sources of the age-related increase in oxidant production. In turn, in the study of Kujoth et al. (2005) the aging process was evaluated in mice expressing a proofreading-deficient version of the mtDNA polymerase γ. Mice accumulated mtDNA mutations while displaying features of accelerated aging. Of note, the accumulation of damage, i.e., mutations and deletions, of mtDNA was not associated with increased markers of oxidative stress in the liver mitochondria (i.e., H_2O_2 production and protein carbonyls, F2-isoprostanes, 8-hydroxy-2′-deoxyguanosine) or a defective cellular proliferation but was associated with the induction of apoptotic markers (i.e., cleaved caspase-3), particularly in tissues characterized by rapid cellular turnover such as the liver. Notably, CR is the only nutritional intervention that retards aging and the accumulation of mtDNA mutations

Table 1 Effect of Aging on Changes of Liver Mass and Function.

Finding(s)	Change(s)	Reference
Liver mass	Decrease by 20%–40%	Schmucker (2005), Wynne et al. (1989)
Perfusion and blood flow	Decrease by up to 50% after age 30	McLean and Le Couteur (2004)
Accumulation of hepatic dense body compartment (lipofuscin)	Increase	Gregg et al. (2012), Schmucker (1998), Cogger et al. (2014), Schmucker (2005)
Hepatocyte size (macrohepatocytes)	Increase	Schmucker (1998), Watanabe et al. (1978)
Polyploidy	Increase	Schmucker (1998), Watanabe et al. (1978)
Pseudocapillarization	Increase	Cogger et al. (2014), McLean et al. (2003)
Functional liver function tests	Decrease	Hall et al. (2005), Rahmioglu et al. (2009)
Albumin synthesis	Decreased (animal studies)	Anantharaju et al. (2002)
	Unchanged (human studies)	Fu and Nair (1998)
Serum albumin	Minor decrease	Tietz, et al. (1992)
Hepatic uptake of HDL1-cholesterol	Decrease	Bravo et al. (1994)
LDL receptor	Decrease	Miller (1984), Schmucker (2005)
serum LDL–cholesterol levels	Increase	Anantharaju et al. (2002), Miller (1984)
Liver steatosis, inflammation, fibrosis, anisokaryosis, cellular senescence	Increase	Gregg et al. (2012)
Accumulation of oxidation products	Increase	Gregg et al. (2012)
Cytochrome P450	Decrease by 30%	Sotaniemi et al. (1997)
Drug metabolic clearance	Slower by 20–40%	Turnheim (2003)
Synthesis of vitamin K–dependent clotting factors	Decrease	Froom et al. (2003)
Gallbladder bile	Increased cholesterol saturation index	Valdivieso et al. (1978), Wang (2002)
Mitogen-activated protein kinase activity	Decrease	Schmucker (2005)
Number of binucleated hepatocytes	Increase	Gan et al. (2011), Premoli et al. (2009)

(Aspnes et al., 1997) with reduction of mitochondria-mediated apoptotic pathways (Cohen et al., 2004; Shelke, Leeuwenburgh, 2003).

ROS production increases with aging in state 3 when mitochondria isolated from old rats are supplemented with succinate. This finding could be explained by the defective suppression of H_2O_2 production, i.e., an example of mitochondrial ROS production, during the energy transition from state 4 to state 3. Also, levels of 8-oxodG in the biological macromolecule mtDNA increase with age in old animals. Notably, this increase is abolished by CR, with a positive effect on the aging rate (Lopez-Torres et al., 2002). In senescent rats from the ages of 28—60 and to 92 weeks, the mitochondrial mass of liver remains unchanged throughout ages, but the aging process is paralleled by increased (36%—45%) content of the oxidation products, i.e., thiobarbituric acid—reactive substances and protein carbonyls. These changes are associated with a progressive decrease in critical enzymes for mitochondrial function, i.e., -47% mitochondrial nitric oxide synthase, -46% SOD2, -30% complex I, and -24% complex IV, in old rats compared with young adult rats. However, liver mitochondria from young to old rats do not differ for fragility and water permeability (Navarro, Boveris, 2004). Age-associated decrease of mtRNA is another event involving also the liver (Anantharaju et al., 2002), whereas damaged mitochondrial proteins increase membrane stiffness (Pamplona et al., 1998). In addition, PUFAs could be easily damaged by ROS (Anantharaju et al., 2002). Thus, the accumulation of oxidized and carboxymethylated proteins in the mitochondrial matrix during senescence concomitant with defective degradation of abnormal matrix proteins affects the ability of aging mitochondria to respond to additional stress (Bakala et al., 2003).

A summary of established damages in the mitochondria of aging liver is depicted in Table 2. These changes include oxidative lesions in the mtDNA (Ames et al., 1993), oxidation of mitochondrial lipids, increased levels of long-chain PUFA, decreased membrane phospholipid peroxidability and decreased $\Delta 9$-desaturase activity coefficient leading to decreased levels of 16:1 and 18:1 fatty acids with less membrane stability. Additional mechanisms include increased apoptotic pathways and increased inner mitochondrial phospholipase A_2 activity (Laganiere, Byung, 1993; Pappu et al., 1978). The content of cytochrome oxidase is associated with loss of enzymatic activity (Wilson, Franks, 1975), whereas malondialdehyde accumulation increases (Von Zglinicki et al., 1991).

In all organs of the body, the aging process develops with progressive unbalanced response of the immune system where proinflammatory

Table 2 Damaging Effects of Aging on Mitochondria of the Hepatocytes

Finding(s)	Change(s)	Reference
mtDNA damage	Oxidative lesions	Ames et al. (1993), Lopez-Torres et al. (2002)
mtDNA mutations	Increased apoptosis Unchanged oxidative stress	Kujoth et al. (2005)
Mitochondrial lipid oxidation	Decreased membrane phospholipid peroxidability. Decreased Δ9-desaturase activity coefficient. Decreased levels of 16:1 and 18:1 fatty acids. Decreased membrane stability Increased amount of polyunsaturated fatty acids (in cardiolipin) Increased inner mitochondrial phospholipase A_2 activity	Laganiere and Byung (1993) Pappu et al. (1978)
Content of cytochrome oxidase	Loss of enzymatic activity	Wilson and Franks (1975)
Malondehaldeide accumulation	Increase	Von Zglinicki et al. (1991)
Mitochondrial number	Decrease	Cogger et al. (2014), Gan et al. (2011), Premoli et al. (2009)
Intracellular oxidant production (dichlorofuorescin)	Increased (whole tissue, mitochondria) Mediated by electron transport chain and NADPH oxidase	Bejma et al. (2000)

cytokines, chemokines, and ROS are not adequately counteracted by antioxidants molecules. The liver is enriched with enzymes with antioxidative functions such as mitochondrial SOD2, SOD1, cytosolic glutathione peroxidase, and peroxisomal catalase (Anantharaju et al., 2002). As already mentioned, aging has been shown to be associated with increased oxidative stress, likely due to decreased capacity to eliminate toxic substrates, i.e., metabolically generated superoxide radical. In particular, aging is paralleled by increased production of superoxide anion, hydrogen peroxide, and hydroxyl radical, with all these molecules driving the oxidative protein damage in the liver. The mitochondria play a key role in this scenario leading the oxidative lesions with age (Ames et al., 1995; Pamplona et al., 1998; Sastre et al., 1996). Few studies have attempted to link the oxidative stress to cell injury (Zhang et al., 2003) examined the effects of hypertermic challenge on levels of ROS, oxidative injury, changes in redox status, and DNA-binding activation of critical stress response transcription factors, i.e., activator protein-1 (AP-1) and nuclear factor-kappa B (NF-κB) in old and young rats. Compared with young rats, old rats show greater oxidative damage with sustained ROS levels through 24 h, higher malondialdehyde (MDA) and 4-hydroxy-2-noneal (4-HNE) levels as marker of lipid peroxidation products, as well as hepatocyte damage, i.e., monocyte infiltration, sinusoidal congestion, hepatocellular vacuolization, and diffuse necrosis. Moreover, the ratio of GSH to glutathione disulfide (GSSG) as a marker of hepatic redox status is significantly lower in old rats than in young rats. The effect on peroxidation of PUFAs in old animals might influence the cellular membrane permeability and membrane leakage. In addition, aging impairs the intracellular redox buffering mechanisms, e.g., antioxidants and glutathione, which, by definition, prevent ROS accumulation. Additional damage includes age-related decrease in DNA base excision repair in mouse hepatocytes (Intano et al., 2003), and increased level of oxidatively damaged DNA in the livers of old mice and rats in comparison to young animals (Hamilton et al., 2001). Increased oxidative stress in old animals appears to be paralleled by the enhanced induction of the antioxidant enzyme heme oxygenase via the transcription factor NF-kB (Lavrovsky et al., 2000). The study by Ikeyama and colleagues (Ikeyama et al., 2003) showed that aging, in rats aged 24–26 months, is associated with elevated basal H_2O_2 and epidermal growth factor (EGF)-induced gadd153 gene expression, as proapoptotic marker, in the livers compared with that in young animals. Moreover, there might be a variable change in the antioxidant family members with aging. Thomas et al. (2002) used gene array analysis to find that both aged rat and human liver samples display increased expression of redox and detoxification enzymes, i.e., the glutathione S-transferase GST, UDP-glucuronosyltransferase, and cytochrome P-450 enzyme families.

The excision repair cross-complementation group 1 *(Ercc1)*$^{-/\Delta}$ murine model hosts a rare human progeroid syndrome caused by inherited defects in DNA repair and premature aging (Gregg et al., 2012). In the 5-month-old *Ercc1*$^{-/\Delta}$ mice that are similarly to the old wild-type group, at the ages of 24–35 months, the livers show architectural and inflammatory damages, elevated liver enzymes, and decreased albumin. Of note, there is a significant increase in oxidative damage in *Ercc1*$^{-/\Delta}$ and old wild-type liver, with a detection of lipid peroxidation and senescence products, i.e., lipofuscin, lipid hydroperoxides, and acrolein (Gregg et al., 2012).

Thus, increased oxidative stress and reduced tolerance to oxidative stress with age can initiate the further signaling pathways that drive cellular dysfunction and reduced stress tolerance in older organisms (Schmucker, 2005; Thomas et al., 2002).

2.1.1 Lipids, Mitochondria, and Aging

Lipid accumulation, i.e., triglycerides and cholesterol, greatly promotes aging in the liver (Ghosh et al., 2012; Petersen et al., 2003; Slawik, Vidal-Puig, 2006), whereas phospholipids remain quantitatively unchanged (Schneeman, Richter, 1993). A typical scenario in this respect is age-related increase in the metabolic syndrome with age (Ford et al., 2002), leading to the progressive redistribution of adipose tissue from subcutaneous sites to visceral ones (Tchkonia et al., 2010), and the increased prevalence of nonalcoholic fatty liver disease (NAFLD) as part of the age-dependent process of fat redistribution in nonadipose tissues. The process of ectopic fat deposition and lipotoxicity, therefore, initiates/perpetuates the damage in the liver, heart, skeletal muscle, and pancreas, thus leading to increased cardiovascular risk, the metabolic syndrome and, in turn, to further enhance liver steatosis (Floreani, 2007; Slawik, Vidal-Puig, 2006; Tchkonia et al., 2010; Tran et al., 2008). Of note, this process also referred to as "inflammaging" can also develop in lean but metabolically obese subjects (Tchkonia et al., 2010; Vecchie et al., 2017). NAFLD refers to the presence of excess fat in the liver, i.e., hepatic steatosis, when more than 5% accumulation of triglycerides occurs in the hepatocytes (Krawczyk et al., 2010). NAFLD encompasses the spectrum of liver abnormalities which range from simple steatosis, i.e., nonalcoholic fatty liver (NAFL), to steatohepatitis (NASH), and to (likely cryptogenic) cirrhosis, and has a potential progress to hepatocellular carcinoma (HCC) (Brunt et al., 2015). NAFLD is one of the most common liver disorders worldwide (Krawczyk et al., 2010), with a 20%–50% prevalence according to ultrasonographic and computed tomography (CT) imaging across population studies (Williams et al., 2011; Zelber-Sagi et al., 2006). Pathogenic factors of NAFLD include a genetic background (Krawczyk et al., 2013), enhanced uptake of FFAs (long-chain fatty acids,

LCFA), increased de novo lipogenesis, decreased fatty acid β-oxidation, and/or decreased synthesis or secretion of very low-density lipoproteins (VLDL) (Cohen et al., 2011).

Overall mechanisms accounting for the age-related increase of hepatic steatosis include accumulation of ROS and DNA damage (Aravinthan et al., 2013), hypercholesterolemia (Bonomini et al., 2013), decreased autophagy (Amir, Czaja, 2011), activation of NF-κB signaling that is the key regulator of inflammatory responses (Franceschi et al., 2000), metabolic dysfunction (Rodriguez et al., 2007), telomere shortening (Tomás-Loba et al., 2013), sedentary lifestyle (Breitling et al., 2009), and cigarette smoking (Booth et al., 2011). Aging also induces liver damage, increases proinflammatory M1 macrophage polarization, and enhances inflammatory response typical of NASH (Fontana et al., 2012). Notably, NAFLD appears to decline in very elderly subjects (Koehler et al., 2012) and this finding becomes more apparent when liver fibrosis and NASH are more advanced (van der Poorten et al., 2013), as observed in elderly patients (Koehler et al., 2012; Noureddin et al., 2013). Different protective mechanisms might be responsible for the age-dependent decrease of liver steatosis and include increased serum adiponectin levels (van der Poorten et al., 2013), the onset of portosystemic shunting (Nosadini et al., 1984), metabolic changes of mitochondria (Caldwell, Crespo, 2004), and inflammatory and catabolic state of cirrhosis (McCullough, Raguso, 1999), as well as collagen deposition in the liver (Noureddin et al., 2013). In the serum, levels of cholesterol, HDL cholesterol, and triglycerides also increase with age. This trend is reverted in individuals older than 90 years old (Tietz et al., 1992). Notably, LDL—cholesterol metabolism rate is decreased by over 30% with age, as a consequence of a decrease in expression of LDL receptors (Miller, 1984). Aging significantly enhances the progression of NAFLD to NASH, and to fibrosis, leading to the conditions that predispose to increase mortality in elderly subjects with NAFLD (Regev, Schiff, 2001). Age-dependent decline of fatty acid β-oxidation and reduced expression of hepatic nuclear receptor peroxisome proliferator—activated receptor may be potential mechanisms (Sanguino et al., 2004). An additional mechanism is the age-dependent increase of the β-adrenergic signaling that is able to drive liver steatosis (Ghosh et al., 2012; Katz et al., 1993). In addition, p300-dependent regulation of chromatin structure during aging is responsible for the activation of five key genes that govern triglyceride synthesis in the liver (Jin et al., 2013). Markers of hepatocyte senescence are also associated with NAFLD (Aravinthan et al., 2013; Park et al., 2010). The link between increased intrahepatic diacylglycerol (DAG) and PKCε activation is essential, as LCFA are oxidized at both mitochondrial and extramitochondrial sites in the hepatocytes. Excessive incorporation of LCFA will increase ROS and mediate hepatocellular and mitochondrial injury (Diogo et al., 2011; Grattagliano et al., 2011, 2013; Zhu et al., 2017).

Mitochondria contribute to the progression of NAFLD (Grattagliano et al., 2004b), a condition characterized by increased predisposition toward prooxidant insults (Pessayre, Fromenty, 2005). One pathway that limits excessive fat accumulation in the liver is the increased mitochondrial oxidation of LCFA, a step associated with an impaired respiration (Fromenty, Pessayre, 1995). Fatty degeneration exposes hepatocytes to a higher risk of oxidative damage, although a number of adaptive metabolic mechanisms have been described (Grattagliano et al., 2003; Yang et al., 2000). Mechanisms include expression of intracellular sensors and signaling molecules for lipid metabolism and oxidative stress pathways (Merriman et al., 2006; Sanyal et al., 2001). Impairment of these systems may have important pathogenic roles in NAFLD progression, including disturbed ATP synthesis (Cortez-Pinto et al., 1999). mtDNA levels, protein expression, and activity of respiratory complexes are also decreased in liver mitochondria (Haque, Sanyal, 2002; Perez-Carreras et al., 2003), pointing to a role for oxidative stress mechanism.

Hepatic steatosis indeed causes cellular damage and ROS production. Our group has shown that rat hepatoma FaO cells loaded with oleate/palmitate to mimic liver steatosis resulted in higher production of ROS and lipid peroxidation, stimulation of catalase activity, and activation of NF-κB. Lipid droplet accumulation also increased levels of peroxisome proliferator–activated receptors (PPARs) and sterol regulatory element–binding protein-1c (SREBP-1c) (Vecchione et al., 2016). In the intact hepatocyte HepG2 cells incubated with saturated fatty acids (a model resembling NASH), mitochondrial function was depressed together with inhibition of mtDNA gene expression and accelerated degradation of respiratory chain subunits (Garcia-Ruiz et al., 2015). In line with these results, we recently showed that sequential exposure of hepatocytes to high concentrations of fatty acids (FAs) and TNF-α mimics in vitro the progression of NAFLD from simple steatosis to steatohepatitis. Several damages were observed at a mitochondrial level and elsewhere in the hepatocyte (reduced hepatocyte viability, increased apoptosis and oxidative stress, reduction in lipid droplet size, and upregulation of I-κB kinase–interacting protein and adipose triglyceride lipase expressions). Notably, silybin, the extract of the milk thistle seeds counteracted the FA-induced mitochondrial damage, increased the mitochondrial size and improved the mitochondrial cristae organization; stimulated mitochondrial FA oxidation; reduced basal and maximal respiration and ATP production in steatohepatitis hepatocytes, stimulated ATP production in steatotic cells, and rescued the FA-induced apoptotic signals and oxidative stress in steatohepatitis hepatocytes (Vecchione et al., 2017). In steatotic livers, ROS formation is increased at the mitochondrial respiratory chain level and determines oxidation of unsaturated lipids (Grattagliano et al., 2003, 2008; Yang et al., 2000). The activity of complex I of the respiratory chain is also reduced (−35%) in mitochondria from fatty livers and is associated with changes in state 3 respiration (Petrosillo et al., 2007); hydrogen peroxide generation and oxidized cardiolipin are significantly increased (Grattagliano et al., 2008; Petrosillo et al., 2007). ROS affect the mitochondrial complex I activity by oxidizing cardiolipin, which is required for the function of this enzyme complex (Paradies et al., 2002).

Oxidation, glutathionylation, and nitrosylation of mitochondrial proteins occur as a response to oxidative stress and result in posttranslational modification of proteins by carbonyl and disulfide formation or by thiol nitrogen exchange. Factors contribute to a block of the electron flow in the respiratory chain resulting in subsequent generation of ROS. This vicious circle involves ROS-mediated antioxidant depletion and the deficient capacity of mitochondria to inactivate ROS (Grattagliano et al., 2003). Ultimately, protein and lipid oxidation, and cytokine production are increased. Hepatocytes react to fat deposition with an early increase of GSH and thioredoxin to prevent lipid and protein oxidation (Grattagliano et al., 2008). Also, increases of protein mixed disulfides (PSSG), nitrates, and nitrosothiols are consistent with both prooxidant protein modifications and increased nitric oxide (NO) synthesis. A critical role for mitochondrial GSH in the development of NASH was in fact proposed (Garcia-Ruizet al., 2006). GSH depletion sensitizes hepatocytes to inflammatory cytokines and TNF-α. Mitochondrial GSH content declines more rapidly than cytosolic GSH, suggesting mitochondria as specific early target for oxidative changes (Grattagliano et al., 2008).

Other pathogenic factors, including nitric oxide, play a role for the progression of liver steatosis and appearance of fibrosis. Thioredoxin, a redox active protein regulates of PSH/PSSG ratio. Thioredoxin is actively involved in the regulation of NO activity via cleavage of nitrosothiols (Nikitovic, Holmgren, 1996; Stoyanovsky et al., 2005), which are formed by conjugation of NO with free thiols and oppose dangerous reactions such as peroxynitrite formation. Nitrosothiols also act as intracellular messengers that control mitochondrial functions (Arnelle, Stamler, 1995; Grattagliano et al., 2004a). Major alterations of thioredoxin levels have been observed with ongoing liver steatosis and have been associated with PSSG and nitrosothiols formation (Garcia-Ruiz et al., 2006). Increased peroxynitrite formation is associated with a variety of interactions, including protein nitration and generation of nitrotyrosine (Sanyal et al., 2001).

Additional mechanisms of mitochondrial damage include increased production of angiotensin II associated with oxidative stress. Animals with elevated endogenous angiotensin II levels display mitochondrial alterations with reduced β-oxidation and consequent decreased mitochondrial palmitate oxidation, decreased enzymatic activities, and expression of mitochondrial proteins, including cytochrome *c*, cytochrome *c* oxidase subunit 1, and TFAM. Administration of angiotensin II receptor blockers or superoxide dismutase/catalase mimetic treatment improves these abnormalities (Wei et al., 2009). Fatty livers show about 35% decrease of catalytic β-F1 subunit of the F0F1-ATP synthase. The process of aging might also influence other pathways in the steatotic liver. Under starvation, mitochondrial oxidative injury is exacerbated to a greater extent in fatty livers. In the steatotic liver, fasting induces a further decrease of the ATP levels, which is accompanied by a 70% fall of the catalytic β-F1 subunit. These changes may account for the observed reduction in the synthesis of ATP (Fernández-Checa et al., 1998). Liver steatosis is also favored by prolonged intake of drugs such as amiodarone or valproate (Berson et al., 1998), and aging is a strong risk factor for medication-induced damage. Amiodarone or valproate accumulates in mitochondria and induces inhibition of fatty acid oxidation and electron transfer chain (Berson et al., 1998). In support of the key role of mitochondria in NAFLD, recent data indicated the critical role of the mitochondrial pyruvate carrier MPC, a heterologous complex made of MPC1 and MPC2 proteins in the inner mitochondrial membrane (Colca et al., 2017). The complex is required for the entry of pyruvate that is synthetized in the cytosol, in the mitochondrial matrix, where it will be further metabolized. MPC might become therefore the target for treatment of several metabolic and inflammatory diseases, namely diabetes (Chen et al., 2012b; Colca et al., 2013; McCommis et al., 2015), and even NASH (McCommis et al., 2016).

Although the abovedescribed observations suggest that fatty livers have compromised mitochondrial function, there are several evidences that the initial stages of the NAFLD condition are associated with an increase in mitochondrial mass, with or without increased mitochondrial fatty acid oxidation, and which serves as an adaptation response to excessive accumulation in the liver. This was elegantly demonstrated in human liver biopsies by Koliaki et al. (2015). This paper showed that when compared with isolated mitochondria from lean individuals, their counterparts from obese humans with or without NAFLD had 4.3- to 5.0-fold higher maximal respiration rates, despite similar mitochondrial content. In opposition to this, and despite the fact that NASH patients featured higher hepatic mitochondrial mass, a 31%—40% lower maximal respiration associated with greater hepatic insulin resistance, mitochondrial uncoupling, and leaking activity was described.

Initial adaptation to a fatty-rich environment contributes to enhance fatty acid oxidation, which means that hypothetically, accelerating the rate of fatty acid burn by mitochondria could be an effective therapeutic strategy. In fact, a liver-targeted mitochondrial protonophore was described to promote a mild depolarization of the inner membrane, contributing to increase electron transfer and fatty acid oxidation. In rodent models for NAFLD and diabetes type 2 (T2D), this approach appears to reverse hypertriglyceridemia, hepatic steatosis, insulin resistance, and hyperglycemia (Perry et al., 2013). Another approach involved adenovirus-mediated liver expression of a malonyl-CoA-insensitive CPT1A (CPT1mt) in a high fat/high sugar animal model, with the ultimate objective of accelerating mitochondrial fatty acid β-oxidation. This approach was able to reverse insulin resistance and glucose intolerance, although not affecting steatosis (Monsenego et al., 2012). The important role of mitochondria in the context of NAFLD and its progression to NASH is well demonstrated by two recent findings. One of them demonstrates that mtDNA, a proinflammatory molecule per se (Boyapati et al., 2017; Zhang et al., 2016), when released from fatty liver hepatocytes, causes liver inflammation by TRL-9 activation (Garcia-Martinez et al., 2016), which can be an important component of the transition between NAFLD and NASH. In another interesting development, it was recently demonstrated that when mitochondria isolated from hepatoma cells were injected into rodents with fatty liver, the phenotype was improved. Because exogenous mitochondria were tagged with green-fluorescence protein (GFP), it was possible to demonstrate accumulation in the mouse liver, lung, brain, muscle, and kidney. How mitochondria entered the different cells and were able to maintain the integrity and restore metabolic activity was not explained.

2.1.2 Bile Mitochondria and Aging

Bile is an aqueous solution containing organic solutes, inorganic electrolytes, and trace amounts of proteins. The three classes of biliary lipids include unesterified cholesterol, phospholipids, and bile acids (i.e., primary bile acids: cholic and chenodeoxycholic acid; and secondary bile acids derived from 7α-dehydroxylation of the primary bile acids in the liver and by intestinal bacteria in the ileum and colon: deoxycholic, lithocholic, ursodeoxycholic, sulfolithocholic, and 7α-oxo-lithocholic acids) (Di Ciaula et al., 2017). Earlier studies have shown that hepatobiliary function as assessed by bile flow and bile acid secretion decline with age in male inbred rats (Schmucker et al., 1985). Moreover, a number of liver enzymes are involved and affected by aging (Schmucker, 2001). Choi et al. (1987) have found a significant decrease in the enzymatic activity of cholesterol 7α-hydroxylase in rats 5—32 weeks of age. In the study by Wang (2002), gallstone-susceptible C57L mice and resistant AKR mice of both genders split into young adult, older adult, and aged groups (8, 36, and 50 weeks of age, respectively) were each fed a lithogenic diet for 8 weeks. Increasing age augments biliary secretion and intestinal absorption of cholesterol, whereas reducing hepatic synthesis and biliary secretion of bile acids. Einarsson et al. (1985) analyzed biliary lipid composition in 60 healthy lean and gallstone-free subjects of various ages and both genders. They found a positive correlation between age and cholesterol saturation index and a negative correlation with bile acid synthesis and the pool size of the cholic acid. A role for a decline in the enzymatic activity of cholesterol 7α-hydroxylase and 7α-hydroxylated cholesterol (as a marker of bile acid synthesis) has been advocated in humans by the study of Bertolotti et al. (1993). By contrast, Valdivieso et al. (1978) found increased proportion of biliary cholesterol and lithogenic index in gallbladder bile of elderly female Chilean patients without significant changes in bile acid metabolism.

Additional age-dependent changes in lipid metabolism include a decrease of hepatic clearance of HDL-cholesterol in older rats (Bravo et al., 1994). Lipophilic and secondary bile acids, moreover, are able to change the membrane composition of hepatocyte mitochondria (where the inner membranes harbor the cytochrome P450 oxidoreductase system). Thus, changes in the membrane fatty acid composition lead to a decrease of the activity of the mitochondrial enzyme system; this step makes enzymes, defective in handling free radicals that originate during the normal process of energy production and detoxification. As a consequence, free radicals may attack membrane PUFAs to initiate and propagate lipid peroxidation, leading to formation of aldehydic lipid peroxidation products (Pandey, Shukla, 2000).

2.1.3 Gallstones, Mitochondria, and Aging

Increasing age is associated with increased risk of gallbladder stones of either types, i.e., cholesterol and pigment stones (Palasciano et al., 1989; Portincasa et al., 2006; Portincasa, Wang, 2015, 2016). It is estimated that the prevalence of gallstone disease is about 50% in females aged 70–75 and in males aged 80–85 (Diehl, 1991; Sama et al., 1990). Pigment stones account for about 20%–25% of all gallstones and form as a consequence of abnormalities in bilirubin metabolism arising in the gut–liver axis. The most common risk factors are hemolytic anemias, liver cirrhosis, Crohn's disease, or extended ileal resection, biliary infection, cystic fibrosis, vitamin B12/folic acid–deficient diets, and genetic factors due to UGT1A1 mutation. Age per se represents a risk factor for such biliary diseases. Cholesterol gallstones represent about 75%–80% of all gallstones in western societies and are due to at least five pathogenetic defects: (1) genetic factors and *LITH* genes; (2) hepatic hypersecretion of biliary cholesterol, leading to nonphysiological and sustained supersaturation of gallbladder bile with cholesterol; (3) accelerated phase transitions of cholesterol in bile; (4) hypomotile gallbladder harvesting the immune-mediated inflammation, as well as hypersecretion and accumulation of mucin gel in the lumen; and (5) increased absorption of cholesterol at the small intestinal enterocyte level (Wang et al., 2017a; Wang et al., 2017b; Wang and Portincasa, 2017, pp. 1–676).

A more detailed analysis of the elegant study by Wang (2002) found that on the lithogenic diet, cholelithiasis prevalence, gallbladder size (i.e., a marker of gallbladder hypomotility and stasis), biliary lipid secretion rate, and HMG-CoA reductase activity (i.e., a marker of cholesterol biosynthesis in the liver) are significantly greater in C57L mice of both genders compared to those in AKR mice. The activity of cholesterol 7α-hydroxylase (i.e., a marker of hepatic bile acid synthesis) is significantly lower in C57L mice than in AKR mice. Of note, increasing age augments biliary secretion and intestinal absorption of cholesterol, while reducing hepatic synthesis and biliary secretion of bile acids, and decreasing gallbladder contractility. Putting these data together, it is clear that all the abovementioned age-dependent factors greatly increase susceptibility to cholesterol cholelithiasis in C57L mice. Mitochondria might play a role in some steps of lithogenesis. Biliary proteins and their redox status have been demonstrated by our group in gallstone patients undergoing cholecystectomy and serial analyses of bile composition and crystallization (Grattagliano et al., 2009). The role of gallbladder contractility during fasting and postprandially is essential in preventing stasis of cholesterol supersaturated bile and precipitation of either cholesterol crystals or bilirubinate pigments (Di Ciaula et al., 2017; Portincasa et al., 1994, 2004, 2006). Mitochondrial Ca^{2++} handling is implicated in spontaneous rhythmic activity in smooth muscle and interstitial cells of Cajal. Indeed, disruption of the MMP (by carbonyl cyanide 3-chlorophenylhydrazone, carbonyl cyanide 4-(trifluoromethoxy) phenylhydrazone, rotenone, and antimycin A) reduced or eliminated action potentials, Ca^{2++} flashes, and Ca^{2++} waves in typical of the gallbladder smooth muscle. Data suggest that mitochondrial Ca^{2++} handling accounts for spontaneous electrical activity, a step involved in gallbladder tone and motility (Balemba et al., 2008). Gallbladder mitochondria are also involved in other pathways such as drug-mediated effects on apoptosis during tumorigenesis in cells and mice (Bao et al., 2014; Li et al., 2015a; Liu et al., 2013; Shu et al., 2014; Wang et al., 2014; Weng et al., 2014) and treatment with oxysterols causing cytochrome *c* release in the dog (Seo et al., 2004). Both aspects are potentially linked to the aging process of the gallbladder with or without stone-formation ability. Accumulation of gut microbiota–derived secondary bile acids in the aging, hypomotile gallbladder, moreover, by virtue of their lipophilic action, might change the membrane composition of and contribute to gallbladder carcinogenesis (Pandey, Shukla, 2000).

Epidemiological and clinical investigations have clearly demonstrated that during the aging process, a person progressively loses the ability to maintain normal physiological functions due to structural alteration of cells or dysfunction of vital organs such as the liver. Aging is a major risk factor for most chronic hepatobiliary disorders because the volume and blood flow of the liver progressively reduce with age. As the elderly population is increasing due to an extended life span, the number of elderly patients with complicated hepatobiliary diseases has grown in the past decades. New and effective strategies for enhancing liver functions should be extensively investigated to improve life quality of human beings.

2.2 Copper Toxicity in Age-Related Diseases and Wilson Disease

The redox active transition metals, copper and iron, are fundamental for vital enzymes (Festa, Thiele, 2011; Winter et al., 2014). However, on overload they might become cell toxic if uncontrolled redox activity occurs. Here, we restrict ourselves to the detailed discussion of copper. Aqueous free copper (and iron) may catalyze the formation of hydroxyl radicals ($^{\bullet}OH$) via Fenton– and Haber–Weiss-based chemistry causing the subsequent damage to proteins, DNA, and lipids that result in cell death

	$Fe^{3+}/\mathbf{Cu^{2+}} + O_2$	$Fe^{2+}/\mathbf{Cu^{1+}} + O_2$
Fenton reaction	$Fe^{2+}/\mathbf{Cu^{1+}} + H_2O_2$	$Fe^{3+}/\mathbf{Cu^{2+}} + OH + OH$
Haber-Weiss reaction	$O_2 + H_2O_2$	$O_2 + OH + OH$

Figure 5 ***Fenton and Haber—Weiss reaction.*** In the presence of catalytic amounts of trace metals (such as iron and copper), highly reactive hydroxyl radicals ($^{\bullet}OH$) are generated from hydrogen peroxide via the Fenton and Haber—Weiss reaction.

(Fig. 5) (Leonard et al., 2004). However, under physiological conditions, this redox activity is prevented by the tight incorporation of copper into the active sites of various vital enzymes that results in the absence of intracellular free copper pools (Kaim, Rall, 1996; Lippard, 1999; Rae et al., 1999; Rubino, Franz, 2012). At present, 54 copper-binding or transporting proteins have been identified in the human proteome. Together they ensure the safe transport of the transition metal to the target enzymes (Blockhuys et al., 2017). Protein copper binding occurs mainly via cysteine residues (e.g., metallothionein) but can also be performed by histidine (e.g., ceruloplasmin) or methionine (e.g., copper transporter Ctr1) residues (Koch et al., 1997). In cells, the most important copper-dependent enzyme resides in mitochondria, the cytochrome *c* oxidase (complex IV) that is responsible for proper cell respiration and oxidative phosphorylation (Wainio et al., 1959). In serum, the transition metal is mainly present tightly bound to the ferroxidase ceruloplasmin (around 70%), with a small part (around 30%) available as so-called "free" or "loosely bound" copper associated to amino acids or albumin (Linder, 2016).

Under pathological situations, the tight regulation of copper uptake, distribution and excretion can be disturbed leading to increased free copper levels that are highly detrimental to cells. Age-related diseases such as AD, PD, diabetes, cardiovascular diseases, and cancer have been suggested to be associated with increased serum free copper levels (Brewer, 2010). For example, in AD, high free copper serum levels (e.g., due to increased copper uptake via drinking water) were linked to an increased risk for cognitive decline (Sparks, Schreurs, 2003; Squitti, Polimanti, 2013). It was shown that copper interacts with the characteristic beta-amyloid plaques resulting in increased ROS production and neuronal death (Huang et al., 1999). Besides this "direct" impact of copper on AD progression, copper was also found to oxidize low-density lipoprotein receptor—related protein responsible for the efflux of beta-amyloid from the brain (Singh et al., 2013).

However, there are also conflicting studies reporting a reduced beta-amyloid production by addition of dietary copper in mice (Bayer et al., 2003) and a correlation between cognitive decline and low copper plasma concentrations in patients with mild to moderate AD (Pajonk et al., 2005). Additionally, high serum copper levels may be associated with an increased risk for several types of cancer, e.g., lung cancer, neoplastic kidney tissue, leukemia, and hepatocellular carcinoma (Hrgovcic et al., 1973; Karcioglu et al., 1978; Mateo et al., 1979; Poo et al., 2003). Whether high copper levels are a cause or "bystander" effect of these malignancies is still under investigation. Thus, at present, future studies have to further substantiate the link between free copper and age-related diseases.

In contrast to the mentioned diseases, copper overload (especially in the liver) is a well-established cell-death causative feature in Wilson disease (WD) (Bearn, 1953; Cumings, 1948). In WD, the hepatocyte demise is linked to copper-induced mitochondrial dysfunction. WD is an autosomal inherited disorder caused by mutation(s) in the gene encoding for the copper transporting ATPase ATP7B (Gitlin, 2003; Tanzi et al., 1993). In hepatocytes, this protein facilitates the transport of excess copper into the bile (Bull et al., 1993). In WD patients, the dysfunction of ATP7B results in the accumulation of massive amounts of copper, primarily in the liver and brain. If untreated, WD is fatal (Liver., 2012).

Mitochondria were identified as first responders in WD patients and in WD animal models. In WD, mitochondria present with dramatically increased copper levels (humans (Sokol et al., 1994), animal models (Lichtmannegger et al., 2016; Zischka et al., 2011)), structural abnormalities ranging from distorted cristae structure to electron-dense inclusions and detachment of the inner and outer membrane (humans (Sternlieb, 1968, 1992; Sternlieb, Feldmann, 1976), animal models (Huster et al., 2006; Lichtmannegger et al., 2016; Roberts et al., 2008; Sternlieb et al., 1995; Yurkova et al., 2011; Zischka et al., 2011)), decreased ETC complex activities (humans (Gu et al., 2000), animal models (Roberts et al., 2008; Sauer et al., 2011)), and a reduced ATP production capacity (animal models (Lichtmannegger et al., 2016)).

These structural and functional alterations are accompanied by the direct attack of copper on mitochondrial protein/peptide thiol residues (Nakamura, Yamazaki, 1972; Zischka et al., 2011). Moreover, treating intact liver mitochondria with copper (in the presence of reducing agents such as DTT or GSH) resulted in WD mitochondrial phenotypes (Zischka et al., 2011). This latter finding questions the hypothesis that ROS (resulting from copper-based Fenton reaction) are primarily causative for mitochondrial structure changes, although such ROS might be relevant in subsequent mitochondrial destruction (Lichtmannegger et al., 2016; Zischka et al., 2011). It rather appears that thiol residues not only in scavenging proteins such as metallothionein but also in vulnerable enzymes such as the ATP synthase (complex V) are prime copper targets (Duncan et al., 2006; Yagi, Hatefi, 1987). Detrimental outcomes by such mitochondrial functional depletion together with persistent copper overload may consequently pave the way for further mitochondrial and subsequently cellular destruction.

Studies in WD patients described a reduced antioxidative defense system (reduced GSH, GST, SOD1/2) (Bruha et al., 2012; Nagasaka et al., 2006; Summer, Eisenburg, 1985), decreased levels of Vit-E in the serum (that correlated with free copper serum level) (von Herbay et al., 1994), and mtDNA deletions (Mansouri et al., 1997). Additionally, aconitase activity, a classical mitochondrial marker for ROS damage to proteins was reduced (Gu et al., 2000). These findings were further confirmed by studies in WD animal models. Here, an increased lipid peroxidation (Kumar et al., 2016; Ohhira et al., 1995; Rui, Suzuki, 1997; Samuele et al., 2005; Yamada et al., 1992; Yamamoto et al., 1999), DNA damage (Chung et al., 1999; Nair et al., 1996; Yamamoto et al., 1993; Yu et al., 2016), decreased antioxidative defense system (GPx, GSH) (Kumar et al., 2016; Samuele et al., 2005; Yamamoto et al., 1999), and reduced free protein thiols were described (Samuele et al., 2005; Zischka et al., 2011). Thus, oxidative damage is clearly associated with WD. However, it needs to be emphasized that these features mostly appear in WD patients or animal models at severe liver damage, i.e., at late disease stages.

Importantly, diverse therapy strategies addressing increased oxidative stress in WD were evaluated in animal models, in particular the LEC rat. Here, a surplus of Vit-E resulted in later onset of hepatitis and reduced lipid peroxidation in male rats compared to control animals (Yamazaki et al., 1993). In contrast, Hawkins et al. described no effect of Vit-E and β-carotene but a delayed onset of jaundice and decreased mortality in animals with enforced administration of proline (80%) and Vit-C (65%) (Hawkins et al., 1995). Another strategy used by Yamashita et al. (1996) was the use of the spin-trapping antioxidant phenyl butyl nitrone that resulted in reduced lipid peroxides and delayed hepatitis and mortality. This finding was further underlined by studies of Asanuma et al. (2007), who described delayed hepatitis, reduced lipid peroxidation, and decreased oxidative DNA damage in treated animals, while the liver copper level was unaltered. Additionally, unsaturated fatty acids, e.g., linolenic acid and linoleic acid, reduced the incidence and onset of hepatitis (Shibata et al., 1999) and increased animal survival by stimulation of bile acid synthesis (Du et al., 2004). However, curcumin (Frank et al., 2003), quercetin, and phytic acid (Kitamura et al., 2005) failed to increase survival or reduce liver damage. In contrast, DL-alpha lipoic acid, a dithiol-containing cofactor of the pyruvate dehydrogenase complex of mitochondria, reduced liver damage and increased the antioxidative defense capacity by upregulation of glutathione peroxidase and reductase (Yamamoto et al., 2001). Additionally, NAC, a prodrug to cysteine and therefore a precursor for glutathione, reduced liver copper and overall liver damage mainly due to metal chelation rather than ROS scavenging (Kitamura et al., 2005). Additionally, fermented brown rice (Shibata et al., 2006) and coffee (Katayama et al., 2014) were tested to reduce liver damage in LEC rats. Both treatment options were able to prolong survival. Whereas fermented brown rice reduced the incidence of hepatitis, coffee delayed the onset of hepatitis in treated animals compared with untreated rats.

In conclusion, the redox-active transition metal copper may be highly detrimental to cells on toxic overload. While such a role is currently under debate in age-related diseases such as AD, there is ample evidence of its destructive power on hepatic mitochondria in WD. With respect to the toxic mode of action of copper on mitochondria, susceptible thiol residues appear as early copper targets. In agreement, treatment regimens that affect the intracellular thiol status (DL-alpha lipoic acid, NAC) were able to increase survival by delaying hepatitis onset or decreasing hepatitis incidence. In contrast, radical scavengers that would counteract a (Fenton chemistry—based) ROS burden (e.g., Vit-E, phenyl butyl nitrone) only showed slight effects on WD progression. Nevertheless, such oxidative stress damage may be present in WD patients and animal models with late-stage disease.

2.3 Mitochondria, Mitochondrial-Associated Membranes, ROS, and Diabetes in Aging

Type 2 diabetes is multifactorial disorder characterized by chronic hyperglycemia due to impaired insulin secretion from pancreatic β-cells and insulin resistance in target tissues. Nowadays, it is estimated that up to 70%—90% of patients with T2D are overweight or obese with aging being an important contributing factor (Al-Goblan et al., 2014). Particularly, excessive lipid intake or age-related prolonged ectopic fat deposition in tissues is believed as the primary reason for insulin resistance development, the islet dysfunction, and disease progression (Janikiewicz et al., 2015; Szymanski et al., 2017). There is now ample evidence that diabetes and obesity-related metabolic dysfunction can accelerate the progression of other age-related diseases and pathologies in both mouse models of disease and normally aging mice (Butterfield et al., 2014; Farr et al., 2008; Morrison et al., 2010). For example, several reports have shown high fat diets, or insulin resistance, can accelerate learning and memory loss as well as neurodegeneration in different mouse models of AD (Kadish et al., 2016; Knight et al., 2014; Maesako et al., 2015; Morrison et al., 2010; Petrov et al., 2015). It is thus possible that obesity-related insulin resistance may be accelerating the aging progression (Salmon, 2016), and mitochondria and oxidative stress may play important role in this process.

Mitochondria-associated membranes are important hubs for insulin signaling due to several proteins that were detected at the MAM location, including protein kinase AKT, mTORC2, phosphatase, and tensin homologue deleted on chromosome 10 (PTEN) (Tubbs, Rieusset, 2017); however, such interconnection awaits further investigations. As T2D is associated with alterations in lipid metabolism and ER—mitochondria contact sites foster lipid species exchange between these organelles, it is a favorable hypothesis that MAM integrity and action participate to lipotoxicity in diabetes. In fact, MAM integrity was required for insulin signaling and it was altered in palmitate-induced insulin-resistant HuH7 hepatic cells, as well as in liver of leptin-deficient ob/ob mice and high-fat and high-sucrose fed mice. Furthermore, disruption of MAM integrity by genetic or pharmacological inhibition of MAM-residing protein cyclophilin D induced insulin resistance in animals and led to aberrant insulin signaling in human primary hepatocytes. Enhancement of MAM formation restored hepatic insulin signaling and action of HuH7 cells and in diabetic mice (Tubbs et al., 2014). In addition, an abnormal increase in MAM formation was reported in livers of ob/ob, and high-fat fed mice, alongside mitochondrial calcium overload, compromised mitochondrial capacity, and augmented oxidative stress (Arruda et al., 2014).

Oxidative stress is considered one of the major risk factors in the onset and progression of T2D. Since the 1960s (Giugliano et al., 1996) hyperglycemia has been hypothesized to contribute to oxidative stress either by direct production of ROS (Nishikawa et al., 2000) or by altering the redox balance. It has been demonstrated that hyperglycemia induces an increased polyol pathway flux and increased intracellular formation of advanced glycation end products and contributes to an oxidative stress environment by activating PKC and causing overproduction of superoxide via mitochondrial ETC (Brownlee, 2001). As emphasized in this review, mitochondria play an important role in the maintenance of cellular redox status, thereby acting as an ROS and redox sink and limiting NADPH oxidase activity. The main sources of mitochondrial ROS under physiological conditions are complexes I and III, which produce $O_2^{\bullet -}$ mainly on the matrix side, where it is rapidly transformed into H_2O_2 by SOD2. Brownlee and colleagues proposed that the mitochondrial transport chain plays a key role in hyperglycemia-induced overproduction of superoxide and in the development of secondary complications such as endothelial dysfunction (Brownlee, 2005). Using ρ^0 cells lacking functional ETC, they demonstrated that the effect of hyperglycemia on ROS production was completely lost. Another study by Nishikawa discovered that T2D alters the primary site of superoxide generation such that complex II becomes the primary source of electrons contributing to superoxide formation under diabetic conditions (Nishikawa et al., 2000). Yu et al. (2006) found that dynamic changes in mitochondrial morphology are an important factor contributing to ROS overproduction under high glucose (HG) conditions. Mitochondria become rapidly fragmented in HG concentrations with a concomitant increase of ROS. These findings suggest that mitochondrial dynamics may influence ROS overproduction in diabetes, obesity, and other related disorders.

Conventional antioxidants neutralize ROS on a one-to-one basis, whereas hyperglycemia-induced overproduction of superoxide is a continuous process. Riley (Salvemini et al., 1999) proposed a novel type of antioxidant, a catalytic antioxidant, such as SOD/catalase mimetic, which works continuously, similar to the enzymes for which these compounds are named. However, SOD is not yet in widespread use in human clinical medicine due to some obstacles: none of the three human SODs possess the necessary pharmacological properties to make it a clinically useful therapeutic agent (McCord, Edeas, 2005).

Another source of ROS in diabetes is NADPH oxidase. This enzyme has been implicated as a major source of ROS generation in the vasculature in response to high glucose and advanced glycation end products (Thallas-Bonke et al., 2008). This study suggests that blockade of NADPH oxidases is a valid intervention to consider for combating established diabetic nephropathy. A study by Li and Shah (2003) supports the theory that NADPH is a mediator of diabetic complications and its action can be suppressed by a variety of PKC inhibitors, implicating this family of kinases in the regulation of hyperglycemia-induced NADPH oxidase activity. Production of ROS from mitochondria has received great attention, and it is now becoming clear that it may be regulated under physiological conditions and plays and important role in redox signaling. A recent work by Darley-Usmar and colleagues (Chacko et al., 2010) hypothesize that MitoQ, the most-studied mitochondria-targeted antioxidant, decreased mitochondrial ROS production and showed beneficial effects in diabetic nephropathy. In this study, $Ins2^{+\setminus -AkitaJ}$ mice were selected as a model of diabetes, showing a dysfunction in pancreatic β-cells, to test the potential benefits of MitoQ therapy. This mouse model develops insulin resistance over time and has many of the characteristics of chronic hyperglycemia (Barber et al., 2005; Bugger et al., 2008). Of the potential compounds available, the ubiquinone analog MitoQ has received particular attention, as it is orally bioavailable, has low toxicity, and reaches concentrations of 200–700 pmol/g of weight in the tissue of a number of organs, including the kidney. The treatment seems to protect mitochondrial redox signaling from hyperglycemia-induced alterations and would thereby ameliorate diabetic nephropathy. Another study by Wang and colleagues proposes the use of vanadium compounds in therapeutic treatment of diabetes and in cancer prevention (Zhao et al., 2010).

Mitochondria also play an important role in β-cell function as insulin release in response to glucose levels and in the sensing of oxygen tension in the carotid body and pulmonary vasculature, two events involved in diabetes pathologies (Duchen, 2004). One of the hypotheses for induction of β-cell dysfunction focuses on changes in the expression and function of UCP2. The inner mitochondrial membrane UCPs are thought to be major facilitators of uncoupled respiration and may modulate the pathophysiology of diabetes (Sack, 2006). It has been proposed that UCP activity and expression contribute to an increase in superoxide formation under diabetic conditions (Krauss et al., 2003). UCP2 is thought to negatively regulate glucose-stimulated insulin secretion by reducing the amount of ATP produced (Fig. 6). Another member of this family, UCP3, seems to have a positive role in T2D. This isoform, very enriched in skeletal muscle, is downregulated in pathologic condition and conversely increased in response to exercise training. UCP3 confers resistance against oxidative stress and the genetic deficiency of UCP3 in primary skeletal myocytes results in excess ROS levels under normoxic and hypoxic conditions (Luet al., 2008). To test whether the induction of UCP3 could modulate insulin sensitivity, a skeletal muscle transgenic mouse line harboring the human UCP3 gene was created (Clapham et al., 2000). These mice are hyperphagic and lean

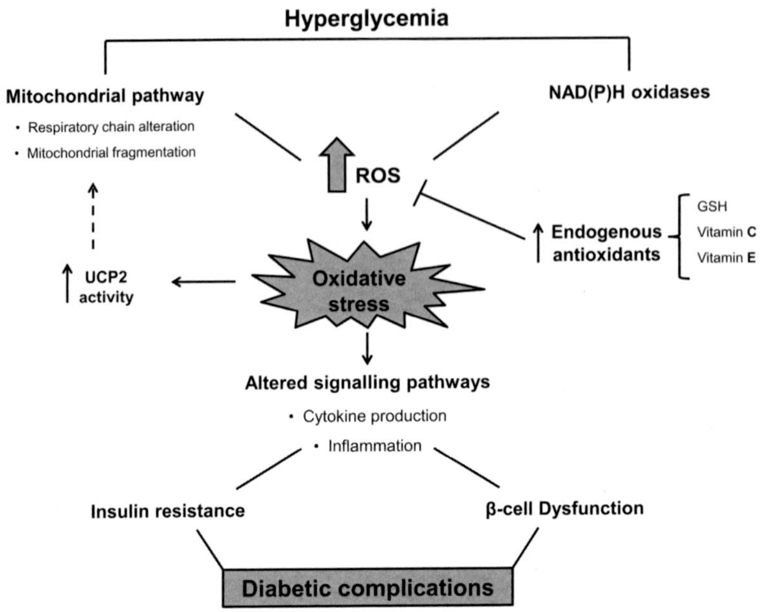

Figure 6 ***Putative pathways linking oxidative stress, mitochondrial dysfunction, and hyperglycemia.*** *GSH*, glutathione; *ROS*, reactive oxygen species; *UCP2*, uncoupling protein-2.

with diminished adipose tissue mass, a phenotype consistent with uncoupled mitochondrial oxidative phosphorylation. Despite this, knockout mice for UCP3 showed some divergent effects, i.e., glucose tolerance improvement, suggesting how the beneficial role of UCP3 may be skeletal muscle specific (for an intriguing discussion on molecular manipulation of mitochondrial metabolism and diabetes, see Pagel-Langenickel et al., 2010).

Cells and tissues contain antioxidant defense mechanisms, which maintain their redox balance and aid in preventing the accumulation of ROS. T2D is associated with reduced levels of antioxidants such as GSH, Vit-C, and Vit-E (Jain, 1998). Thus, antioxidant therapy has been of great interest as a means to combat oxidative stress in diabetic patients. However, these treatments lack broad therapeutic effect across the patient population (Ceriello, Testa, 2009). Nevertheless, the current data rise an important, and rather unexplored, question of whether antioxidant treatment could be used in obese patients with metabolic dysfunction to prevent the progression of additional comorbidities or slow the acceleration of the aging process caused by insulin resistance and chronic inflammation.

2.4 Mitochondria, ROS, Cardiovascular Pathology, and Aging

Cardiovascular diseases (CVDs) are multifactorial, but several lines of evidence suggest that mitochondrial dysfunction contributes to their pathophysiology. The underlying mechanisms appear to involve not only damage to the organelle and loss of bioenergetic function but also disruption of mitochondrion-dependent redox-signaling pathways. Mitochondria are considered as a major source of ROS, and loss of control of their formation leads to mitochondrial oxidative damage and dysfunction; mitochondria are also important targets for ROS. ROS can lead to the activation of pathways that control cell differentiation and apoptosis, both of which are mechanisms of particular relevance to CVDs. Indeed, mitochondrial oxidative damage has been implicated in a range of degenerative conditions that include CVDs, namely atherosclerosis, hypertension, heart failure, and ischemia/reperfusion (I/R) injury (Fig. 7) (Delles et al., 2008; Di Lisa, Bernardi, 2006; Madamanchi, Runge, 2007; Misra et al., 2009).

Some of the most compelling evidence that mitochondrial ROS is a causative agent in the development of CVDs in vivo comes from experiments using transgenic mice to alter expression of mitochondrial antioxidant proteins. Initial experiments using genetic knockouts showed that mice lacking SOD2 produce huge amounts of mitochondrial ROS and develop cardiomyopathy within the first weeks of birth (Li et al., 1995; Schriner et al., 2005). Nowadays, it is widely accepted that deficiencies in mitochondrial antioxidants and/or regulatory proteins that modulate mitochondrial oxidant production promote the onset of CVDs. In a recent work Nox4, a member of the NADPH oxidases (Nox) family expressed primarily in the mitochondria in cardiac myocytes, was reported to be a major source of superoxide production in the cardiovascular system. Nox4 mediates cardiac hypertrophy and heart failure in response to pressure overload. Upregulation of Nox4 increased mitochondrial superoxide thereby directly mediating oxidative stress, mitochondrial dysfunction, and myocardial cell death during pressure overload-induced cardiac hypertrophy. Because expression of Nox4 is upregulated by cardiac stress, including pressure overload, heart failure, and aging, it could become an ideal target for pharmacological interventions to in the heart (Kuroda et al., 2010). Another study shows sex differences in the phosphorylation of mitochondrial proteins, including aldehyde dehydrogenase 2 (an enzyme that detoxifies ROS-generated aldehyde adducts) and α-ketoglutarate dehydrogenase

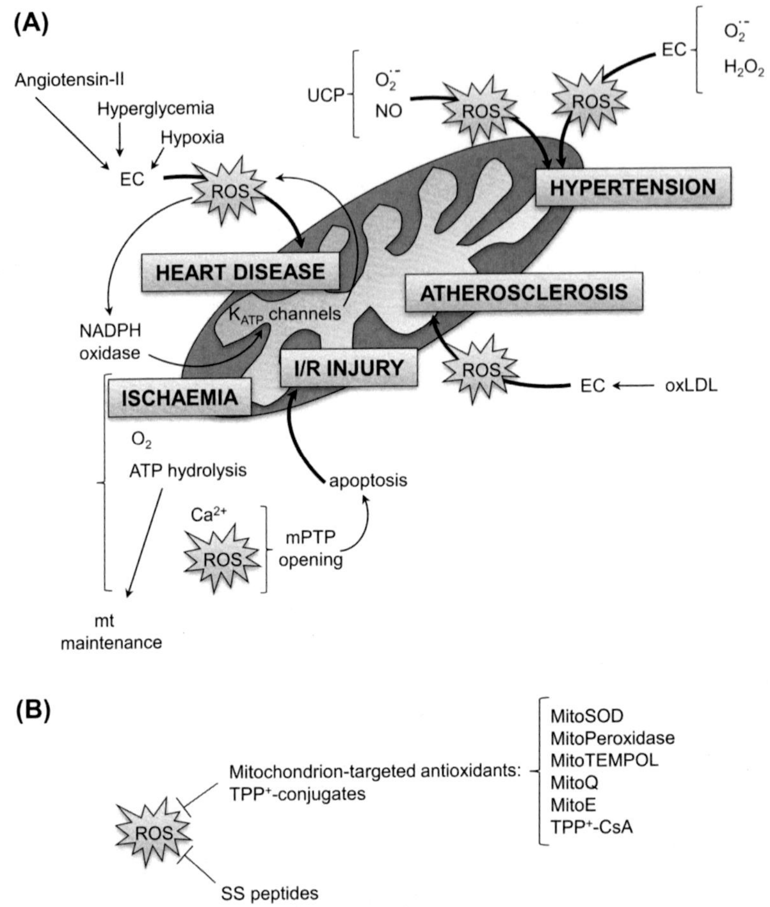

Figure 7 ***ROS as causative agents in cardiovascular diseases (CVDs) and therapeutic approaches targeting mitochondrial oxidative damage.*** (A) Principal mechanisms of ROS production and subsequent mitochondrial dysfunction leading to CVDs. Mitochondria participate in the pathogenesis of hypertension: increased blood pressure has been associated with an excessive endothelial cells (EC) production of superoxide ($O_2^{\cdot-}$) and H_2O_2; UCP1 expression also increases $O_2^{\cdot-}$ production and decreases the availability of NO. ROS production by endothelial mitochondria contribute to heart disease: angiotensin-II, hyperglycemia, or hypoxia increases mitochondrial ROS production in EC, which then stimulates the NADPH oxidase; moreover ROS, produced by the NADPH oxidase, activate mitochondrial KATP channels, suggesting a possible feedback amplification system. Exposure of endothelial cells to oxidized lipids (oxLDL) induces ROS formation, which has a pivotal role in atherogenesis. During ischemia O_2 is lacking and mitochondria hydrolyze ATP to maintain the mitochondrial membrane potential ($\Delta\Psi$); the cardiac cell strives to maintain ATP production, this eventually results in mitochondrial Ca^{2+} overload, mitochondrial depolarization, and increases the generation of ROS. Restoration of blood flow will help to restore ATP levels, but the damaged mitochondria generate enormous amounts of ROS during reperfusion and promote

(one of the major sources of ROS generation), resulting in a reduced production of ROS and cardioprotection in females (Lagranha et al., 2010). In addition, mtDNA damage is increased in cardiovascular tissues of CVD patients, and it has been established that mtDNA mutations lead to increased ROS production (Aliev et al., 2002).

ROS may be both a cause and an effect of hypertension because increased blood pressure has been associated with an excessive endothelial production of superoxide and H_2O_2 in both animal models and humans with increased blood pressure (Puddu et al., 2008). The participation of mitochondria in the pathogenesis of hypertension is also suggested by the involvement of UCPs in experimental and human hypertensive states (Table 2). In mice with doxycycline-inducible expression of UCP1 in arterial walls, UCP1 expression increases superoxide production and decrease the biological availability NO, causing increased blood pressure (Bernal-Mizrachi et al., 2005). In addition, a common polymorphism of the UCP2 gene has been associated with hypertension (Ji et al., 2004); however, as the hypertension-associated allele was reported to increase transcription of the UCP2 gene, and UCP2 is known to diminish ROS production and emission from mitochondria (Teshima et al., 2003), oxidative stress seems unlikely to be a link between the polymorphism and hypertension. ROS also underlie much of the endothelial cell (EC) damage related to heart disease (Davidson, 2010). Damage to the endothelium contributes to the development of atherosclerosis and hence to possible myocardial infarction and subsequent heart failure. EC has relatively little dependence on oxidative phosphorylation for ATP production; however, endothelial mitochondria are centrally involved in maintaining the fine regulatory balance between mitochondrial Ca^{2+} concentration, ROS production, and NO. Moreover, a general principle appears to be emerging in which mitochondrial ROS

mitochondrial permeability transition pore (mPTP) opening and activation of apoptosis, triggering ischemia/reperfusion (I/R) injury. (B) Approaches to deliver drugs to the mitochondria and prevent ROS-induced mitochondrial oxidative damage. Compounds conjugated to the lipophilic triphenylphosphonium cation (TPP^+) can be delivered selectively into the mitochondrial matrix in a potential-driven manner; a series of mitochondria-targeted antioxidants have been designed to decrease superoxide (MitoSOD), hydrogen peroxide (MitoPeroxidase), ferrous iron (MitoTEMPOL), lipid peroxidation (MitoQ, MitoE), and preventing mPTP opening (TPP^+-CsA). Recently, Szeto and Schiller (SS) peptides targeting the inner mitochondrial membrane have been developed; they concentrate in a potential independent manner and possess intrinsic mitoprotective activities.

signals to other cellular sources, triggering ROS production from these. For example, angiotensin-II, hyperglycemia, or hypoxia, each increases mitochondrial ROS production in EC, which then stimulates NADPH oxidase via activation of mitogen-activated protein kinase (MAPK). Furthermore, a "reverse" pathway, in which ROS produced by NADPH oxidase leads to increased mitochondrial ROS production (via activation of mitochondrial K_{ATP} channels, matrix swelling, and alkalization) may also exist, suggesting a feedback amplification system (Daiber, 2010). These data demonstrate that ROS production by endothelial mitochondria contribute to heart disease, therefore targeted scavenging of ROS may have various protective and beneficial effects on the heart. Interestingly, ROS play a pivotal role in atherogenesis, which is one of the major factors in the development of heart failure. Oliveira et al. (2005) have shown that mitochondria from atherosclerosis-prone hypercholesterolemic LDL receptor knockout mice have oxidative phosphorylation efficiency similar to that of control mice, while their net production of ROS and susceptibility to developing membrane permeability transition are both higher. Indeed, it has been demonstrated that exposure of endothelial cells to oxidized lipids such as oxLDL induces ROS/reactive nitrogen species (RNS) formation (Zmijewski et al., 2005). Recently, overexpression of mitochondrial thioredoxin (Trx2) in the endothelium has been proven to protect from atherosclerosis (Zhang et al., 2007).

ROS production is also a feature of ischemia and is implicated in the pathogenesis of I/R injury (Misra et al., 2009). In acute myocardial infarction, two distinct types of damage occur to the heart: ischemic injury and reperfusion injury. The first results from the initial loss of blood flow and the second from the restoration of oxygenated blood flow. To maintain the MMP when O_2 is lacking, mitochondria hydrolyze ATP; this activity destroys any available ATP, favors Ca^{2+} accumulation, and increases the generation of ROS. During ischemia, ROS are generated in the myocardium, at complexes I and III of the ETC, and primarily formed by the degradation of adenosine. Increased ROS production (initiated during ischemia and exacerbated upon reperfusion) coupled with increased cellular $[Ca^{2+}]$ are thought to be the main causes of reperfusion injury. The combined effects of ROS and elevated $[Ca^{2+}]$ lead to the opening of the PTP, which ultimately induces apoptosis (Solaini, Harris, 2005). Ischemic preconditioning (IPC) is a very effective way of protecting the heart from reperfusion injury. This process involves one or more short nonlethal cycles of I/R that protect the heart against a subsequent prolonged period of ischemia. It seems clear that IPC protects the heart by reducing oxidative stress during

I/R, and that this decreases PTP opening. However, the signaling pathways involved in mediating these effects have yet to be elucidated and several possibilities exist. For example, UCP2 overexpression in rat neonatal cardiomyocytes confers tolerance to oxidative stress by diminishing mitochondrial Ca^{2+} overload and reducing ROS generation, suggesting that UCP2 may mitigate ischemia-reperfusion injury and be a mechanism of cardioprotection (Teshima et al., 2003). Other data demonstrate that nitrite (NO_2^-), a stable endocrine pool of nitric oxide that is selectively reduced to NO in ischemic conditions, mediates cardioprotection after I/R. The mechanism involves the inhibition of mitochondrial complex I by S-nitrosation, leading to the inhibition of electron transport and subsequent decrease in mitochondrial ROS generation, which limits apoptosis and cytotoxicity at reperfusion (Shiva et al., 2007). Another possible mechanism involved mitochondrial ROS formation catalyzed by p66Shc; hearts from p66Shc knockout mice display decreased ROS production and decreased myocardial injury caused by postischemic reperfusion. In particular the absence of this protein involved in mitochondrial ROS formation prevented the oxidative attack of structural components of cardiomyocytes, such as lipids and proteins (Carpi et al., 2009).

Today, oxidative stress remains an attractive target for cardiovascular prevention and therapy. A general therapy for decreasing mitochondrial oxidative damage should be effective as a future therapy for CVDs. As the mitochondrial respiratory chain in the IMM is considered as important intracellular source of ROS, a specific action of antioxidants on the mitochondrial respiratory chain may constitute an important mechanism of cardiovascular protection. For example, uncouplers such as dinitrophenol are protective, as they decrease MMP and thus decrease the activity of the mitochondrial Ca^{2+} uniporter, whereas F1FO inhibitors such as oligomycin can decrease wasteful ATP hydrolysis. Unfortunately, these two agents show complementary deleterious effects: uncouplers increase ATP hydrolysis and F1FO inhibitors increase MMP and, presumably, increase mitochondrial Ca^{2+} uptake, thus they are used only as research tools. Other studies seeking to counteract the deleterious effects of ROS have shown that antioxidants such as Vit-E, CoQ10, and NAC decrease mitochondrial oxidative damage in different models. Despite, a number of preclinical and clinical lines of evidence, studies testing the effects of classical antioxidants such as Vit-C, Vit-E, or folic acid in combination with Vit-E have been disappointing. Vit-E or CoQ is quite lipophilic and tends to be retained in cell membranes and subsequently fail to achieve significant intracellular concentrations. NAC fails to provide significant antioxidant effect, presumably due to its low lipid solubility and tissue distribution. As these compounds do not significantly accumulate within mitochondria, their effectiveness remains limited. A number of low-molecular weight catalytic antioxidants, generally called SOD mimetics, have been shown to provide some protection against I/R injury (Gianello et al., 1996). Unfortunately, again, while these SOD mimetics are cell permeable, they do not selectively target mitochondria.

Recent drug development efforts have focused their attention on reducing mitochondrial oxidative stress using mitochondrion-targeted antioxidants, which show potential as future therapies for CVDs (Fig. 7). MitoVit-E, one of the first TPP^+ (triphenylphosphonium cation) conjugates, has been shown to successfully decrease ROS production and apoptosis in bovine aortic EC exposed to oxidative stress (Dhanasekaran et al., 2004) but was ineffective against hypoxic–ischemic striatal injury in neonatal rats (Covey et al., 2006). One disadvantage of Vit-E is that it is not a catalytic antioxidant and its scavenging activity is not regenerated. In contrast, MitoQ consists of a TPP^+ covalently attached via an aliphatic linker to a ubiquinone derivative; after detoxifying an oxidant species, it is regenerated by the respiratory chain. MitoQ concentrates several 100-fold within the mitochondria, is orally bioavailable, distributes to various organs, including the heart, without adverse effects in rats or humans, and decreases mitochondrial oxidative damage in rodent models of cardiac I/R injury (Adlam et al., 2005). Administration of MitoQ to stroke-prone hypertensive rats improves endothelial function and attenuates cardiac hypertrophy (Graham et al., 2009). Also a cyclosporine A (CsA) derivate, specifically targeted to mitochondria by conjugation to a TPP^+, has been designed (Malouitre et al., 2010); CsA is well known to protect the heart from reperfusion injury by preventing PTP opening; however, side effects associated with PTP inhibitors limit their therapeutic potential. An alternative compound, called SkQ1, when perfused through isolated heart preparations or fed to rats, was able to reduce ischemia-induced arrhythmia and infarct size, despite the use of a concentration that was an astounding six orders of magnitude lower than that of MitoQ (Bakeeva et al., 2008). On the other hand, the accumulation of these lipophilic cations in the mitochondrial matrix can disrupt MMP and inhibit mitochondrial respiration and ATP production. As a result, the therapeutic index of these molecules is rather low, with toxic concentrations being only $\sim$10-fold greater than effective concentrations. The utility of TPP^+-conjugated antioxidants may also be limited by their requirement of MMP for mitochondrial uptake, especially considering that diseased mitochondria are unlikely to have normal MMP. The SS peptides, a novel class of mitochondrial-targeted antioxidant (see above, Section vii), appear to hold great promise in the setting of anticipated ischemic intervals and may be used for minimizing I/R injury during angioplasty, coronary bypass surgery, cardiac surgery, and organ transplantation. SS-02, SS-31, and SS-20 have been reported to reduce myocardial ischemia—reperfusion injury in ex vivo and in vivo studies, and in vivo studies showed that SS-02 and SS-31 both reduced cardiac infarct size. A recent study showed that mitochondrial targeted peptide SS-31 ameliorates angiotensin-induced cardiomyopathy through the reduction of mitochondrial ROS and provided a strong rationale for investigating the clinical application of SS-31 for treatment or prevention of hypertensive cardiovascular diseases (Dai et al., 2011).

Furthermore, physical activity is another condition that increases antioxidant capacity in the heart by augmenting ROS scavenging enzymes such as catalase, superoxide dismutase, and glutathione peroxidase (Linke et al., 2005; Powers et al., 1994; Roh et al., 2016; Somani et al., 1995). Heat-shock proteins (HSPs) play important role in cellular defense against oxidative stress. It was shown that physical training induced HSP70 and HSP27 expression in trained old rats compared with sedentary old and young rats partially counterbalanced the heart age-related effects in the antioxidant system without altering peroxidation levels (Rinaldi et al., 2006). The antioxidant action of exercise seems to be dependent on exercise protocol. While endurance exercise exacerbates oxidative stress and cardiac remodeling, acute or moderate aerobic exercise training preserves cardiac health and prevents remodeling by maintaining myocardial defense system through stabilizing NRF2-antioxidant signaling (Narasimhan, Rajasekaran, 2016). Although aging hearts exhibit reduced NRF2-dependent antioxidant mechanisms, exercise training in aged mice increases NRF2 activity and induces of its electrophile-responsive/antioxidant-responsive elements (EpRE/ARE) target pathway to near-normal levels seen in young counterparts (Gounder et al., 2012; Narasimhan, Rajasekaran, 2016). NRF2 also *trans*-activates genes of the antioxidant response (Kang et al., 2005) and is coactivated by PGC-1α during oxidative stress (Aquilano et al., 2013). Long- and short-term endurance exercise increases PGC-1α expression in cardiac muscle (Bayod et al., 2012; Safdar et al., 2011) and induces mitochondrial biogenesis (Chinsomboon et al., 2009; Safdar et al., 2011). Besides these, cardioprotective effects of PGC-1 are mediated through its

ROS-lowering effects because PGC-1α induces glutathione peroxidase 1 and superoxide dismutase 2 and protects neural cells in culture from oxidative-stressor-mediated death (St-Pierre et al., 2006). Moreover, Ferrara et al., 2008 demonstrated that exercise training, which significantly increases activity of SIRT1, a factor that plays important role in antioxidant pathways, could counteract age-related systems impairment, by activation of antioxidant systems and DNA repair and disability (Cacciatore et al., 2004), suggesting therefore a possible role of exercise training in conditioning life span. Like SIRT1, SIRT3 protects against oxidative stress, in large part through FOXO3a-dependent mechanisms that induce superoxide dismutase and catalase (Sundaresan et al., 2009). Notably, these two sirtuins are upregulated during exercise in the heart and are positive modulators of PGC-1α activity (Palacios et al., 2009; Planavila et al., 2011). The enriched oxidative status caused by exercise training leads to number of beneficial effects, e.g., decreased arterial stiffness, improved endothelial function and metabolic and clotting setting, and reduced body weight (Corbi et al., 2012). Other findings derived from clinical studies show that regular physical activity decreases cardiovascular comorbidity and mortality in adult and in elderly by restoring the protective effect of ischemic preconditioning and partially contrasting loss of antioxidant defense in the aged heart (Corbi et al., 2012).

2.5 Mitochondria, ROS, Inflammation, and Aging

Recent studies highlight the important role of mitochondria, especially mitochondrial ROS (mtROS) production, as an upstream messenger in the induction of proinflammatory signaling. Inflammation initiates as a defensive immune response to pathogenic stimuli. This inflammatory response is a protective mechanism induced by the organism to remove the injurious stimuli and to promote the repair of damaged tissues. Dysregulation of the inflammatory response is observed in a variety of human diseases, including diabetes, neurodegeneration, and cancer. Activation of inflammatory processes starts through the formation of the inflammasome: a high-molecular weight multiprotein complex consisting of scaffold proteins and the adaptor protein ASC that recruits and binds caspase-1 (the initiator inflammatory caspase-1. There are four subfamilies of inflammasomes depending on the sensor molecule: NLRP3, NLRP1, NLRC4 (NLR family, CARD domain containing 4), and AIM2 (absent in melanoma 2) (Schroder, Tschopp, 2010). On activation, inflammasomes trigger the proteolytic maturation of proinflammatory cytokines, such as the potent

IL-1β, to engage the immune defenses. Due to its association with numerous inflammatory diseases, the NLRP3 inflammasome is currently the most fully characterized (Martinon et al., 2009).

A variety of danger signals, exogenous as well as endogenous, can activate the NLRP3 inflammasome, although the mechanisms of activation are poorly understood (Baroja-Mazo et al., 2014; Tschopp, 2011). Recent evidence suggests that mitochondria may integrate these distinct signals and deliver information to NLRP3 inflammasomes. Zhou et al. (2011) found that mtROS are key signals that directly trigger NLRP3 inflammasome activation. These authors artificially induced mtROS production, either by blocking key enzymes of the respiratory chain (complex I or complex III) or by inhibiting mitophagy/autophagy, and resulted in an increased amount of IL-1β secretion. In accordance with these findings, Nakahira et al. (2011) observed the fundamental role of autophagy in regulating mtROS production, demonstrating that decreased autophagy impairs mitochondrial homeostasis, increases mtROS levels and leads to NLRP3 inflammasome activation and IL-1β production. Moreover, treatment with the antioxidant 4-amino-2,4-pyrrolidinedicarboxylic acid (APDC) blocked NLRP3 inflammasome activation and IL-1β secretion (Zhou et al., 2011). Consistent with these observations, defective mitophagy increased inflammasome activation. In fact, macrophages treated with 3-methyladenine (3MA) or silenced of the autophagy regulator beclin 1 and autophagy protein 5 (ATG5) showed a hyperactivation of NLRP3 and IL-1β release on stimulation with monosodium urate (MSU) crystals and nigericin due to the accumulation of damaged mitochondria and increased ROS generation, while resveratrol treatment attenuated this effect (Wu et al., 2016).

The link between NLRP3 inflammasomes and mitochondria is further confirmed by the subcellular localization of NLRP3 and ASC. Under resting conditions, NLRP3 is localized in the ER but relocates to MAMs and mitochondria after inflammasome stimulation. Similarly, ASC proteins translocate from the cytosolic fraction to mitochondria and MAM after stimulation (Zhou et al., 2011). Based on these findings, Zhou et al. proposed a model in which NLRP3 acts as a sensor of mitochondrial status. This model was confirmed by evidence that NLRP3 and ASC are required for the release of mtDNA into the cytosol, a key step for caspase-1 activation and IL-1β secretion (Nakahira et al., 2011). Additional evidence of the intimate relationship between mitochondria and inflammasome is sustained by the role of mitochondrial Ca^{2+} uniporter (MCU). Modulation of MCU expression has been shown to regulate the activation of NLRP3 inflammasome induced by complement complex (Triantafilou et al., 2013) or by *Pseudomonas aeruginosa* in lung epithelial cells, preventing mitochondrial ROS production (Rimessi et al., 2015).

In contrast, other data reveal a distinct role for ROS in inducing inflammatory cytokines through a transcriptional-dependent mechanism, rather than through direct activation (van de Veerdonk et al., 2010). Following this theory, Bulua et al. demonstrated that mtROS influence the transcription of proinflammatory cytokines through the regulation of the MAPK pathway (Bulua et al., 2011; Kamata et al., 2005). It was found that patients with an autoinflammatory syndrome (TRAPS, TNF receptor-associated periodic syndrome) possessed increased basal levels of mtROS in cells that are responsible for production of IL-6 and TNF-α, independent of the NLRP3 inflammasome activation. Indeed, increased levels of IL-6 and TNF-α were observed after inflammation induction in the absence of NLRP3, caspase-1, or IL-1R. All of these findings underscore a common pathway identifying mitochondria as a crucial source of ROS, driving inflammation through mechanisms either inflammasome-dependent or inflammasome-independent. MtROS in particular, but also other key factors in inflammation induction such as NLRP3 and IL-1β, might be therapeutically targeted for treating inflammatory diseases. It has already been demonstrated that blocking excessive ROS with antioxidants reduces inflammation. Treatments with mitochondria-targeted antioxidants such as Mito-TEMPO (Trnka et al., 2009) or MitoQ (Murphy, Smith, 2007; Villalba et al., 2010), scavengers specific for mtROS, have been shown to reduce the secretion of IL-1β or IL-6 in vitro (Bulua et al., 2011; Nakahira et al., 2011). Such findings are promising and warrant further study in controlled clinical trials to confirm the in vivo efficacy of antioxidants.

Chronic inflammation is associated with physiological and pathological aging (Franceschi et al., 2007). Jurk et al. demonstrated that *nfkb1* −/− mice show premature aging. In particular, their data suggest that chronic low-grade inflammation can accelerate aging via ROS-mediated exacerbation of telomere dysfunction and cell senescence (Jurk et al., 2014). Recently, the NLRP3 inflammasome has been shown to be implicated in several aging-related diseases such as gout, T2D, obesity, and cancer. AMP-activated protein kinase (AMPK) plays a pivotal role in the control of metabolic events involved in the pathophysiology of aging. Nevertheless, it has emerged as an important integrator of inflammation signaling (Cordero et al., 2018). In fact, many of the various AMPK-dependent pathways regulate NLRP3 inflammasome activation during aging. Aging has a negative effect on the quality of immune responses, which increases the frequency and severity of infectious diseases. Recent results demonstrate the role of inflammation and oxidative stress in age-related changes of immune cell survival factors in the bone marrow with an accumulation of IFN-y, TNF, and ROS, which induce IL-15 and IL-6 expression. The treatment with the ROS scavengers NAC and Vit-C reduces cytokine levels, suggesting that antioxidants may be beneficial in counteracting immunosenescence by improving immunological memory in old age (Pangrazzi et al., 2017). Furthermore, osteoarthritis, a whole-joint degenerative multifactorial disorder, is influenced by oxidative stress and aging. Cartilage of osteoarthritis patients has significantly more ROS-induced DNA damage than normal cartilage and this damage is mediated by IL-1β. The overproduction of ROS in osteoarthritis regulates intracellular signaling processes, such as chondrocyte senescence and apoptosis, extracellular matrix synthesis, and degradation along with synovial inflammation and dysfunction of the subchondral bone (Lepetsos, Papavassiliou, 2016).

All these findings identify the ROS-mediated inflammation as a possible target in the treatment of a wide range of aging-related diseases, suggesting the use of antioxidants, both natural and synthetic.

2.6 Mitochondria, ROS, Cell Death, and Aging

Apoptosis, the process that allows multicellular organisms to eliminate unnecessary, dangerous, or damaged cells without evoking inflammation or tissue damage, takes place in a wide number of physiological and pathological events (Green, Kroemer, 2004). Its efficacy for removing hazardous cells is circumvented in viral diseases and neoplasia by specific molecular routes. On the other hand, the progressive occurrence of apoptosis in nonproliferating cells has been proposed as the basis for a number of degenerative disorders, as well as in progressive loss of organ function during aging (Green et al., 2011). Thus, understanding the control mechanisms of apoptosis is a major goal for the development of new therapeutic approaches. Mitochondria, traditionally considered the powerhouses that supply energy to the cells, are also considered vessels filled with an array of weaponry that can be unleashed to promote the apoptotic signaling cascade, resulting in the demise of the cell. Apoptotic stimuli cause the release of proapoptotic mitochondrial mediators into the cytoplasm (cytochrome *c*, AIF, Smac/DIABLO). Their assembly with cytosolic proteins forms a complex (apoptosome) that recruits and activates caspases leading the cell into apoptotic death (Adams, Cory, 2002). The molecular mechanism of this release is unclear but most likely requires the activity of a large-conductance channel, known as the PTP, which remains again an elusive molecular entity (Lepetsos, Papavassiliou, 2016). Main regulators of the PTP include voltage-dependent anion channel (VDAC) in the outer mitochondrial membrane (OMM), adenine nucleotide translocase (ANT) in the inner mitochondrial membrane (IMM), cyclophilin D in the mitochondrial matrix, and more recently the mitochondrial phosphate carrier (PiC) (Leung et al., 2008). Whereas, hexokinase II (HKII), mitochondrial creatine kinase (CK), benzodiazepine receptor (PBR), and Bcl-2-family members (Bcl-2, Bcl-x_L, and Bax) are currently included as putative regulatory components. However, despite the use of multiple methodologies, the identity and the number of PTP's components are still elusive, and it remains to be clarified whether ANT and VDAC are the main components of pore or accessory. As showed using isolated mitochondria from mice knockout for ANT or VDAC and CypD, where a permeability transition response to the classical inducer calcium (Ca^{2+}) was monitored, suggested a limited role for these proteins (Baines et al., 2007; Basso et al., 2005; Kokoszka et al., 2004; Nakagawa et al., 2005). Could be, that the structure of pore depends on the tissue as well as the pathophysiological state (Halestrap, 2009). In fact, four homologous genes, whose expression is not only tissue specific but also vary according to the pathophysiological state of the cell, encode ANT. Similarly VDAC, where the different isoforms appear to rely on their ability to engage protein—protein interactions with different partners complicating VDAC's contribution to cell death, strictly dependent from isoform and stimulus (Cheng et al., 2003; Rapizzi et al., 2002). These and other PTP proteins are targets for ROS, and oxidative modifications of those proteins containing thiol groups will significantly impact PTP activation and thus mitochondrial ions fluxes, as mentioned later (Zoratti, Szabo, 1995) (Costantini et al., 1996; Kowaltowski et al., 2001).

In response to proapoptotic stimuli, including ROS and Ca^{2+} overload, the PTP assumes a high-conductance state that allows the deregulated entry of small solutes into the mitochondrial matrix along their electrochemical gradient (Morciano et al., 2015). The opening of the PTP induces mitochondrial swelling, and these large-scale alterations of organelle morphology may allow the release of the caspase cofactors into the cytosol (Yang, Cortopassi, 1998). Evidences by which PTP opening may accelerate aging can represent the link between PTP-driven apoptosis, ROS, and senescence. It is reported that PTP may be the culprit of the progressive

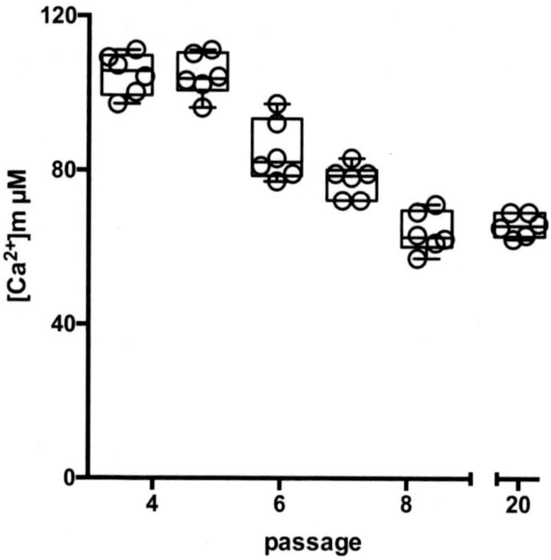

Figure 8 ***Mitochondrial calcium uptake as a function of cell culture passage number.*** Representative quantitation of mitochondrial Ca^{2+} uptake elicited by ATP 100 µM n mouse embryonic fibroblasts (MEFs) from low and high passage rates. Mitochondrial calcium concentration ($[Ca^{2+}]_m$).

age-dependent NAD^+ depletion inside mitochondrial matrix (Schriewer et al., 2013) affecting DNA repair mechanisms and the protective sirtuin pathway (Merksamer et al., 2013) prompting the cell to senescence and death. This relationship is not univocal; indeed also aging could favor PTP opening (Rottenberg, Hoek, 2017) giving rise to an intricate loop. Interestingly, an extracellular agonist–stimulated Ca^{2+} uptake by mitochondria in mouse embryonic fibroblasts (MEFs) is gradually decreased with culture time (Fig. 8). Moreover, many reports described dysregulations in Ca^{2+} homeostasis and MMP in aged cells in terms of calcium buffering thresholds, which significantly decrease (Gant et al., 2015; Pandya et al., 2015) and a lowered MMP (Sugrue, Tatton, 2001). These changes intensify PTP opening events in senescent cells (Paradies et al., 2013) and the liaison with age-dependent mitochondrial ROS accumulation is very strong as the oxidation of channels and transporters such as MCU (Dong et al., 2017) in which the oxidation of the reactive thiol represented by Cys-97 showed constitutive channel activity and a consequent mitochondrial calcium overload), the mitochondrial calcium uptake 1 (MICU1) and phospholipids contribute to the progressive damage and dysregulation of proteins and to the exchange of radical species between mitochondria and cytosol, where during aging, it could become the main exchange route.

Recently, it has been proposed a multistep nature of the PTP complex opening involving first, a disassembly of ATP synthase dimers and second, a correct rearrangement of the C-ring in the IMM (Bonora et al., 2017; Morciano et al., 2017). This is coherent with previous findings in which a reduced dimerization status of ATP synthase was detected in aging cells, thus prompting them to cell death (Daum et al., 2013). Indeed, it is reported that aging is intimately related to a decline of mitochondrial functions (Sun et al., 2016) and young, middle-aged and senescent cells own important mitochondrial differences in terms of IMM integrity and organization, as well as the oxidative state of proteins, mtDNA, and phospholipids; for instances, it has to be remarked that ATP synthase in young cells is mainly detected in dimeric and oligomeric structures (Daum et al., 2013) to fully support the energetic requirement of the cell and the correct curvature of the IMM. Recent findings award the synthasome assembly to PTP modulators, such as CypD and generally, to conditions preventing mitochondrial permeability transition (MPT) (Beutner et al., 2017); on the other hand, in senescent mitochondria, a highly dynamic transition from dimers to monomers occur (Daum et al., 2013).

Investigation of the molecular routes of apoptosis has revealed the important role of mitochondria in decoding oxidative insults (Fig. 2). In fact, oxidative stress and altered mitochondrial function have consistently been proposed to be major determinants of life span, as shown by several studies with transgenic overexpression—antioxidants systems (Parkes et al., 1998; Sun et al., 2008), or as shown by experiments with fundamental genes for mitochondrial redox regulation thioredoxins and glutaredoxins (Chen et al., 2002; Choksi et al., 2011; Diotte et al., 2009; Enoksson et al., 2005; Kim et al., 2010; Nagy et al., 2008; Stanley et al., 2011). Indeed, mouse embryonic fibroblast cells (MEFs), from which p66shc has been completely depleted, are resistant to oxidative stress induced apoptotic death (Migliaccio et al., 1999). This death is p53-dependent, and knockout of either p53 or p66shc causes resistance, suggesting that p66shc is downstream of p53 in the pathway (Trinei et al., 2002). In agreement with this finding, mice with p66shc completely knocked out are resistant to paraquat and live about 30% longer than controls. The relationship between mitochondria and p66shc emerged by the effective localization of the protein to the organelle: part of the cytosolic pool translocates to mitochondria (Orsini et al., 2004). In the intermembrane space (IMS), p66shc binds cytochrome *c* and acts as a redox enzyme, generating H_2O_2, in turn inducing the opening of the PTP and cellular apoptosis. ROS production by p66shc appears to be a specialized function whereby electrons are subtracted from the ETC to catalyze the partial reduction of molecular oxygen (Giorgio et al., 2005). The redox activity of p66shc explains the decrease in ROS levels observed in p66shc knockout cells (Migliaccio et al., 2006) and is also responsible for an altered mitochondrial metabolism under basal conditions, characterized by lower oxygen consumption (Nemoto et al., 2006). Clear and specific references about p66shc and its aging activity will be addressed in detail in a next section, emphasizing pivotal molecular proteins involved in p66shc's signal transduction as new targets of pharmacological therapy.

The tumor suppressor Fhit protein characterizes another pivotal mitochondrial proapoptotic route (Rimessi et al., 2009) associated with oxidative-stress induction. It is absent or reduced in many types of human tumors including lung, esophagus, stomach, kidney, and cervical carcinomas (Croce et al., 1999). Fhit is encoded by the FHIT gene, located within a fragile region of chromosome 3 of the human genome, and is frequently altered in cancers and inactivated in cancer-derived cell lines (Inoue et al., 1997). Direct evidence from Fhit-deficient cancer cells shows that chemotherapy-induced cell death is accompanied by a mild response to production of ROS, suggesting that Fhit-deficiency could negatively influence treatment outcome. In addition, Fhit loss has been considered as a mechanism to delay cell aging due to its ability to overcome oncogene-induced senescence (Waters et al., 2014); indeed, it is known that oncogene activation generates, most of the time, irreversible DNA damage allowing senescence or apoptosis pathways in preneoplastic cells. In this key passage, observations made in $Fhit^{+/+}$ and $Fhit^{-/-}$ MEF cells, Fhit function is highly dependent on its DNA "caretaker" role prompting p53 deregulation and avoiding senescence once lost (Miuma et al., 2013). Although Fhit is identified as a cytosolic protein, it may sort to mitochondria where it interacts with ferredoxin reductase (fdxr), a flavoprotein transactivated by p53 (Trapasso et al., 2008). Through interaction with chaperones Hsp60 and Hsp10, mitochondrial Fhit modulates electron-transfer from NADPH via the activity of fdxr. The Fhit/fdxr complex generates ROS and increases Ca^{2+} uptake into mitochondria, potentiating the effects of apoptotic agents (Rimessi et al., 2009). The sorting of Fhit to mitochondria is now recognized as essential to its tumor suppressor actions in apoptosis induction. Together, these results identify a mitochondrial signaling step at the center of the mechanisms of the anticancer action of Fhit and draw attention to the importance of pharmacologically regulating its intracellular sorting and organelle activity, as confirmed by the overexpression of mitochondrial-targeted Fhit-chimera in tumor cells (Rimessi et al., 2009).

In recent years, an interesting role in regulating cell destiny has been identified for autophagy (the catabolic pathway for degradation of intracellular proteins and organelles via the lysosome) (Klionsky et al., 2016; Scherz-Shouval et al., 2007). Autophagy is mainly activated by nutrient starvation and it plays a dual role: not only it is primarily a survival mechanism but it also leads to cell death, thus possibly acting as an alternative to apoptosis (Levine, Kroemer, 2008). In contrast to the well-documented prosurvival function of autophagy, some examples of ROS-induced autophagy linked to cell death have been described in tumor cells. Treatments with rotenone and TTFA (inhibitors of complex I and II, respectively) in cancer cell lines induced autophagy-dependent cell death, displaying increased mitochondrial ROS production. Use of siRNA against autophagy genes or the autophagy inhibitor 3-MA produces reversal of the effects of autophagic cell death (Chen et al., 2007b). Today, the precise role of autophagy in cancer is strongly debated, but it remains a potentially important area of development for new therapeutic interventions. Scientific evidence supports both tumor promoting and suppressive functions for autophagy. The exact role of it during cancer progression depends on tumor type, context, and stage. Although genetic evidences confirm a role for autophagy as a tumor suppressor mechanism, it can also promote the maintenance of models of ovarian cancer and gastrointestinal stromal tumor (Guo et al., 2011; Gupta et al., 2010; Lu et al., 2008). The requirement for autophagy becomes more apparent in later stages as tumor cells cope with microenvironmental stresses encountered during progression and metastasis (Roy, Debnath, 2010). The tumor suppressive functions are most apparent during tumor initiation, where H-ras^{12v} induces different autophagic responses depending on the duration of oncogene overexpression. After 48 h of Ras overexpression, autophagy inhibits cell proliferation, whereas a longer time of oncogene overexpression, cell proliferation was enhanced by autophagy (Wu et al., 2011). Growing evidence suggests that in different pathological contexts, cross talk between apoptosis and autophagy takes place; this is intimately interconnected during stress responses, and anti- or prosurvival effects have been shown in cancer or neurodegenerative diseases, respectively (Codogno, Meijer, 2005; Kroemer et al., 2007). For this reason, researchers have a marked interest in developing pharmacological regulators of autophagy as alternative therapeutic approaches to counter the dysfunctional apoptotic response during pathological conditions. On the other hand, in healthy conditions, growing (but not yet direct) evidences link autophagy to aging revealing this pathway as cure-all for longevity and thus, a landmark for therapeutic manipulations. These observations were made in many systems such as yeast, *S. Cerevisiae*, *Drosophila* and ultimately in mammals with concordant findings that describe a reduced age-dependent activation of autophagy in mammals (Kaushik et al., 2012; Vittorini et al., 1999), in in vitro senescent cultures and isolated organs from old rodents (Cuervo, Dice, 2000). In addition, the disruption of a type of autophagy, classified as chaperone-mediated autophagy, promoted cell sensitivity to stress, a condition dramatically associated to decreased longevity (Massey et al., 2006). Another proof by which autophagy could be deeply involved in an extended life span derive from protein overexpression studies in which increased levels of ATG5, ATG7, ATG8 and SIRT1 autophagy proteins prevented ER stress and oxidative insult-induced cell death, improved glucose tolerance and clearance, and preserved motor function (Lee et al., 2008; Pyo et al., 2013; Yang et al., 2010). Although all these observations support the hypothesis that some forms of basal autophagy could ameliorate aging-associated dysregulated pathways, further efforts are required to get direct and final evidences.

Redox regulation by moderate levels of ROS is also observed in autophagy. It is generally accepted that mitochondria play a fundamental role in ROS-mediated autophagy regulation (Azad et al., 2009; Kim et al., 2007a; Kirkland et al., 2002; Kissova et al., 2006; Xu et al., 2006). A specialized form of the autophagy process, called mitophagy, degrades defective mitochondria. Mitophagy helps to maintain a healthy population of mitochondria in the cell (Rimessi et al., 2013) and thus reduces oxidative damage (Scherz-Shouvalet al., 2007). Increases in cellular ROS lead to loss of MMP, which is considered a trigger for mitophagy (Kim et al., 2007b). Under serum deprivation, a typical decrease in MMP is observed in hepatocytes prior to engulfment by autophagosomes, whereas cyclosporin A (PTP inhibitor) prevents this depolarization and the autophagosomal proliferation (Elmore et al., 2001; Rodriguez-Enriquez et al., 2009). Indeed, a recent study showed that mitochondria provide the membrane source for autophagosome biogenesis during the autophagy process (Hailey et al., 2010). Tracking photolabeled mitochondria showed that fusion and fission events permitted the segregation of abnormal mitochondria, which were then degraded by mitophagy (Twig et al., 2008), and that the profission protein Fis1 triggered mitophagy only when associated with mitochondrial dysfunction (Gomes, Scorrano, 2008).

2.7 Mitochondrial Dysfunction and Oxidative Stress in Age-Related Neurodegenerative Diseases

Links between mitochondria and neurodegeneration have been reported for several years, with the first hints stemming from studies of classical mitochondrial disorders. Interestingly, the incidence of neurodegenerative diseases increases with aging, which means that some of the mitochondrial phenotypes found in aged individuals are similar to some mitochondrial hallmarks found in neurodegeneration. In this type of disease, mtDNA mutations lead to impairment of mitochondrial respiration, ATP synthesis, reduction in MMP, and increased ROS production (DiMauro, Hirano, 2009). These defects usually cause a bioenergetic deficit in tissues with the highest energy demand, particularly the central nervous system and muscular tissue, causing neurological disorders, ataxia, muscle weakness, stroke-like episodes, epilepsy, and other disease types. Similarly, deregulation of mitochondrial physiology and increased oxidative stress, often at an early stage, have both been correlated with the most common neurodegenerative diseases including AD, PD, Huntington's disease (HD), or Amyotrophic Lateral Sclerosis (ALS), among others (Lin, Beal, 2006). In all of these pathologies, mitochondria appear to be central hubs of the pathophysiological process, due to their impaired ability for ATP synthesis and for an increased production of ROS (Szeto, 2006). In fact, alterations of mitochondrial function and increased oxidative stress were found in patients affected by these pathologies.

Loss of neurons in the hippocampus and neocortex combines with two brain alterations in AD: the accumulation of senile plaques (composed of amyloid-β, Aβ) and neurofibrillary tangles (made of the hyperphosphorylated protein tau). The principal risk factor for AD is age, but there are also cases of autosomal dominant familial forms caused by mutations in amyloid precursor protein (APP), presenilin-1 (PS1), or −2 (PS2). Mitochondrial abnormalities, such as alterations of TCA cycle enzymes, and reduced activity of respiratory complexes I, III, and IV were found in postmortem brains of patients with AD (Bosetti et al., 2002; Gibson et al., 1998; Lin, Beal, 2006; Reddy, 2006; Reddy, Beal, 2008). Direct evidence of oxidative stress was obtained in the brains of early-stage AD animal models by using in vivo electron paramagnetic resonance (EPR) imaging with methoxycarbamyl-proxyl (MCP) as a redox-sensitive probe (Fang et al., 2016), with these defects being associated not only with cognitive defects but also with mitochondrial dysfunction. It has been demonstrated that Aβ25-35 and Aβ1-40 induce apoptotic cell death in cerebral endothelial cells, and Aβ-neurotoxicity is associated with mitochondrial dysfunction (increased generation of ROS, excitotoxicity, apoptosis, and inflammation) (Mattson, 2000). Moreover, AD-associated metabolic alterations have been described to alter mitochondrial dynamics in the neurons, causing interference with local ATP and calcium gradients (Correia et al., 2016), as well as alterations in cardiolipin species, which may explain alterations in mitochondrial respiratory chain activities and increased ROS generation (Monteiro-Cardoso et al., 2015). SIRT3, a mitochondrial sirtuin has recently been shown to be decreased in the context of AD, leading to p53-mediated negative effects on mtDNA transcription and resulting in inhibition of respiration and increased oxidative stress (Lee et al., 2017). As referred above, SIRT3 is involved in the regulation of mitochondrial respiration and oxidative stress through regulation of protein acetylation and SOD2 expression/activity (Pereira et al., 2012).

Rhein and colleagues showed that convergence of Aβ and tau on mitochondria with associated defects in mitochondrial complexes I and IV cause disturbances in the respiratory and energy system of ^{triple}AD mice. The same work describes that age-related oxidative stress leads to dysfunctional energy metabolism and consequently to neuronal death (David et al., 2005; Rhein et al., 2009). Other in vitro and in vivo studies by Keil et al. (2004) demonstrated that Aβ induces mitochondrial adaptation and failure in a vulnerable and dose-dependent pattern, with NO being involved in these processes.

APP has an unidentified mitochondrial-signal sequence that targets it to mitochondria. Aβ interacts with Aβ-binding alcohol dehydrogenase (ABAD), and this interaction leads to release of cytochrome *c* and an increase of ROS generation (Lustbader et al., 2004). Yet, it has not been revealed if APP could be processed to Aβ directly within mitochondria or if processing precedes Aβ translocation. The finding that AD patients display impaired mitochondrial APP translocation and accumulation between OMM and IMM as well as evidences showing the presence of γ secretase in the mitochondrial matrix suggest that mitochondria could be a processing site for APP and that this event is fundamental for the onset of pathology and increased oxidative stress (Devi et al., 2006; Pavlov et al., 2011). Recently, it was proposed that overexpression of Aβ modifies the activity mitochondrial fusion and fission proteins. In particular, an increase of Fis1, a decrease of DLP1 (fission proteins), and a decrease of the fusion protein OPA1 were measured. All these alterations lead to mitochondrial fragmentation, reduction of MMP and increase in ROS generation, most likely resulting from deficient quality control mechanisms (de Moura et al., 2010; Wang et al., 2008b).

It was proposed that the "mitochondrial cascade" might provoke AD development due to accumulation of mutations in mitochondrial genes, which causes insufficient respiratory chain accompanied with increased ROS production, and results in more and more severe mtDNA damage that leads to severe oxidative stress, stimulating AB toxicity (Swerdlow et al., 2014). Although it was not possible to identify any mitochondrial mutations causing AD, an increased number of point mutations in mtDNA isolated from brains of AD patients, relative to controls, has been reported (Lin et al., 2002). However, recent studies indicated such mutations appear to be effects of errors during the mtDNA replication process rather than be caused by ROS (Hoekstra et al., 2016). In contrast to AD, PD is characterized by tremor, rigidity, postural instability, and bradykinesia. PD is caused by degeneration of the dopaminergic neurons of the substantia nigra, combined with accumulation of α-synuclein- and ubiquitin-containing inclusions, called Lewy bodies, in the surviving neurons (de Moura et al., 2010). Disordered protein handling/degradation, mitochondrial dysfunction, and increased oxidative stress have been shown to be correlated with sporadic and familial PD as well as Parkinsonism due to exposure to toxins or pesticides (Dagda et al., 2009).

Several neurotoxins that affect dopaminergic neurons act at the mitochondrial level and induce oxidative stress in that organelle. One toxin is the heroin contaminant 1-methyl-4-phenyl-1,2,3,6-tetrahydroxypyridine (MPTP), which produces a Parkinsonian phenotype (Przedborski, Jackson-Lewis, 1998). MPTP is converted to $MPDP^{+}$ and then to the active metabolite MPP^{+} by astrocytes and acts as an inhibitor of complex I of the mitochondrial ETC (Langston et al., 1983), promoting cytochrome *c* release from the IMM (Banerjee et al., 2009). Interestingly, both MPTP and its metabolites, $MPDP^{+}$ and MPP^{+}, reveal also mutagenic properties (Ulanowska, Wegrzyn, 2006) (Ulanowska et al., 2007). Thus, one can speculate that they might potentially cause also mutations in mtDNA, accelerating the disease development. Other compounds that inhibit complex I and contribute to PD development include the pesticides rotenone and paraquat, commonly used in farming. The latter acts by causing degeneration of dopaminergic neurons and increased oxidative stress (Cicchetti et al., 2005). Chronic exposure to rotenone leads to dopaminergic degeneration and formation and aggregation of α-synuclein and ubiquitin, as well as oxidative damage and caspase-dependent cell death (Rego, Oliveira, 2003). Moreover, rotenone itself is widely used as a tool for the generation of PD models in *Drosophila*, mice, and rats. The appearance of cytoplasmic inclusions containing alpha synuclein and ubiquitin is reported in the different biological models, as well as dopaminergic neuron degeneration and impairment of locomotor activity in animal models, resembling the human pathology (Blandini, Armentero, 2012; Coulom, Birman, 2004; Inden et al., 2011). It was reported how rotenone administration induces ATP depletion, increased oxidative stress and death in a neuroblastoma model. Interestingly, ATP depletion in that model, induced by exposure to 2-deoxyglucose did not induce cell death, suggesting that ROS production is the critical event in rotenone-induced toxicity. Moreover, these effects were abolished by overexpression of the rotenone insensitive respiratory complex I subunit ND1 derived from yeast (Sherer et al., 2007). Nonetheless, it has been shown that NDUFS4 KO mice were completely insensitive to rotenone-induced neurodegeneration (Choi et al., 2008). The results obtained appear to suggest that there are more than one binding site for rotenone in mitochondrial Complex I (Fendel et al., 2008), with the most relevant appearing to be ND1 (Murai et al., 2007). Moreover, a redox-active dopamine analogue, 6-hydroxy-dopamine (6-OHDA), induces the death of dopaminergic neurons and leads to an increase of free radicals, abolishing α-synuclein's role.

A new player which has been described to have an important role in an early stage of PD is the mitochondrial LON protease. It was recently determined that Lon protease expression increased in the ventral mesencephalon of MPTP-treated animals, in the same time frame as the appearance of oxidized proteins and dopaminergic cell loss. The authors also observed a loss of Lon activity by ROS and carbonylation in α-ketoglutarate dehydrogenase (KGDH), aconitase, or subunits of respiratory chain complexes (Bulteau et al., 2017). Interestingly, not only Lon operates as a mtDNA and protein quality control protein (Pinti et al., 2016; Sepuri et al., 2017) but it also was recently shown to relocate to mitochondrial-associated membranes on different stimuli (Polo et al., 2017), suggesting that ROS may affect mitochondrial—ER interactions in PD through inactivation of Lon, besides avoid its quality control activity.

Recent studies demonstrated that mutations of different genes involved in mitochondrial function or with antioxidant activities cause familial PD. The leucine-rich repeat kinase 2 (LRRK2) is the most commonly mutated gene in the familial and sporadic type of PD (Kachergus et al., 2005), whereas DJ-1, parkin, and PINK-1 are involved in autosomal recessive Parkinsonism (Dagda et al., 2009; Kitada et al., 1999).

Relatively high levels of mutant mtDNA, particularly with deletions, were reported in dopaminergic neurons of elderly people and in PD patients (Bender et al., 2006) (Kraytsberg et al., 2006; Lin et al., 2012). Studies on animals gave quite similar results, i.e., accumulation of deletions in mtDNA in brains, particularly in substantia nigra, of PD models (Dolle et al., 2016; Parkinson et al., 2014; Tzoulis et al., 2016). Moreover, depletion of mtDNA was also reported in this disease (Grunewald et al., 2016). Thus, it is possible that mtDNA dysfunction might contribute to development of PD.

In HD, the abnormal expansion of polyglutamine repeats in the Huntingtin (HTT) protein (above 40 repeats) causes aggregation of the unfolded protein that leads to neuronal degeneration in the cortex and striatum. Similarly to PD, alterations of mitochondrial respiratory complexes (especially II and III) were found in *postmortem* brain samples (Damiano et al., 2010). Furthermore, altered MMP in lymphoblasts from patients and augmented lactate production in the brain suggest a connection between mitochondrial dysfunction and HD. In addition, inhibitors of respiratory complex II, such as 3-nitropropionic acid and malonate, induce an HD pathological phenotype in animals (Beal, 1994). Mutated HTT is sufficient to decrease ATP synthesis and impair respiratory chain activity, whereas overexpression of respiratory complex II subunits is enough to recover mitochondrial function and sensitivity to apoptosis in neurons expressing an 82 glutamine HTT (Benchoua et al., 2006). The physiological role of HTT is still under debate, although it has been shown to colocate with the OMM and to regulate mitochondrial trafficking along axons (Chang et al., 2006). Interestingly, HTT facilitates PTP opening during Ca^{2+} stimulation in neurons and promote ROS production on 3-NP exposure in cybrids (Ferreira et al., 2010), suggesting that mutated HTT could have a role in the regulation of ROS metabolism. In fact, the PTP has been proposed as a major factor for mitochondrial damage in HD (Quintanilla et al., 2017).

Increased levels of DNA lesions and mutations, both base pair substitutions and deletions of larger DNA fragments, were reported in nuclear and mitochondrial genomes of HD patients and in animal models of this disease (summarized in Ayala-Pena, 2013). This suggests that mutations in DNA, including those in mtDNA, might be important factors in development of HD. The problem of mtDNA depletion is more complicated, as different groups reported either decreased (Liu et al., 2008; Petersen et al., 2014) or increased (Chen et al., 2007a) levels of mtDNA in HD patients relative to controls. Recent studies, based on testing biological material from a relatively large population, indicated higher levels of mtDNA in leukocytes, but depletion of mtDNA in fibroblasts of HD patients relative to healthy controls (Jędrak et al., 2017). Therefore, it was suggested that both size of the study group, and particularly the kind of investigated tissue, as well as some methodological and technical details important for adequate measurement of mtDNA levels, might be responsible for differences in results published by various groups (Jedrak et al., 2017). Nevertheless, it appears that levels of mtDNA might be changed in HD patients, while either decreased or increased, depending on the tissue. Because mtDNA depletion has been suggested to occur in the brain of HD patients due to ROS-generated DNA damage, and due to evident accumulation of mutations in mtDNA and nDNA, a model of progression of HD, driven by accumulation of mtDNA lesions caused by toxicity of mutant HTT, and resultant enhanced production of ROS that cause mtDNA depletion, has been proposed (Ayala-Pena, 2013). According to this model, mtDNA dysfunction leads to deficiency in mitochondrial functions and subsequent neurodegeneration.

Similar connections between mitochondria and neurodegeneration are present in ALS. Alterations of mitochondrial structures and number, as well as defects, in respiratory chain complexes have been observed in postmortem samples of spinal cord. Genetic causes of ALS are still poorly understood. Approximately 90% of ALS cases are sporadic, and of the remaining 10% (familial ALS) only 20% are attributed to genetic disorders. These cases are linked to mutations that alter SOD1 activity. SOD1 is considered one of the most important cellular scavengers of the cytoplasmic superoxide anion. Recently, SOD1 presence has been confirmed in mitochondria, particularly in the OMM and the IMS, where it may exert a scavenging activity. Apparently, its localization to the IMS is fundamental for its correct maturation (Reddehase et al., 2009) and this observation gains special relevance in the context of ALS. It is well known that mutant SOD1 found in patients generates toxic aggregates (Bruijn et al., 1998). Recently, mutant SOD1 was discovered to specifically aggregate in mitochondria and induce mitochondrial impairment and induction of apoptosis (Cozzolino et al., 2009). Although SOD1 mutations affect only a small fraction of ALS patients, it is plausible that similar mechanisms involving mitochondria and oxidative stress are responsible for the induction of pathogenesis in most of the patients.

2.8 Is Preventing Mitochondrial Oxidative Stress With Antioxidants Effective?—The Example of Neurodegenerative Diseases

As a natural consequence of the multiple connections between mitochondria, oxidative stress, and neurodegeneration, several strategies have been developed directed at reducing oxidative stress and recovering from the pathological phenotype. Several compounds with antioxidant properties have been shown to reduce oxidative stress and increase cell survival in in vitro systems (Fig. 9).

Lipoic acid (LA), a cofactor for pyruvate dehydrogenase and α-ketoglutarate dehydrogenase, and N-acetyl-cysteine decrease mitochondrial oxidative stress in fibroblasts derived from AD patients. LA used in a combination with acetyl-L-carnitine (ALCAR, a membrane permeable form of the mitochondrial acetyl carrier carnitine) protected primary cortical neurons against apoptosis induced by 4-hydroxy-2 nonenal (Abdul, Butterfield, 2007).

Coenzyme Q 10 (CoQ10), a fundamental cofactor in the respiratory chain with elevated antioxidant properties, was able to protect human neuroblastoma cells from paraquat- and rotenone-induced mitochondrial dysfunction and cell death. Similarly, the mitochondria-targeted form, MitoQ, prevented cell death in fibroblasts from patients affected by Friedreich's Ataxia (FA). This molecule possesses elevated antioxidant properties, preserves mitochondrial functions, and reduces ROS formation even in Rho zero cells (lacking mtDNA).

Other mitochondria-targeted antioxidants are potentially interesting. MitoE, an analogue of Vit-E, shows elevated scavenging activity in fibroblasts from FA patients. MitoQ and LPBNAH, a derivative of PBN, protected neuroblastoma cells from H_2O_2-induced oxidative stress and Aβ1-42 toxicity.

Antioxidant-based strategies also appear widely effective in animal models of neurodegenerative disease. Natural antioxidant-like CoQ, Vit-E, creatine, and green tea polyphenols showed protective effects in mouse models of both PD and AD (Beal, 2003). LA significantly increased survival in HD mice models N171-82Q and R6/2 (Andreassen et al., 2001) and also reduces oxidative stress in aged rats and reduces memory impairment in aged mice (Quinn et al., 2007). The combination of LA and ALCAR reduced aging-related mitochondrial damage in rats and promotes neuron survival during glutamate-induced toxicity (Nagesh Babu et al., 2011).

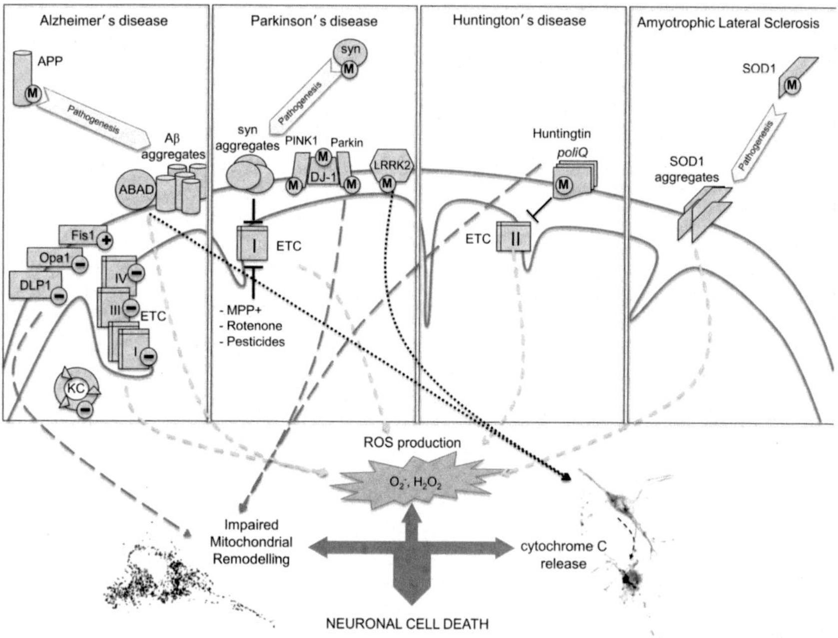

Figure 9 ***Mitochondrial-related alterations in principal neurodegenerative disorders.*** In Alzheimer's disease amyloid peptides (Ab) can aggregate to mitochondria and interact with Ab alcohol dehydrogenase (ABAD) to induce cytochrome C release and ROS production. Contemporary altered activity of Krebs cycle enzymes (KC), complexes of respiratory chain, and proteins involved in mitochondria fusion and fission leads to impaired mitochondrial remodeling and ROS production. Similar readouts on mitochondrial physiology are observed in Parkinson's disease. Aggregation of mutant synuclein (syn) impair activity of respiratory complex I similarly as many pro-parkinsonian compounds. In this scenario should be added the activity of mutant proteins such as PINK1, DJ-1, Parkin, LRRK2 that impair mitochondrial modeling and recycling as promoting induction of cytochrome C release. In Huntington disease mutant huntingtin (carrying poliQ expansion) can impair activity of II respiratory complex (with consequent ROS production) and mitochondria transport along filaments in axons. Also superoxide dismutase 1 (SOD1) mutations causing ALS can induce appearance of toxic mitochondrial aggregates that lead to improved toxic ROS production. Altered protein or OXPHOS complexes activities are shown with a plus or minus symbol in a circle, mutant proteins instead are marked by an M in a circle.

Ginkgo biloba extract seems to be of therapeutic benefit in the treatment of mild dementia of different etiology, especially as regards AD (Janssen et al., 2010). Leuner and colleagues suggest that mitochondrial protection and reduction of oxidative stress are important components of the neuroprotective activity of Ginkgo biloba extract (Leuner et al., 2007).

A novel class of cell-permeable small peptides that selectively partition into the inner mitochondrial membrane and possess intrinsic mito-protective activities have been developed and proposed as a novel class of mitochondria-targeted antioxidants. These novel peptides were originally designed by Szeto and Schiller and have been designated SS peptides. This class of molecules is characterized by a structural motif that alternates aromatic residues and basic amino acids. Contrary to MitoQ and MitoE, these aromatic-cationic peptides are taken up by mitochondria but are not delivered into the matrix. Furthermore, uptake is not dependent on MMP (the extent of uptake was only reduced by ~10–15% in mitochondria that were depolarized with FCCP) and is not limited to mitochondria with normal MMP. This is a major advantage when dealing with diseased mitochondria with a reduced MMP (Zhao et al., 2004). Animal studies indicated that these novel mitochondria-targeted peptides have excellent pharmacokinetic properties and are relatively free of toxicity, suggesting that they may have enormous therapeutic potential. Several different compounds have been tested and most of them display antioxidant properties (SS-02, SS-19, SS-31) and have been demonstrated to inhibit mitochondrial ROS production, prevent mitochondrial swelling, and neuronal cell loss in animal models of ALS (Petri et al., 2006) and PD (Yang et al., 2009). Interestingly, the SS-20 peptide that does not show antioxidant properties appears to be potent in protecting neuronal loss in a mouse model for Parkinson's disease.

Despite the collective success of antioxidant strategies in recovering the pathological phenotype in cells and animal systems, convincing clinical results are still lacking. In 1997, Sano et al. published one of the first clinical studies testing compounds that protect against mitochondria-mediated oxidative stress (Sano et al., 1997). This study was a double blind, placebo-controlled, randomized, multicenter trial in patients with moderate severity AD. This study tested both selegiline (an inhibitor of monoamine oxidase) and α-tocopherol (Vit-E) for 2 years in a total of 341 patients. Unfortunately, no significant benefits were observed.

More recently, MitoQ, considered one of the most promising mitochondria-targeted antioxidants, was tested in a Phase 2 trial on PD patients. The double blind, placebo-controlled study was conducted for 12 months in 128 patients, but failed to provide evidence that MitoQ could ameliorate the pathological condition (Snow et al., 2010).

Clinical failure for the tested antioxidant therapies could be explained by many reasons. First, the inability to measure mitochondrial damage and contributors to oxidative damage may be due to low bioavailability of the compound within the brain because of difficulties in crossing the blood–brain barrier. Second, the therapy is usually administered to patients in advanced stages of the pathology, in a condition under which the scavenging activity of the compound might not be sufficient to reverse the phenotype (as this may be due to a different, specific mechanism, at this stage already independent of oxidative stress).

Together with mitochondrial dysfunction and oxidative stress, another characteristic common in most neurodegenerative diseases is the presence of aggregated proteins. Examples include Aβ in AD, α-synuclein and ubiquitin in PD, mutant SOD1 in some forms of ALS, and mutant HTT in HD. As already cited, in most of the cases, protein aggregates accumulate also in mitochondria, promoting the increase of oxidative stress and impairment of mitochondrial functions.

Interestingly, in rats, administration of epoxomicin or PSI, two proteasomal inhibitors, leads to generation of an animal model of Parkinson's disease (McNaught et al., 2004). It is well known that proteasomal degradation requires ATP, while elevated levels of oxidative stress impair correct protein folding and, in the case of AD, promote BACE overexpression and accumulation of Aβ (Tamagno et al., 2005). A model could be envisioned in which the presence of protein aggregates can promote mitochondrial dysfunction and ROS generation. This would promote protein misfolding and impairment of proteasomal activity, initiating a cycle leading to neuronal death and progression of the pathology.

It should be considered that even if altered oxidative stress is a common feature of neurodegeneration, ROS might not be generalized as toxic components, but rather act as proper signaling molecules. The pathways of MAPK (mitogen-activated kinase), PI3K (phosphoinositide 3-kinase), and PKCs are ROS-sensitive and could promote cell proliferation in the presence of some oxidants (Giorgi et al., 2010; Hole et al., 2010). In the presence of ROS, the PI3K pathway can induce the nuclear respiratory factor 2, whereas PKCs can sustain brain remodeling and synaptogenesis after stroke (Sun et al., 2008). Moreover, ROS also mediate cell survival through the activation of HIF1a and NF-κB. HIF1a results in glycolytic switch and reduction of respiratory protection during stroke or hypoxia conditions (Siddiq et al., 2005) or in HD neurodegeneration animal models, whereas NF-κB maintains the expression of antiapoptotic factors such as GADD45B and XIAP (Pahl, 1999).

Thus, it should be considered that in an oxidative environment, such as found in brain cells from patients with neurodegenerative disease, systems are likely adapting in an attempt to survive. Administration of antioxidants could help cells to recover their equilibrium but could also be helpful for stimulating pathways that manage ROS metabolism. From this point of view, the stimulation of Nrf2 is a promising target. This protein is normally present at low levels in the cytosol, bound to the Cul3-based E3 ligase adaptor KEAP1. Under oxidative stress, KEAP1 undergoes conformational changes, leaving Nrf2 free to move within the nucleus. In the nucleus, it binds to ARE and exerts its transcriptional activity to promote the expression of antioxidant genes such as NQO1, SOD1, and GST (Nguyen et al., 2009).

To date, sulforaphane is considered one of the most potent compounds for promoting Nrf2 activation. Sulforaphane is usually obtained from cruciferous vegetables, converted from glucosinolate or glucoraphanin. Sulforaphane is able to oxidize cysteine residues on KEAP1, leading to Nrf2 activation. In an animal model of stroke, administration of sulforaphane or carnosic acid (another Nrf2 activator) led to neuroprotection and improvement of neurological functions (Satoh et al., 2008). Furthermore, modulation of Nrf2 improved neurological impairment in animal models of AD, HD, or after administration of the pro-Parkinsonian drug MPTP (Calkins et al., 2005; Kanninen et al., 2009).

Investigating antioxidant strategies as therapeutic interventions in neurodegenerative disease are a promising area of inquiry, but convincing clinical results remain elusive. The development of such strategies should involve compounds that act on pathways controlling intracellular ROS metabolism, including adjuvants of antioxidant compounds, as well as compounds modulating proteasomal activity (Fig. 10).

3. CONCLUSIONS

In addition to the well-characterized energy-producing functions, mitochondria are an important intracellular source of ROS. Multiple mitochondrial functions and interconnections exist between aerobic energy metabolism, generation of ROS, activation of the apoptotic pathways, and other fundamental homeostatic and signaling pathways (e.g., Ca^{2+} homeostasis, lipid and nucleotide synthesis). Thus, mitochondrial impairment determines various degrees of energy failure and deregulation of ROS production. An in-depth investigation of all these effects will be the prerequisite to identify effective strategies to counteract the deleterious

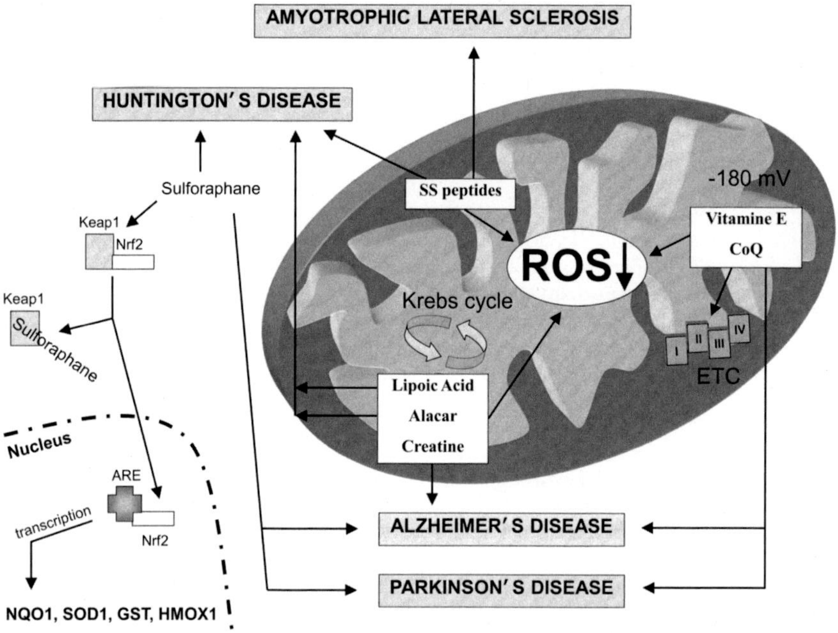

Figure 10 ***Schematic view of various compounds that reduce oxidative stress in diseases mentioned in the text and their correlation with neurodegenerative disorders.*** Mitochondrial selective compounds can act in directly buffering ROS production, such as Szeto–Shiller peptides (SS), vitamine E or coenzyme Q (CoQ), and derivatives. Buffering of ROS production could by obtained by inhibiting interaction between Keap1 and the nuclear respiratory factor 2 (NRF2). Restoration of mitochondrial functions has been also obtained by the use of drugs targeting Kreb's cycle, such as lipoic acid, acetyl-l-carnitine (ALCAR), or creatine.

and multiple consequences of mitochondrial malfunctioning. The comprehension of the mechanisms regulating mitochondrial physiology and homeostasis, and in particular the control of mitochondrial ROS production, may have a significant impact for the development of novel therapies for the treatment of a wide variety of human diseases.

However, it should be noted that many clinical trials using antioxidants (in conditions such as cardiovascular diseases, cancer, diabetes, and neurological degenerative diseases, as well as to slow the aging process) have provided contradictory results. Thus, although the experimental evidence for an antioxidant therapy is quite promising, further validation work is required. Despite the unresolved questions about the parallel role of ROS in oxidative damage or as signaling molecules, the causal link of mitochondria impairment in aging and associated diseases is unequivocal.

ABBREVIATIONS

3MA	3-Methyladenine
4-HNE1	4-Hydroxy-2-noneal
6-OHDA	6-Hydroxy-dopamine
8-oxodG	8-Oxo-7,8-dihydro-2′-deoxyguanosine
Aβ	Amyloid-beta
ABAD	Ab-binding alcohol dehydrogenase
AD	Alzheimer's disease
ANT	Adenine nucleotide translocase
AIF	Apoptosis-inducing factor
ALS	Amyotrophic Lateral Sclerosis
AHL	Age-related hearing loss
ALCAR	Acetyl-1-carnitine
AMPK	AMP-activated protein kinase
AP-1	Activator protein-1
APDC	4-Amino-2,4-pyrrolidinedicarboxylic acid
APP	Amyloid precursor protein
ARE	Nrf2/antioxidant response element
Ca^{2+}	Calcium ion
$[Ca^{2+}]c$	Cytosolic calcium concentration
$[Ca^{2+}]m$	Mitochondrial calcium concentration
CoQ10	Coenzyme Q 10
CPT1mt	Malonyl-CoA-insensitive CPT1A
CR	Caloric restriction
CRTC 1	CREB-regulated transcription coactivator 1
CsA	Cyclosporine A
CT	Computed tomography
CVDs	Cardiovascular diseases
DA	Dopamine
DAG	Intrahepatic diacylglycerol
Dnm1p	Dynamin-related protein 1
Drp1	Dynamin-related protein 1
DTT	Dithiothreitol
EC	Endothelial cell
EDRF	Endothelium-derived relaxing factor
EGF	Epidermal growth factor
EPR	Electron paramagnetic resonance imaging
ER	Endoplasmic reticulum
***(Ercc 1)*$^{-/\Delta}$**	The excision repair cross-complementation group 1
ETC	Electron transport chain
Fdxr	Ferredoxin reductase
FA	Friedreich's Ataxia
FAs	Fatty acids
FFA	Free fatty acids
Fis1	Fission protein 1
FRTA	Free Radical Theory of Aging
GFP	Green-fluorescence protein
GPx	Glutathione peroxidase
GSH	Glutathione
GST	Glutathione S-transferase

HCC	Hepatocellular carcinoma
HD	Huntington's disease
HUVECs	Human umbilical vein endothelial cells
HG	High glucose
HK	Hexokinase
HO-1	Heme oxygenase-1
H_2O_2,	Hydrogen peroxide
HSPs	Heat-shock proteins
HTT	Huntingtin
IDH2	$NADP^+$-dependent isocitrate dehydrogenase 2
IIS	Insulin signaling
IMM	Inner mitochondrial membrane
IMS	Intermembrane space
IPC	Ischemic preconditioning
iPSC	Induced pluripotent stem cells
I/R	Ischemia/reperfusion
ITPR2	Inositol 1,4,5-trisphosphate receptor type 2
JNK	c-Jun NH2-terminal kinase
KO	Knockout
LA	Lipoic acid
LC3	MAP1 light chain 3
LCFA	Long-chain fatty acids
LRRK2	Leucine-rich repeat kinase 2
MAMs	Mitochondria-associated membranes
MAPK	Mitogen-activated protein kinase
(Marf)	Mitochondrial assembly regulatory factor
MDA	Malondialdehyde
MEFs	Mouse embryonic fibroblast cells
Mff	Mitochondrial fission factor
Mfn1	Mitofusin 1
Mfn2	Mitofusin 2
MFRTA	Mitochondrial Free Radical Theory of Aging
MGM1	Dynamin-like GTPas
Mid49	Mitochondrial dynamics protein of 49 kDa
Mid51	Mitochondrial dynamics protein of 51 kDa
MMP	Mitochondrial membrane potential
MnSOD	Manganese-dependent superoxide dismutase
MPC	Mitochondrial pyruvate carrier
MPP^+	1-Metyl-4-phenylpyridinium
MPTP	1-Methyl-4-phenyl-1,2,3,6-tetrahydropyridine
mtDNA	Mitochondrial DNA
Mtfp1	Mitochondrial fission process 1
mTOR	Mammalian target of rapamycin kinase
NAC	N-acetylcysteine
(NAFL)	Nonalcoholic fatty liver
NAFLD	Nonalcoholic fatty liver disease
NASH	Steatohepatitis
NF-κB	Nuclear factor-kappa B
NO	Nitric oxide
Nox	NADPH oxidases
$O_2^{\bullet -}$	Superoxide
$^{\bullet}OH$	Hydroxyl radical

OMM	Outer mitochondrial membrane
ONOO⁻	Peroxynitrite
Opa1	Optic atrophy 1
PGC-1α	Peroxisome proliferator–activated receptor gamma coactivator 1-alpha
PD	Parkinson's disease
PBR	Peripheral benzodiazepine receptor
Pin1	Peptidyl–prolyl isomerase
PINK1	PTEN-induced putative kinase protein 1
PKC	Protein kinase C
PiC	Mitochondrial phosphate carrier
POLG	DNA polymerase γ
PPARs	Peroxisome proliferator–activated receptors
Prx	Peroxiredoxins
PS	Presenilin
PSSG	Protein mixed disulfide
PTP	Permeability transition pore
PUFAs	Polyunsaturated fatty acids
RNS	Reactive nitrogen species
ROS	Reactive oxygen species
SIRT	Mitochondrial NAD^+–dependent deacetylase sirtuin
SOD1	Zn–Cu superoxide dismutase
SOD2	Manganese superoxide dismutase
SOD3	Superoxide dismutase 3
SREBP-1c	Sterol regulatory element–binding protein-1c
SS peptides	Szeto–Schiller peptides
STOML2	Stomatin-like protein 2
T2D	Type 2 diabetes
TCA	Tricarboxylic acid
TFAM	Transcriptor factor A
TMEM135	Transmembrane protein 135
TPP^+	Triphenylphosphonium cation
Trx2	Thioredoxin
UCP	Uncoupling proteins
Vit-C	Vitamin C
Vit-E	Vitamin E
VLDL	Very low-density lipoproteins
VDAC	Voltage-dependent anion channel
WD	Wilson disease

ACKNOWLEDGMENTS

CG is supported by local funds from the University of Ferrara, the Italian Association for Cancer Research (AIRC: IG-19803), the Italian Ministry of Health, and by a Fondazione Cariplo grant. I.C.M.S, H.Z., P.J.O., V.S, P.P and M.R.W. gratefully acknowledge the financial support for this research from the FOIE GRAS and mtFOIE GRAS projects. These projects received funding from the European Union's Horizon 2020 Research and Innovation programme under the Marie Skłodowska-Curie Grant Agreement No. 722619 (FOIE GRAS) and Grant Agreement No. 734719 (mtFOIE GRAS). This research was also supported by the Polish National Science Centre grants (UMO-2014/15/B/NZ1/00490) for M.R.W., (UMO-2013/08/W/NZ1/00687) for J.D., J.S. and P.P.K., (2015/17/N/NZ2/04267) for P.J. and G.W., (UMO-2011/03/B/NZ3/00693 and UMO-2013/10/E/NZ3/

00670) for A.D. as well as The National Centre for Research and Development, grant STRATEGMED 3/305813/2/NCBR/2017 for A.D., (UMO-2014/13/B/NZ4/00199 and UMO-2016/22/E/NZ4/00650) for P.D. and by the European Union's Horizon 2020 research and innovation program under Marie Sklodowska-Curie grant agreement no. 665735 (Bio4Med) and by funding from the Polish Ministry of Science and Higher Education as part of 2016–2020 funds for the implementation of international projects (agreement no. 3548/H2020/COFUND/2016/2) for A.D., P.D., J.D. This work was also supported by the Internal Projects of the Children's Memorial Health Institute No S125/2012 for A.K.W.

P.P. is grateful to Camilla degli Scrovegni for continuous support. P.P. is supported by the Italian Ministry of Education, University and Research; the Italian Ministry of Health; Telethon (GGP15219/B); the Italian Association for Cancer Research (AIRC: IG-18624); and by local funds from the University of Ferrara. S.M. is supported by "Fondazione Umberto Veronesi" and the Italian Ministry of Health. A.R. was supported by local funds from the University of Ferrara (FIR 2017), the Italian Ministry of Health and Italian Cystic Fibrosis Foundation (FFC # 20/2015). G.S. was supported by the grants Telethon (GEP12066 and GGP16026), Ricerca Finalizzata, Italian Ministry of Health (RF-2013-02359065), Association for Cancer Research (AIRC, IG13447); Biochemical and Biophysical Research Council (grants BB/L020874/1; BB/P018726/1) and the Wellcome Trust (097815/Z/11/Z). P.J.O. is also thankful to the Foundation for Science and Technology (Portugal), through FEDER/COMPETE funding (Grants POCI-01-0145-FEDER-007440, POCI-01-0145-FEDER-016659, and PTDC/DTP-FTO/2433/2014). V.A.S. acknowledges support from the Foundation for Science and Technology (Portugal), IF/01182/2015. The authors are deeply indebted to past and present collaborators.

REFERENCES

Abdul, H.M., Butterfield, D.A., 2007. Involvement of PI3K/PKG/ERK1/2 signaling pathways in cortical neurons to trigger protection by cotreatment of acetyl-L-carnitine and alpha-lipoic acid against HNE-mediated oxidative stress and neurotoxicity: implications for Alzheimer's disease. Free Radic. Biol. Med. 42, 371–384.

Abete, P., Testa, G., Cacciatore, F., Della-Morte, D., Galizia, G., Langellotto, A., Rengo, F., 2011. Ischemic preconditioning in the younger and aged heart. Aging Dis. 2, 138–148.

Abeti, R., Abramov, A.Y., Duchen, M.R., 2011. β-amyloid activates PARP causing astrocytic metabolic failure and neuronal death. Brain 134, 1658–1672.

Adam-Vizi, V., Chinopoulos, C., 2006. Bioenergetics and the formation of mitochondrial reactive oxygen species. Trends Pharmacol. Sci. 27, 639–645.

Adams, J.M., Cory, S., 2002. Apoptosomes: engines for caspase activation. Curr. Opin. Cell Biol. 14, 715–720.

Adlam, V.J., Harrison, J.C., Porteous, C.M., James, A.M., Smith, R.A., Murphy, M.P., Sammut, I.A., 2005. Targeting an antioxidant to mitochondria decreases cardiac ischemia-reperfusion injury. FASEB J. 19, 1088–1095.

Al-Goblan, A.S., Al-Alfi, M.A., Khan, M.Z., 2014. Mechanism linking diabetes mellitus and obesity. Diabetes Metab. Syndr. Obes. 7, 587–591.

Aliev, G., Seyidova, D., Neal, M.L., Shi, J., Lamb, B.T., Siedlak, S.L., Vinters, H.V., Head, E., Perry, G., Lamanna, J.C., Friedland, R.P., Cotman, C.W., 2002. Atherosclerotic lesions and mitochondria DNA deletions in brain microvessels as a central target for the development of human AD and AD-like pathology in aged transgenic mice. Ann. N. Y. Acad. Sci. 977, 45–64.

Altieri, D.C., 2017. Mitochondria on the move: emerging paradigms of organelle trafficking in tumour plasticity and metastasis. British J. Cancer 117, 301–305.
Amara, C.E., Shankland, E.G., Jubrias, S.A., Marcinek, D.J., Kushmerick, M.J., Conley, K.E., 2007. Mild mitochondrial uncoupling impacts cellular aging in human muscles in vivo. Proc. Natl. Acad. Sci. U.S.A 104, 1057–1062.
Ames, B.N., Shigenaga, M.K., Hagen, T.M., 1993. Oxidants, antioxidants, and the degenerative diseases of aging. Proc. Natl. Acad. Sci. Unit. States Am. 90, 7915–7922.
Ames, B.N., Shigenaga, M.K., Hagen, T.M., 1995. Mitochondrial decay in aging. Biochim. Biophys. Acta 1271, 165–170.
Amir, M., Czaja, M.J., 2011. Autophagy in nonalcoholic steatohepatitis. Expert Rev. Gastroenterol. Hepatol. 5, 159–166.
Anantharaju, A., Feller, A., Chedid, A., 2002. Aging Liver. A review. Gerontology 48, 343–353.
Anderson, A., Bowman, A., Boulton, S.J., Manning, P., Birch-Machin, M.A., 2014. A role for human mitochondrial complex II in the production of reactive oxygen species in human skin. Redox. Biol. 2, 1016–1022.
Andreassen, O.A., Ferrante, R.J., Dedeoglu, A., Beal, M.F., 2001. Lipoic acid improves survival in transgenic mouse models of Huntington's disease. Neuroreport 12, 3371–3373.
Angelova, P.R., Abramov, A.Y., 2018. Role of mitochondrial ROS in the brain: from physiology to neurodegeneration. FEBS Lett.
Anson, R.M., Hudson, E., Bohr, V.A., 2000. Mitochondrial endogenous oxidative damage has been overestimated. FASEB J. 14, 355–360.
Aon, M.A., Cortassa, S., O'Rourke, B., 2010. Redox-optimized ROS balance: a unifying hypothesis. Biochim. Biophys. Acta 1797, 865–877.
Appelhans, T., Richter, C.P., Wilkens, V., Hess, S.T., Piehler, J., Busch, K.B., 2012. Nanoscale organization of mitochondrial microcompartments revealed by combining tracking and localization microscopy. Nano Lett. 12, 610–616.
Aquilano, K., Baldelli, S., Pagliei, B., Cannata, S.M., Rotilio, G., Ciriolo, M.R., 2013. p53 orchestrates the PGC-1alpha-mediated antioxidant response upon mild redox and metabolic imbalance. Antioxid. Redox Signal. 18, 386–399.
Aravinthan, A., Scarpini, C., Tachtatzis, P., Verma, S., Penrhyn-Lowe, S., Harvey, R., Davies, S.E., Allison, M., Coleman, N., Alexander, G., 2013. Hepatocyte senescence predicts progression in non-alcohol-related fatty liver disease. J. Hepatol. 58, 549–556.
Area-Gomez, E., Schon, E.A., 2017. On the pathogenesis of Alzheimer's disease: the MAM hypothesis. Faseb. J. 31, 864–867.
Arnelle, D.R., Stamler, J.S., 1995. NO+, NO., and NO– donation by S-nitrosothiols: implications for regulation of physiological functions by S-nitrosylation and acceleration of disulfide formation. Arch. Biochem. Biophys. 318, 279–285.
Arruda, A.P., Pers, B.M., Parlakgül, G., Güney, E., Inouye, K., Hotamisligil, G.S., 2014. Chronic enrichment of hepatic endoplasmic reticulum–mitochondria contact leads to mitochondrial dysfunction in obesity. Nat. Med. 20, 1427.
Asanuma, T., Yasui, H., Inanami, O., Waki, K., Takahashi, M., Iizuka, D., Uemura, T., Durand, G., Polidori, A., Kon, Y., Pucci, B., Kuwabara, M., 2007. A new amphiphilic derivative, N-{[4-(lactobionamido)methyl]benzylidene}-1,1-dimethyl-2-(octylsulfanyl)ethylamin e N-oxide, has a protective effect against copper-induced fulminant hepatitis in Long-Evans Cinnamon rats at an extremely low concentration compared with its original form alpha-phenyl-N-(tert-butyl) nitrone. Chem. Biodivers. 4, 2253–2267.
Ascensao, A., Martins, M.J., Santos-Alves, E., Goncalves, I.O., Portincasa, P., Oliveira, P.J., Magalhaes, J., 2013. Modulation of hepatic redox status and mitochondrial metabolism by exercise: therapeutic strategy for liver diseases. Mitochondrion 13, 862–870.

Asencio, C., Rodriguez-Aguilera, J.C., Ruiz-Ferrer, M., Vela, J., Navas, P., 2003. Silencing of ubiquinone biosynthesis genes extends life span in *Caenorhabditis elegans*. FASEB J. 17, 1135–1137.
Aspnes, L.E., Lee, C.M., Weindruch, R., Chung, S.S., Roecker, E.B., Aiken, J.M., 1997. Caloric restriction reduces fiber loss and mitochondrial abnormalities in aged rat muscle. FASEB J. 11, 573–581.
Ayala-Pena, S., 2013. Role of oxidative DNA damage in mitochondrial dysfunction and Huntington's disease pathogenesis. Free Radic. Biol. Med. 62, 102–110.
Ayala, V., Naudi, A., Sanz, A., Caro, P., Portero-Otin, M., Barja, G., Pamplona, R., 2007. Dietary protein restriction decreases oxidative protein damage, peroxidizability index, and mitochondrial complex I content in rat liver. J. Gerontol. A Biol. Sci. Med. Sci. 62, 352–360.
Azad, M.B., Chen, Y., Gibson, S.B., 2009. Regulation of autophagy by reactive oxygen species (ROS): implications for cancer progression and treatment. Antioxid. Redox Signal. 11, 777–790.
Bahonar, A., Saadatnia, M., Khorvash, F., Maracy, M., Khosravi, A., 2017. Carotenoids as potential antioxidant agents in stroke prevention: A systematic review. Int. J. Prev. Med. 8, 70.
Baines, C.P., Kaiser, R.A., Sheiko, T., Craigen, W.J., Molkentin, J.D., 2007. Voltage-dependent anion channels are dispensable for mitochondrial-dependent cell death. Nat. Cell Biol. 9, 550–555.
Bakala, H., Delaval, E., Hamelin, M., Bismuth, J., Borot-Laloi, C., Corman, B., Friguet, B., 2003. Changes in rat liver mitochondria with aging. Eur. J. Biochem. 270, 2295–2302.
Bakeeva, L.E., Barskov, I.V., Egorov, M.V., Isaev, N.K., Kapelko, V.I., Kazachenko, A.V., Kirpatovsky, V.I., Kozlovsky, S.V., Lakomkin, V.L., Levina, S.B., Pisarenko, O.I., Plotnikov, E.Y., Saprunova, V.B., Serebryakova, L.I., Skulachev, M.V., Stelmashook, E.V., Studneva, I.M., Tskitishvili, O.V., Vasilyeva, A.K., Victorov, I.V., Zorov, D.B., Skulachev, V.P., 2008. Mitochondria-targeted plastoquinone derivatives as tools to interrupt execution of the aging program. 2. Treatment of some ROS- and age-related diseases (heart arrhythmia, heart infarctions, kidney ischemia, and stroke). Biochemistry (Mosc.) 73, 1288–1299.
Baker, D.J., Peleg, S., 2017. Biphasic modeling of mitochondrial metabolism dysregulation during aging. Trends Biochem. Sci. 42, 702–711.
Balemba, O.B., Bartoo, A.C., Nelson, M.T., Mawe, G.M., 2008. Role of mitochondria in spontaneous rhythmic activity and intracellular calcium waves in the guinea pig gallbladder smooth muscle. Am. J. Physiol. Gastrointest. Liver Physiol. 294. G467–476.
Banerjee, R., Starkov, A.A., Beal, M.F., Thomas, B., 2009. Mitochondrial dysfunction in the limelight of Parkinson's disease pathogenesis. Biochim. Biophys. Acta 1792, 651–663.
Bao, R., Shu, Y., Wu, X., Weng, H., Ding, Q., Cao, Y., Li, M., Mu, J., Wu, W., Ding, Q., 2014. Oridonin induces apoptosis and cell cycle arrest of gallbladder cancer cells via the mitochondrial pathway. BMC Cancer 14, 217.
Baranano, D.E., Rao, M., Ferris, C.D., Snyder, S.H., 2002. Biliverdin reductase: a major physiologic cytoprotectant. Proc. Natl. Acad. Sci. U.S.A 99, 16093–16098.
Barber, A.J., Antonetti, D.A., Kern, T.S., Reiter, C.E., Soans, R.S., Krady, J.K., Levison, S.W., Gardner, T.W., Bronson, S.K., 2005. The Ins2Akita mouse as a model of early retinal complications in diabetes. Invest. Ophthalmol. Vis. Sci. 46, 2210–2218.
Barnhart, E.L., 2016. Mechanics of mitochondrial motility in neurons. Cuurr. Opin. Cell Biol. 38, 90–99.
Baroja-Mazo, A., Martin-Sanchez, F., Gomez, A.I., Martinez, C.M., Amores-Iniesta, J., Compan, V., Barbera-Cremades, M., Yague, J., Ruiz-Ortiz, E., Anton, J., Bujan, S., Couillin, I., Brough, D., Arostegui, J.I., Pelegrin, P., 2014. The NLRP3 inflammasome is released as a particulate danger signal that amplifies the inflammatory response. Nat. Immunol. 15, 738–748.

Basisty, N., Dai, D.F., Gagnidze, A., Gitari, L., Fredrickson, J., Maina, Y., Beyer, R.P., Emond, M.J., Hsieh, E.J., MacCoss, M.J., Martin, G.M., Rabinovitch, P.S., 2016. Mitochondrial-targeted catalase is good for the old mouse proteome, but not for the young: 'reverse' antagonistic pleiotropy? Aging Cell 15, 634–645.

Basso, E., Fante, L., Fowlkes, J., Petronilli, V., Forte, M.A., Bernardi, P., 2005. Properties of the permeability transition pore in mitochondria devoid of Cyclophilin D. J. Biol. Chem. 280, 18558–18561.

Bautista-Niño, P.K., Portilla-Fernandez, E., Vaughan, D.E., Danser, A.H.J., Roks, A.J.M., 2016. DNA damage: A main determinant of vascular aging. Int. J. Mol. Sci. 17, 748.

Bayer, T.A., Schafer, S., Simons, A., Kemmling, A., Kamer, T., Tepest, R., Eckert, A., Schussel, K., Eikenberg, O., Sturchler-Pierrat, C., Abramowski, D., Staufenbiel, M., Multhaup, G., 2003. Dietary Cu stabilizes brain superoxide dismutase 1 activity and reduces amyloid Abeta production in APP23 transgenic mice. Proc. Natl. Acad. Sci. U.S.A 100, 14187–14192.

Bayod, S., Del Valle, J., Lalanza, J.F., Sanchez-Roige, S., de Luxan-Delgado, B., Coto-Montes, A., Canudas, A.M., Camins, A., Escorihuela, R.M., Pallas, M., 2012. Long-term physical exercise induces changes in sirtuin 1 pathway and oxidative parameters in adult rat tissues. Exp. Gerontol. 47, 925–935.

Beal, M.F., 1994. Neurochemistry and toxin models in Huntington's disease. Curr. Opin. Neurol. 7, 542–547.

Beal, M.F., 2003. Mitochondria, oxidative damage, and inflammation in Parkinson's disease. Ann. N. Y. Acad. Sci. 991, 120–131.

Bearn, A.G., 1953. Genetic and biochemical aspects of Wilson's disease. Am. J. Med. 15, 442–449.

Bedard, K., Krause, K.H., 2007. The NOX family of ROS-generating NADPH oxidases: physiology and pathophysiology. Physiol. Rev. 87, 245–313.

Bejma, J., Ramires, P., Ji, L., 2000. Free radical generation and oxidative stress with ageing and exercise: differential effects in the myocardium and liver. Acta Physiologica Scandinavica 169, 343–351.

Benard, G., Rossignol, R., 2008. Ultrastructure of the mitochondrion and its bearing on function and bioenergetics. Antioxid. Redox Signal. 10, 1313–1342.

Benchoua, A., Trioulier, Y., Zala, D., Gaillard, M.C., Lefort, N., Dufour, N., Saudou, F., Elalouf, J.M., Hirsch, E., Hantraye, P., Deglon, N., Brouillet, E., 2006. Involvement of mitochondrial complex II defects in neuronal death produced by N-terminus fragment of mutated huntingtin. Mol. Biol. Cell 17, 1652–1663.

Bender, A., Krishnan, K.J., Morris, C.M., Taylor, G.A., Reeve, A.K., Perry, R.H., Jaros, E., Hersheson, J.S., Betts, J., Klopstock, T., Taylor, R.W., Turnbull, D.M., 2006. High levels of mitochondrial DNA deletions in substantia nigra neurons in aging and Parkinson disease. Nat. Genet. 38, 515–517.

Bernal-Mizrachi, C., Gates, A.C., Weng, S., Imamura, T., Knutsen, R.H., DeSantis, P., Coleman, T., Townsend, R.R., Muglia, L.J., Semenkovich, C.F., 2005. Vascular respiratory uncoupling increases blood pressure and atherosclerosis. Nature 435, 502–506.

Bernhardt, D., Müller, M., Reichert, A.S., Osiewacz, H.D., 2015. Simultaneous impairment of mitochondrial fission and fusion reduces mitophagy and shortens replicative lifespan. Sci. Rep. 5, 7885.

Berson, A., De Beco, V., Letteron, P., Robin, M.A., Moreau, C., El Kahwaji, J., Verthier, N., Feldmann, G., Fromenty, B., Pessayre, D., 1998. Steatohepatitis-inducing drugs cause mitochondrial dysfunction and lipid peroxidation in rat hepatocytes. Gastroenterology 114, 764–774.

Bertolotti, M., Abate, N., Bertolotti, S., Loria, P., Concari, M., Messora, R., Carubbi, F., Pinetti, A., Carulli, N., 1993. Effect of aging on cholesterol 7 alpha-hydroxylation in humans. J. Lipid Res. 34, 1001–1007.

Beutner, G., Alanzalon, R.E., Porter Jr., G.A., 2017. Cyclophilin D regulates the dynamic assembly of mitochondrial ATP synthase into synthasomes. Sci. Rep. 7, 14488.

Bevilacqua, L., Ramsey, J.J., Hagopian, K., Weindruch, R., Harper, M.E., 2004. Effects of short- and medium-term calorie restriction on muscle mitochondrial proton leak and reactive oxygen species production. Am. J. Physiol. Endocrinol. Metab. 286. E852–861.

Bhattacharyya, S., Sinha, K., Sil, P.C., 2014. Cytochrome P450s: mechanisms and biological implications in drug metabolism and its interaction with oxidative stress. Curr. Drug Metabol. 15, 719–742.

Birket, M.J., Passos, J.F., von Zglinicki, T., Birch-Machin, M.A., 2009. The relationship between the aging- and photo-dependent T414G mitochondrial DNA mutation with cellular senescence and reactive oxygen species production in cultured skin fibroblasts. J. Invest. Dermatol. 129, 1361–1366.

Blandini, F., Armentero, M.T., 2012. Animal models of Parkinson's disease. FEBS J. 279, 1156–1166.

Blockhuys, S., Celauro, E., Hildesjo, C., Feizi, A., Stal, O., Fierro-Gonzalez, J.C., Wittung-Stafshede, P., 2017. Defining the human copper proteome and analysis of its expression variation in cancers. Metallomics 9, 112–123.

Bokov, A., Chaudhuri, A., Richardson, A., 2004. The role of oxidative damage and stress in aging. Mech. Ageing Dev. 125, 811–826.

Bonekamp, N.A., Volkl, A., Fahimi, H.D., Schrader, M., 2009. Reactive oxygen species and peroxisomes: struggling for balance. Biofactors 35, 346–355.

Bonomini, F., Rodella, L., Moghadasian, M., Lonati, C., Rezzani, R., 2013. Apolipoprotein E deficiency and a mouse model of accelerated liver aging. Biogerontology 14, 209–220.

Bononi, A., Giorgi, C., Patergnani, S., Larson, D., Verbruggen, K., Tanji, M., Pellegrini, L., Signorato, V., Olivetto, F., Pastorino, S., Nasu, M., Napolitano, A., Gaudino, G., Morris, P., Sakamoto, G., Ferris, L.K., Danese, A., Raimondi, A., Tacchetti, C., Kuchay, S., Pass, H.I., Affar, E.B., Yang, H., Pinton, P., Carbone, M., 2017. BAP1 regulates IP3R3-mediated Ca(2+) flux to mitochondria suppressing cell transformation. Nature 546, 549–553.

Bonora, M., Morganti, C., Morciano, G., Pedriali, G., Lebiedzinska-Arciszewska, M., Aquila, G., Giorgi, C., Rizzo, P., Campo, G., Ferrari, R., Kroemer, G., Wieckowski, M.R., Galluzzi, L., Pinton, P., 2017. Mitochondrial permeability transition involves dissociation of F1FO ATP synthase dimers and C-ring conformation. EMBO Rep. 18, 1077–1089.

Booth, F.W., Laye, M.J., Roberts, M.D., 2011. Lifetime sedentary living accelerates some aspects of secondary aging. J. Appl. Physiol. 111, 1497–1504.

Bosetti, F., Brizzi, F., Barogi, S., Mancuso, M., Siciliano, G., Tendi, E.A., Murri, L., Rapoport, S.I., Solaini, G., 2002. Cytochrome c oxidase and mitochondrial F1F0-ATPase (ATP synthase) activities in platelets and brain from patients with Alzheimer's disease. Neurobiol. Aging 23, 371–376.

Boyapati, R.K., Tamborska, A., Dorward, D.A., Ho, G.T., 2017. Advances in the understanding of mitochondrial DNA as a pathogenic factor in inflammatory diseases. F1000Res 6, 169.

Brand, M.D., 2010. The sites and topology of mitochondrial superoxide production. Exp. Gerontol. 45, 466–472.

Bravo, E., Pignatelli, E., Masella, R., Verna, R., Cantafora, A., 1994. Influence of age on hepatic uptake of HDL1-cholesterol in male Wistar rats with bile duct cannulation. J. Biochem. 115, 833–836.

Breitling, L.P., Raum, E., Müller, H., Rothenbacher, D., Brenner, H., 2009. Synergism between smoking and alcohol consumption with respect to serum gamma-glutamyltransferase. Hepatology 49, 802–808.

Brewer, G.J., 2010. Risks of copper and iron toxicity during aging in humans. Chem. Res. Toxicol. 23, 319–326.

Broche, B., Ben Fradj, S., Aguilar, E., Sancerni, T., Benard, M., Makaci, F., Berthault, C., Scharfmann, R., Alves-Guerra, M.C., Duvillie, B., 2018. Mitochondrial protein UCP2 controls pancreas development. Diabetes 67, 78–84.

Brown, G.C., Borutaite, V., 2011. There is no evidence that mitochondria are the main source of reactive oxygen species in mammalian cells. Mitochondrion.

Brownlee, M., 2001. Biochemistry and molecular cell biology of diabetic complications. Nature 414, 813–820.

Brownlee, M., 2005. The pathobiology of diabetic complications: a unifying mechanism. Diabetes 54, 1615–1625.

Bruha, R., Vitek, L., Marecek, Z., Pospisilova, L., Nevsimalova, S., Martasek, P., Petrtyl, J., Urbanek, P., Jiraskova, A., Malikova, I., Haluzik, M., Ferenci, P., 2012. Decreased serum antioxidant capacity in patients with Wilson disease is associated with neurological symptoms. J. Inherit. Metab. Dis. 35, 541–548.

Bruijn, L.I., Houseweart, M.K., Kato, S., Anderson, K.L., Anderson, S.D., Ohama, E., Reaume, A.G., Scott, R.W., Cleveland, D.W., 1998. Aggregation and motor neuron toxicity of an ALS-linked SOD1 mutant independent from wild-type SOD1. Science 281, 1851–1854.

Brunt, E.M., Wong, V.W., Nobili, V., Day, C.P., Sookoian, S., Maher, J.J., Bugianesi, E., Sirlin, C.B., Neuschwander-Tetri, B.A., Rinella, M.E., 2015. Nonalcoholic fatty liver disease. Nat. Rev. Dis. Primers 1, 15080.

Bugger, H., Boudina, S., Hu, X.X., Tuinei, J., Zaha, V.G., Theobald, H.A., Yun, U.J., McQueen, A.P., Wayment, B., Litwin, S.E., Abel, E.D., 2008. Type 1 diabetic akita mouse hearts are insulin sensitive but manifest structurally abnormal mitochondria that remain coupled despite increased uncoupling protein 3. Diabetes 57, 2924–2932.

Bugger, H., Guzman, C., Zechner, C., Palmeri, M., Russell, K.S., Russell 3rd, R.R., 2011. Uncoupling protein downregulation in doxorubicin-induced heart failure improves mitochondrial coupling but increases reactive oxygen species generation. Cancer Chemother Pharmacol. 67, 1381–1388.

Bull, P.C., Thomas, G.R., Rommens, J.M., Forbes, J.R., Cox, D.W., 1993. The Wilson disease gene is a putative copper transporting P-type ATPase similar to the Menkes gene. Nat. Genet. 5, 327–337.

Bulteau, A.L., Mena, N.P., Auchere, F., Lee, I., Prigent, A., Lobsiger, C.S., Camadro, J.M., Hirsch, E.C., 2017. Dysfunction of mitochondrial Lon protease and identification of oxidized protein in mouse brain following exposure to MPTP: implications for Parkinson disease. Free Radic. Biol. Med. 108, 236–246.

Bulua, A.C., Simon, A., Maddipati, R., Pelletier, M., Park, H., Kim, K.Y., Sack, M.N., Kastner, D.L., Siegel, R.M., 2011. Mitochondrial reactive oxygen species promote production of proinflammatory cytokines and are elevated in TNFR1-associated periodic syndrome (TRAPS). J. Exp. Med. 208, 519–533.

Burkewitz, K., Morantte, I., Weir, H.J.M., Yeo, R., Zhang, Y., Huynh, F.K., Ilkayeva, O.R., Hirschey, M.D., Grant, A.R., Mair, W.B., 2015. Neuronal CRTC-1 governs systemic mitochondrial metabolism and lifespan via a catecholamine signal. Cell 160, 842–855.

Butterfield, D.A., Di Domenico, F., Barone, E., 2014. Elevated risk of type 2 diabetes for development of Alzheimer disease: a key role for oxidative stress in brain. Biochim. Biophys. Acta 1842, 1693–1706.

Cacciatore, F., Abete, P., Maggi, S., Luchetti, G., Calabrese, C., Viati, L., Leosco, D., Ferrara, N., Vitale, D.F., Rengo, F., 2004. Disability and 6-year mortality in elderly population. role of visual impairment. Aging Clin. Exp. Res. 16, 382–388.

Cadenas, E., Davies, K.J., 2000. Mitochondrial free radical generation, oxidative stress, and aging. Free Radic. Biol. Med. 29, 222–230.
Cahill, A., Hershman, S., Davies, A., Sykora, P., 2005. Ethanol feeding enhances age-related deterioration of the rat hepatic mitochondrion. Am. J. Physiol. Gastrointest. Liver Physiol. 289. G1115–1123.
Caino, M.C., Seo, J.H., Aguinaldo, A., Wait, E., Bryant, K.G., Kossenkov, A.V., Hayden, J.E., Vaira, V., Morotti, A., Ferrero, S., Bosari, S., Gabrilovich, D.I., Languino, L.R., Cohen, A.R., Altieri, D.C., 2016. A neuronal network of mitochondrial dynamics regulates metastasis. Nat. Commun. 7, 13730.
Caldeira da Silva, C.C., Cerqueira, F.M., Barbosa, L.F., Medeiros, M.H., Kowaltowski, A.J., 2008. Mild mitochondrial uncoupling in mice affects energy metabolism, redox balance and longevity. Aging Cell 7, 552–560.
Caldwell, S.H., Crespo, D.M., 2004. The spectrum expanded: cryptogenic cirrhosis and the natural history of non-alcoholic fatty liver disease. Powell EE, Cooksley WGE, Hanson R, Searle J, Halliday JW, Powell LW. The natural history of nonalcoholic steatohepatitis: a follow-up study of forty-two patients for up to 21 years [Hepatology 1990; 11: 74–80] J. Hepatol. 40, 578–584.
Calkins, M.J., Jakel, R.J., Johnson, D.A., Chan, K., Kan, Y.W., Johnson, J.A., 2005. Protection from mitochondrial complex II inhibition in vitro and in vivo by Nrf2-mediated transcription. Proc. Natl. Acad. Sci. U.S.A 102, 244–249.
Calkins, M.J., Manczak, M., Mao, P., Shirendeb, U., Reddy, P.H., 2011. Impaired mitochondrial biogenesis, defective axonal transport of mitochondria, abnormal mitochondrial dynamics and synaptic degeneration in a mouse model of Alzheimer's disease. Hum. Mol. Genet. 20, 4515–4529.
Cao, Z., Lindsay, J.G., Isaacs, N.W., 2007. Mitochondrial peroxiredoxins. Subcell. Biochem. 44, 295–315.
Carpi, A., Menabo, R., Kaludercic, N., Pelicci, P., Di Lisa, F., Giorgio, M., 2009. The cardioprotective effects elicited by p66(Shc) ablation demonstrate the crucial role of mitochondrial ROS formation in ischemia/reperfusion injury. Biochim. Biophys. Acta 1787, 774–780.
Cecchi, C., Fiorillo, C., Sorbi, S., Latorraca, S., Nacmias, B., Bagnoli, S., Nassi, P., Liguri, G., 2002. Oxidative stress and reduced antioxidant defenses in peripheral cells from familial Alzheimer's patients. Free Radic. Biol. Med. 33, 1372–1379.
Cereghetti, G.M., Scorrano, L., 2006. The many shapes of mitochondrial death. Oncogene 25, 4717–4724.
Ceriello, A., Testa, R., 2009. Antioxidant anti-inflammatory treatment in type 2 diabetes. Diabetes Care 32 (Suppl 2). S232–236.
Chacko, B.K., Reily, C., Srivastava, A., Johnson, M.S., Ye, Y., Ulasova, E., Agarwal, A., Zinn, K.R., Murphy, M.P., Kalyanaraman, B., Darley-Usmar, V., 2010. Prevention of diabetic nephropathy in Ins2(+/)(AkitaJ) mice by the mitochondria-targeted therapy MitoQ. Biochem. J. 432, 9–19.
Chakrabarti, R., Ji, W.-K., Stan, R.V., de Juan Sanz, J., Ryan, T.A., Higgs, H.N., 2017. INF2-mediated actin polymerization at the ER stimulates mitochondrial calcium uptake, inner membrane constriction, and division. J. Cell Biol.
Chalmers, S., Saunter, C.D., Girkin, J.M., McCarron, J.G., 2016. Age decreases mitochondrial motility and increases mitochondrial size in vascular smooth muscle. J. Physiol. 594, 4283–4295.
Chance, B., Sies, H., Boveris, A., 1979. Hydroperoxide metabolism in mammalian organs. Physiol. Rev. 59, 527–605.
Chang, D.T., Rintoul, G.L., Pandipati, S., Reynolds, I.J., 2006. Mutant huntingtin aggregates impair mitochondrial movement and trafficking in cortical neurons. Neurobiol. Dis. 22, 388–400.

Chen, C.-M., Wu, Y.-R., Cheng, M.-L., Liu, J.-L., Lee, Y.-M., Lee, P.-W., Soong, B.-W., Chiu, D.T.-Y., 2007a. Increased oxidative damage and mitochondrial abnormalities in the peripheral blood of Huntington's disease patients. Biochem. Biophys. Res. Commun. 359, 335–340.

Chen, H., Chan, D.C., 2009. Mitochondrial dynamics—fusion, fission, movement, and mitophagy—in neurodegenerative diseases. Hum. Mol. Genet. 18, R169–R176.

Chen, H., Detmer, S.A., Ewald, A.J., Griffin, E.E., Fraser, S.E., Chan, D.C., 2003. Mitofusins Mfn1 and Mfn2 coordinately regulate mitochondrial fusion and are essential for embryonic development. J. Cell Biol. 160, 189–200.

Chen, L., Gong, Q., Stice, J.P., Knowlton, A.A., 2009. Mitochondrial OPA1, apoptosis, and heart failure. Cardiovasc. Res. 84, 91–99.

Chen, L., Liu, T., Tran, A., Lu, X., Tomilov, A.A., Davies, V., Cortopassi, G., Chiamvimonvat, N., Bers, D.M., Votruba, M., Knowlton, A.A., 2012a. OPA1 mutation and late-onset cardiomyopathy: mitochondrial dysfunction and mtDNA instability. J. Am. Heart Assoc. 1.

Chen, Y., Cai, J., Murphy, T.J., Jones, D.P., 2002. Overexpressed human mitochondrial thioredoxin confers resistance to oxidant-induced apoptosis in human osteosarcoma cells. J. Biol. Chem. 277, 33242–33248.

Chen, Y., McMillan-Ward, E., Kong, J., Israels, S.J., Gibson, S.B., 2007b. Mitochondrial electron-transport-chain inhibitors of complexes I and II induce autophagic cell death mediated by reactive oxygen species. J. Cell Sci. 120, 4155–4166.

Chen, Z., Vigueira, P.A., Chambers, K.T., Hall, A.M., Mitra, M.S., Qi, N., McDonald, W.G., Colca, J.R., Kletzien, R.F., Finck, B.N., 2012b. Insulin resistance and metabolic derangements in obese mice are ameliorated by a novel peroxisome proliferator-activated receptor gamma-sparing thiazolidinedione. J. Biol. Chem. 287, 23537–23548.

Cheng, E.H., Sheiko, T.V., Fisher, J.K., Craigen, W.J., Korsmeyer, S.J., 2003. VDAC2 inhibits BAK activation and mitochondrial apoptosis. Science 301, 513–517.

Chih-Hao, W., Shi-Bei, W., Yu-Ting, W., Yau-Huei, W., 2013. Oxidative stress response elicited by mitochondrial dysfunction: implication in the pathophysiology of aging. Exp. Biol. Med. 238, 450–460.

Chinsomboon, J., Ruas, J., Gupta, R.K., Thom, R., Shoag, J., Rowe, G.C., Sawada, N., Raghuram, S., Arany, Z., 2009. The transcriptional coactivator PGC-1alpha mediates exercise-induced angiogenesis in skeletal muscle. Proc. Natl. Acad. Sci. U.S.A 106, 21401–21406.

Cho, B., Cho, H.M., Jo, Y., Kim, H.D., Song, M., Moon, C., Kim, H., Kim, K., Sesaki, H., Rhyu, I.J., Kim, H., Sun, W., 2017a. Constriction of the mitochondrial inner compartment is a priming event for mitochondrial division. Nat. Commun. 8, 15754.

Cho, S., Namkoong, K., Shin, M., Park, J., Yang, E., Ihm, J., Thu, V.T., Kim, H.K., Han, J., 2017b. Cardiovascular protective effects and clinical applications of resveratrol. J. Med. Food 20, 323–334.

Choi, W.S., Kruse, S.E., Palmiter, R.D., Xia, Z., 2008. Mitochondrial complex I inhibition is not required for dopaminergic neuron death induced by rotenone, MPP+, or paraquat. Proc. Natl. Acad. Sci. U.S.A 105, 15136–15141.

Choi, Y.S., Ide, T., Sugano, M., 1987. Age-related changes in the regulation of cholesterol metabolism in rats. Exp. Gerontol. 22, 339–349.

Choksi, S., Lin, Y., Pobezinskaya, Y., Chen, L., Park, C., Morgan, M., Li, T., Jitkaew, S., Cao, X., Kim, Y.S., Kim, H.S., Levitt, P., Shih, G., Birre, M., Deng, C.X., Liu, Z.G., 2011. A HIF-1 target, ATIA, protects cells from apoptosis by modulating the mitochondrial thioredoxin, TRX2. Mol. Cell 42, 597–609.

Chung, F.L., Nath, R.G., Nagao, M., Nishikawa, A., Zhou, G.D., Randerath, K., 1999. Endogenous formation and significance of 1,N2-propanodeoxyguanosine adducts. Mutat. Res. 424, 71–81.

Cicchetti, F., Lapointe, N., Roberge-Tremblay, A., Saint-Pierre, M., Jimenez, L., Ficke, B.W., Gross, R.E., 2005. Systemic exposure to paraquat and maneb models early Parkinson's disease in young adult rats. Neurobiol. Dis. 20, 360–371.

Cipolat, S., de Brito, O.M., Dal Zilio, B., Scorrano, L., 2004. OPA1 requires mitofusin 1 to promote mitochondrial fusion. Proc. Natl. Acad. Sci. U.S.A 101, 15927–15932.

Clapham, J.C., Arch, J.R., Chapman, H., Haynes, A., Lister, C., Moore, G.B., Piercy, V., Carter, S.A., Lehner, I., Smith, S.A., Beeley, L.J., Godden, R.J., Herrity, N., Skehel, M., Changani, K.K., Hockings, P.D., Reid, D.G., Squires, S.M., Hatcher, J., Trail, B., Latcham, J., Rastan, S., Harper, A.J., Cadenas, S., Buckingham, J.A., Brand, M.D., Abuin, A., 2000. Mice overexpressing human uncoupling protein-3 in skeletal muscle are hyperphagic and lean. Nature 406, 415–418.

Cocheme, H.M., Quin, C., McQuaker, S.J., Cabreiro, F., Logan, A., Prime, T.A., Abakumova, I., Patel, J.V., Fearnley, I.M., James, A.M., Porteous, C.M., Smith, R.A., Saeed, S., Carre, J.E., Singer, M., Gems, D., Hartley, R.C., Partridge, L., Murphy, M.P., 2011. Measurement of H_2O_2 within living *Drosophila* during aging using a ratiometric mass spectrometry probe targeted to the mitochondrial matrix. Cell Metab. 13, 340–350.

Codogno, P., Meijer, A.J., 2005. Autophagy and signaling: their role in cell survival and cell death. Cell Death Differ. 12 (Suppl 2), 1509–1518.

Cogger, V.C., Svistounov, D., Warren, A., Zykova, S., Melvin, R.G., Solon-Biet, S.M., O'Reilly, J.N., McMahon, A.C., Ballard, J.W., De Cabo, R., Le Couteur, D.G., Lebel, M., 2014. Liver aging and pseudocapillarization in a Werner syndrome mouse model. J. Gerontol. A Biol. Sci. Med. Sci. 69, 1076–1086.

Cohen, H.Y., Miller, C., Bitterman, K.J., Wall, N.R., Hekking, B., Kessler, B., Howitz, K.T., Gorospe, M., de Cabo, R., Sinclair, D.A., 2004. Calorie restriction promotes mammalian cell survival by inducing the SIRT1 deacetylase. Science 305, 390–392.

Cohen, J.C., Horton, J.D., Hobbs, H.H., 2011. Human fatty liver disease: old questions and new insights. Science 332, 1519–1523.

Colca, J.R., McDonald, W.G., McCommis, K.S., Finck, B.N., 2017. Treating fatty liver disease by modulating mitochondrial pyruvate metabolism. Hepatol. Commun. 1, 193–197.

Colca, J.R., VanderLugt, J.T., Adams, W.J., Shashlo, A., McDonald, W.G., Liang, J., Zhou, R., Orloff, D.G., 2013. Clinical proof-of-concept study with MSDC-0160, a prototype mTOT-modulating insulin sensitizer. Clin. Pharmacol. Ther. 93, 352–359.

Collins, J.J., Evason, K., Kornfeld, K., 2006. Pharmacology of delayed aging and extended lifespan of *Caenorhabditis elegans*. Exp. Gerontol. 41, 1032–1039.

Colman, R.J., Anderson, R.M., Johnson, S.C., Kastman, E.K., Kosmatka, K.J., Beasley, T.M., Allison, D.B., Cruzen, C., Simmons, H.A., Kemnitz, J.W., Weindruch, R., 2009. Caloric restriction delays disease onset and mortality in rhesus monkeys. Science 325, 201–204.

Copeland, J.M., Cho, J., Lo, T., Hur, J.H., Bahadorani, S., Arabyan, T., Rabie, J., Soh, J., Walker, D.W., 2009. Extension of *Drosophila* life span by RNAi of the mitochondrial respiratory chain. Curr. Biol. 19, 1591–1598.

Corbi, G., Conti, V., Russomanno, G., Rengo, G., Vitulli, P., Ciccarelli, A.L., Filippelli, A., Ferrara, N., 2012. Is physical activity able to modify oxidative damage in cardiovascular aging? Oxid. Med. Cell Longev. 2012, 728547.

Cordero, M.D., Williams, M.R., Ryffel, B., 2018. AMP-Activated protein kinase regulation of the NLRP3 inflammasome during aging. Trends Endocrinol. Metab. 29, 8–17.

Corral-Debrinski, M., Horton, T., Lott, M.T., Shoffner, J.M., Flint Beal, M., Wallace, D.C., 1992. Mitochondrial DNA deletions in human brain: regional variability and increase with advanced age. Nat. Genet. 2, 324.

Correia, S.C., Perry, G., Moreira, P.I., 2016. Mitochondrial traffic jams in Alzheimer's disease - pinpointing the roadblocks. Biochim. Biophys. Acta 1862, 1909–1917.

Cortez-Pinto, H., Chatham, J., Chacko, V.P., Arnold, C., Rashid, A., Diehl, A.M., 1999. Alterations in liver ATP homeostasis in human nonalcoholic steatohepatitis: a pilot study. J. Am. Med. Assoc. 282, 1659–1664.

Costantini, P., Chernyak, B.V., Petronilli, V., Bernardi, P., 1996. Modulation of the mitochondrial permeability transition pore by pyridine nucleotides and dithiol oxidation at two separate sites. J. Biol. Chem. 271, 6746–6751.

Coulom, H., Birman, S., 2004. Chronic exposure to rotenone models sporadic Parkinson's disease in *Drosophila melanogaster*. J. Neurosci. 24, 10993–10998.

Covey, M.V., Murphy, M.P., Hobbs, C.E., Smith, R.A., Oorschot, D.E., 2006. Effect of the mitochondrial antioxidant, mito vitamin E, on hypoxic-ischemic striatal injury in neonatal rats: a dose-response and stereological study. Exp. Neurol. 199, 513–519.

Cozzolino, M., Pesaresi, M.G., Amori, I., Crosio, C., Ferri, A., Nencini, M., Carri, M.T., 2009. Oligomerization of mutant SOD1 in mitochondria of motoneuronal cells drives mitochondrial damage and cell toxicity. Antioxid. Redox Signal. 11, 1547–1558.

Croce, C.M., Sozzi, G., Huebner, K., 1999. Role of FHIT in human cancer. J. Clin. Oncol. 17, 1618–1624.

Csiszar, A., Labinskyy, N., Perez, V., Recchia, F.A., Podlutsky, A., Mukhopadhyay, P., Losonczy, G., Pacher, P., Austad, S.N., Bartke, A., Ungvari, Z., 2008. Endothelial function and vascular oxidative stress in long-lived GH/IGF-deficient Ames dwarf mice. Am. J. Physiol. Heart Circ. Physiol. 295. H1882–1894.

Cuervo, A.M., Dice, J.F., 2000. Age-related decline in chaperone-mediated autophagy. J. Biol. Chem. 275, 31505–31513.

Cumings, J.N., 1948. The copper and iron content of brain and liver in the normal and in hepato-lenticular degeneration. Brain 71, 410–415.

D'Autreaux, B., Toledano, M.B., 2007. ROS as signalling molecules: mechanisms that generate specificity in ROS homeostasis. Nat. Rev. Mol. Cell Biol. 8, 813–824.

Dagda, R.K., Zhu, J., Chu, C.T., 2009. Mitochondrial kinases in Parkinson's disease: converging insights from neurotoxin and genetic models. Mitochondrion 9, 289–298.

Dai, D.F., Chen, T., Szeto, H., Nieves-Cintron, M., Kutyavin, V., Santana, L.F., Rabinovitch, P.S., 2011. Mitochondrial targeted antioxidant Peptide ameliorates hypertensive cardiomyopathy. J. Am. Coll. Cardiol. 58, 73–82.

Daiber, A., 2010. Redox signaling (cross-talk) from and to mitochondria involves mitochondrial pores and reactive oxygen species. Biochim. Biophys. Acta 1797, 897–906.

Damiano, M., Galvan, L., Deglon, N., Brouillet, E., 2010. Mitochondria in Huntington's disease. Biochim. Biophys. Acta 1802, 52–61.

Dani, D., Shimokawa, I., Komatsu, T., Higami, Y., Warnken, U., Schokraie, E., Schnolzer, M., Krause, F., Sugawa, M.D., Dencher, N.A., 2010. Modulation of oxidative phosphorylation machinery signifies a prime mode of anti-ageing mechanism of calorie restriction in male rat liver mitochondria. Biogerontology 11, 321–334.

Danz, E.D., Skramsted, J., Henry, N., Bennett, J.A., Keller, R.S., 2009. Resveratrol prevents doxorubicin cardiotoxicity through mitochondrial stabilization and the Sirt1 pathway. Free Radic. Biol. Med. 46, 1589–1597.

Daum, B., Walter, A., Horst, A., Osiewacz, H.D., Kuhlbrandt, W., 2013. Age-dependent dissociation of ATP synthase dimers and loss of inner-membrane cristae in mitochondria. Proc. Natl. Acad. Sci. U.S.A 110, 15301–15306.

David, D.C., Hauptmann, S., Scherping, I., Schuessel, K., Keil, U., Rizzu, P., Ravid, R., Drose, S., Brandt, U., Muller, W.E., Eckert, A., Gotz, J., 2005. Proteomic and functional analyses reveal a mitochondrial dysfunction in P301L tau transgenic mice. J. Biol. Chem. 280, 23802–23814.
Davidson, S.M., 2010. Endothelial mitochondria and heart disease. Cardiovasc. Res. 88, 58–66.
de Almeida, A.F., Curi, R., Newsholme, P., Newsholme, E.A., 1989. Maximal activities of key enzymes of glutaminolysis, glycolysis, Krebs cycle and pentose-phosphate pathway of several tissues in mature and aged rats. Int. J. Biochem. 21, 937–940.
de Moura, M.B., dos Santos, L.S., Van Houten, B., 2010. Mitochondrial dysfunction in neurodegenerative diseases and cancer. Environ. Mol. Mutagen. 51, 391–405.
De Paepe, B., Van Coster, R., 2017. A critical assessment of the therapeutic potential of resveratrol supplements for treating mitochondrial disorders. Nutrients 9.
DeBalsi, K.L., Hoff, K.E., Copeland, W.C., 2017. Role of the mitochondrial DNA replication machinery in mitochondrial DNA mutagenesis, aging and age-related diseases. Ageing Res. Rev. 33, 89–104.
Debattisti, V., Gerencser, A.A., Saotome, M., Das, S., Hajnóczky, G., 2017. ROS control mitochondrial motility through p38 and the motor adaptor Miro/Trak. Cell Rep. 21, 1667–1680.
Delles, C., Miller, W.H., Dominiczak, A.F., 2008. Targeting reactive oxygen species in hypertension. Antioxid. Redox Signal. 10, 1061–1077.
Deus, C.M., Serafim, T.L., Magalhaes-Novais, S., Vilaca, A., Moreira, A.C., Sardao, V.A., Cardoso, S.M., Oliveira, P.J., 2017. Sirtuin 1-dependent resveratrol cytotoxicity and pro-differentiation activity on breast cancer cells. Arch. Toxicol. 91, 1261–1278.
Devi, L., Prabhu, B.M., Galati, D.F., Avadhani, N.G., Anandatheerthavarada, H.K., 2006. Accumulation of amyloid precursor protein in the mitochondrial import channels of human Alzheimer's disease brain is associated with mitochondrial dysfunction. J. Neurosci. 26, 9057–9068.
Dhanasekaran, A., Kotamraju, S., Kalivendi, S.V., Matsunaga, T., Shang, T., Keszler, A., Joseph, J., Kalyanaraman, B., 2004. Supplementation of endothelial cells with mitochondria-targeted antioxidants inhibit peroxide-induced mitochondrial iron uptake, oxidative damage, and apoptosis. J. Biol. Chem. 279, 37575–37587.
Di Ciaula, A., Garruti, G., Baccetto, R.L., Molina-Molina, E., Bonfrate, L., Wang, D.Q., Portincasa, P., 2017. Bile acid physiology. Ann. Hepatol. 16, 4–14.
Di Lisa, F., Bernardi, P., 2006. Mitochondria and ischemia-reperfusion injury of the heart: fixing a hole. Cardiovasc. Res. 70, 191–199.
Diao, J., Allister, E.M., Koshkin, V., Lee, S.C., Bhattacharjee, A., Tang, C., Giacca, A., Chan, C.B., Wheeler, M.B., 2008. UCP2 is highly expressed in pancreatic alpha-cells and influences secretion and survival. Proc. Natl. Acad. Sci. U.S.A 105, 12057–12062.
Diehl, A.K., 1991. Epidemiology and natural history of gallstone disease. Gastroenterol. Clin. North Am. 20, 1–19.
DiMauro, S., Hirano, M., 2009. Pathogenesis and treatment of mitochondrial disorders. Adv. Exp. Med. Biol. 652, 139–170.
Diogo, C.V., Grattagliano, I., Oliveira, P.J., Bonfrate, L., Portincasa, P., 2011. Re-wiring the circuit: mitochondria as a pharmacological target in liver disease. Curr. Med. Chem. 18, 5448–5465.
Diotte, N.M., Xiong, Y., Gao, J., Chua, B.H., Ho, Y.S., 2009. Attenuation of doxorubicin-induced cardiac injury by mitochondrial glutaredoxin 2. Biochim. Biophys. Acta 1793, 427–438.
Dolle, C., Flones, I., Nido, G.S., Miletic, H., Osuagwu, N., Kristoffersen, S., Lilleng, P.K., Larsen, J.P., Tysnes, O.B., Haugarvoll, K., Bindoff, L.A., Tzoulis, C., 2016. Defective mitochondrial DNA homeostasis in the substantia nigra in Parkinson disease. Na.t Commun. 7, 13548.

Dong, Z., Shanmughapriya, S., Tomar, D., Siddiqui, N., Lynch, S., Nemani, N., Breves, S.L., Zhang, X., Tripathi, A., Palaniappan, P., Riitano, M.F., Worth, A.M., Seelam, A., Carvalho, E., Subbiah, R., Jana, F., Soboloff, J., Peng, Y., Cheung, J.Y., Joseph, S.K., Caplan, J., Rajan, S., Stathopulos, P.B., Madesh, M., 2017. Mitochondrial Ca(2+) uniporter is a mitochondrial luminal redox sensor that augments MCU channel activity. Mol Cell 65, 1014–1028 e1017.

Dorn, G.W., Scorrano, L., 2010. Two close, too close: sarcoplasmic reticulum-mitochondrial cross-talk and cardiomyocyte fate. Circulation Res. 107, 689–699.

Droge, W., 2002. Free radicals in the physiological control of cell function. Physiol. Rev. 82, 47–95.

Du, C., Fujii, Y., Ito, M., Harada, M., Moriyama, E., Shimada, R., Ikemoto, A., Okuyama, H., 2004. Dietary polyunsaturated fatty acids suppress acute hepatitis, alter gene expression and prolong survival of female Long-Evans Cinnamon rats, a model of Wilson disease. J. Nutr. Biochem. 15, 273–280.

Duchen, M.R., 2004. Roles of mitochondria in health and disease. Diabetes 53 (Suppl 1). S96–102.

Duncan, K.E., Ngu, T.T., Chan, J., Salgado, M.T., Merrifield, M.E., Stillman, M.J., 2006. Peptide folding, metal-binding mechanisms, and binding site structures in metallothioneins. Exp. Biol. Med. 231, 1488–1499.

Echtay, K.S., Roussel, D., St-Pierre, J., Jekabsons, M.B., Cadenas, S., Stuart, J.A., Harper, J.A., Roebuck, S.J., Morrison, A., Pickering, S., Clapham, J.C., Brand, M.D., 2002. Superoxide activates mitochondrial uncoupling proteins. Nature 415, 96–99.

Edgar, D., Trifunovic, A., 2009. The mtDNA mutator mouse: dissecting mitochondrial involvement in aging. Aging (Albany NY) 1, 1028–1032.

Einarsson, K., Nilsell, K., Leijd, B., Angelin, B., 1985. Influence of age on secretion of cholesterol and synthesis of bile acids by the liver. N. Engl. J. Med. 313, 277–282.

Elmore, S.P., Qian, T., Grissom, S.F., Lemasters, J.J., 2001. The mitochondrial permeability transition initiates autophagy in rat hepatocytes. FASEB J. 15, 2286–2287.

Emelyanova, L., Preston, C., Gupta, A., Viqar, M., Negmadjanov, U., Edwards, S., Kraft, K., Devana, K., Holmuhamedov, E., O'Hair, D., Tajik, A.J., Jahangir, A., 2017. Effect of aging on mitochondrial energetics in the human atria. J. Gerontol. A Biol. Sci. Med. Sci.

Enoksson, M., Fernandes, A.P., Prast, S., Lillig, C.H., Holmgren, A., Orrenius, S., 2005. Overexpression of glutaredoxin 2 attenuates apoptosis by preventing cytochrome c release. Biochem. Biophy.s Res. Commun. 327, 774–779.

Epperly, M.W., Wang, H., Jones, J.A., Dixon, T., Montesinos, C.A., Greenberger, J.S., 2011. Antioxidant-chemoprevention diet ameliorates late effects of total-body irradiation and supplements radioprotection by MnSOD-plasmid liposome administration. Radiat. Res. 175, 759–765.

Facchini, F.S., Hua, N., Abbasi, F., Reaven, G.M., 2001. Insulin resistance as a predictor of age-related diseases. J. Clinical Endocrinol. Metabol. 86, 3574–3578.

Fang, D., Zhang, Z., Li, H., Yu, Q., Douglas, J.T., Bratasz, A., Kuppusamy, P., Yan, S.S., 2016. Increased electron paramagnetic resonance signal correlates with mitochondrial dysfunction and oxidative stress in an Alzheimer's disease mouse brain. J. Alzheimers Dis. 51, 571–580.

Farr, S.A., Yamada, K.A., Butterfield, D.A., Abdul, H.M., Xu, L., Miller, N.E., Banks, W.A., Morley, J.E., 2008. Obesity and hypertriglyceridemia produce cognitive impairment. Endocrinology 149, 2628–2636.

Fayet, G., Jansson, M., Sternberg, D., Moslemi, A.R., Blondy, P., Lombes, A., Fardeau, M., Oldfors, A., 2002. Ageing muscle: clonal expansions of mitochondrial DNA point mutations and deletions cause focal impairment of mitochondrial function. Neuromuscul. Disord. 12, 484–493.

Fendel, U., Tocilescu, M.A., Kerscher, S., Brandt, U., 2008. Exploring the inhibitor binding pocket of respiratory complex I. Biochim. Biophys. Acta 1777, 660–665.
Fernández-Checa, J.C., Kaplowitz, N., García-Ruiz, C., Colell, A., 1998. Mitochondrial glutathione: importance and transport. In: Seminars in Liver Disease. Thieme Medical Publishers, Inc, pp. 389–401.
Fernandez-Sanz, C., Ruiz-Meana, M., Miro-Casas, E., Nunez, E., Castellano, J., Loureiro, M., Barba, I., Poncelas, M., Rodriguez-Sinovas, A., Vazquez, J., Garcia-Dorado, D., 2014. Defective sarcoplasmic reticulum-mitochondria calcium exchange in aged mouse myocardium. Cell Death Dis. 5, e1573.
Ferrara, N., Rinaldi, B., Corbi, G., Conti, V., Stiuso, P., Boccuti, S., Rengo, G., Rossi, F., Filippelli, A., 2008. Exercise training promotes SIRT1 activity in aged rats. Rejuvenation Res. 11, 139–150.
Ferreira, I.L., Nascimento, M.V., Ribeiro, M., Almeida, S., Cardoso, S.M., Grazina, M., Pratas, J., Santos, M.J., Januario, C., Oliveira, C.R., Rego, A.C., 2010. Mitochondrial-dependent apoptosis in Huntington's disease human cybrids. Exp. Neurol. 222, 243–255.
Festa, R.A., Thiele, D.J., 2011. Copper: an essential metal in biology. Curr. Biol. 21. R877–883.
Filadi, R., Theurey, P., Pizzo, P., 2017. The endoplasmic reticulum-mitochondria coupling in health and disease: molecules, functions and significance. Cell Calcium 62, 1–15.
Fjell, A.M., McEvoy, L., Holland, D., Dale, A.M., Walhovd, K.B., 2014. What is normal in normal aging? effects of aging, amyloid and alzheimer's disease on the cerebral cortex and the hippocampus. Progress Neurobiol. 117, 20–40.
Floreani, A., 2007. Liver diseases in the elderly: an update. Digestive Dis. 25, 138–143.
Fontana, L., Zhao, E., Tanaka, K., Czaja, M.J., 2012. Aging promotes the development of diet-induced murine steatohepatitis but not steatosis. Hepatology 56, 317A.
Ford, E.S., Giles, W.H., Dietz, W.H., 2002. Prevalence of the metabolic syndrome among US adults: findings from the third national health and nutrition examination survey. Jama 287, 356–359.
Franceschi, C., Bonafè, M., Valensin, S., Olivieri, F., De Luca, M., Ottaviani, E., De Benedictis, G., 2000. Inflamm-aging: an evolutionary perspective on immunosenescence. Ann. N. Y. Acad. Sci. 908, 244–254.
Franceschi, C., Capri, M., Monti, D., Giunta, S., Olivieri, F., Sevini, F., Panourgia, M.P., Invidia, L., Celani, L., Scurti, M., Cevenini, E., Castellani, G.C., Salvioli, S., 2007. Inflammaging and anti-inflammaging: a systemic perspective on aging and longevity emerged from studies in humans. Mech. Ageing Dev. 128, 92–105.
Frandsen, J.R., Narayanasamy, P., 2018. Neuroprotection through flavonoid: enhancement of the glyoxalase pathway. Redox Biol. 14, 465–473.
Frank, N., Knauft, J., Amelung, F., Nair, J., Wesch, H., Bartsch, H., 2003. No prevention of liver and kidney tumors in Long-Evans Cinnamon rats by dietary curcumin, but inhibition at other sites and of metastases. Mutat. Res. 523–524, 127–135.
Frederick, R.L., Shaw, J.M., 2007. Moving mitochondria: establishing distribution of an essential organelle. Traffic (Copenhagen, Denmark) 8, 1668–1675.
Freitas, I., Boncompagni, E., Tarantola, E., Gruppi, C., Bertone, V., Ferrigno, A., Milanesi, G., Vaccarone, R., Tira, M.E., Vairetti, M., 2016. In situ evaluation of oxidative stress in rat fatty liver induced by a methionine- and choline-deficient diet. Oxid. Med. Cell Longev. 2016, 9307064.
Fridovich, I., 2004. Mitochondria: are they the seat of senescence? Aging Cell 3, 13–16.
Friedman, J.R., Lackner, L.L., West, M., DiBenedetto, J.R., Nunnari, J., Voeltz, G.K., 2011. ER tubules mark sites of mitochondrial division. Science 334, 358–362.
Froom, P., Miron, E., Barak, M., 2003. Oral anticoagulants in the elderly. British J. Haematol. 120, 526–528.

Fromenty, B., Pessayre, D., 1995. Inhibition of mitochondrial beta-oxidation as a mechanism of hepatotoxicity. Pharmacol. Ther. 67, 101–154.

Fu, A., Nair, K.S., 1998. Age effect on fibrinogen and albumin synthesis in humans. Am. J. Physiol. 275. E1023–1030.

Gan, L., Chitturi, S., Farrell, G.C., 2011. Mechanisms and implications of age-related changes in the liver: nonalcoholic Fatty liver disease in the elderly. Curr. Gerontol. Geriatr. Res. 2011, 831536.

Gant, J.C., Chen, K.C., Kadish, I., Blalock, E.M., Thibault, O., Porter, N.M., Landfield, P.W., 2015. Reversal of aging-related neuronal Ca2+ dysregulation and cognitive impairment by delivery of a transgene encoding FK506-binding protein 12.6/1b to the hippocampus. J. Neurosci. 35, 10878–10887.

Garcia-Macia, M., Vega-Naredo, I., De Gonzalo-Calvo, D., Rodriguez-Gonzalez, S.M., Camello, P.J., Camello-Almaraz, C., Martin-Cano, F.E., Rodriguez-Colunga, M.J., Pozo, M.J., Coto-Montes, A.M., 2011. Melatonin induces neural SOD2 expression independent of the NF-kappaB pathway and improves the mitochondrial population and function in old mice. J. Pineal. Res. 50, 54–63.

Garcia-Martinez, I., Santoro, N., Chen, Y., Hoque, R., Ouyang, X., Caprio, S., Shlomchik, M.J., Coffman, R.L., Candia, A., Mehal, W.Z., 2016. Hepatocyte mitochondrial DNA drives nonalcoholic steatohepatitis by activation of TLR9. J. Clin. Invest. 126, 859–864.

Garcia-Ruiz, C., Fernandez-Checa, J.C., 2006. Mitochondrial glutathione: hepatocellular survival-death switch. J. Gastroenterol. Hepatol. 21 (Suppl. 3). S3–6.

Garcia-Ruiz, I., Rodriguez-Juan, C., Diaz-Sanjuan, T., del Hoyo, P., Colina, F., Munoz-Yague, T., Solis-Herruzo, J.A., 2006. Uric acid and anti-TNF antibody improve mitochondrial dysfunction in ob/ob mice. Hepatology 44, 581–591.

Garcia-Ruiz, I., Solis-Munoz, P., Fernandez-Moreira, D., Munoz-Yague, T., Solis-Herruzo, J.A., 2015. In vitro treatment of HepG2 cells with saturated fatty acids reproduces mitochondrial dysfunction found in nonalcoholic steatohepatitis. Dis. Model. Mech. 8, 183–191.

Gasparre, G., Rossignol, R., Sonveaux, P., 2017. Mitochondria in cancer. Biochimica et Biophysica Acta 1858, 553–555.

Gems, D., Partridge, L., 2008. Stress-response hormesis and aging: "that which does not kill us makes us stronger". Cell Metab. 7, 200–203.

Ghosh, P.M., Shu, Z.-J., Zhu, B., Lu, Z., Ikeno, Y., Barnes, J.L., Yeh, C.-K., Zhang, B.-X., Katz, M.S., Kamat, A., 2012. Role of β-adrenergic receptors in regulation of hepatic fat accumulation during aging. J. Endocrinol. 213, 251–261.

Gianello, P., Saliez, A., Bufkens, X., Pettinger, R., Misseleyn, D., Hori, S., Malfroy, B., 1996. EUK-134, a synthetic superoxide dismutase and catalase mimetic, protects rat kidneys from ischemia-reperfusion-induced damage. Transplantation 62, 1664–1666.

Gibson, G.E., Sheu, K.F., Blass, J.P., 1998. Abnormalities of mitochondrial enzymes in Alzheimer disease. J. Neural. Transm. (Vienna) 105, 855–870.

Giorgi, C., Agnoletto, C., Baldini, C., Bononi, A., Bonora, M., Marchi, S., Missiroli, S., Patergnani, S., Poletti, F., Rimessi, A., Zavan, B., Pinton, P., 2010. Redox control of protein kinase C: cell- and disease-specific aspects. Antioxid Redox Signal 13, 1051–1085.

Giorgio, M., Migliaccio, E., Orsini, F., Paolucci, D., Moroni, M., Contursi, C., Pelliccia, G., Luzi, L., Minucci, S., Marcaccio, M., Pinton, P., Rizzuto, R., Bernardi, P., Paolucci, F., Pelicci, P.G., 2005. Electron transfer between cytochrome c and p66Shc generates reactive oxygen species that trigger mitochondrial apoptosis. Cell 122, 221–233.

Gitlin, J.D., 2003. Wilson disease. Gastroenterology 125, 1868–1877.

Giugliano, D., Ceriello, A., Paolisso, G., 1996. Oxidative stress and diabetic vascular complications. Diabetes Care 19, 257–267.

Gledhill, J.R., Walker, J.E., 2005. Inhibition sites in F1-ATPase from bovine heart mitochondria. Biochem. J. 386, 591–598.
Gomes, L.C., Scorrano, L., 2008. High levels of Fis1, a pro-fission mitochondrial protein, trigger autophagy. Biochim. Biophys. Acta 1777, 860–866.
Goncalves, I.O., Oliveira, P.J., Ascensao, A., Magalhaes, J., 2013. Exercise as a therapeutic tool to prevent mitochondrial degeneration in nonalcoholic steatohepatitis. Eur. J. Clin. Invest. 43, 1184–1194.
Gounder, S.S., Kannan, S., Devadoss, D., Miller, C.J., Whitehead, K.J., Odelberg, S.J., Firpo, M.A., Paine 3rd, R., Hoidal, J.R., Abel, E.D., Rajasekaran, N.S., 2012. Impaired transcriptional activity of Nrf2 in age-related myocardial oxidative stress is reversible by moderate exercise training. PLoS One 7, e45697.
Graham, D., Huynh, N.N., Hamilton, C.A., Beattie, E., Smith, R.A., Cocheme, H.M., Murphy, M.P., Dominiczak, A.F., 2009. Mitochondria-targeted antioxidant MitoQ10 improves endothelial function and attenuates cardiac hypertrophy. Hypertension 54, 322–328.
Grattagliano, I., Caraceni, P., Calamita, G., Ferri, D., Gargano, I., Palasciano, G., Portincasa, P., 2008. Severe liver steatosis correlates with nitrosative and oxidative stress in rats. Eur. J. Clin. Invest. 38, 523–530.
Grattagliano, I., Caraceni, P., Portincasa, P., Domenicali, M., Palmieri, V.O., Trevisani, F., Bernardi, M., Palasciano, G., 2003. Adaptation of subcellular glutathione detoxification system to stress conditions in choline-deficient diet induced rat fatty liver. Cell Biol. Toxicol. 19, 355–366.
Grattagliano, I., Diogo, C.V., Mastrodonato, M., de Bari, O., Persichella, M., Wang, D.Q., Liquori, A., Ferri, D., Carratu, M.R., Oliveira, P.J., Portincasa, P., 2013. A silybin-phospholipids complex counteracts rat fatty liver degeneration and mitochondrial oxidative changes. World J. Gastroenterol. 19, 3007–3017.
Grattagliano, I., Portincasa, P., Cocco, T., Moschetta, A., Di Paola, M., Palmieri, V.O., Palasciano, G., 2004a. Effect of dietary restriction and N-acetylcysteine supplementation on intestinal mucosa and liver mitochondrial redox status and function in aged rats. Exp. Gerontol. 39, 1323–1332.
Grattagliano, I., Russmann, S., Diogo, C., Bonfrate, L., Oliveira, P.J., Wang, D.Q., Portincasa, P., 2011. Mitochondria in chronic liver disease. Curr. Drug Targets 12, 879–893.
Grattagliano, I., Russmann, S., Palmieri, V.O., Juni, P., Bihl, F., Portincasa, P., Palasciano, G., Lauterburg, B.H., 2004b. Low membrane protein sulfhydrils but not G6PD deficiency predict ribavirin-induced hemolysis in hepatitis C. Hepatology 39, 1248–1255.
Grattagliano, I., Wang, D.Q., Di, C.A., Diogo, C.V., Palasciano, G., Portincasa, P., 2009. Biliary proteins and their redox status changes in gallstone patients. Eur.J. Clin. Invest. 39, 986–992.
Green, D.R., Galluzzi, L., Kroemer, G., 2011. Mitochondria and the autophagy-inflammation-cell death axis in organismal aging. Science 333, 1109–1112.
Green, D.R., Kroemer, G., 2004. The pathophysiology of mitochondrial cell death. Science 305, 626–629.
Gregg, S.Q., Gutierrez, V., Robinson, A.R., Woodell, T., Nakao, A., Ross, M.A., Michalopoulos, G.K., Rigatti, L., Rothermel, C.E., Kamileri, I., Garinis, G.A., Stolz, D.B., Niedernhofer, L.J., 2012. A mouse model of accelerated liver aging caused by a defect in DNA repair. Hepatology 55, 609–621.
Grunewald, A., Rygiel, K.A., Hepplewhite, P.D., Morris, C.M., Picard, M., Turnbull, D.M., 2016. Mitochondrial DNA depletion in respiratory chain-deficient Parkinson disease neurons. Ann. Neurol. 79, 366–378.

Gu, M., Cooper, J.M., Butler, P., Walker, A.P., Mistry, P.K., Dooley, J.S., Schapira, A.H., 2000. Oxidative-phosphorylation defects in liver of patients with Wilson's disease. Lancet 356, 469–474.

Guest, J.R., Russell, G.C., 1992. Complexes and complexities of the citric acid cycle in *Escherichia coli*. Curr. Top Cell Regul. 33, 231–247.

Guo, J.Y., Chen, H.Y., Mathew, R., Fan, J., Strohecker, A.M., Karsli-Uzunbas, G., Kamphorst, J.J., Chen, G., Lemons, J.M., Karantza, V., Coller, H.A., Dipaola, R.S., Gelinas, C., Rabinowitz, J.D., White, E., 2011. Activated ras requires autophagy to maintain oxidative metabolism and tumorigenesis. Genes Dev. 25, 460–470.

Gupta, A., Roy, S., Lazar, A.J., Wang, W.L., McAuliffe, J.C., Reynoso, D., McMahon, J., Taguchi, T., Floris, G., Debiec-Rychter, M., Schoffski, P., Trent, J.A., Debnath, J., Rubin, B.P., 2010. Autophagy inhibition and antimalarials promote cell death in gastrointestinal stromal tumor (GIST). Proc. Natl. Acad. Sci. U.S.A 107, 14333–14338.

Guzman, J.N., Sanchez-Padilla, J., Wokosin, D., Kondapalli, J., Ilijic, E., Schumacker, P.T., Surmeier, D.J., 2010. Oxidant stress evoked by pacemaking in dopaminergic neurons is attenuated by DJ-1. Nature 468, 696.

Haga, S., Terui, K., Fukai, M., Oikawa, Y., Irani, K., Furukawa, H., Todo, S., Ozaki, M., 2008. Preventing hypoxia/reoxygenation damage to hepatocytes by p66(shc) ablation: up-regulation of anti-oxidant and anti-apoptotic proteins. J. Hepatol. 48, 422–432.

Hagopian, K., Harper, M.E., Ram, J.J., Humble, S.J., Weindruch, R., Ramsey, J.J., 2005. Long-term calorie restriction reduces proton leak and hydrogen peroxide production in liver mitochondria. Am. J. Physiol. Endocrinol. Metab. 288. E674–684.

Hailey, D.W., Rambold, A.S., Satpute-Krishnan, P., Mitra, K., Sougrat, R., Kim, P.K., Lippincott-Schwartz, J., 2010. Mitochondria supply membranes for autophagosome biogenesis during starvation. Cell 141, 656–667.

Halestrap, A.P., 2009. What is the mitochondrial permeability transition pore? J. Mol. Cell Cardiol. 46, 821–831.

Hall, K.E., Proctor, D.D., Fisher, L., Rose, S., 2005. American gastroenterological association future trends committee report: effects of aging of the population on gastroenterology practice, education, and research. Gastroenterology 129, 1305–1338.

Hamilton, M.L., Van Remmen, H., Drake, J.A., Yang, H., Guo, Z.M., Kewitt, K., Walter, C.A., Richardson, A., 2001. Does oxidative damage to DNA increase with age? Proc. Natl. Acad. Sci. 98, 10469–10474.

Haque, M., Sanyal, A.J., 2002. The metabolic abnormalities associated with non-alcoholic fatty liver disease. Best practice & research. Clin. Gastroenterol. 16, 709–731.

Harman, D., 1956. Aging: a theory based on free radical and radiation chemistry. J. Gerontol. 11, 298–300.

Harman, D., 1972. The biologic clock: the mitochondria? J. Am. Geriatr. Soc. 20, 145–147.

Harrington, L.A., Harley, C.B., 1988. Effect of vitamin E on lifespan and reproduction in *Caenorhabditis elegans*. Mech. Ageing Dev. 43, 71–78.

Hashizume, O., Ohnishi, S., Mito, T., Shimizu, A., Ishikawa, K., Nakada, K., Soda, M., Mano, H., Togayachi, S., Miyoshi, H., Okita, K., Hayashi, J.-I., 2015. Epigenetic regulation of the nuclear-coded GCAT and SHMT2 genes confers human age-associated mitochondrial respiration defects. Sci. Rep. 5, 10434.

Hatch, A.L., Ji, W.-K., Merrill, R.A., Strack, S., Higgs, H.N., 2016. Actin filaments as dynamic reservoirs for Drp1 recruitment. Mol. Biol. Cell 27, 3109–3121.

Hawkins, R.L., Mori, M., Inoue, M., Torii, K., 1995. Proline, ascorbic acid, or thioredoxin affect jaundice and mortality in long evans cinnamon rats. Pharmacol. Biochem. Behav. 52, 509–515.

Hayashi, J.-I., Hashizume, O., Ishikawa, K., Shimizu, A., 2016. Mutations in mitochondrial DNA regulate mitochondrial diseases and metastasis but do not regulate aging. Curr. Opin. Genet. Dev. 38, 63–67.

Hekimi, S., Guarente, L., 2003. Genetics and the specificity of the aging process. Science 299, 1351–1354.

Hekimi, S., Lapointe, J., Wen, Y., 2011. Taking a "good" look at free radicals in the aging process. Trends Cell Biol. 21, 569–576.

Herbener, G.H., 1976. A Morphometric study of age-dependent changes in mitochondrial populations of mouse liver and heart1. J. Gerontol. 31, 8–12.

Hirschey, M.D., Shimazu, T., Goetzman, E., Jing, E., Schwer, B., Lombard, D.B., Grueter, C.A., Harris, C., Biddinger, S., Ilkayeva, O.R., Stevens, R.D., Li, Y., Saha, A.K., Ruderman, N.B., Bain, J.R., Newgard, C.B., Farese Jr., R.V., Alt, F.W., Kahn, C.R., Verdin, E., 2010. SIRT3 regulates mitochondrial fatty-acid oxidation by reversible enzyme deacetylation. Nature 464, 121–125.

Hoekstra, J.G., Hipp, M.J., Montine, T.J., Kennedy, S.R., 2016. Mitochondrial DNA mutations increase in early stage Alzheimer disease and are inconsistent with oxidative damage. Ann. Neurol. 80, 301–306.

Hole, P.S., Pearn, L., Tonks, A.J., James, P.E., Burnett, A.K., Darley, R.L., Tonks, A., 2010. Ras-induced reactive oxygen species promote growth factor-independent proliferation in human CD34+ hematopoietic progenitor cells. Blood 115, 1238–1246.

Hrgovcic, M., Tessmer, C.F., Thomas, F.B., Ong, P.S., Gamble, J.F., Shullenberger, C.C., 1973. Serum copper observations in patients with malignant lymphoma. Cancer 32, 1512–1524.

Huang, X., Atwood, C.S., Hartshorn, M.A., Multhaup, G., Goldstein, L.E., Scarpa, R.C., Cuajungco, M.P., Gray, D.N., Lim, J., Moir, R.D., Tanzi, R.E., Bush, A.I., 1999. The A beta peptide of Alzheimer's disease directly produces hydrogen peroxide through metal ion reduction. Biochemistry 38, 7609–7616.

Huang, Y., Hong, H., Li, M., Liu, J., Jiang, C., Zhang, H., Ye, L., Zheng, J., 2017. Age-Dependent oxidative DNA damage does not correlate with reduced proliferation of cardiomyocytes in humans. PLoS One 12, e0170351.

Huster, D., Finegold, M.J., Morgan, C.T., Burkhead, J.L., Nixon, R., Vanderwerf, S.M., Gilliam, C.T., Lutsenko, S., 2006. Consequences of copper accumulation in the livers of the Atp7b-/- (Wilson disease gene) knockout mice. Am. J. Pathol. 168, 423–434.

Hwang, I., Lee, J., Huh, J.Y., Park, J., Lee, H.B., Ho, Y.S., Ha, H., 2012. Catalase deficiency accelerates diabetic renal injury through peroxisomal dysfunction. Diabetes 61, 728–738.

Ide, T., Tsutsui, H., Kinugawa, S., Utsumi, H., Kang, D., Hattori, N., Uchida, K., Arimura, K., Egashira, K., Takeshita, A., 1999. Mitochondrial electron transport complex I is a potential source of oxygen free radicals in the failing myocardium. Circ Res. 85, 357–363.

Ikeyama, S., Wang, X.-T., Li, J., Podlutsky, A., Martindale, J.L., Kokkonen, G., van Huizen, R., Gorospe, M., Holbrook, N.J., 2003. Expression of the pro-apoptotic gen-egadd153/chop is elevated in liver with aging and sensitizes cells to oxidant Injury. J. Biological. Chem. 278, 16726–16731.

Inden, M., Kitamura, Y., Abe, M., Tamaki, A., Takata, K., Taniguchi, T., 2011. Parkinsonian rotenone mouse model: reevaluation of long-term administration of rotenone in C57BL/6 mice. Biol. Pharm. Bull. 34, 92–96.

Inoue, H., Ishii, H., Alder, H., Snyder, E., Druck, T., Huebner, K., Croce, C.M., 1997. Sequence of the FRA3B common fragile region: implications for the mechanism of FHIT deletion. Proc. Natl. Acad. Sci. U.S.A 94, 14584–14589.

Intano, G.W., Cho, E.J., McMahan, C.A., Walter, C.A., 2003. Age-related base excision repair activity in mouse brain and liver nuclear extracts. J. Gerontol. A Biol. Sci. Med. Sci. 58, 205–211.

Ishii, T., Miyazawa, M., Hartman, P.S., Ishii, N., 2011. Mitochondrial superoxide anion (O(2)(-)) inducible "mev-1" animal models for aging research. BMB Rep. 44, 298–305.

Ishii, T., Yasuda, K., Akatsuka, A., Hino, O., Hartman, P.S., Ishii, N., 2005. A mutation in the SDHC gene of complex II increases oxidative stress, resulting in apoptosis and tumorigenesis. Cancer Res. 65, 203–209.

Jain, S.K., 1998. Glutathione and glucose-6-phosphate dehydrogenase deficiency can increase protein glycosylation. Free Radic. Biol. Med. 24, 197–201.

Janikiewicz, J., Hanzelka, K., Kozinski, K., Kolczynska, K., Dobrzyn, A., 2015. Islet beta-cell failure in type 2 diabetes—Within the network of toxic lipids. Biochem. Biophys. Res. Commun. 460, 491–496.

Janssen, I.M., Sturtz, S., Skipka, G., Zentner, A., Velasco Garrido, M., Busse, R., 2010. Ginkgo biloba in Alzheimer's disease: a systematic review. Wien. Med. Wochenschr. 160, 539–546.

Jędrak, P., Krygier, M., Tońska, K., Drozd, M., Kaliszewska, M., Bartnik, E., Sołtan, W., Sitek, E.J., Stanisławska-Sachadyn, A., Limon, J., Sławek, J., Węgrzyn, G., Barańska, S., 2017. Mitochondrial DNA levels in Huntington disease leukocytes and dermal fibroblasts. Metabolic Brain Disease 32, 1237–1247.

Jedrak, P., Sowa, N., Baranska, S., Wegrzyn, G., 2017. Characterization of conditions and determination of practical tips for mtDNA level estimation in various human cells. Acta Biochimica. Polonica. 64, 699–704.

Jendrach, M., Pohl, S., Vöth, M., Kowald, A., Hammerstein, P., Bereiter-Hahn, J., 2005. Morpho-dynamic changes of mitochondria during ageing of human endothelial cells. Mech. Ageing Dev. 126, 813–821.

Ji, L.L., 1993. Antioxidant enzyme response to exercise and aging. Med. Sci. Sports Exerc. 25, 225–231.

Ji, Q., Ikegami, H., Fujisawa, T., Kawabata, Y., Ono, M., Nishino, M., Ohishi, M., Katsuya, T., Rakugi, H., Ogihara, T., 2004. A common polymorphism of uncoupling protein 2 gene is associated with hypertension. J. Hypertens. 22, 97–102.

Jin, J., Iakova, P., Breaux, M., Sullivan, E., Jawanmardi, N., Chen, D., Jiang, Y., Medrano, E.M., Timchenko, N.A., 2013. Increased expression of enzymes of triglyceride synthesis is essential for the development of hepatic steatosis. Cell Rep. 3, 831–843.

Johnson, S.C., Rabinovitch, P.S., Kaeberlein, M., 2013. mTOR is a key modulator of ageing and age-related disease. Nature 493, 338.

Joseph, A.-M., Adhihetty, P.J., Buford, T.W., Wohlgemuth, S.E., Lees, H.A., Nguyen, L.M.D., Aranda, J.M., Sandesara, B.D., Pahor, M., Manini, T.M., Marzetti, E., Leeuwenburgh, C., 2012. The impact of aging on mitochondrial function and biogenesis pathways in skeletal muscle of sedentary high- and low-functioning elderly individuals. Aging Cell 11, 801–809.

Jurk, D., Wilson, C., Passos, J.F., Oakley, F., Correia-Melo, C., Greaves, L., Saretzki, G., Fox, C., Lawless, C., Anderson, R., Hewitt, G., Pender, S.L., Fullard, N., Nelson, G., Mann, J., van de Sluis, B., Mann, D.A., von Zglinicki, T., 2014. Chronic inflammation induces telomere dysfunction and accelerates ageing in mice. Nat. Commun. 2, 4172.

Kachergus, J., Mata, I.F., Hulihan, M., Taylor, J.P., Lincoln, S., Aasly, J., Gibson, J.M., Ross, O.A., Lynch, T., Wiley, J., Payami, H., Nutt, J., Maraganore, D.M., Czyzewski, K., Styczynska, M., Wszolek, Z.K., Farrer, M.J., Toft, M., 2005. Identification of a novel LRRK2 mutation linked to autosomal dominant parkinsonism: evidence of a common founder across European populations. Am. J. Hum. Genet. 76, 672–680.

Kadish, I., Kumar, A., Beitnere, U., Jennings, E., McGilberry, W., van Groen, T., 2016. Dietary composition affects the development of cognitive deficits in WT and Tg AD model mice. Exp. Gerontol. 86, 39–49.

Kaim, W., Rall, J., 1996. Copper - A 'modern' bioelement. Angew. Chem. Int. Edit. 35, 43–60.

Kalyani, R.R., Corriere, M., Ferrucci, L., 2014. Age-related and disease-related muscle loss: the effect of diabetes, obesity, and other diseases. Lancet. Diabetes Endocrinol. 2, 819–829.

Kamata, H., Honda, S., Maeda, S., Chang, L., Hirata, H., Karin, M., 2005. Reactive oxygen species promote TNFalpha-induced death and sustained JNK activation by inhibiting MAP kinase phosphatases. Cell 120, 649–661.

Kamp, F., Exner, N., Lutz, A.K., Wender, N., Hegermann, J., Brunner, B., Nuscher, B., Bartels, T., Giese, A., Beyer, K., Eimer, S., Winklhofer, K.F., Haass, C., 2010. Inhibition of mitochondrial fusion by α-synuclein is rescued by PINK1, Parkin and DJ-1. EMBO J. 29, 3571–3589.

Kandimalla, R., Manczak, M., Yin, X., Wang, R., Reddy, P.H., 2018. Hippocampal phosphorylated tau induced cognitive decline, dendritic spine loss and mitochondrial abnormalities in a mouse model of Alzheimer's disease. Hum. Mol. Genet. 27, 30–40.

Kanfer, G., Kornmann, B., 2016. Dynamics of the mitochondrial network during mitosis. Biochem. Soc. Trans. 44, 510–516.

Kang, K.W., Lee, S.J., Kim, S.G., 2005. Molecular mechanism of nrf2 activation by oxidative stress. Antioxid. Redox Signal. 7, 1664–1673.

Kanninen, K., Heikkinen, R., Malm, T., Rolova, T., Kuhmonen, S., Leinonen, H., Yla-Herttuala, S., Tanila, H., Levonen, A.L., Koistinaho, M., Koistinaho, J., 2009. Intrahippocampal injection of a lentiviral vector expressing Nrf2 improves spatial learning in a mouse model of Alzheimer's disease. Proc. Natl. Acad. Sci. U.S.A 106, 16505–16510.

Karcioglu, Z.A., Sarper, R.M., Van Rinsvelt, H.A., Guffey, J.A., Fink, R.W., 1978. Trace element concentrations in renal cell carcinoma. Cancer 42, 1330–1340.

Kasahara, A., Cipolat, S., Chen, Y., Dorn, G.W., Scorrano, L., 2013. Mitochondrial fusion directs cardiomyocyte differentiation via calcineurin and notch signaling. Science 342, 734–737.

Katayama, M., Donai, K., Sakakibara, H., Ohtomo, Y., Miyagawa, M., Kuroda, K., Kodama, H., Suzuki, K., Kasai, N., Nishimori, K., Uchida, T., Watanabe, K., Aso, H., Isogai, E., Sone, H., Fukuda, T., 2014. Coffee consumption delays the hepatitis and suppresses the inflammation related gene expression in the Long-Evans Cinnamon rat. Clin. Nutr. 33, 302–310.

Katz, M.S., Dax, E.M., Gregerman, R.I., 1993. Beta adrenergic regulation of rat liver glycogenolysis during aging. Exp. Gerontol. 28, 329–340.

Kaushik, S., Arias, E., Kwon, H., Lopez, N.M., Athonvarangkul, D., Sahu, S., Schwartz, G.J., Pessin, J.E., Singh, R., 2012. Loss of autophagy in hypothalamic POMC neurons impairs lipolysis. EMBO Rep. 13, 258–265.

Keil, U., Bonert, A., Marques, C.A., Scherping, I., Weyermann, J., Strosznajder, J.B., Muller-Spahn, F., Haass, C., Czech, C., Pradier, L., Muller, W.E., Eckert, A., 2004. Amyloid beta-induced changes in nitric oxide production and mitochondrial activity lead to apoptosis. J. Biol. Chem. 279, 50310–50320.

Kennedy, S.R., Salk, J.J., Schmitt, M.W., Loeb, L.A., 2013. Ultra-sensitive sequencing reveals an age-related increase in somatic mitochondrial mutations that are inconsistent with oxidative damage. PLoS Genet. 9, e1003794.

Kim, A., Joseph, S., Khan, A., Epstein, C.J., Sobel, R., Huang, T.T., 2010. Enhanced expression of mitochondrial superoxide dismutase leads to prolonged in vivo cell cycle progression and up-regulation of mitochondrial thioredoxin. Free Radic. Biol. Med. 48, 1501–1512.

Kim, E.H., Sohn, S., Kwon, H.J., Kim, S.U., Kim, M.J., Lee, S.J., Choi, K.S., 2007a. Sodium selenite induces superoxide-mediated mitochondrial damage and subsequent autophagic cell death in malignant glioma cells. Cancer Res. 67, 6314–6324.

Kim, H.R., Won, S.J., Fabian, C., Kang, M.G., Szardenings, M., Shin, M.G., 2015. Mitochondrial DNA aberrations and pathophysiological implications in hematopoietic diseases, chronic inflammatory diseases, and cancers. Ann. Lab. Med. 35, 1–14.

Kim, I., Rodriguez-Enriquez, S., Lemasters, J.J., 2007b. Selective degradation of mitochondria by mitophagy. Arch. Biochem. Biophys. 462, 245–253.

Kirkland, R.A., Adibhatla, R.M., Hatcher, J.F., Franklin, J.L., 2002. Loss of cardiolipin and mitochondria during programmed neuronal death: evidence of a role for lipid peroxidation and autophagy. Neuroscience 115, 587–602.
Kirkwood, T.B.L., Kowald, A., 2012. The free-radical theory of ageing – older, wiser and still alive. BioEssays 34, 692–700.
Kissova, I., Deffieu, M., Samokhvalov, V., Velours, G., Bessoule, J.J., Manon, S., Camougrand, N., 2006. Lipid oxidation and autophagy in yeast. Free Radic. Biol. Med. 41, 1655–1661.
Kitada, T., Asakawa, S., Matsumine, H., Hattori, N., Minoshima, S., Shimizu, N., Mizuno, Y., 1999. Positional cloning of the autosomal recessive juvenile parkinsonism (AR-JP) gene and its diversity in deletion mutations. Parkinsonism Relat. Disord. 5, 163–168.
Kitamura, Y., Nishikawa, A., Nakamura, H., Furukawa, F., Imazawa, T., Umemura, T., Uchida, K., Hirose, M., 2005. Effects of N-acetylcysteine, quercetin, and phytic acid on spontaneous hepatic and renal lesions in LEC rats. Toxicol Pathol 33, 584–592.
Klionsky, D.J., Abdelmohsen, K., Abe, A., Abedin, M.J., Abeliovich, H., Acevedo Arozena, A., et al., 2016. Guidelines for the use and interpretation of assays for monitoring autophagy (3rd edition). Autophagy 12, 1–222.
Knight, E.M., Martins, I.V.A., Gümüsgöz, S., Allan, S.M., Lawrence, C.B., 2014. High-fat diet-induced memory impairment in triple-transgenic Alzheimer's disease (3xTgAD) mice is independent of changes in amyloid and tau pathology. Neurobiol. Aging 35, 1821–1832.
Ko, J.H., Sethi, G., Um, J.Y., Shanmugam, M.K., Arfuso, F., Kumar, A.P., Bishayee, A., Ahn, K.S., 2017. The role of resveratrol in cancer therapy. Int. J. Mol. Sci. 18.
Koch, K.A., Pena, M.M., Thiele, D.J., 1997. Copper-binding motifs in catalysis, transport, detoxification and signaling. Chem. Biol. 4, 549–560.
Koehler, E.M., Schouten, J.N., Hansen, B.E., van Rooij, F.J., Hofman, A., Stricker, B.H., Janssen, H.L., 2012. Prevalence and risk factors of non-alcoholic fatty liver disease in the elderly: results from the Rotterdam study. J. Hepatol. 57, 1305–1311.
Kokoszka, J.E., Waymire, K.G., Levy, S.E., Sligh, J.E., Cai, J., Jones, D.P., MacGregor, G.R., Wallace, D.C., 2004. The ADP/ATP translocator is not essential for the mitochondrial permeability transition pore. Nature 427, 461–465.
Koliaki, C., Szendroedi, J., Kaul, K., Jelenik, T., Nowotny, P., Jankowiak, F., Herder, C., Carstensen, M., Krausch, M., Knoefel, W.T., Schlensak, M., Roden, M., 2015. Adaptation of hepatic mitochondrial function in humans with non-alcoholic fatty liver is lost in steatohepatitis. Cell Metab. 21, 739–746.
Korbecki, J., Baranowska-Bosiacka, I., Gutowska, I., Chlubek, D., 2013. The effect of reactive oxygen species on the synthesis of prostanoids from arachidonic acid. J. Physiol. Pharmacol. 64, 409–421.
Korobova, F., Gauvin, T.J., Higgs, H.N., 2014. A role for myosin II in mammalian mitochondrial fission. Curr. Biol. 24, 409–414.
Korobova, F., Ramabhadran, V., Higgs, H.N., 2013. An actin-dependent step in mitochondrial fission mediated by the ER-associated Formin INF2. Science 339. https://doi.org/10.1126/science.1228360.
Korshunov, S.S., Korkina, O.V., Ruuge, E.K., Skulachev, V.P., Starkov, A.A., 1998. Fatty acids as natural uncouplers preventing generation of $O_2^{\cdot-}$ and H_2O_2 by mitochondria in the resting state. FEBS Lett. 435, 215–218.
Korshunov, S.S., Skulachev, V.P., Starkov, A.A., 1997. High protonic potential actuates a mechanism of production of reactive oxygen species in mitochondria. FEBS Lett. 416, 15–18.
Kowaltowski, A.J., Castilho, R.F., Vercesi, A.E., 2001. Mitochondrial permeability transition and oxidative stress. FEBS Lett. 495, 12–15.

Kraus, F., Ryan, M.T., 2017. The constriction and scission machineries involved in mitochondrial fission. J. Cell Sci. 130, 2953–2960.

Krauss, S., Zhang, C.Y., Scorrano, L., Dalgaard, L.T., St-Pierre, J., Grey, S.T., Lowell, B.B., 2003. Superoxide-mediated activation of uncoupling protein 2 causes pancreatic beta cell dysfunction. J Clin Invest 112, 1831–1842.

Krawczyk, M., Bonfrate, L., Portincasa, P., 2010. Nonalcoholic fatty liver disease. Best practice & research. Clin. Gastroenterol. 24, 695–708.

Krawczyk, M., Portincasa, P., Lammert, F., 2013. PNPLA3-associated steatohepatitis: toward a gene-based classification of fatty liver disease. Semin. Liver Dis. 33, 369–379.

Kraytsberg, Y., Kudryavtseva, E., McKee, A.C., Geula, C., Kowall, N.W., Khrapko, K., 2006. Mitochondrial DNA deletions are abundant and cause functional impairment in aged human substantia nigra neurons. Nat. Genet. 38, 518–520.

Kroemer, G., Galluzzi, L., Brenner, C., 2007. Mitochondrial membrane permeabilization in cell death. Physiol. Rev. 87, 99–163.

Kroemer, G., Marino, G., Levine, B., 2010. Autophagy and the integrated stress response. Mol Cell 40, 280–293.

Kuchay, S., Giorgi, C., Simoneschi, D., Pagan, J., Missiroli, S., Saraf, A., Florens, L., Washburn, M.P., Collazo-Lorduy, A., Castillo-Martin, M., Cordon-Cardo, C., Sebti, S.M., Pinton, P., Pagano, M., 2017. PTEN counteracts FBXL2 to promote IP3R3- and Ca(2+)-mediated apoptosis limiting tumour growth. Nature 546, 554–558.

Kujoth, G.C., Hiona, A., Pugh, T.D., Someya, S., Panzer, K., Wohlgemuth, S.E., Hofer, T., Seo, A.Y., Sullivan, R., Jobling, W.A., Morrow, J.D., Van Remmen, H., Sedivy, J.M., Yamasoba, T., Tanokura, M., Weindruch, R., Leeuwenburgh, C., Prolla, T.A., 2005. Mitochondrial DNA mutations, oxidative stress, and apoptosis in mammalian aging. Science 309, 481–484.

Kukat, C., Wurm, C.A., Spahr, H., Falkenberg, M., Larsson, N.G., Jakobs, S., 2011. Super-resolution microscopy reveals that mammalian mitochondrial nucleoids have a uniform size and frequently contain a single copy of mtDNA. Proc. Natl. Acad. Sci. U.S.A 108, 13534–13539.

Kumar, V., Kalita, J., Bora, H.K., Misra, U.K., 2016. Relationship of antioxidant and oxidative stress markers in different organs following copper toxicity in a rat model. Toxicol. Appl. Pharmacol. 293, 37–43.

Kuroda, J., Ago, T., Matsushima, S., Zhai, P., Schneider, M.D., Sadoshima, J., 2010. NADPH oxidase 4 (Nox4) is a major source of oxidative stress in the failing heart. Proc. Natl. Acad. Sci. U.S.A 107, 15565–15570.

Kushnareva, Y., Murphy, A.N., Andreyev, A., 2002. Complex I-mediated reactive oxygen species generation: modulation by cytochrome c and NAD(P)+ oxidation-reduction state. Biochem. J. 368, 545–553.

Kwon, S., Seok, S., Yau, P., Li, X., Kemper, B., Kemper, J.K., 2017. Obesity and aging diminish sirtuin 1 (SIRT1)-mediated deacetylation of SIRT3, leading to hyperacetylation and decreased activity and stability of SIRT3. J. Biol. Chem. 292, 17312–17323.

Laganiere, S., Byung, P.Y., 1993. Modulation of membrane phospholipid fatty acid composition by age and food restriction. Gerontology 39, 7–18.

Lagouge, M., Argmann, C., Gerhart-Hines, Z., Meziane, H., Lerin, C., Daussin, F., Messadeq, N., Milne, J., Lambert, P., Elliott, P., Geny, B., Laakso, M., Puigserver, P., Auwerx, J., 2006. Resveratrol improves mitochondrial function and protects against metabolic disease by activating SIRT1 and PGC-1alpha. Cell 127, 1109–1122.

Lagouge, M., Larsson, N.G., 2013. The role of mitochondrial DNA mutations and free radicals in disease and ageing. J. Internal Med. 273, 529–543.

Lagranha, C.J., Deschamps, A., Aponte, A., Steenbergen, C., Murphy, E., 2010. Sex differences in the phosphorylation of mitochondrial proteins result in reduced production of reactive oxygen species and cardioprotection in females. Circ. Res. 106, 1681–1691.

Lambert, A.J., Boysen, H.M., Buckingham, J.A., Yang, T., Podlutsky, A., Austad, S.N., Kunz, T.H., Buffenstein, R., Brand, M.D., 2007. Low rates of hydrogen peroxide production by isolated heart mitochondria associate with long maximum lifespan in vertebrate homeotherms. Aging Cell 6, 607–618.

Lambertucci, R.H., Hirabara, S.M., Silveira Ldos, R., Levada-Pires, A.C., Curi, R., Pithon-Curi, T.C., 2008. Palmitate increases superoxide production through mitochondrial electron transport chain and NADPH oxidase activity in skeletal muscle cells. J. Cell. Physiol. 216, 796–804.

Lange, K.W., Li, S., 2017. Resveratrol, pterostilbene, and dementia. Biofactors.

Langston, J.W., Ballard, P., Tetrud, J.W., Irwin, I., 1983. Chronic Parkinsonism in humans due to a product of meperidine-analog synthesis. Science 219, 979–980.

Lanza, I.R., Short, D.K., Short, K.R., Raghavakaimal, S., Basu, R., Joyner, M.J., McConnell, J.P., Nair, K.S., 2008. Endurance exercise as a countermeasure for aging. Diabetes 57, 2933–2942.

Lapasset, L., Milhavet, O., Prieur, A., Besnard, E., Babled, A., Aït-Hamou, N., Leschik, J., Pellestor, F., Ramirez, J.-M., De Vos, J., Lehmann, S., Lemaitre, J.-M., 2011. Rejuvenating senescent and centenarian human cells by reprogramming through the pluripotent state. Genes Dev. 25, 2248–2253.

Lapointe, J., Hekimi, S., 2008. Early mitochondrial dysfunction in long-lived Mclk1+/- mice. J. Biol. Chem. 283, 26217–26227.

Larsen, P.L., Clarke, C.F., 2002. Extension of life-span in *Caenorhabditis elegans* by a diet lacking coenzyme Q. Science 295, 120–123.

Laughlin, M.H., Simpson, T., Sexton, W.L., Brown, O.R., Smith, J.K., Korthuis, R.J., 1990. Skeletal muscle oxidative capacity, antioxidant enzymes, and exercise training. J. Appl. Physiol. 68, 2337–2343.

Lauri, A., Pompilio, G., Capogrossi, M.C., 2014. The mitochondrial genome in aging and senescence. Ageing Res. Rev. 18, 1–15.

Lavrovsky, Y., Song, C.S., Chatterjee, B., Roy, A.K., 2000. Age-dependent increase of heme oxygenase–1 gene expression in the liver mediated by NFκB. Mech. Ageing Dev. 114, 49–60.

Leduc-Gaudet, J.-P., Picard, M., Pelletier, F.S.-J., Sgarioto, N., Auger, M.-J., Vallée, J., Robitaille, R., St-Pierre, D.H., Gouspillou, G., 2015. Mitochondrial morphology is altered in atrophied skeletal muscle of aged mice. Oncotarget 6, 17923–17937.

Lee, H.C., Wei, Y.H., 2012. Mitochondria and aging. In: Scatena, R., Bottoni, P., Giardina, B. (Eds.), Advances in Mitochondrial Medicine. Springer Netherlands, Dordrecht, pp. 311–327.

Lee, H.C., Wei, Y.H., 1997. Mutation and oxidative damage of mitochondrial DNA and defective turnover of mitochondria in human aging. J. Formos. Med. Assoc. 96, 770–778.

Lee, I.H., Cao, L., Mostoslavsky, R., Lombard, D.B., Liu, J., Bruns, N.E., Tsokos, M., Alt, F.W., Finkel, T., 2008. A role for the NAD-dependent deacetylase Sirt1 in the regulation of autophagy. Proc. Natl. Acad. Sci. U.S.A 105, 3374–3379.

Lee, J., Kim, Y., Liu, T., Hwang, Y.J., Hyeon, S.J., Im, H., Lee, K., Alvarez, V.E., McKee, A.C., Um, S.J., Hur, M., Mook-Jung, I., Kowall, N.W., Ryu, H., 2017. SIRT3 deregulation is linked to mitochondrial dysfunction in Alzheimer's disease. Aging Cell.

Lee, J.E., Westrate, L.M., Wu, H., Page, C., Voeltz, G.K., 2016a. Multiple Dynamin family members collaborate to drive mitochondrial division. Nature 540, 139–143.

Lee, W.-H., Higuchi, H., Ikeda, S., Macke, E.L., Takimoto, T., Pattnaik, B.R., Liu, C., Chu, L.-F., Siepka, S.M., Krentz, K.J., Rubinstein, C.D., Kalejta, R.F., Thomson, J.A., Mullins, R.F., Takahashi, J.S., Pinto, L.H., Ikeda, A., 2016b. Mouse Tmem135 mutation reveals a mechanism involving mitochondrial dynamics that leads to age-dependent retinal pathologies. eLife 5, e19264.

Lenaz, G., 2001. The mitochondrial production of reactive oxygen species: mechanisms and implications in human pathology. IUBMB Life 52, 159–164.
Leonard, S.S., Harris, G.K., Shi, X., 2004. Metal-induced oxidative stress and signal transduction. Free Radic. Biol. Med. 37, 1921–1942.
Lepetsos, P., Papavassiliou, A.G., 2016. ROS/oxidative stress signaling in osteoarthritis. Biochim. Biophys. Acta 1862, 576–591.
Leuner, K., Hauptmann, S., Abdel-Kader, R., Scherping, I., Keil, U., Strosznajder, J.B., Eckert, A., Muller, W.E., 2007. Mitochondrial dysfunction: the first domino in brain aging and Alzheimer's disease? Antioxid. Redox Signal. 9, 1659–1675.
Leung, A.W., Varanyuwatana, P., Halestrap, A.P., 2008. The mitochondrial phosphate carrier interacts with cyclophilin D and may play a key role in the permeability transition. J. Biol. Chem. 283, 26312–26323.
Levine, B., Kroemer, G., 2008. Autophagy in the pathogenesis of disease. Cell 132, 27–42.
Li, H.F., Wang, X.A., Xiang, S.S., Hu, Y.P., Jiang, L., Shu, Y.J., Li, M.L., Wu, X.S., Zhang, F., Ye, Y.Y., 2015a. Oleanolic acid induces mitochondrial-dependent apoptosis and G0/G1 phase arrest in gallbladder cancer cells. Drug Des. Devel. Ther. 9, 3017.
Li, J.M., Shah, A.M., 2003. ROS generation by nonphagocytic NADPH oxidase: potential relevance in diabetic nephropathy. J. Am. Soc. Nephrol. 14. S221–226.
Li, S., Xu, S., Roelofs, B.A., Boyman, L., Lederer, W.J., Sesaki, H., Karbowski, M., 2015b. Transient assembly of F-actin on the outer mitochondrial membrane contributes to mitochondrial fission. J. Cell Biol. 208, 109–123.
Li, Y., Huang, T.T., Carlson, E.J., Melov, S., Ursell, P.C., Olson, J.L., Noble, L.J., Yoshimura, M.P., Berger, C., Chan, P.H., Wallace, D.C., Epstein, C.J., 1995. Dilated cardiomyopathy and neonatal lethality in mutant mice lacking manganese superoxide dismutase. Nat. Genet. 11, 376–381.
Li, Y.R., Li, S., Lin, C.C., 2017. Effect of resveratrol and pterostilbene on aging and longevity. Biofactors.
Liang, H., Masoro, E.J., Nelson, J.F., Strong, R., McMahan, C.A., Richardson, A., 2003. Genetic mouse models of extended lifespan. Exp. Gerontol. 38, 1353–1364.
Lichtmannegger, J., Leitzinger, C., Wimmer, R., Schmitt, S., Schulz, S., Kabiri, Y., Eberhagen, C., Rieder, T., Janik, D., Neff, F., Straub, B.K., Schirmacher, P., DiSpirito, A.A., Bandow, N., Baral, B.S., Flatley, A., Kremmer, E., Denk, G., Reiter, F.P., Hohenester, S., Eckardt-Schupp, F., Dencher, N.A., Adamski, J., Sauer, V., Niemietz, C., Schmidt, H.H., Merle, U., Gotthardt, D.N., Kroemer, G., Weiss, K.H., Zischka, H., 2016. Methanobactin reverses acute liver failure in a rat model of Wilson disease. J. Clin. Invest. 126, 2721–2735.
Lim, K.S., Jeyaseelan, K., Whiteman, M., Jenner, A., Halliwell, B., 2005. Oxidative damage in mitochondrial DNA is not extensive. Ann. N. Y. Acad. Sci. 1042, 210–220.
Lin, M.T., Beal, M.F., 2006. Mitochondrial dysfunction and oxidative stress in neurodegenerative diseases. Nature 443, 787–795.
Lin, M.T., Cantuti-Castelvetri, I., Zheng, K., Jackson, K.E., Tan, Y.B., Arzberger, T., Lees, A.J., Betensky, R.A., Beal, M.F., Simon, D.K., 2012. Somatic mitochondrial DNA mutations in early Parkinson and incidental Lewy body disease. Ann. Neurol. 71, 850–854.
Lin, M.T., Simon, D.K., Ahn, C.H., Kim, L.M., Beal, M.F., 2002. High aggregate burden of somatic mtDNA point mutations in aging and Alzheimer's disease brain. Hum. Mol. Genet. 11, 133–145.
Linder, M.C., 2016. Ceruloplasmin and other copper binding components of blood plasma and their functions: an update. Metallomics 8, 887–905.
Linke, A., Adams, V., Schulze, P.C., Erbs, S., Gielen, S., Fiehn, E., Mobius-Winkler, S., Schubert, A., Schuler, G., Hambrecht, R., 2005. Antioxidative effects of exercise training in patients with chronic heart failure: increase in radical scavenger enzyme activity in skeletal muscle. Circulation 111, 1763–1770.

Linnane, A.W., Marzuki, S., Ozawa, T., Tanaka, M., 1989. Mitochondrial DNA mutations as an important contributor to ageing and degenerative diseases. Lancet 1, 642–645.
Lippard, S.J., 1999. Free copper ions in the cell? Science 284, 748–749.
Liu, C.S., Cheng, W.L., Kuo, S.J., Li, J.Y., Soong, B.W., Wei, Y.H., 2008. Depletion of mitochondrial DNA in leukocytes of patients with poly-Q diseases. J. Neurol. Sci. 264, 18–21.
Liu, T.Y., Tan, Z.J., Jiang, L., Gu, J.F., Wu, X.S., Cao, Y., Li, M.L., Wu, K.J., Liu, Y.B., 2013. Curcumin induces apoptosis in gallbladder carcinoma cell line GBC-SD cells. Cancer Cell. Int. 13, 64.
Liu, X., Zhang, L., Wang, P., Li, X., Qiu, D., Li, L., Zhang, J., Hou, X., Han, L., Ge, J., Li, M., Gu, L., Wang, Q., 2017. Sirt3-dependent deacetylation of SOD2 plays a protective role against oxidative stress in oocytes from diabetic mice. Cell Cycle 16, 1302–1308.
Liu, Y., Fiskum, G., Schubert, D., 2002. Generation of reactive oxygen species by the mitochondrial electron transport chain. J. Neurochem. 80, 780–787.
Liu, Y., Liu, J., Tetzlaff, W., Paty, D.W., Cynader, M.S., 2006. Biliverdin reductase, a major physiologic cytoprotectant, suppresses experimental autoimmune encephalomyelitis. Free Radic. Biol. Med. 40, 960–967.
Liver., E. A. f. S. o, 2012. EASL Clinical Practice Guidelines: wilson's disease. J. Hepatol. 56, 671–685.
Lloret, A., Badia, M.C., Mora, N.J., Pallardo, F.V., Alonso, M.D., Vina, J., 2009. Vitamin E paradox in Alzheimer's disease: it does not prevent loss of cognition and may even be detrimental. J. Alzheimers Dis. 17, 143–149.
Lonnrot, K., Holm, P., Lagerstedt, A., Huhtala, H., Alho, H., 1998. The effects of lifelong ubiquinone Q10 supplementation on the Q9 and Q10 tissue concentrations and life span of male rats and mice. Biochem. Mol. Biol. Int. 44, 727–737.
Lopez-Lluch, G., Rodriguez-Aguilera, J.C., Santos-Ocana, C., Navas, P., 2010. Is coenzyme Q a key factor in aging? Mech. Ageing Dev. 131, 225–235.
Lopez-Torres, M., Gredilla, R., Sanz, A., Barja, G., 2002. Influence of aging and long-term caloric restriction on oxygen radical generation and oxidative DNA damage in rat liver mitochondria. Free Radic. Biol. Med. 32, 882–889.
Losón, O.C., Song, Z., Chen, H., Chan, D.C., 2013. Fis1, Mff, MiD49, and MiD51 mediate Drp1 recruitment in mitochondrial fission. Mol. Biol. Cell 24, 659–667.
Lu, J.M., Lin, P.H., Yao, Q., Chen, C., 2010. Chemical and molecular mechanisms of antioxidants: experimental approaches and model systems. J. Cell. Mol. Med. 14, 840–860.
Lu, Z., Luo, R.Z., Lu, Y., Zhang, X., Yu, Q., Khare, S., Kondo, S., Kondo, Y., Yu, Y., Mills, G.B., Liao, W.S., Bast Jr., R.C., 2008. The tumor suppressor gene ARHI regulates autophagy and tumor dormancy in human ovarian cancer cells. J. Clin. Invest. 118, 3917–3929.
Lu, Z., Sack, M.N., 2008. ATF-1 is a hypoxia-responsive transcriptional activator of skeletal muscle mitochondrial-uncoupling protein 3. J. Biol. Chem. 283, 23410–23418.
Lustbader, J.W., Cirilli, M., Lin, C., Xu, H.W., Takuma, K., Wang, N., Caspersen, C., Chen, X., Pollak, S., Chaney, M., Trinchese, F., Liu, S., Gunn-Moore, F., Lue, L.F., Walker, D.G., Kuppusamy, P., Zewier, Z.L., Arancio, O., Stern, D., Yan, S.S., Wu, H., 2004. ABAD directly links Abeta to mitochondrial toxicity in Alzheimer's disease. Science 304, 448–452.
McLean, A.J., Cogger, V.C., Chong, G.C., Warren, A., Markus, A.M., Dahlstrom, J.E., Le Couteur, D.G., 2003. Age-related pseudocapillarization of the human liver. J. Pathol. 200, 112–117.
McLean, A.J., Le Couteur, D.G., 2004. Aging biology and geriatric clinical pharmacology. Pharmacol. Rev. 56, 163–184.

Madamanchi, N.R., Runge, M.S., 2007. Mitochondrial dysfunction in atherosclerosis. Circ Res. 100, 460–473.

Maesako, M., Uemura, M., Tashiro, Y., Sasaki, K., Watanabe, K., Noda, Y., Ueda, K., Asada-Utsugi, M., Kubota, M., Okawa, K., Ihara, M., Shimohama, S., Uemura, K., Kinoshita, A., 2015. High fat diet enhances β-site cleavage of amyloid precursor protein (APP) via promoting β-site APP cleaving enzyme 1/adaptor protein 2/clathrin complex formation. PLoS One 10, e0131199.

Magwere, T., West, M., Riyahi, K., Murphy, M.P., Smith, R.A., Partridge, L., 2006. The effects of exogenous antioxidants on lifespan and oxidative stress resistance in *Drosophila melanogaster*. Mech. Ageing Dev. 127, 356–370.

Mai, S., Klinkenberg, M., Auburger, G., Bereiter-Hahn, J., Jendrach, M., 2010. Decreased expression of Drp1 and Fis1 mediates mitochondrial elongation in senescent cells and enhances resistance to oxidative stress through PINK1. J. Cell Sci. 123, 917–926.

Mailloux, R.J., Harper, M.E., 2011. Uncoupling proteins and the control of mitochondrial reactive oxygen species production. Free Radic. Biol. Med. 51, 1106–1115.

Malouitre, S., Dube, H., Selwood, D., Crompton, M., 2010. Mitochondrial targeting of cyclosporin A enables selective inhibition of cyclophilin-D and enhanced cytoprotection after glucose and oxygen deprivation. Biochem. J. 425, 137–148.

Manczak, M., Calkins, M.J., Reddy, P.H., 2011. Impaired mitochondrial dynamics and abnormal interaction of amyloid beta with mitochondrial protein Drp1 in neurons from patients with Alzheimer's disease: implications for neuronal damage. Hum. Mol. Genet. 20, 2495–2509.

Manczak, M., Kandimalla, R., Fry, D., Sesaki, H., Reddy, P.H., 2016. Protective effects of reduced dynamin-related protein 1 against amyloid beta-induced mitochondrial dysfunction and synaptic damage in Alzheimer's disease. Hum. Mol. Genet. 25, 5148–5166.

Manczak, M., Reddy, P.H., 2012. Abnormal interaction between the mitochondrial fission protein Drp1 and hyperphosphorylated tau in Alzheimer's disease neurons: implications for mitochondrial dysfunction and neuronal damage. Hum. Mol. Genet. 21, 2538–2547.

Manor, U., Bartholomew, S., Golani, G., Christenson, E., Kozlov, M., Higgs, H., Spudich, J., Lippincott-Schwartz, J., 2015. A mitochondria-anchored isoform of the actin-nucleating spire protein regulates mitochondrial division. eLife 4, e08828.

Mansouri, A., Gaou, I., Fromenty, B., Berson, A., Letteron, P., Degott, C., Erlinger, S., Pessayre, D., 1997. Premature oxidative aging of hepatic mitochondrial DNA in Wilson's disease. Gastroenterology 113, 599–605.

Marchi, S., Patergnani, S., Missiroli, S., Morciano, G., Rimessi, A., Wieckowski, M.R., Giorgi, C., Pinton, P., 2018. Mitochondrial and endoplasmic reticulum calcium homeostasis and cell death. Cell Calcium. 69, 62–72.

Marinho, H.S., Real, C., Cyrne, L., Soares, H., Antunes, F., 2014. Hydrogen peroxide sensing, signaling and regulation of transcription factors. Redox Biol. 2, 535–562.

Martín-Maestro, P., Gargini, R., García, E., Perry, G., Avila, J., García-Escudero, V., 2017. Slower dynamics and aged mitochondria in Sporadic Alzheimer's disease. Oxid. Med. Cell Longev. 2017, 14.

Martinet, W., Agostinis, P., Vanhoecke, B., Dewaele, M., De Meyer, G.R., 2009. Autophagy in disease: a double-edged sword with therapeutic potential. Clin. Sci. (Lond.) 116, 697–712.

Martinon, F., Mayor, A., Tschopp, J., 2009. The inflammasomes: guardians of the body. Annu. Rev. Immunol. 27, 229–265.

Massey, A.C., Kaushik, S., Sovak, G., Kiffin, R., Cuervo, A.M., 2006. Consequences of the selective blockage of chaperone-mediated autophagy. Proc. Natl. Acad. Sci. U.S.A 103, 5805–5810.

Mateo, M.C.M., Bustamante, J.B., Arellano, I.F., 1979. Serum copper, ceruloplasmin, lactic-desydrogenase and alpha-2-globulin in lung-cancer. Biomed. Express 31, 66–68.
Mattson, M.P., 2000. Apoptosis in neurodegenerative disorders. Nat. Rev. Mol. Cell Biol. 1, 120–129.
McCommis, K., Hodges, W., Brunt, E., Nalbantoglu, I., McDonald, W., Holley, C., Fujiwara, H., Schaffer, J., Colca, J., Finck, B., 2016. Targeting the mitochondrial pyruvate carrier attenuates fibrosis in a mouse model of nonalcoholic steatohepatitis. Hepatology.
McCommis, K.S., Chen, Z., Fu, X., McDonald, W.G., Colca, J.R., Kletzien, R.F., Burgess, S.C., Finck, B.N., 2015. Loss of mitochondrial pyruvate carrier 2 in the liver leads to defects in gluconeogenesis and compensation via pyruvate-Alanine cycling. Cell Metab. 22, 682–694.
McCord, J.M., Edeas, M.A., 2005. SOD, oxidative stress and human pathologies: a brief history and a future vision. Biomed. Pharmacother 59, 139–142.
McCullough, A., Raguso, C., 1999. Effect of cirrhosis on energy expenditure. Am. J. Clinical Nutr. 69, 1066–1068.
McNaught, K.S., Perl, D.P., Brownell, A.L., Olanow, C.W., 2004. Systemic exposure to proteasome inhibitors causes a progressive model of Parkinson's disease. Ann. Neurol. 56, 149–162.
Meigs, J.B., Muller, D.C., Nathan, D.M., Blake, D.R., Andres, R., 2003. The Natural history of progression from normal glucose tolerance to type 2 diabetes in the Baltimore longitudinal study of aging. Diabetes 52, 1475–1484.
Melkov, A., Abdu, U., 2018. Regulation of long-distance transport of mitochondria along microtubules. Cell. Mol. Life Sci. 75, 163–176.
Merksamer, P.I., Liu, Y., He, W., Hirschey, M.D., Chen, D., Verdin, E., 2013. The sirtuins, oxidative stress and aging: an emerging link. Aging (Albany NY) 5, 144–150.
Merriman, R.B., Aouizerat, B.E., Bass, N.M., 2006. Genetic influences in nonalcoholic fatty liver disease. J. Clin. Gastroenterol. 40 (Suppl. 1). S30–33.
Migliaccio, E., Giorgio, M., Mele, S., Pelicci, G., Reboldi, P., Pandolfi, P.P., Lanfrancone, L., Pelicci, P.G., 1999. The p66shc adaptor protein controls oxidative stress response and life span in mammals. Nature 402, 309–313.
Migliaccio, E., Giorgio, M., Pelicci, P.G., 2006. Apoptosis and aging: role of p66Shc redox protein. Antioxid. Redox Signal. 8, 600–608.
Miller, N., 1984. Why does plasma low density lipoprotein concentration in adults increase with age? Lancet 323, 263–267.
Miquel, J., Economos, A.C., Fleming, J., Johnson .r, J.E., 1980. Mitochondrial role in cell aging. Exp. Gerontol. 15, 575–591.
Miquel, J., Ferrandiz, M.L., De Juan, E., Sevila, I., Martinez, M., 1995. N-acetylcysteine protects against age-related decline of oxidative phosphorylation in liver mitochondria. Eur. J. Pharmacol. 292, 333–335.
Misgeld, T., Schwarz, T., 2017. Mitostasis in neurons: Maintaining mitochondria in an extended cellular architecture. Neuron 96, 651–66.
Mishra, P., Chan, D.C., 2014. Mitochondrial dynamics and inheritance during cell division, development and disease. Nature Rev. Mol. Cell Biol. 15, 634–646.
Misra, M.K., Sarwat, M., Bhakuni, P., Tuteja, R., Tuteja, N., 2009. Oxidative stress and ischemic myocardial syndromes. Med. Sci. Monit. 15. RA209–219.
Miuma, S., Saldivar, J.C., Karras, J.R., Waters, C.E., Paisie, C.A., Wang, Y., Jin, V., Sun, J., Druck, T., Zhang, J., Huebner, K., 2013. Fhit deficiency-induced global genome instability promotes mutation and clonal expansion. PLoS One 8, e80730.
Monsenego, J., Mansouri, A., Akkaoui, M., Lenoir, V., Esnous, C., Fauveau, V., Tavernier, V., Girard, J., Prip-Buus, C., 2012. Enhancing liver mitochondrial fatty acid oxidation capacity in obese mice improves insulin sensitivity independently of hepatic steatosis. J. Hepatol. 56, 632–639.

Monteiro-Cardoso, V.F., Oliveira, M.M., Melo, T., Domingues, M.R., Moreira, P.I., Ferreiro, E., Peixoto, F., Videira, R.A., 2015. Cardiolipin profile changes are associated to the early synaptic mitochondrial dysfunction in Alzheimer's disease. J. Alzheimers Dis. 43, 1375–1392.

Montero, R., Pineda, M., Aracil, A., Vilaseca, M.A., Briones, P., Sanchez-Alcazar, J.A., Navas, P., Artuch, R., 2007. Clinical, biochemical and molecular aspects of cerebellar ataxia and Coenzyme Q10 deficiency. Cerebellum 6, 118–122.

Moore, A.S., Wong, Y.C., Simpson, C.L., Holzbaur, E.L., 2016. Dynamic actin cycling through mitochondrial subpopulations locally regulates the fission–fusion balance within mitochondrial networks. Nat. Commun. 7, 12886.

Moosmann, B., Behl, C., 2008. Mitochondrially encoded cysteine predicts animal lifespan. Aging Cell 7, 32–46.

Mootha, V.K., Lindgren, C.M., Eriksson, K.-F., Subramanian, A., Sihag, S., Lehar, J., Puigserver, P., Carlsson, E., Ridderstråle, M., Laurila, E., Houstis, N., Daly, M.J., Patterson, N., Mesirov, J.P., Golub, T.R., Tamayo, P., Spiegelman, B., Lander, E.S., Hirschhorn, J.N., Altshuler, D., Groop, L.C., 2003. PGC-1α-responsive genes involved in oxidative phosphorylation are coordinately downregulated in human diabetes. Nat. Genetics 34, 267.

Morciano, G., Bonora, M., Giorgi, C., Pinton, P., 2017. Other bricks for the correct construction of the mitochondrial permeability transition pore complex. Cell Death Dis. 8, e2698.

Morciano, G., Giorgi, C., Bonora, M., Punzetti, S., Pavasini, R., Wieckowski, M.R., Campo, G., Pinton, P., 2015. Molecular identity of the mitochondrial permeability transition pore and its role in ischemia-reperfusion injury. J. Mol Cell. Cardiol. 78, 142–153.

Morita, M., Prudent, J., Basu, K., Goyon, V., Katsumura, S., Hulea, L., Pearl, D., Siddiqui, N., Strack, S., McGuirk, S., St-Pierre, J., Larsson, O., Topisirovic, I., Vali, H., McBride, H.M., Bergeron, J.J., Sonenberg, N., 2017. mTOR controls mitochondrial dynamics and cell survival via MTFP1. Molecular Cell 67, 922–935.e925.

Morrison, C.D., Pistell, P.J., Ingram, D.K., Johnson, W.D., Liu, Y., Fernandez-Kim, S.O., White, C.L., Purpera, M.N., Uranga, R.M., Bruce-Keller, A.J., Keller, J.N., 2010. High fat diet increases hippocampal oxidative stress and cognitive impairment in aged mice: implications for decreased Nrf2 signaling. J. Neurochem. 114, 1581–1589.

Morsci, N.S., Hall, D.H., Driscoll, M., Sheng, Z.H., 2016. Age-Related phasic patterns of mitochondrial maintenance in adult *Caenorhabditis elegans* neurons. J. Neurosci. 36, 1373–1385.

Morse, D., Choi, A.M., 2005. Heme oxygenase-1: from bench to bedside. Am. J. Respir. Crit. Care Med. 172, 660–670.

Morselli, E., Maiuri, M.C., Markaki, M., Megalou, E., Pasparaki, A., Palikaras, K., Criollo, A., Galluzzi, L., Malik, S.A., Vitale, I., Michaud, M., Madeo, F., Tavernarakis, N., Kroemer, G., 2010. Caloric restriction and resveratrol promote longevity through the Sirtuin-1-dependent induction of autophagy. Cell Death Dis. 1, e10.

Muhammad, M.H., Allam, M.M., 2017. Resveratrol and/or exercise training counteract aging-associated decline of physical endurance in aged mice; targeting mitochondrial biogenesis and function. J. Physiol. Sci.

Murai, M., Ishihara, A., Nishioka, T., Yagi, T., Miyoshi, H., 2007. The ND1 subunit constructs the inhibitor binding domain in bovine heart mitochondrial complex I. Biochemistry 46, 6409–6416.

Murakami, K., Murata, N., Noda, Y., Tahara, S., Kaneko, T., Kinoshita, N., Hatsuta, H., Murayama, S., Barnham, K.J., Irie, K., Shirasawa, T., Shimizu, T., 2011. SOD1 (copper/zinc superoxide dismutase) deficiency drives amyloid beta protein oligomerization and memory loss in mouse model of Alzheimer disease. J. Biol. Chem. 286, 44557–44568.

Murphy, M.P., Smith, R.A., 2007. Targeting antioxidants to mitochondria by conjugation to lipophilic cations. Annu. Rev. Pharmacol. Toxicol. 47, 629–656.

Murry, C.E., Jennings, R.B., Reimer, K.A., 1986. Preconditioning with ischemia: a delay of lethal cell injury in ischemic myocardium. Circulation 74, 1124–1136.

Muscari, C., Giaccari, A., Stefanelli, C., Viticchi, C., Giordano, E., Guarnieri, C., Caldarera, C.M., 1996. Presence of a DNA-4236 bp deletion and 8-hydroxy-deoxyguanosine in mouse cardiac mitochondrial DNA during aging. Aging (Milano) 8, 429–433.

Nagasaka, H., Inoue, I., Inui, A., Komatsu, H., Sogo, T., Murayama, K., Murakami, T., Yorifuji, T., Asayama, K., Katayama, S., Uemoto, S., Kobayashi, K., Takayanagi, M., Fujisawa, T., Tsukahara, H., 2006. Relationship between oxidative stress and antioxidant systems in the liver of patients with Wilson disease: hepatic manifestation in Wilson disease as a consequence of augmented oxidative stress. Pediatr. Res. 60, 472–477.

Nagesh Babu, G., Kumar, A., Singh, R.L., 2011. Chronic pretreatment with acetyl-L-carnitine and +/-DL-alpha-lipoic acid protects against acute glutamate-induced neurotoxicity in rat brain by altering mitochondrial function. Neurotox Res. 19, 319–329.

Nagy, N., Malik, G., Tosaki, A., Ho, Y.S., Maulik, N., Das, D.K., 2008. Overexpression of glutaredoxin-2 reduces myocardial cell death by preventing both apoptosis and necrosis. J. Mol Cell Cardiol. 44, 252–260.

Nair, J., Sone, H., Nagao, M., Barbin, A., Bartsch, H., 1996. Copper-dependent formation of miscoding etheno-DNA adducts in the liver of Long Evans cinnamon (LEC) rats developing hereditary hepatitis and hepatocellular carcinoma. Cancer Res. 56, 1267–1271.

Nakagawa, T., Shimizu, S., Watanabe, T., Yamaguchi, O., Otsu, K., Yamagata, H., Inohara, H., Kubo, T., Tsujimoto, Y., 2005. Cyclophilin D-dependent mitochondrial permeability transition regulates some necrotic but not apoptotic cell death. Nature 434, 652–658.

Nakahira, K., Haspel, J.A., Rathinam, V.A., Lee, S.J., Dolinay, T., Lam, H.C., Englert, J.A., Rabinovitch, M., Cernadas, M., Kim, H.P., Fitzgerald, K.A., Ryter, S.W., Choi, A.M., 2011. Autophagy proteins regulate innate immune responses by inhibiting the release of mitochondrial DNA mediated by the NALP3 inflammasome. Nat. Immunol. 12, 222–230.

Nakamura, M., Yamazaki, I., 1972. One-electron transfer reactions in biochemical systems. VI. Changes in electron transfer mechanism of lipoamide dehydrogenase by modification of sulfhydryl groups. Biochim. Biophys. Acta 267, 249–257.

Narasimhan, M., Rajasekaran, N.S., 2016. Exercise, Nrf2 and antioxidant signaling in cardiac aging. Front. Physiol. 7, 241.

Narendra, D.P., Jin, S.M., Tanaka, A., Suen, D.F., Gautier, C.A., Shen, J., Cookson, M.R., Youle, R.J., 2010. PINK1 is selectively stabilized on impaired mitochondria to activate Parkin. PLoS Biol. 8, e1000298.

Navarro-Cruz, A.R., Ramirez, Y.A., Ochoa-Velasco, C., Brambila, E., Avila-Sosa, R., Perez-Fernandez, S., Morales-Medina, J.C., Aguilar-Alonso, P., 2017. Effect of chronic administration of resveratrol on cognitive performance during aging process in rats. Oxid. Med. Cell Longev. 2017, 8510761.

Navarro, A., Boveris, A., 2004. Rat brain and liver mitochondria develop oxidative stress and lose enzymatic activities on aging. Am. J. Physiol. Regul. Integr. Comp. Physiol. 287. R1244–1249.

Navarro, A., Gomez, C., Sanchez-Pino, M.J., Gonzalez, H., Bandez, M.J., Boveris, A.D., Boveris, A., 2005. Vitamin E at high doses improves survival, neurological performance, and brain mitochondrial function in aging male mice. Am. J. Physiol. Regul. Integr. Comp. Physiol. 289. R1392–1399.

Nemoto, S., Combs, C.A., French, S., Ahn, B.H., Fergusson, M.M., Balaban, R.S., Finkel, T., 2006. The mammalian longevity-associated gene product p66shc regulates mitochondrial metabolism. J. Biol. Chem. 281, 10555–10560.

Nguyen, T., Nioi, P., Pickett, C.B., 2009. The Nrf2-antioxidant response element signaling pathway and its activation by oxidative stress. J. Biol. Chem. 284, 13291–13295.

Nikitovic, D., Holmgren, A., 1996. S-nitrosoglutathione is cleaved by the thioredoxin system with liberation of glutathione and redox regulating nitric oxide. J. Biological. Chem. 271, 19180–19185.

Nimse, S.B., Pal, D., 2015. Free radicals, natural antioxidants, and their reaction mechanisms. RSC Adv. 5, 27986–28006.

Nishikawa, T., Edelstein, D., Du, X.L., Yamagishi, S., Matsumura, T., Kaneda, Y., Yorek, M.A., Beebe, D., Oates, P.J., Hammes, H.P., Giardino, I., Brownlee, M., 2000. Normalizing mitochondrial superoxide production blocks three pathways of hyperglycaemic damage. Nature 404, 787–790.

Nisoli, E., Tonello, C., Cardile, A., Cozzi, V., Bracale, R., Tedesco, L., Falcone, S., Valerio, A., Cantoni, O., Clementi, E., Moncada, S., Carruba, M.O., 2005. Calorie restriction promotes mitochondrial biogenesis by inducing the expression of eNOS. Science 310, 314–317.

Norton, M., Ng, A.C.-H., Baird, S., Dumoulin, A., Shutt, T., Mah, N., Andrade-Navarro, M.A., McBride, H.M., Screaton, R.A., 2014. ROMO1 Is an essential redox-dependent regulator of mitochondrial dynamics. Sci. Signal. 7, ra10.

Nosadini, R., Avogaro, A., Mollo, F., Marescotti, C., Tiengo, A., Duner, E., MERKEL, A., Gatta, A., Zuin, R., De Kreutzenberg, S., 1984. Carbohydrate and lipid metabolism in cirrhosis. Evidence that hepatic uptake of gluconeogenic precursors and of free fatty acids depends on effective hepatic flow. J. Clinical Endocrinol. Metabol. 58, 1125–1132.

Noureddin, M., Yates, K.P., Vaughn, I.A., Neuschwander-Tetri, B.A., Sanyal, A.J., McCullough, A., Merriman, R., Hameed, B., Doo, E., Kleiner, D.E., Behling, C., Loomba, R., Nash, C.R., 2013. Clinical and histological determinants of nonalcoholic steatohepatitis and advanced fibrosis in elderly patients. Hepatology 58, 1644–1654.

Ohhira, M., Ono, M., Ohhira, M., Sekiya, C., Namiki, M., Fujimoto, Y., Nagao, M., Mori, M., 1995. Changes in free radical-metabolizing enzymes and lipid peroxides in the liver of long-evans with cinnamon-like coat color rats. J. Gastroenterol. 30, 619–623.

Okawara, M., Katsuki, H., Kurimoto, E., Shibata, H., Kume, T., Akaike, A., 2007. Resveratrol protects dopaminergic neurons in midbrain slice culture from multiple insults. Biochem. Pharmacol. 73, 550–560.

Oliveira, H.C., Cosso, R.G., Alberici, L.C., Maciel, E.N., Salerno, A.G., Dorighello, G.G., Velho, J.A., de Faria, E.C., Vercesi, A.E., 2005. Oxidative stress in atherosclerosis-prone mouse is due to low antioxidant capacity of mitochondria. FASEB J. 19, 278–280.

Orr, W.C., Sohal, R.S., 1994. Extension of life-span by overexpression of superoxide dismutase and catalase in *Drosophila melanogaster*. Science 263, 1128–1130.

Orsini, F., Migliaccio, E., Moroni, M., Contursi, C., Raker, V.A., Piccini, D., Martin-Padura, I., Pelliccia, G., Trinei, M., Bono, M., Puri, C., Tacchetti, C., Ferrini, M., Mannucci, R., Nicoletti, I., Lanfrancone, L., Giorgio, M., Pelicci, P.G., 2004. The life span determinant p66Shc localizes to mitochondria where it associates with mitochondrial heat shock protein 70 and regulates trans-membrane potential. J. Biol. Chem. 279, 25689–25695.

Otten, A.B., Smeets, H.J., 2015. Evolutionary defined role of the mitochondrial DNA in fertility, disease and ageing. Hum. Reprod. Update 21, 671–689.

Otterbein, L.E., Choi, A.M., 2000. Heme oxygenase: colors of defense against cellular stress. Am. J. Physiol. Lung Cell Mol. Physiol. 279. L1029–1037.

Paddenberg, R., Goldenberg, A., Faulhammer, P., Braun-Dullaeus, R.C., Kummer, W., 2003a. Mitochondrial complex II is essential for hypoxia-induced ROS generation and vasoconstriction in the pulmonary vasculature. Adv. Exp. Med. Biol. 536, 163–169.

Paddenberg, R., Ishaq, B., Goldenberg, A., Faulhammer, P., Rose, F., Weissmann, N., Braun-Dullaeus, R.C., Kummer, W., 2003b. Essential role of complex II of the respiratory chain in hypoxia-induced ROS generation in the pulmonary vasculature. Am. J. Physiol. Lung Cell Mol. Physiol. 284. L710–719.

Page, M.M., Robb, E.L., Salway, K.D., Stuart, J.A., 2010. Mitochondrial redox metabolism: aging, longevity and dietary effects. Mech. Ageing Dev. 131, 242–252.

Pagel-Langenickel, I., Bao, J., Pang, L., Sack, M.N., 2010. The role of mitochondria in the pathophysiology of skeletal muscle insulin resistance. Endocr. Rev. 31, 25–51.

Pahl, H.L., 1999. Activators and target genes of Rel/NF-kappaB transcription factors. Oncogene 18, 6853–6866.

Pajonk, F.G., Kessler, H., Supprian, T., Hamzei, P., Bach, D., Schweickhardt, J., Herrmann, W., Obeid, R., Simons, A., Falkai, P., Multhaup, G., Bayer, T.A., 2005. Cognitive decline correlates with low plasma concentrations of copper in patients with mild to moderate Alzheimer's disease. J. Alzheimers Dis. 8, 23–27.

Palacios, O.M., Carmona, J.J., Michan, S., Chen, K.Y., Manabe, Y., Ward 3rd, J.L., Goodyear, L.J., Tong, Q., 2009. Diet and exercise signals regulate SIRT3 and activate AMPK and PGC-1alpha in skeletal muscle. Aging (Albany NY) 1, 771–783.

Palasciano, G., Portincasa, P., Vinciguerra, V., Velardi, A., Tardi, S., Baldassarre, G., Albano, O., 1989. Gallstone prevalence and gallbladder volume in children and adolescents: an epidemiological ultrasonographic survey and relationship to body mass index. Am. J. Gastroenterol. 84, 1378–1382.

Pamplona, R., 2011. Mitochondrial DNA damage and animal longevity: insights from comparative studies. J. Aging Res. 2011, 807108.

Pamplona, R., Barja, G., Portero-Otin, M., 2002. Membrane fatty acid unsaturation, protection against oxidative stress, and maximum life span: a homeoviscous-longevity adaptation? Ann. N. Y. Acad. Sci. 959, 475–490.

Pamplona, R., Portero-Otin, M., Riba, D., Ruiz, C., Prat, J., Bellmunt, M.J., Barja, G., 1998. Mitochondrial membrane peroxidizability index is inversely related to maximum life span in mammals. J. Lipid. Res. 39, 1989–1994.

Pandey, M., Shukla, V., 2000. Fatty acids, biliary bile acids, lipid peroxidation products and gallbladder carcinogenesis. Europ. J. Cancer Prev. 9, 165–172.

Pandya, J.D., Grondin, R., Yonutas, H.M., Haghnazar, H., Gash, D.M., Zhang, Z., Sullivan, P.G., 2015. Decreased mitochondrial bioenergetics and calcium buffering capacity in the basal ganglia correlates with motor deficits in a nonhuman primate model of aging. Neurobiol. Aging 36, 1903–1913.

Pangrazzi, L., Meryk, A., Naismith, E., Koziel, R., Lair, J., Krismer, M., Trieb, K., Grubeck-Loebenstein, B., 2017. "Inflamm-aging" influences immune cell survival factors in human bone marrow. Eur. J. Immunol. 47, 481–492.

Pappu, A., Fatterpaker, P., Sreenivasan, A., 1978. Phospholipase A2 of rat liver mitochondria in vitamin E deficiency. Biochem. J. 172, 349–352.

Paradies, G., Paradies, V., Ruggiero, F.M., Petrosillo, G., 2013. Changes in the mitochondrial permeability transition pore in aging and age-associated diseases. Mech. Ageing Dev. 134, 1–9.

Paradies, G., Paradies, V., Ruggiero, F.M., Petrosillo, G., 2017. Mitochondrial bioenergetics decay in aging: beneficial effect of melatonin. Cell Mol. Life Sci. 74, 3897–3911.

Paradies, G., Petrosillo, G., Paradies, V., Reiter, R.J., Ruggiero, F.M., 2010. Melatonin, cardiolipin and mitochondrial bioenergetics in health and disease. J. Pineal. Res. 48, 297–310.

Paradies, G., Petrosillo, G., Pistolese, M., Ruggiero, F.M., 2002. Reactive oxygen species affect mitochondrial electron transport complex I activity through oxidative cardiolipin damage. Gene 286, 135–141.

Park, G., Sim, Y., Lee, W., Sung, S.H., Oh, M.S., 2016. Protection on skin aging mediated by antiapoptosis effects of the Water Lily (Nymphaea Tetragona Georgi) via reactive oxygen species scavenging in human epidermal keratinocytes. Pharmacology 97, 282–293.
Park, H., Ishigami, A., Shima, T., Mizuno, M., Maruyama, N., Yamaguchi, K., Mitsuyoshi, H., Minami, M., Yasui, K., Itoh, Y., 2010. Hepatic senescence marker protein-30 is involved in the progression of nonalcoholic fatty liver disease. J. Gastroenterol. 45, 426–434.
Parkes, T.L., Elia, A.J., Dickinson, D., Hilliker, A.J., Phillips, J.P., Boulianne, G.L., 1998. Extension of Drosophila lifespan by overexpression of human SOD1 in motorneurons. Nat. Genet. 19, 171–174.
Parkinson, G.M., Dayas, C.V., Smith, D.W., 2014. Increased mitochondrial DNA deletions in substantia nigra dopamine neurons of the aged rat. Curr. Aging Sci. 7, 155–160.
Parkinson Study, G., 1993. Effects of tocopherol and deprenyl on the progression of disability in early Parkinson's disease. N. Engl. J. Med. 328, 176–183.
Pathak, D., Sepp, K.J., Hollenbeck, P.J., 2010. Evidence that myosin activity opposes microtubule-based axonal transport of mitochondria. J. Neurosci 30, 8984–8992.
Paul, A., Belton, A., Nag, S., Martin, I., Grotewiel, M.S., Duttaroy, A., 2007. Reduced mitochondrial SOD displays mortality characteristics reminiscent of natural aging. Mech. Ageing Dev. 128, 706–716.
Pavlov, P.F., Wiehager, B., Sakai, J., Frykman, S., Behbahani, H., Winblad, B., Ankarcrona, M., 2011. Mitochondrial gamma-secretase participates in the metabolism of mitochondria-associated amyloid precursor protein. FASEB J. 25, 78–88.
Pereira, C.V., Lebiedzinska, M., Wieckowski, M.R., Oliveira, P.J., 2012. Regulation and protection of mitochondrial physiology by sirtuins. Mitochondrion 12, 66–76.
Perez-Carreras, M., Del Hoyo, P., Martin, M.A., Rubio, J.C., Martin, A., Castellano, G., Colina, F., Arenas, J., Solis-Herruzo, J.A., 2003. Defective hepatic mitochondrial respiratory chain in patients with nonalcoholic steatohepatitis. Hepatology 38, 999–1007.
Perez, V.I., Bokov, A., Van Remmen, H., Mele, J., Ran, Q., Ikeno, Y., Richardson, A., 2009a. Is the oxidative stress theory of aging dead? Biochim. Biophys. Acta 1790, 1005–1014.
Perez, V.I., Van Remmen, H., Bokov, A., Epstein, C.J., Vijg, J., Richardson, A., 2009b. The overexpression of major antioxidant enzymes does not extend the lifespan of mice. Aging Cell 8, 73–75.
Pernas, L., Scorrano, L., 2016. Mito-Morphosis: mitochondrial fusion, fission, and cristae remodeling as key mediators of cellular function. Annu. Rev. Physiol. 78, 505–531.
Perry, R.J., Kim, T., Zhang, X.M., Lee, H.Y., Pesta, D., Popov, V.B., Zhang, D., Rahimi, Y., Jurczak, M.J., Cline, G.W., Spiegel, D.A., Shulman, G.I., 2013. Reversal of hypertriglyceridemia, fatty liver disease, and insulin resistance by a liver-targeted mitochondrial uncoupler. Cell Metab. 18, 740–748.
Pessayre, D., Fromenty, B., 2005. NASH: a mitochondrial disease. J. Hepatol. 42, 928–940.
Petersen, K.F., Befroy, D., Dufour, S., Dziura, J., Ariyan, C., Rothman, D.L., DiPietro, L., Cline, G.W., Shulman, G.I., 2003. Mitochondrial dysfunction in the elderly: possible role in insulin resistance. Science 300, 1140–1142.
Petersen, M.H., Budtz-Jorgensen, E., Sorensen, S.A., Nielsen, J.E., Hjermind, L.E., Vinther-Jensen, T., Nielsen, S.M., Norremolle, A., 2014. Reduction in mitochondrial DNA copy number in peripheral leukocytes after onset of Huntington's disease. Mitochondrion 17, 14–21.
Petri, S., Kiaei, M., Damiano, M., Hiller, A., Wille, E., Manfredi, G., Calingasan, N.Y., Szeto, H.H., Beal, M.F., 2006. Cell-permeable peptide antioxidants as a novel therapeutic approach in a mouse model of amyotrophic lateral sclerosis. J. Neurochem. 98, 1141–1148.

Petrosillo, G., Fattoretti, P., Matera, M., Ruggiero, F.M., Bertoni-Freddari, C., Paradies, G., 2008. Melatonin prevents age-related mitochondrial dysfunction in rat brain via cardiolipin protection. Rejuvenation Res. 11, 935–943.

Petrosillo, G., Matera, M., Moro, N., Ruggiero, F.M., Paradies, G., 2009. Mitochondrial complex I dysfunction in rat heart with aging: critical role of reactive oxygen species and cardiolipin. Free Radic. Biol. Med. 46, 88–94.

Petrosillo, G., Portincasa, P., Grattagliano, I., Casanova, G., Matera, M., Ruggiero, F.M., Ferri, D., Paradies, G., 2007. Mitochondrial dysfunction in rat with nonalcoholic fatty liver Involvement of complex I, reactive oxygen species and cardiolipin. Biochim. Biophys. Acta 1767, 1260–1267.

Petrov, D., Pedrós, I., Artiach, G., Sureda, F.X., Barroso, E., Pallàs, M., Casadesús, G., Beas-Zarate, C., Carro, E., Ferrer, I., Vazquez-Carrera, M., Folch, J., Camins, A., 2015. High-fat diet-induced deregulation of hippocampal insulin signaling and mitochondrial homeostasis deficiences contribute to Alzheimer disease pathology in rodents. Biochimica et Biophysica Acta 1852, 1687–1699.

Pinti, M., Gibellini, L., Nasi, M., De Biasi, S., Bortolotti, C.A., Iannone, A., Cossarizza, A., 2016. Emerging role of Lon protease as a master regulator of mitochondrial functions. Biochim. Biophys. Acta 1857, 1300–1306.

Pinton, P., Rimessi, A., Marchi, S., Orsini, F., Migliaccio, E., Giorgio, M., Contursi, C., Minucci, S., Mantovani, F., Wieckowski, M.R., Del Sal, G., Pelicci, P.G., Rizzuto, R., 2007. Protein kinase C beta and prolyl isomerase 1 regulate mitochondrial effects of the life-span determinant p66Shc. Science 315, 659–663.

Pinton, P., Rizzuto, R., 2008. p66Shc, oxidative stress and aging: importing a lifespan determinant into mitochondria. Cell Cycle 7, 304–308.

Planavila, A., Iglesias, R., Giralt, M., Villarroya, F., 2011. Sirt1 acts in association with PPARalpha to protect the heart from hypertrophy, metabolic dysregulation, and inflammation. Cardiovasc. Res. 90, 276–284.

Plecita-Hlavata, L., Jezek, J., Jezek, P., 2009. Pro-oxidant mitochondrial matrix-targeted ubiquinone MitoQ10 acts as anti-oxidant at retarded electron transport or proton pumping within Complex I. Int. J. Biochem. Cell Biol. 41, 1697–1707.

Polo, M., Alegre, F., Moragrega, A.B., Gibellini, L., Marti-Rodrigo, A., Blas-Garcia, A., Esplugues, J.V., Apostolova, N., 2017. Lon protease: a novel mitochondrial matrix protein in the interconnection between drug-induced mitochondrial dysfunction and endoplasmic reticulum stress. Br. J. Pharmacol. 174, 4409–4429.

Poo, J.L., Rosas-Romero, R., Montemayor, A.C., Isoard, F., Uribe, M., 2003. Diagnostic value of the copper/zinc ratio in hepatocellular carcinoma: a case control study. J. Gastroenterol. 38, 45–51.

Portincasa, P., Di Ciaula, A., Baldassarre, G., Palmieri, V., Gentile, A., Cimmino, A., Palasciano, G., 1994. Gallbladder motor function in gallstone patients: sonographic and in vitro studies on the role of gallstones, smooth muscle function and gallbladder wall inflammation. J. Hepatol. 21, 430–440.

Portincasa, P., Moschetta, A., Berardino, M., Di-Ciaula, A., Vacca, M., Baldassarre, G., Pietrapertosa, A., Cammarota, R., Tannoia, N., Palasciano, G., 2004. Impaired gallbladder motility and delayed orocecal transit contribute to pigment gallstone and biliary sludge formation in beta-thalassemia major adults. World J. Gastroenterol. 10, 2383–2390.

Portincasa, P., Moschetta, A., Palasciano, G., 2006. Cholesterol gallstone disease. Lancet 368, 230–239.

Portincasa, P., Wang, D.Q., 2015. Gallstones. In: Podolsky, K.D., Camilleri, M., Fitz, J.G., Kalloo, A.N., Shanahan, F., Wang, T.C. (Eds.), Yamada's Textbook of Gastroenterology. Wiley-Blackwell, UK, pp. 1808–1834.

Portincasa, P., Wang, D.Q., 2016. Gallstones. In: Podolsky, K.D., Camilleri, M., Fitz, J.G., Kalloo, A.N., Shanahan, F., Wang, T.C. (Eds.), Yamada's Atlas of Gastroenterology. Wiley-Blackwell, UK, pp. 335–353.

Powers, S.K., Criswell, D., Lawler, J., Ji, L.L., Martin, D., Herb, R.A., Dudley, G., 1994. Influence of exercise and fiber type on antioxidant enzyme activity in rat skeletal muscle. Am. J. Physiol. 266. R375–380.

Powers, S.K., Ji, L.L., Leeuwenburgh, C., 1999. Exercise training-induced alterations in skeletal muscle antioxidant capacity: a brief review. Med. Sci. Sports Exerc. 31, 987–997.

Prakash, C., Soni, M., Kumar, V., 2015. Biochemical and molecular alterations following arsenic-induced oxidative stress and mitochondrial dysfunction in rat brain. Biol. Trace Elem. Res. 167, 121–129.

Premoli, A., Paschetta, E., Hvalryg, M., Spandre, M., Bo, S., Durazzo, M., 2009. Characteristics of liver diseases in the elderly: a review. Minerva Gastroenterologica e Dietologica 55, 71–78.

Prudent, J., McBride, H.M., 2016. Mitochondrial dynamics: ER actin tightens the Drp1 noose. Curr. Biol. 26, R207–R209.

Przedborski, S., Jackson-Lewis, V., 1998. Mechanisms of MPTP toxicity. Mov. Disord. 13 (Suppl. 1), 35–38.

Ptak, R.G., Gallay, P.A., Jochmans, D., Halestrap, A.P., Ruegg, U.T., Pallansch, L.A., Bobardt, M.D., de Bethune, M.P., Neyts, J., De Clercq, E., Dumont, J.M., Scalfaro, P., Besseghir, K., Wenger, R.M., Rosenwirth, B., 2008. Inhibition of human immunodeficiency virus type 1 replication in human cells by Debio-025, a novel cyclophilin binding agent. Antimicrob. Agents Chemother. 52, 1302–1317.

Puddu, P., Puddu, G.M., Cravero, E., Rosati, M., Muscari, A., 2008. The molecular sources of reactive oxygen species in hypertension. Blood Press 17, 70–77.

Putti, R., Migliaccio, V., Sica, R., Lionetti, L., 2015. Skeletal muscle mitochondrial bioenergetics and morphology in high fat diet induced obesity and insulin resistance: focus on dietary fat source. Front. Physiol. 6, 426.

Pyo, J.O., Yoo, S.M., Ahn, H.H., Nah, J., Hong, S.H., Kam, T.I., Jung, S., Jung, Y.K., 2013. Overexpression of Atg5 in mice activates autophagy and extends lifespan. Nat. Commun. 4, 2300.

Qiu, X., Brown, K., Hirschey, M.D., Verdin, E., Chen, D., 2010. Calorie restriction reduces oxidative stress by SIRT3-mediated SOD2 activation. Cell Metab. 12, 662–667.

Quan, C., Cho, M.K., Perry, D., Quan, T., 2015. Age-associated reduction of cell spreading induces mitochondrial DNA common deletion by oxidative stress in human skin dermal fibroblasts: implication for human skin connective tissue aging. J. Biomed. Sci. 22, 62.

Quiles, J.L., Ochoa, J.J., Huertas, J.R., Mataix, J., 2004. Coenzyme Q supplementation protects from age-related DNA double-strand breaks and increases lifespan in rats fed on a PUFA-rich diet. Exp. Gerontol. 39, 189–194.

Quinlan, C.L., Perevoshchikova, I.V., Hey-Mogensen, M., Orr, A.L., Brand, M.D., 2013. Sites of reactive oxygen species generation by mitochondria oxidizing different substrates. Redox Biol. 1, 304–312.

Quinn, J.F., Bussiere, J.R., Hammond, R.S., Montine, T.J., Henson, E., Jones, R.E., Stackman Jr., R.W., 2007. Chronic dietary alpha-lipoic acid reduces deficits in hippocampal memory of aged Tg2576 mice. Neurobiol. Aging 28, 213–225.

Quintanilla, R.A., Tapia, C., Perez, M.J., 2017. Possible role of mitochondrial permeability transition pore in the pathogenesis of Huntington disease. Biochem. Biophys. Res. Commun. 483, 1078–1083.

Rae, T.D., Schmidt, P.J., Pufahl, R.A., Culotta, V.C., O'Halloran, T.V., 1999. Undetectable intracellular free copper: the requirement of a copper chaperone for superoxide dismutase. Science 284, 805–808.

Raha, S., McEachern, G.E., Myint, A.T., Robinson, B.H., 2000. Superoxides from mitochondrial complex III: the role of manganese superoxide dismutase. Free Radic. Biol. Med. 29, 170–180.

Raha, S., Robinson, B.H., 2000. Mitochondria, oxygen free radicals, disease and ageing. Trends Biochem. Sci 25, 502–508.

Rahman, M., Halade, G.V., Bhattacharya, A., Fernandes, G., 2009. The fat-1 transgene in mice increases antioxidant potential, reduces pro-inflammatory cytokine levels, and enhances PPAR-gamma and SIRT-1 expression on a calorie restricted diet. Oxid. Med. Cell Longev. 2, 307–316.

Rahmioglu, N., Andrew, T., Cherkas, L., Surdulescu, G., Swaminathan, R., Spector, T., Ahmadi, K.R., 2009. Epidemiology and genetic epidemiology of the liver function test proteins. PLoS One 4, e4435.

Rana, A., Oliveira, M.P., Khamoui, A.V., Aparicio, R., Rera, M., Rossiter, H.B., Walker, D.W., 2017. Promoting Drp1-mediated mitochondrial fission in midlife prolongs healthy lifespan of *Drosophila melanogaster.* Nat. Commun. 8, 448.

Rana, A., Rera, M., Walker, D.W., 2013. Parkin overexpression during aging reduces proteotoxicity, alters mitochondrial dynamics, and extends lifespan. Proc. Natl. Acad. Sci. 110, 8638–8643.

Rapizzi, E., Pinton, P., Szabadkai, G., Wieckowski, M.R., Vandecasteele, G., Baird, G., Tuft, R.A., Fogarty, K.E., Rizzuto, R., 2002. Recombinant expression of the voltage-dependent anion channel enhances the transfer of Ca2+ microdomains to mitochondria. J. Cell. Biol. 159, 613–624.

Rapsomaniki, E., Timmis, A., George, J., Pujades-Rodriguez, M., Shah, A.D., Denaxas, S., White, I.R., Caulfield, M.J., Deanfield, J.E., Smeeth, L., Williams, B., Hingorani, A., Hemingway, H., 2014. Blood pressure and incidence of twelve cardiovascular diseases: lifetime risks, healthy life-years lost, and age-specific associations in 1·25 million people. Lancet 383, 1899–1911.

Rebrin, I., Sohal, R.S., 2004. Comparison of thiol redox state of mitochondria and homogenates of various tissues between two strains of mice with different longevities. Exp. Gerontol. 39, 1513–1519.

Reddehase, S., Grumbt, B., Neupert, W., Hell, K., 2009. The disulfide relay system of mitochondria is required for the biogenesis of mitochondrial Ccs1 and Sod1. J. Mol. Biol. 385, 331–338.

Reddy, P.H., 2006. Amyloid precursor protein-mediated free radicals and oxidative damage: implications for the development and progression of Alzheimer's disease. J. Neurochem. 96, 1–13.

Reddy, P.H., Beal, M.F., 2008. Amyloid beta, mitochondrial dysfunction and synaptic damage: implications for cognitive decline in aging and Alzheimer's disease. Trends Mol. Med. 14, 45–53.

Reddy, P.H., Manczak, M., Yin, X., 2017. Mitochondria-division inhibitor 1 protects against amyloid-β induced mitochondrial fragmentation and synaptic damage in Alzheimer's disease. J. Alzheimer's Dis. 58, 147–162.

Regev, A., Schiff, E.R., 2001. Liver disease in the elderly. Gastroenterol. Clin. North Am. 30, 547–563.

Regmi, S.G., Rolland, S.G., Conradt, B., 2014. Age-dependent changes in mitochondrial morphology and volume are not predictors of lifespan. Aging (Albany NY) 6, 118–130.

Rego, A.C., Oliveira, C.R., 2003. Mitochondrial dysfunction and reactive oxygen species in excitotoxicity and apoptosis: implications for the pathogenesis of neurodegenerative diseases. Neurochem. Res. 28, 1563–1574.

Rhein, V., Song, X., Wiesner, A., Ittner, L.M., Baysang, G., Meier, F., Ozmen, L., Bluethmann, H., Drose, S., Brandt, U., Savaskan, E., Czech, C., Gotz, J., Eckert, A., 2009. Amyloid-beta and tau synergistically impair the oxidative phosphorylation system in triple transgenic Alzheimer's disease mice. Proc. Natl. Acad. Sci. U.S.A 106, 20057–20062.
Rimessi, A., Bezzerri, V., Patergnani, S., Marchi, S., Cabrini, G., Pinton, P., 2015. Mitochondrial Ca2+-dependent NLRP3 activation exacerbates the Pseudomonas aeruginosa-driven inflammatory response in cystic fibrosis. Nat. Commun. 6, 6201.
Rimessi, A., Bonora, M., Marchi, S., Patergnani, S., Marobbio, C.M., Lasorsa, F.M., Pinton, P., 2013. Perturbed mitochondrial Ca2+ signals as causes or consequences of mitophagy induction. Autophagy 9, 1677–1686.
Rimessi, A., Marchi, S., Fotino, C., Romagnoli, A., Huebner, K., Croce, C.M., Pinton, P., Rizzuto, R., 2009. Intramitochondrial calcium regulation by the FHIT gene product sensitizes to apoptosis. Proc. Natl. Acad. Sci. U.S.A 106, 12753–12758.
Rinaldi, B., Corbi, G., Boccuti, S., Filippelli, W., Rengo, G., Leosco, D., Rossi, F., Filippelli, A., Ferrara, N., 2006. Exercise training affects age-induced changes in SOD and heat shock protein expression in rat heart. Exp. Gerontol. 41, 764–770.
Ristow, M., Schmeisser, S., 2011. Extending life span by increasing oxidative stress. Free Radic. Biol. Med. 51, 327–336.
Ristow, M., Zarse, K., 2010. How increased oxidative stress promotes longevity and metabolic health: The concept of mitochondrial hormesis (mitohormesis). Exp. Gerontol. 45, 410–418.
Rizzuto, R., Pinton, P., Carrington, W., Fay, F.S., Fogarty, K.E., Lifshitz, L.M., Tuft, R.A., Pozzan, T., 1998. Close contacts with the endoplasmic reticulum as determinants of mitochondrial Ca2+ responses. Science 280, 1763–1766.
Robb, E.L., Winkelmolen, L., Visanji, N., Brotchie, J., Stuart, J.A., 2008. Dietary resveratrol administration increases MnSOD expression and activity in mouse brain. Biochem. Biophys. Res. Commun. 372, 254–259.
Roberts, E.A., Robinson, B.H., Yang, S., 2008. Mitochondrial structure and function in the untreated Jackson toxic milk (tx-j) mouse, a model for Wilson disease. Mol. Genet. Metab. 93, 54–65.
Robson-Doucette, C.A., Sultan, S., Allister, E.M., Wikstrom, J.D., Koshkin, V., Bhattacharjee, A., Prentice, K.J., Sereda, S.B., Shirihai, O.S., Wheeler, M.B., 2011. Beta-cell uncoupling protein 2 regulates reactive oxygen species production, which influences both insulin and glucagon secretion. Diabetes 60, 2710–2719.
Rodriguez-Enriquez, S., Kai, Y., Maldonado, E., Currin, R.T., Lemasters, J.J., 2009. Roles of mitophagy and the mitochondrial permeability transition in remodeling of cultured rat hepatocytes. Autophagy 5, 1099–1106.
Rodriguez, A., Muller, D.C., Metter, E.J., Maggio, M., Harman, S.M., Blackman, M.R., Andres, R., 2007. Aging, androgens, and the metabolic syndrome in a longitudinal study of aging. J. Clinical. Endocrinol. Metabol. 92, 3568–3572.
Rodriguez, M., Rodriguez-Sabate, C., Morales, I., Sanchez, A., Sabate, M., 2015. Parkinson's disease as a result of aging. Aging Cell 14, 293–308.
Rogina, B., Helfand, S.L., 2004. Sir2 mediates longevity in the fly through a pathway related to calorie restriction. Proc. Natl. Acad. Sci. U.S.A 101, 15998–16003.
Roh, J., Rhee, J., Chaudhari, V., Rosenzweig, A., 2016. The role of exercise in cardiac aging: from physiology to molecular mechanisms. Circ Res. 118, 279–295.
Romanello, V., Guadagnin, E., Gomes, L., Roder, I., Sandri, C., Petersen, Y., Milan, G., Masiero, E., Del Piccolo, P., Foretz, M., Scorrano, L., Rudolf, R., Sandri, M., 2010. Mitochondrial fission and remodelling contributes to muscle atrophy. EMBO J. 29, 1774–1785.

Rossignol, R., Faustin, B., Rocher, C., Malgat, M., Mazat, J.P., Letellier, T., 2003. Mitochondrial threshold effects. Biochem. J. 370, 751–762.

Röth, D., Krammer, P.H., Gülow, K., 2014. Dynamin related protein 1-dependent mitochondrial fission regulates oxidative signalling in T cells. FEBS Lett. 588, 1749–1754.

Rottenberg, H., Hoek, J.B., 2017. The path from mitochondrial ROS to aging runs through the mitochondrial permeability transition pore. Aging Cell 16, 943–955.

Roy, S., Debnath, J., 2010. Autophagy and tumorigenesis. Semin Immunopathol. 32, 383–396.

Rubino, J.T., Franz, K.J., 2012. Coordination chemistry of copper proteins: how nature handles a toxic cargo for essential function. J. Inorg. Biochem. 107, 129–143.

Rui, M., Suzuki, K.T., 1997. Copper in plasma reflects its status and subsequent toxicity in the liver of LEC rats. Res. Commun. Mol. Pathol. Pharmacol. 98, 335–346.

Ryan, M.J., Dudash, H.J., Docherty, M., Geronilla, K.B., Baker, B.A., Haff, G.G., Cutlip, R.G., Alway, S.E., 2010. Vitamin E and C supplementation reduces oxidative stress, improves antioxidant enzymes and positive muscle work in chronically loaded muscles of aged rats. Exp. Gerontol. 45, 882–895.

Sack, M.N., 2006. Mitochondrial depolarization and the role of uncoupling proteins in ischemia tolerance. Cardiovasc. Res. 72, 210–219.

Safdar, A., Little, J.P., Stokl, A.J., Hettinga, B.P., Akhtar, M., Tarnopolsky, M.A., 2011. Exercise increases mitochondrial PGC-1alpha content and promotes nuclear-mitochondrial cross-talk to coordinate mitochondrial biogenesis. J. Biol. Chem. 286, 10605–10617.

Salmon, A.B., 2016. Beyond diabetes: does obesity-induced oxidative stress drive the aging process? Antioxidants 5, 24.

Salvemini, D., Wang, Z.Q., Zweier, J.L., Samouilov, A., Macarthur, H., Misko, T.P., Currie, M.G., Cuzzocrea, S., Sikorski, J.A., Riley, D.P., 1999. A nonpeptidyl mimic of superoxide dismutase with therapeutic activity in rats. Science 286, 304–306.

Sama, C., Labate, A.M., Taroni, F., Barbara, L., 1990. Epidemiology and natural history of gallstone disease. Semin. Liver Dis. 10, 149–158.

Samuele, A., Mangiagalli, A., Armentero, M.T., Fancellu, R., Bazzini, E., Vairetti, M., Ferrigno, A., Richelmi, P., Nappi, G., Blandini, F., 2005. Oxidative stress and pro-apoptotic conditions in a rodent model of Wilson's disease. Biochim. Biophys. Acta 1741, 325–330.

Sanguino, E., Bejarano, R., Alegret, M., Sánchez, R., Vázquez-Carrera, M., Laguna, J., 2004. Sexual dimorphism in lipid metabolic phenotype associated with old age in Sprague–Dawley rats. Exp. Gerontol. 39, 1295–1306.

Sano, M., Ernesto, C., Thomas, R.G., Klauber, M.R., Schafer, K., Grundman, M., Woodbury, P., Growdon, J., Cotman, C.W., Pfeiffer, E., Schneider, L.S., Thal, L.J., 1997. A controlled trial of selegiline, alpha-tocopherol, or both as treatment for Alzheimer's disease. The Alzheimer's disease cooperative study. N. Engl. J. Med. 336, 1216–1222.

Santos, R.X., Correia, S.C., Zhu, X., Smith, M.A., Moreira, P.I., Castellani, R.J., Nunomura, A., Perry, G., 2013. Mitochondrial DNA oxidative damage and repair in aging and Alzheimer's disease. Antioxid. Redox Signal. 18, 2444–2457.

Sanyal, A.J., Campbell-Sargent, C., Mirshahi, F., Rizzo, W.B., Contos, M.J., Sterling, R.K., Luketic, V.A., Shiffman, M.L., Clore, J.N., 2001. Nonalcoholic steatohepatitis: association of insulin resistance and mitochondrial abnormalities. Gastroenterology 120, 1183–1192.

Sanz, A., Caro, P., Ibanez, J., Gomez, J., Gredilla, R., Barja, G., 2005. Dietary restriction at old age lowers mitochondrial oxygen radical production and leak at complex I and oxidative DNA damage in rat brain. J. Bioenerg. Biomembr. 37, 83–90.

Sastre, J., Pallardo, F.V., Pla, R., Pellin, A., Juan, G., O'Connor, J.E., Estrela, J.M., Miquel, J., Vina, J., 1996. Aging of the liver: age-associated mitochondrial damage in intact hepatocytes. Hepatology 24, 1199–1205.

Satoh, T., Kosaka, K., Itoh, K., Kobayashi, A., Yamamoto, M., Shimojo, Y., Kitajima, C., Cui, J., Kamins, J., Okamoto, S., Izumi, M., Shirasawa, T., Lipton, S.A., 2008. Carnosic acid, a catechol-type electrophilic compound, protects neurons both in vitro and in vivo through activation of the Keap1/Nrf2 pathway via S-alkylation of targeted cysteines on Keap1. J. Neurochem. 104, 1116–1131.

Sauer, S.W., Merle, U., Opp, S., Haas, D., Hoffmann, G.F., Stremmel, W., Okun, J.G., 2011. Severe dysfunction of respiratory chain and cholesterol metabolism in Atp7b(-/-) mice as a model for Wilson disease. Biochim. Biophys. Acta 1812, 1607–1615.

Sawda, C., Moussa, C., Turner, R.S., 2017. Resveratrol for Alzheimer's disease. Ann. N. Y. Acad. Sci. 1403, 142–149.

Schaffer, S., Muller, W.E., Eckert, G.P., 2005. Tocotrienols: constitutional effects in aging and disease. J. Nutr. 135, 151–154.

Scheckhuber, C.Q., Erjavec, N., Tinazli, A., Hamann, A., Nyström, T., Osiewacz, H.D., 2006. Reducing mitochondrial fission results in increased life span and fitness of two fungal ageing models. Nat. Cell Biol. 9, 99.

Scherz-Shouval, R., Elazar, Z., 2007. ROS, mitochondria and the regulation of autophagy. Trends Cell Biol. 17, 422–427.

Scherz-Shouval, R., Shvets, E., Fass, E., Shorer, H., Gil, L., Elazar, Z., 2007. Reactive oxygen species are essential for autophagy and specifically regulate the activity of Atg4. EMBO J. 26, 1749–1760.

Schlicker, C., Gertz, M., Papatheodorou, P., Kachholz, B., Becker, C.F., Steegborn, C., 2008. Substrates and regulation mechanisms for the human mitochondrial sirtuins Sirt3 and Sirt5. J. Mol. Biol. 382, 790–801.

Schmucker, D.L., 1998. Aging and the liver: an update. J. Gerontol. A Biol. Sci. Med. Sci. 53. B315–320.

Schmucker, D.L., 2001. Liver function and phase I drug metabolism in the elderly. Drugs Aging 18, 837–851.

Schmucker, D.L., 2005. Age-related changes in liver structure and function: Implications for disease. Exp. Gerontol. 40, 650–659.

Schmucker, D.L., Gilbert, R., Jones, A.L., Hradek, G.T., Bazin, H., 1985. Effect of aging on the hepatobiliary transport of dimeric immunoglobulin A in the male Fischer rat. Gastroenterology 88, 436–443.

Schneeman, B.O., Richter, D., 1993. Changes in plasma and hepatic lipids, small intestinal histology and pancreatic enzyme activity due to aging and dietary fiber in rats. J. Nutr. 123, 1328–1337.

Schriewer, J.M., Peek, C.B., Bass, J., Schumacker, P.T., 2013. ROS-mediated PARP activity undermines mitochondrial function after permeability transition pore opening during myocardial ischemia-reperfusion. J. Am. Heart Assoc. 2, e000159.

Schriner, S.E., Linford, N.J., Martin, G.M., Treuting, P., Ogburn, C.E., Emond, M., Coskun, P.E., Ladiges, W., Wolf, N., Van Remmen, H., Wallace, D.C., Rabinovitch, P.S., 2005. Extension of murine life span by overexpression of catalase targeted to mitochondria. Science 308, 1909–1911.

Schroder, K., Tschopp, J., 2010. The inflammasomes. Cell 140, 821–832.

Senft, D., Ronai, Z. e. A., 2016. Regulators of mitochondrial dynamics in cancer. Curr. Opin. Cell Biol. 39, 43–52.

Seo, D.W., Choi, H.-S., Lee, S.P., Kuver, R., 2004. Oxysterols from human bile induce apoptosis of canine gallbladder epithelial cells in monolayer culture. Am. J. Physiol. Gastrointest. Liver Physiol. 287, G1247–G1256.

Sepuri, N.B., Angireddy, R., Srinivasan, S., Guha, M., Spear, J., Lu, B., Anandatheerthavarada, H.K., Suzuki, C.K., Avadhani, N.G., 2017. Mitochondrial LON protease-dependent degradation of cytochrome c oxidase subunits under hypoxia and myocardial ischemia. Biochim. Biophys. Acta 1858, 519–528.

Serviddio, G., Romano, A.D., Cassano, T., Bellanti, F., Altomare, E., Vendemiale, G., 2011. Principles and therapeutic relevance for targeting mitochondria in aging and neurodegenerative diseases. Curr. Pharm. Des. 17, 2036–2055.
Shabalina, I.G., Nedergaard, J., 2011. Mitochondrial ('mild') uncoupling and ROS production: physiologically relevant or not? Biochem. Soc. Trans. 39, 1305–1309.
Shao, X., Li, Q., Mogilner, A., Bershadsky, A.D., Shivashankar, G.V., 2015. Mechanical stimulation induces formin-dependent assembly of a perinuclear actin rim. Proc. Natl. Acad. Sci.U.S.A 112, E2595–E2601.
Shelke, R.R., Leeuwenburgh, C., 2003. Lifelong caloric restriction increases expression of apoptosis repressor with a caspase recruitment domain (ARC) in the brain. FASEB J. 17, 494–496.
Shenouda, S.M., Widlansky, M.E., Chen, K., Xu, G., Holbrook, M., Tabit, C.E., Hamburg, N.M., Frame, A.A., Caiano, T.L., Kluge, M.A., Duess, M.-A., Levit, A., Kim, B., Hartman, M.-L., Joseph, L., Shirihai, O.S., Vita, J.A., 2011. Altered mitochondrial dynamics contributes to endothelial dysfunction in diabetes mellitus. Circulation 124, 444–453.
Sherer, T.B., Richardson, J.R., Testa, C.M., Seo, B.B., Panov, A.V., Yagi, T., Matsuno-Yagi, A., Miller, G.W., Greenamyre, J.T., 2007. Mechanism of toxicity of pesticides acting at complex I: relevance to environmental etiologies of Parkinson's disease. J. Neurochem. 100, 1469–1479.
Shi, R., Guberman, M., Kirshenbaum, L.A., 2017. Mitochondrial quality control: The role of mitophagy in aging. Trends Cardiovasc. Med.
Shibata, T., Nagayasu, H., Kawano, T., Kitajo, H., Hamada, J., Moriuchi, T., Okada, F., Watanabe, S., Yasuda, S., Okuyama, H., Kobayashi, H., Arisue, M., 1999. Unsaturated fatty acid feeding prevents the development of acute hepatitis in Long-Evans Cinnamon (LEC) rats. Anticancer Res. 19, 5169–5174.
Shibata, T., Nagayasu, H., Kitajo, H., Arisue, M., Yamashita, T., Hatakeyama, D., Iwasaki, T., Kobayashi, H., 2006. Inhibitory effects of fermented brown rice and rice bran on the development of acute hepatitis in Long-Evans Cinnamon rats. Oncol. Rep. 15, 869–874.
Shintoku, R., Takigawa, Y., Yamada, K., Kubota, C., Yoshimoto, Y., Takeuchi, T., Koshiishi, I., Torii, S., 2017. Lipoxygenase-mediated generation of lipid peroxides enhances ferroptosis induced by erastin and RSL3. Cancer Sci. 108, 2187–2194.
Shiva, S., Sack, M.N., Greer, J.J., Duranski, M., Ringwood, L.A., Burwell, L., Wang, X., MacArthur, P.H., Shoja, A., Raghavachari, N., Calvert, J.W., Brookes, P.S., Lefer, D.J., Gladwin, M.T., 2007. Nitrite augments tolerance to ischemia/reperfusion injury via the modulation of mitochondrial electron transfer. J. Exp. Med. 204, 2089–2102.
Shpilka, T., Haynes, C.M., 2017. The mitochondrial UPR: mechanisms, physiological functions and implications in ageing. Nat. Rev. Mol. Cell Biol.
Shu, Y.J., Bao, R.F., Wu, X.S., Weng, H., Ding, Q., Cao, Y., Li, M.L., Mu, J.S., Wu, W.G., Ding, Q.C., 2014. Baicalin induces apoptosis of gallbladder carcinoma cells in vitro via a mitochondrial-mediated pathway and suppresses tumor growth in vivo. Anti-Cancer Agents Med. Chem. 14, 1136–1145.
Siddiq, A., Ayoub, I.A., Chavez, J.C., Aminova, L., Shah, S., LaManna, J.C., Patton, S.M., Connor, J.R., Cherny, R.A., Volitakis, I., Bush, A.I., Langsetmo, I., Seeley, T., Gunzler, V., Ratan, R.R., 2005. Hypoxia-inducible factor prolyl 4-hydroxylase inhibition. A target for neuroprotection in the central nervous system. J. Biol. Chem. 280, 41732–41743.
Sies, H., 2017. Hydrogen peroxide as a central redox signaling molecule in physiological oxidative stress: oxidative eustress. Redox Biol. 11, 613–619.

Siler-Marsiglio, K.I., Pan, Q., Paiva, M., Madorsky, I., Khurana, N.C., Heaton, M.B., 2005. Mitochondrially targeted vitamin E and vitamin E mitigate ethanol-mediated effects on cerebellar granule cell antioxidant defense systems. Brain Res. 1052, 202–211.
Silva, J.P., Shabalina, I.G., Dufour, E., Petrovic, N., Backlund, E.C., Hultenby, K., Wibom, R., Nedergaard, J., Cannon, B., Larsson, N.G., 2005. SOD2 overexpression: enhanced mitochondrial tolerance but absence of effect on UCP activity. EMBO J. 24, 4061–4070.
Singh, I., Sagare, A.P., Coma, M., Perlmutter, D., Gelein, R., Bell, R.D., Deane, R.J., Zhong, E., Parisi, M., Ciszewski, J., Kasper, R.T., Deane, R., 2013. Low levels of copper disrupt brain amyloid-beta homeostasis by altering its production and clearance. Proc. Natl. Acad. Sci. U.S.A 110, 14771–14776.
Skulachev, V.P., Anisimov, V.N., Antonenko, Y.N., Bakeeva, L.E., Chernyak, B.V., Erichev, V.P., Filenko, O.F., Kalinina, N.I., Kapelko, V.I., Kolosova, N.G., Kopnin, B.P., Korshunova, G.A., Lichinitser, M.R., Obukhova, L.A., Pasyukova, E.G., Pisarenko, O.I., Roginsky, V.A., Ruuge, E.K., Senin, I.I., S.everina, I.I., S.kulachev, M.V., Spivak, I.M., Tashlitsky, V.N., Tkachuk, V.A., Vyssokikh, M.Y., Yaguzhinsky, L.S., Zorov, D.B., 2009. An attempt to prevent senescence: a mitochondrial approach. Biochim. Biophys. Acta 1787, 437–461.
Slawik, M., Vidal-Puig, A.J., 2006. Lipotoxicity, overnutrition and energy metabolism in aging. Ageing Res. Rev. 5, 144–164.
Smith, R.A., Porteous, C.M., Gane, A.M., Murphy, M.P., 2003. Delivery of bioactive molecules to mitochondria in vivo. Proc. Natl. Acad. Sci. U.S.A 100, 5407–5412.
Snow, B.J., Rolfe, F.L., Lockhart, M.M., Frampton, C.M., O'Sullivan, J.D., Fung, V., Smith, R.A., Murphy, M.P., Taylor, K.M., 2010. A double-blind, placebo-controlled study to assess the mitochondria-targeted antioxidant MitoQ as a disease-modifying therapy in Parkinson's disease. Mov. Disord. 25, 1670–1674.
Sohal, R.S., Weindruch, R., 1996. Oxidative stress, caloric restriction, and aging. Science 273, 59–63.
Sokol, R.J., Twedt, D., McKim Jr., J.M., Devereaux, M.W., Karrer, F.M., Kam, I., von Steigman, G., Narkewicz, M.R., Bacon, B.R., Britton, R.S., et al., 1994. Oxidant injury to hepatic mitochondria in patients with Wilson's disease and Bedlington terriers with copper toxicosis. Gastroenterology 107, 1788–1798.
Solaini, G., Harris, D.A., 2005. Biochemical dysfunction in heart mitochondria exposed to ischaemia and reperfusion. Biochem. J. 390, 377–394.
Somani, S.M., Frank, S., Rybak, L.P., 1995. Responses of antioxidant system to acute and trained exercise in rat heart subcellular fractions. Pharmacol. Biochem. Behav. 51, 627–634.
Someya, S., Yu, W., Hallows, W.C., Xu, J., Vann, J.M., Leeuwenburgh, C., Tanokura, M., Denu, J.M., Prolla, T.A., 2010. Sirt3 mediates reduction of oxidative damage and prevention of age-related hearing loss under caloric restriction. Cell 143, 802–812.
Soong, N.W., Hinton, D.R., Cortopassi, G., Arnheim, N., 1992. Mosaicism for a specific somatic mitochondrial DNA mutation in adult human brain. Nat. Genet. 2, 318–323.
Sorato, E., Menazza, S., Zulian, A., Sabatelli, P., Gualandi, F., Merlini, L., Bonaldo, P., Canton, M., Bernardi, P., Di Lisa, F., 2014. Monoamine oxidase inhibition prevents mitochondrial dysfunction and apoptosis in myoblasts from patients with collagen VI myopathies. Free Radic. Biol. Med. 75, 40–47.
Sotaniemi, E.A., Arranto, A.J., Pelkonen, O., Pasanen, M., 1997. Age and cytochrome P450-linked drug metabolism in humans: an analysis of 226 subjects with equal histopathologic conditions. Clin. Pharmacol. Therap. 61, 331–339.
Sparks, D.L., Schreurs, B.G., 2003. Trace amounts of copper in water induce beta-amyloid plaques and learning deficits in a rabbit model of Alzheimer's disease. Proc. Natl. Acad. Sci. U.S.A 100, 11065–11069.

Squitti, R., Polimanti, R., 2013. Copper phenotype in Alzheimer's disease: dissecting the pathway. Am. J. Neurodegener. Dis. 2, 46–56.

St-Pierre, J., Buckingham, J.A., Roebuck, S.J., Brand, M.D., 2002. Topology of superoxide production from different sites in the mitochondrial electron transport chain. J. Biol. Chem. 277, 44784–44790.

St-Pierre, J., Drori, S., Uldry, M., Silvaggi, J.M., Rhee, J., Jager, S., Handschin, C., Zheng, K., Lin, J., Yang, W., Simon, D.K., Bachoo, R., Spiegelman, B.M., 2006. Suppression of reactive oxygen species and neurodegeneration by the PGC-1 transcriptional coactivators. Cell 127, 397–408.

Stadtman, E.R., 1992. Protein oxidation and aging. Science 257, 1220–1224.

Stadtman, E.R., 2006. Protein oxidation and aging. Free Radic. Res. 40, 1250–1258.

Stanley, B.A., Sivakumaran, V., Shi, S., McDonald, I., Lloyd, D., Watson, W.H., Aon, M.A., Paolocci, N., 2011. Thioredoxin reductase-2 is essential for keeping low levels of H(2) O(2) emission from isolated heart mitochondria. J. Biol. Chem. 286, 33669–33677.

Starkov, A.A., Fiskum, G., Chinopoulos, C., Lorenzo, B.J., Browne, S.E., Patel, M.S., Beal, M.F., 2004. Mitochondrial alpha-ketoglutarate dehydrogenase complex generates reactive oxygen species. J. Neurosci. 24, 7779–7788.

Stefanatos, R., Sanz, A., 2011. Mitochondrial complex I: a central regulator of the aging process. Cell Cycle 10, 1528–1532.

Sterky, F.H., Lee, S., Wibom, R., Olson, L., Larsson, N.G., 2011. Impaired mitochondrial transport and Parkin-independent degeneration of respiratory chain-deficient dopamine neurons in vivo. Proc. Natl. Acad. Sci. U.S.A 108, 12937–12942.

Sternlieb, I., 1968. Mitochondrial and fatty changes in hepatocytes of patients with Wilson's disease. Gastroenterology 55, 354–367.

Sternlieb, I., 1992. Fraternal concordance of types of abnormal hepatocellular mitochondria in Wilson's disease. Hepatology 16, 728–732.

Sternlieb, I., Feldmann, G., 1976. Effects of anticopper therapy on hepatocellular mitochondria in patients with Wilson's disease: an ultrastructural and stereological study. Gastroenterology 71, 457–461.

Sternlieb, I., Quintana, N., Volenberg, I., Schilsky, M.L., 1995. An array of mitochondrial alterations in the hepatocytes of Long-Evans Cinnamon rats. Hepatology 22, 1782–1787.

Stockl, P., Zankl, C., Hutter, E., Unterluggauer, H., Laun, P., Heeren, G., Bogengruber, E., Herndler-Brandstetter, D., Breitenbach, M., Jansen-Durr, P., 2007. Partial uncoupling of oxidative phosphorylation induces premature senescence in human fibroblasts and yeast mother cells. Free Radic. Biol. Med. 43, 947–958.

Stoyanovsky, D.A., Tyurina, Y.Y., Tyurin, V.A., Anand, D., Mandavia, D.N., Gius, D., Ivanova, J., Pitt, B., Billiar, T.R., Kagan, V.E., 2005. Thioredoxin and lipoic acid catalyze the denitrosation of low molecular weight and protein S-nitrosothiols. J. Am. Chem. Soc. 127, 15815–15823.

Sturtz, L.A., Diekert, K., Jensen, L.T., Lill, R., Culotta, V.C., 2001. A fraction of yeast Cu,Zn-superoxide dismutase and its metallochaperone, CCS, localize to the intermembrane space of mitochondria. A physiological role for SOD1 in guarding against mitochondrial oxidative damage. J. Biol. Chem. 276, 38084–38089.

Sugrue, M.M., Tatton, W.G., 2001. Mitochondrial membrane potential in aging cells. Biol. Signals Recept. 10, 176–188.

Summer, K.H., Eisenburg, J., 1985. Low content of hepatic reduced glutathione in patients with Wilson's disease. Biochem. Med. 34, 107–111.

Sun, M.K., Hongpaisan, J., Nelson, T.J., Alkon, D.L., 2008. Poststroke neuronal rescue and synaptogenesis mediated in vivo by protein kinase C in adult brains. Proc. Natl. Acad. Sci. U.S.A 105, 13620–13625.

Sun, N., Youle, R.J., Finkel, T., 2016. The mitochondrial basis of aging. Mol Cell 61, 654–666.

Sundaresan, N.R., Gupta, M., Kim, G., Rajamohan, S.B., Isbatan, A., Gupta, M.P., 2009. Sirt3 blocks the cardiac hypertrophic response by augmenting Foxo3a-dependent antioxidant defense mechanisms in mice. J. Clin. Invest. 119, 2758–2771.

Suski, J., Lebiedzinska, M., Machado, N.G., Oliveira, P.J., Pinton, P., Duszynski, J., Wieckowski, M.R., 2011. Mitochondrial tolerance to drugs and toxic agents in ageing and disease. Curr. Drug Targets 12, 827–849.

Swerdlow, R.H., Burns, J.M., Khan, S.M., 2014. The Alzheimer's disease mitochondrial cascade hypothesis: progress and perspectives. Biochimica et Biophysica Acta 1842, 1219–1231.

Szczepanowska, K., Trifunovic, A., 2017. Origins of mtDNA mutations in ageing. Essays Biochem. 61, 325–337.

Szeto, H.H., 2006. Mitochondria-targeted peptide antioxidants: novel neuroprotective agents. AAPS J. 8. E521–531.

Szymanski, J., Janikiewicz, J., Michalska, B., Patalas-Krawczyk, P., Perrone, M., Ziolkowski, W., Duszynski, J., Pinton, P., Dobrzyn, A., Wieckowski, M.R., 2017. Interaction of mitochondria with the endoplasmic reticulum and plasma membrane in calcium homeostasis, lipid trafficking and mitochondrial structure. Int. J. Mol. Sci. 18.

Tahara, E.B., Barros, M.H., Oliveira, G.A., Netto, L.E., Kowaltowski, A.J., 2007. Dihydrolipoyl dehydrogenase as a source of reactive oxygen species inhibited by caloric restriction and involved in *Saccharomyces cerevisiae* aging. FASEB J. 21, 274–283.

Tamagno, E., Parola, M., Bardini, P., Piccini, A., Borghi, R., Guglielmotto, M., Santoro, G., Davit, A., Danni, O., Smith, M.A., Perry, G., Tabaton, M., 2005. Beta-site APP cleaving enzyme up-regulation induced by 4-hydroxynonenal is mediated by stress-activated protein kinases pathways. J. Neurochem. 92, 628–636.

Tanzi, R.E., Petrukhin, K., Chernov, I., Pellequer, J.L., Wasco, W., Ross, B., Romano, D.M., Parano, E., Pavone, L., Brzustowicz, L.M., et al., 1993. The Wilson disease gene is a copper transporting ATPase with homology to the Menkes disease gene. Nat. Genet. 5, 344–350.

Tchkonia, T., Morbeck, D.E., Von Zglinicki, T., Van Deursen, J., Lustgarten, J., Scrable, H., Khosla, S., Jensen, M.D., Kirkland, J.L., 2010. Fat tissue, aging, and cellular senescence. Aging Cell 9, 667–684.

Teshima, Y., Akao, M., Jones, S.P., Marban, E., 2003. Uncoupling protein-2 overexpression inhibits mitochondrial death pathway in cardiomyocytes. Circ Res. 93, 192–200.

Thallas-Bonke, V., Thorpe, S.R., Coughlan, M.T., Fukami, K., Yap, F.Y., Sourris, K.C., Penfold, S.A., Bach, L.A., Cooper, M.E., Forbes, J.M., 2008. Inhibition of NADPH oxidase prevents advanced glycation end product-mediated damage in diabetic nephropathy through a protein kinase C-alpha-dependent pathway. Diabetes 57, 460–469.

Thomas, R.P., Guigneaux, M., Wood, T., Evers, B.M., 2002. Age-associated changes in gene expression patterns in the liver. J. Gastrointest. Surg. 6, 445–453. Discussion 454.

Tiepolo, T., Angelin, A., Palma, E., Sabatelli, P., Merlini, L., Nicolosi, L., Finetti, F., Braghetta, P., Vuagniaux, G., Dumont, J.M., Baldari, C.T., Bonaldo, P., Bernardi, P., 2009. The cyclophilin inhibitor Debio 025 normalizes mitochondrial function, muscle apoptosis and ultrastructural defects in Col6a1-/- myopathic mice. Br. J. Pharmacol. 157, 1045–1052.

Tietz, N.W., Shuey, D.F., Wekstein, D.R., 1992. Laboratory values in fit aging individuals—sexagenarians through centenarians. Clin. Chem. 38, 1167–1185.

Tomás-Loba, A., Bernardes de Jesus, B., Mato, J.M., Blasco, M.A., 2013. A metabolic signature predicts biological age in mice. Aging Cell 12, 93–101.

Tomasetti, M., Alleva, R., Borghi, B., Collins, A.R., 2001. In vivo supplementation with coenzyme Q10 enhances the recovery of human lymphocytes from oxidative DNA damage. FASEB J. 15, 1425–1427.

Tran, T.T., Yamamoto, Y., Gesta, S., Kahn, C.R., 2008. Beneficial effects of subcutaneous fat transplantation on metabolism. Cell Metab. 7, 410–420.

Trapasso, F., Pichiorri, F., Gaspari, M., Palumbo, T., Aqeilan, R.I., Gaudio, E., Okumura, H., Iuliano, R., Di Leva, G., Fabbri, M., Birk, D.E., Raso, C., Green-Church, K., Spagnoli, L.G., Venuta, S., Huebner, K., Croce, C.M., 2008. Fhit interaction with ferredoxin reductase triggers generation of reactive oxygen species and apoptosis of cancer cells. J. Biol. Chem. 283, 13736–13744.

Triantafilou, K., Hughes, T.R., Triantafilou, M., Morgan, B.P., 2013. The complement membrane attack complex triggers intracellular Ca2+ fluxes leading to NLRP3 inflammasome activation. J. Cell Sci. 126, 2903–2913.

Trifunovic, A., Hansson, A., Wredenberg, A., Rovio, A.T., Dufour, E., Khvorostov, I., Spelbrink, J.N., Wibom, R., Jacobs, H.T., Larsson, N.G., 2005. Somatic mtDNA mutations cause aging phenotypes without affecting reactive oxygen species production. Proc. Natl. Acad. Sci. U.S.A 102, 17993–17998.

Trifunovic, A., Wredenberg, A., Falkenberg, M., Spelbrink, J.N., Rovio, A.T., Bruder, C.E., Bohlooly, Y.M., Gidlof, S., Oldfors, A., Wibom, R., Tornell, J., Jacobs, H.T., Larsson, N.G., 2004. Premature ageing in mice expressing defective mitochondrial DNA polymerase. Nature 429, 417–423.

Trinei, M., Giorgio, M., Cicalese, A., Barozzi, S., Ventura, A., Migliaccio, E., Milia, E., Padura, I.M., Raker, V.A., Maccarana, M., Petronilli, V., Minucci, S., Bernardi, P., Lanfrancone, L., Pelicci, P.G., 2002. A p53-p66Shc signalling pathway controls intracellular redox status, levels of oxidation-damaged DNA and oxidative stress-induced apoptosis. Oncogene 21, 3872–3878.

Trnka, J., Blaikie, F.H., Logan, A., Smith, R.A., Murphy, M.P., 2009. Antioxidant properties of MitoTEMPOL and its hydroxylamine. Free Radic. Res. 43, 4–12.

Truong, V.L., Jun, M., Jeong, W.S., 2017. Role of resveratrol in regulation of cellular defense systems against oxidative stress. Biofactors.

Tschopp, J., 2011. Mitochondria: sovereign of inflammation? Eur. J. Immunol. 41, 1196–1202.

Tubbs, E., Rieusset, J., 2017. Metabolic signaling functions of ER–mitochondria contact sites: role in metabolic diseases. J. Mol. Endocrinol. 58, R87–R106.

Tubbs, E., Theurey, P., Vial, G., Bendridi, N., Bravard, A., Chauvin, M.A., Ji-Cao, J., Zoulim, F., Bartosch, B., Ovize, M., Vidal, H., Rieusset, J., 2014. Mitochondria-associated endoplasmic reticulum membrane (MAM) integrity is required for insulin signaling and is implicated in hepatic insulin resistance. Diabetes 63, 3279–3294.

Turnheim, K., 2003. When drug therapy gets old: pharmacokinetics and pharmacodynamics in the elderly. Exp. Gerontol. 38, 843–853.

Twig, G., Elorza, A., Molina, A.J., Mohamed, H., Wikstrom, J.D., Walzer, G., Stiles, L., Haigh, S.E., Katz, S., Las, G., Alroy, J., Wu, M., Py, B.F., Yuan, J., Deeney, J.T., Corkey, B.E., Shirihai, O.S., 2008. Fission and selective fusion govern mitochondrial segregation and elimination by autophagy. EMBO J. 27, 433–446.

Tzoulis, C., Schwarzlmuller, T., Biermann, M., Haugarvoll, K., Bindoff, L.A., 2016. Mitochondrial DNA homeostasis is essential for nigrostriatal integrity. Mitochondrion 28, 33–37.

Ulanowska, K., Piosik, J., Gwizdek-Wisniewska, A., Wegrzyn, G., 2007. Impaired mutagenic activities of MPDP(+) (1-methyl-4-phenyl-2,3-dihydropyridinium) and MPP(+) (1-methyl-4-phenylpyridinium) due to their interactions with methylxanthines. Bioorg. Med. Chem. 15, 5150–5157.

Ulanowska, K., Wegrzyn, G., 2006. Mutagenic activity of 1-methyl-4-phenyl-1,2,3,6-tetrahydropyridine. J. Appl. Genet. 47, 85–87.

Ungvari, Z., Parrado-Fernandez, C., Csiszar, A., de Cabo, R., 2008. Mechanisms underlying caloric restriction and lifespan regulation: implications for vascular aging. Circ Res. 102, 519–528.

Uttara, B., Singh, A.V., Zamboni, P., Mahajan, R.T., 2009. Oxidative stress and neurodegenerative diseases: a review of upstream and downstream antioxidant therapeutic options. Curr. Neuropharmacol. 7, 65–74.
Uzhachenko, R., Boyd, K., Olivares-Villagomez, D., Zhu, Y., Goodwin, J.S., Rana, T., Shanker, A., Tan, W.J., Bondar, T., Medzhitov, R., Ivanova, A.V., 2017. Mitochondrial protein Fus1/Tusc2 in premature aging and age-related pathologies: critical roles of calcium and energy homeostasis. Aging (Albany NY) 9, 627–648.
Valcarcel-Jimenez, L., Gaude, E., Torrano, V., Frezza, C., Carracedo, A., 2017. Mitochondrial metabolism: Yin and Yang for tumor progression. Trends Endocrinol. Metabol. 28, 748–757.
Valdivieso, V., Palma, R., Wunkhaus, R., Antezana, C., Severin, C., Contreras, A., 1978. Effect of aging on biliary lipid composition and bile acid metabolism in normal Chilean women. Gastroenterology 74, 871–874.
Valle, A., Guevara, R., Garcia-Palmer, F.J., Roca, P., Oliver, J., 2007. Sexual dimorphism in liver mitochondrial oxidative capacity is conserved under caloric restriction conditions. Am. J. Physiol. Cell Physiol. 293. C1302–1308.
van de Veerdonk, F.L., Smeekens, S.P., Joosten, L.A., Kullberg, B.J., Dinarello, C.A., van der Meer, J.W., Netea, M.G., 2010. Reactive oxygen species-independent activation of the IL-1beta inflammasome in cells from patients with chronic granulomatous disease. Proc. Natl. Acad. Sci. U.S.A 107, 3030–3033.
van der Poorten, D., Samer, C.F., Ramezani-Moghadam, M., Coulter, S., Kacevska, M., Schrijnders, D., Wu, L.E., McLeod, D., Bugianesi, E., Komuta, M., 2013. Hepatic fat loss in advanced nonalcoholic steatohepatitis: are alterations in serum adiponectin the cause? Hepatology 57, 2180–2188.
Vecchie, A., Dallegri, F., Carbone, F., Bonaventura, A., Liberale, L., Portincasa, P., Fruhbeck, G., Montecucco, F., 2017. Obesity phenotypes and their paradoxical association with cardiovascular diseases. Europ. J. Internal Med.
Vecchione, G., Grasselli, E., Cioffi, F., Baldini, F., Oliveira, P.J., Sardao, V.A., Cortese, K., Lanni, A., Voci, A., Portincasa, P., Vergani, L., 2017. The nutraceutic silybin counteracts excess lipid accumulation and ongoing oxidative stress in an In vitro model of non-alcoholic fatty liver disease progression. Front. Nutr. 4, 42.
Vecchione, G., Grasselli, E., Voci, A., Baldini, F., Grattagliano, I., Wang, D.Q., Portincasa, P., Vergani, L., 2016. Silybin counteracts lipid excess and oxidative stress in cultured steatotic hepatic cells. World J. Gastroenterol. 22, 6016–6026.
Verfaillie, T., Rubio, N., Garg, A.D., Bultynck, G., Rizzuto, R., Decuypere, J.P., Piette, J., Linehan, C., Gupta, S., Samali, A., Agostinis, P., 2012. PERK is required at the ER-mitochondrial contact sites to convey apoptosis after ROS-based ER stress. Cell Death Differ. 19, 1880–1891.
Vermulst, M., Bielas, J.H., Kujoth, G.C., Ladiges, W.C., Rabinovitch, P.S., Prolla, T.A., Loeb, L.A., 2007. Mitochondrial point mutations do not limit the natural lifespan of mice. Nat. Genet. 39, 540–543.
Villalba, J.M., Parrado, C., Santos-Gonzalez, M., Alcain, F.J., 2010. Therapeutic use of coenzyme Q10 and coenzyme Q10-related compounds and formulations. Expert. Opin. Investig. Drugs 19, 535–554.
Vittorini, S., Paradiso, C., Donati, A., Cavallini, G., Masini, M., Gori, Z., Pollera, M., Bergamini, E., 1999. The age-related accumulation of protein carbonyl in rat liver correlates with the age-related decline in liver proteolytic activities. J. Gerontol. A Biol. Sci. Med. Sci. 54. B318–323.
von Herbay, A., de Groot, H., Hegi, U., Stremmel, W., Strohmeyer, G., Sies, H., 1994. Low vitamin E content in plasma of patients with alcoholic liver disease, hemochromatosis and Wilson's disease. J. Hepatol. 20, 41–46.

Von Zglinicki, T., Marzabadi, M., Roomans, G., 1991. Water and ion distributions in myocytes cultured under oxidative stress mimic changes found in the process of aging. Mech. Ageing Dev. 58, 49–60.

Wai, T., García-Prieto, J., Baker, M.J., Merkwirth, C., Benit, P., Rustin, P., Rupérez, F.J., Barbas, C., Ibañez, B., Langer, T., 2015. Imbalanced OPA1 processing and mitochondrial fragmentation cause heart failure in mice. Science 350.

Wainio, W.W., Vanderwende, C., Shimp, N.F., 1959. Copper in Cytochrome-C oxidase. J. Biol. Chem. 234, 2433–2436.

Wales, P., Schuberth, C.E., Aufschnaiter, R., Fels, J., García-Aguilar, I., Janning, A., Dlugos, C.P., Schäfer-Herte, M., Klingner, C., Wälte, M., Kuhlmann, J., Menis, E., Hockaday Kang, L., Maier, K.C., Hou, W., Russo, A., Higgs, H.N., Pavenstädt, H., Vogl, T., Roth, J., Qualmann, B., Kessels, M.M., Martin, D.E., Mulder, B., Wedlich-Söldner, R., 2016. Calcium-mediated actin reset (CaAR) mediates acute cell adaptations. eLife 5, e19850.

Wang, D.Q., 2002. Aging per se is an independent risk factor for cholesterol gallstone formation in gallstone susceptible mice. J. Lipid. Res. 43, 1950–1959.

Wang, D.Q., Portincasa, P. (Eds.), 2017. Gallstones. Recent advances in epidemiology, pathogenesis, diagnosis and management, The first edition. Nova Science Publisher Inc., New York, NY, pp. 1–676.

Wang, D.Q., Portincasa, P., Tso, P., 2017a. Transintestinal cholesterol excretion: A secondary, nonbiliary pathway contributing to reverse cholesterol transport. Hepatology 66, 1337–1340.

Wang, D.Q., Neuschwander-Tetri, B.A., Portincasa, P., 2017b. The Biliary System In: Colloquium Series on Integrated Systems Physiology: From Molecule to Function, 2nd ed., 8 pp. i–178.

Wang, X., Su, B., Fujioka, H., Zhu, X., 2008a. Dynamin-like protein 1 reduction underlies mitochondrial morphology and distribution abnormalities in fibroblasts from Sporadic Alzheimer's disease patients. Am. J. Pathol. 173, 470–482.

Wang, X., Su, B., Siedlak, S.L., Moreira, P.I., Fujioka, H., Wang, Y., Casadesus, G., Zhu, X., 2008b. Amyloid-β overproduction causes abnormal mitochondrial dynamics via differential modulation of mitochondrial fission/fusion proteins. Proc. Natl. Acad. Sci. U.S.A 105, 19318–19323.

Wang, X.A., Xiang, S.S., Li, H.F., Wu, X.S., Li, M.L., Shu, Y.J., Zhang, F., Cao, Y., Ye, Y.Y., Bao, R.F., Weng, H., Wu, W.G., Mu, J.S., Hu, Y.P., Jiang, L., Tan, Z.J., Lu, W., Wang, P., Liu, Y.B., 2014. Cordycepin induces S phase arrest and apoptosis in human gallbladder cancer cells. Molecules 19, 11350–11365.

Watanabe, T., Shimada, H., Tanaka, Y., 1978. Human hepatocytes and aging: a cytophotometrical analysis in 35 sudden-death cases. Virchows. Archiv. B. Cell Pathol. 27, 307–316.

Waters, C.E., Saldivar, J.C., Hosseini, S.A., Huebner, K., 2014. The FHIT gene product: tumor suppressor and genome "caretaker". Cell Mol. Life Sci. 71, 4577–4587.

Webster, B.R., Lu, Z., Sack, M.N., Scott, I., 2012. The role of sirtuins in modulating redox stressors. Free Radic. Biol. Med. 52, 281–290.

Weglewska, A., Jakobkiewicz-Banecka, J., Wegrzyn, G., 2005. A modified procedure for quantitative analysis of mtDNA, detecting mtDNA depletion. J. Appl. Genet. 46, 423–428.

Wei, Y., Clark, S.E., Thyfault, J.P., Uptergrove, G.M., Li, W., Whaley-Connell, A.T., Ferrario, C.M., Sowers, J.R., Ibdah, J.A., 2009. Oxidative stress-mediated mitochondrial dysfunction contributes to angiotensin II-induced nonalcoholic fatty liver disease in transgenic Ren2 rats. Am. J. Pathol. 174, 1329–1337.

Wei, Y.H., 1992. Mitochondrial DNA alterations as ageing-associated molecular events. Mutat. Res. 275, 145–155.

Wei, Y.H., 1998. Oxidative stress and mitochondrial DNA mutations in human aging. Proc. Soc. Exp. Biol. Med. 217, 53–63.
Wei, Y.H., Ma, Y.S., Lee, H.C., Lee, C.F., Lu, C.Y., 2001. Mitochondrial theory of aging matures—roles of mtDNA mutation and oxidative stress in human aging. Zhonghua Yi Xue Za Zhi (Taipei) 64, 259–270.
Weng, H., Tan, Z.J., Hu, Y.P., Shu, Y.J., Bao, R.F., Jiang, L., Wu, X.S., Li, M.L., Ding, Q., Wang, X.A., Xiang, S.S., Li, H.F., Cao, Y., Tao, F., Liu, Y.B., 2014. Ursolic acid induces cell cycle arrest and apoptosis of gallbladder carcinoma cells. Cancer Cell Int. 14, 96.
Wieckowski, M.R., Wojtczak, L., 1997. Involvement of the dicarboxylate carrier in the protonophoric action of long-chain fatty acids in mitochondria. Biochem. Biophys. Res. Commun. 232, 414–417.
Wiel, C., Lallet-Daher, H., Gitenay, D., Gras, B., Le Calvé, B., Augert, A., Ferrand, M., Prevarskaya, N., Simonnet, H., Vindrieux, D., Bernard, D., 2014. Endoplasmic reticulum calcium release through ITPR2 channels leads to mitochondrial calcium accumulation and senescence. Nat. Commun. 5, 3792.
Wiesner, R.J., Zsurka, G., Kunz, W.S., 2006. Mitochondrial DNA damage and the aging process: facts and imaginations. Free Radic. Res. 40, 1284–1294.
Willems, P.H., Rossignol, R., Dieteren, C.E., Murphy, M.P., Koopman, W.J., 2015. Redox homeostasis and mitochondrial dynamics. Cell Metab. 22, 207–218.
Williams, C.D., Stengel, J., Asike, M.I., Torres, D.M., Shaw, J., Contreras, M., Landt, C.L., Harrison, S.A., 2011. Prevalence of nonalcoholic fatty liver disease and nonalcoholic steatohepatitis among a largely middle-aged population utilizing ultrasound and liver biopsy: a prospective study. Gastroenterology 140, 124–131.
Wilson, P., Franks, L., 1975. The effect of age on mitochondrial ultrastructure and enzymes. In: Cell impairment in aging and development. Springer, pp. 171–183.
Winter, W.E., Bazydlo, L.A., Harris, N.S., 2014. The molecular biology of human iron metabolism. Lab. Med. 45, 92–102.
Wojtczak, L., Lebiedzinska, M., Suski, J.M., Wieckowski, M.R., Schonfeld, P., 2011. Inhibition by purine nucleotides of the release of reactive oxygen species from muscle mitochondria: indication for a function of uncoupling proteins as superoxide anion transporters. Biochem. Biophys. Res. Commun. 407, 772–776.
Wu, J., Li, X., Zhu, G., Zhang, Y., He, M., Zhang, J., 2016. The role of Resveratrol-induced mitophagy/autophagy in peritoneal mesothelial cells inflammatory injury via NLRP3 inflammasome activation triggered by mitochondrial ROS. Exp. Cell Res. 341, 42–53.
Wu, S.Y., Lan, S.H., Cheng, D.E., Chen, W.K., Shen, C.H., Lee, Y.R., Zuchini, R., Liu, H.S., 2011. Ras-related tumorigenesis is suppressed by BNIP3-mediated autophagy through inhibition of cell proliferation. Neoplasia 13, 1171–1182.
Wynne, H.A., Cope, L.H., Mutch, E., Rawlins, M.D., Woodhouse, K.W., James, O.F., 1989. The effect of age upon liver volume and apparent liver blood flow in healthy man. Hepatology 9, 297–301.
Xu, Y., Kim, S.O., Li, Y., Han, J., 2006. Autophagy contributes to caspase-independent macrophage cell death. J. Biol. Chem. 281, 19179–19187.
Yagi, T., Hatefi, Y., 1987. Thiols in oxidative phosphorylation: thiols in the F0 of ATP synthase essential for ATPase activity. Arch. Biochem. Biophys. 254, 102–109.
Yamada, T., Adachi, Y., Iijima, M., Sesaki, H., 2016. Making a division apparatus on mitochondria. Trends Biochem. Sci. 41, 209–210.
Yamada, T., Sogawa, K., Suzuki, Y., Izumi, K., Agui, T., Matsumoto, K., 1992. Elevation of the level of lipid-peroxidation associated with hepatic-Injury in Lec Mutant Rat. Res. Commun. Chem. Path. 77, 121–124.

Yamamoto, F., Kasai, H., Togashi, Y., Takeichi, N., Hori, T., Nishimura, S., 1993. Elevated level of 8-Hydroxydeoxyguanosine in DNA of liver, kidneys, and brain of Long-Evans Cinnamon rats. Jpn. J. Cancer Res. 84, 508–511.
Yamamoto, H., Hirose, K., Hayasaki, Y., Masuda, M., Kazusaka, A., Fujita, S., 1999. Mechanism of enhanced lipid peroxidation in the liver of Long-Evans cinnamon (LEC) rats. Arch. Toxicol. 73, 457–464.
Yamamoto, H., Watanabe, T., Mizuno, H., Endo, K., Fukushige, J., Hosokawa, T., Kazusaka, A., Fujita, S., 2001. The antioxidant effect of DL-alpha-lipoic acid on copper-induced acute hepatitis in Long-Evans Cinnamon (LEC) rats. Free Radic. Res. 34, 69–80.
Yamashita, T., Ohshima, H., Asanuma, T., Inukai, N., Miyoshi, I., Kasai, N., Kon, Y., Watanabe, T., Sato, F., Kuwabara, M., 1996. The effects of alpha-phenyl-tert-butyl nitrone (PBN) on copper-induced rat fulminant hepatitis with jaundice. Free Radic. Biol. Med. 21, 755–761.
Yamazaki, K., Ohyama, H., Kurata, K., Wakabayashi, T., 1993. Effects of dietary vitamin E on clinical course and plasma glutamic oxaloacetic transaminase and glutamic pyruvic transaminase activities in hereditary hepatitis of LEC rats. Lab. Anim. Sci. 43, 61–67.
Yang, C.C., Chen, D., Lee, S.S., Walter, L., 2011. The dynamin-related protein DRP-1 and the insulin signaling pathway cooperate to modulate *C. elegans* longevity. Aging Cell 10, 724–728.
Yang, J.C., Cortopassi, G.A., 1998. Induction of the mitochondrial permeability transition causes release of the apoptogenic factor cytochrome c. Free Radic. Biol. Med. 24, 624–631.
Yang, L., Li, P., Fu, S., Calay, E.S., Hotamisligil, G.S., 2010. Defective hepatic autophagy in obesity promotes ER stress and causes insulin resistance. Cell Metab. 11, 467–478.
Yang, L., Zhao, K., Calingasan, N.Y., Luo, G., Szeto, H.H., Beal, M.F., 2009. Mitochondria targeted peptides protect against 1-methyl-4-phenyl-1,2,3,6-tetrahydropyridine neurotoxicity. Antioxid. Redox Signal. 11, 2095–2104.
Yang, S., Zhu, H., Li, Y., Lin, H., Gabrielson, K., Trush, M.A., Diehl, A.M., 2000. Mitochondrial adaptations to obesity-related oxidant stress. Arch. Biochem. Biophys. 378, 259–268.
Yang, W., Hekimi, S., 2010. A mitochondrial superoxide signal triggers increased longevity in *Caenorhabditis elegans*. PLoS Biol. 8, e1000556.
Yankovskaya, V., Horsefield, R., Tornroth, S., Luna-Chavez, C., Miyoshi, H., Leger, C., Byrne, B., Cecchini, G., Iwata, S., 2003. Architecture of succinate dehydrogenase and reactive oxygen species generation. Science 299, 700–704.
Yen, T.C., Chen, Y.S., King, K.L., Yeh, S.H., Wei, Y.H., 1989. Liver mitochondrial respiratory functions decline with age. Biochem. Biophys. Res. Commun. 165, 994–1003.
Yen, T.C., Su, J.H., King, K.L., Wei, Y.H., 1991. Ageing-associated 5 kb deletion in human liver mitochondrial DNA. Biochem. Biophys. Res. Commun. 178, 124–131.
Youle, R.J., van der Bliek, A.M., 2012. Mitochondrial fission, fusion, and stress. Science 337, 1062–1065.
Yu, T., Robotham, J.L., Yoon, Y., 2006. Increased production of reactive oxygen species in hyperglycemic conditions requires dynamic change of mitochondrial morphology. Proc. Natl. Acad. Sci. U.S.A 103, 2653–2658.
Yu, Y., Guerrero, C.R., Liu, S., Amato, N.J., Sharma, Y., Gupta, S., Wang, Y., 2016. Comprehensive assessment of oxidatively induced modifications of DNA in a rat model of human Wilson's Disease. Mol. Cell Proteomics 15, 810–817.
Yun, J., Finkel, T., 2014. Mitohormesis. Cell Metab. 19, 757–766.

Yurkova, I.L., Arnhold, J., Fitzl, G., Huster, D., 2011. Fragmentation of mitochondrial cardiolipin by copper ions in the Atp7b-/- mouse model of Wilson's disease. Chem. Phys. Lipids 164, 393–400.

Zapico, S.C., Ubelaker, D.H., 2013. mtDNA mutations and their role in aging, diseases and forensic sciences. Aging Dis. 4, 364–380.

Zelber-Sagi, S., Nitzan-Kaluski, D., Halpern, Z., Oren, R., 2006. Prevalence of primary non-alcoholic fatty liver disease in a population-based study and its association with biochemical and anthropometric measures. Liver Int. 26, 856–863.

Zhang, H., Luo, Y., Zhang, W., He, Y., Dai, S., Zhang, R., Huang, Y., Bernatchez, P., Giordano, F.J., Shadel, G., Sessa, W.C., Min, W., 2007. Endothelial-specific expression of mitochondrial thioredoxin improves endothelial cell function and reduces atherosclerotic lesions. Am. J. Pathol. 170, 1108–1120.

Zhang, H.J., Xu, L.J., Drake, V.J., Xie, L.T., Oberley, L.W., Kregel, K.C., 2003. Heat-induced liver injury in old rats is associated with exaggerated oxidative stress and altered transcription factor activation. Faseb. J. 17, 2293–2295.

Zhang, L., Deng, S., Zhao, S., Ai, Y., Zhang, L., Pan, P., Su, X., Tan, H., Wu, D., 2016. Intra-peritoneal administration of mitochondrial DNA provokes acute lung injury and systemic inflammation via toll-like receptor 9. Int. J. Mol. Sci. 17.

Zhang, Y., Ikeno, Y., Qi, W., Chaudhuri, A., Li, Y., Bokov, A., Thorpe, S.R., Baynes, J.W., Epstein, C., Richardson, A., Van Remmen, H., 2009. Mice deficient in both Mn superoxide dismutase and glutathione peroxidase-1 have increased oxidative damage and a greater incidence of pathology but no reduction in longevity. J. Gerontol. A Biol. Sci. Med. Sci. 64, 1212–1220.

Zhao, K., Zhao, G.M., Wu, D., Soong, Y., Birk, A.V., Schiller, P.W., Szeto, H.H., 2004. Cell-permeable peptide antioxidants targeted to inner mitochondrial membrane inhibit mitochondrial swelling, oxidative cell death, and reperfusion injury. J. Biol. Chem. 279, 34682–34690.

Zhao, Y., Ye, L., Liu, H., Xia, Q., Zhang, Y., Yang, X., Wang, K., 2010. Vanadium compounds induced mitochondria permeability transition pore (PTP) opening related to oxidative stress. J. Inorg. Biochem. 104, 371–378.

Zhou, R., Yazdi, A.S., Menu, P., Tschopp, J., 2011. A role for mitochondria in NLRP3 inflammasome activation. Nature 469, 221–225.

Zhu, C.G., Liu, Y.X., Wang, H., Wang, B.P., Qu, H.Q., Wang, B.L., Zhu, M., 2017. Active form of vitamin D ameliorates non-alcoholic fatty liver disease by alleviating oxidative stress in a high-fat diet rat model. Endocr. J. 64, 663–673.

Zini, R., Morin, C., Bertelli, A., Bertelli, A.A., Tillement, J.P., 1999. Effects of resveratrol on the rat brain respiratory chain. Drugs Exp. Clin. Res. 25, 87–97.

Zischka, H., Lichtmannegger, J., Schmitt, S., Jagemann, N., Schulz, S., Wartini, D., Jennen, L., Rust, C., Larochette, N., Galluzzi, L., Chajes, V., Bandow, N., Gilles, V.S., DiSpirito, A.A., Esposito, I., Goettlicher, M., Summer, K.H., Kroemer, G., 2011. Liver mitochondrial membrane crosslinking and destruction in a rat model of Wilson disease. J. Clin. Invest. 121, 1508–1518.

Zmijewski, J.W., Moellering, D.R., Le Goffe, C., Landar, A., Ramachandran, A., Darley-Usmar, V.M., 2005. Oxidized LDL induces mitochondrially associated reactive oxygen/nitrogen species formation in endothelial cells. Am. J. Physiol. Heart Circ. Physiol. 289. H852–861.

Zoratti, M., Szabo, I., 1995. The mitochondrial permeability transition. Biochim. Biophys. Acta 1241, 139–176.

FURTHER READING

Di Ciaula, A., Wang, D.Q., Portincasa, P., 2018. An update on the pathogenesis of cholesterol gallstone disease. Curr. Opin. Gastroenterol. 34, 71–80.

Han, B., Sivaramakrishnan, P., Lin, C.-C.J., Neve, I.A.A., He, J., Tay, L.W.R., Sowa, J.N., Sizovs, A., Du, G., Wang, J., Herman, C., Wang, M.C., 2017. Microbial genetic composition tunes host longevity. Cell 169, 1249–1262.e1213.

Printed in the United States
By Bookmasters